A NEW ECOLOGY

A NEW ECOLOGY

Novel Approaches
to Interactive Systems

Edited by
PETER W. PRICE, C. N. SLOBODCHIKOFF,
and WILLIAM S. GAUD

Department of Biological Sciences
Northern Arizona University
Flagstaff, Arizona

A Wiley-Interscience Publication

JOHN WILEY & SONS

New York   Chichester   Brisbane   Toronto   Singapore

Library of Congress Cataloging in Publication Data:

Main entry under title:

A new ecology.

   Based on a conference held at Northern Arizona
University, Aug. 11–13, 1982.
   "A Wiley-Interscience publication."
   Includes index.
   1. Ecology—Congresses. I. Price, Peter W.
II. Slobodchikoff, C. N. III. Gaud, William S.
IV. Northern Arizona University.

QH540.N395 1984   574.5   83-14603
ISBN 0-471-89670-5

Printed in the United States of America

10 9 8 7 6 5 4 3 2 1

LIST OF CONTRIBUTORS

John F. Addicott, Department of Zoology, University of Alberta, Edmonton, Alberta, Canada, and Rocky Mountain Biological Laboratory, Crested Butte, Colorado

Thomas Caraco, Department of Biology, University of Rochester, Rochester, New York

Robert K. Colwell, Department of Zoology, University of California, Berkeley, California

Paul K. Dayton, Scripps Institution of Oceanography, La Jolla, California

Hugh Dingle, Department of Entomology, University of California at Davis, Davis, California

Gordon W. Frankie, Department of Entomological Sciences, University of California, Berkeley, California

William S. Gaud, Department of Biological Sciences, Northern Arizona University, Flagstaff, Arizona

Conrad A. Istock, Department of Biology, University of Rochester, Rochester, New York

Clive G. Jones, Department of Chemical Ecology, New York Botanical Garden, Cary Arboretum, Millbrook, New York

John H. Lawton, Department of Biology, University of York, Heslington, York, England

Richard E. Michod, Department of Ecology and Evolutionary Biology, University of Arizona, Tucson, Arizona

David L. Morgan, Research and Extension Center, Texas A & M University, Dallas, Texas

Peter W. Price, Department of Biological Sciences, Northern Arizona University, Flagstaff, Arizona

H. Ronald Pulliam, Department of Biological Sciences, State University of New York, Albany, New York

Anne M. Robinson, Department of Biological Sciences, Northern Arizona University, Flagstaff, Arizona

C. N. Slobodchikoff, Department of Biological Sciences, Northern Arizona University, Flagstaff, Arizona

Donald R. Strong, Department of Biological Science, Florida State University, Tallahassee, Florida

Mia J. Tegner, Scripps Institution of Oceanography, University of California, La Jolla, California

Thomas G. Whitham, Department of Biological Sciences, Northern Arizona University, Flagstaff, Arizona, and Biology Department, Museum of Northern Arizona, Flagstaff, Arizona

John A. Wiens, Department of Biology, University of New Mexico, Albuquerque, New Mexico

Henry M. Wilbur, Department of Zoology, Duke University, Durham, North Carolina

Alan G. Williams, Department of Biological Sciences, Northern Arizona University, Flagstaff, Arizona

PREFACE

We designed this volume specifically for students of ecology who are beginning a research career. Here they can find some of the excitement and ferment in this field and readily assess how to make a major contribution to its development. We asked each author to comment specifically on future developments and research needs in the area they address. We anticipate that this volume will be used in undergraduate and graduate ecology seminars, and as a source of discussion topics in general ecology courses. We expect, also, that established researchers will find enough new research, synthesis, and discussion of future ecology to make this a valuable reference volume.

The book is derived from a conference of the same title held at Northern Arizona University, August 11–13, 1982. The conference was organized as a celebration of the increased research capacity on this campus in the form of the new Ralph M. Bilby Research Center building and the establishment of the Center for Ecological Research.

We selected four areas of ecology in which we saw rapid development: the relationship between resources and populations; life history strategies; ecology of social behavior; organization of communities. The need for synthesis among these areas resulted in a fifth part to the book. We do not pretend to cover the full range of exciting areas in ecology, and the topics we chose reflect our personal interests, as well as those of a large body of researchers in the ecological community. Authors were chosen as much for their demonstrated ability for integration and synthesis as for their expertise in the particular area of ecology they addressed. At the conference in which these topics were presented there was much lively discussion and a refreshing sense of rediscovery of areas and points of view resulting from interaction between scientists working in disparate subdisciplines of ecology. We resisted the temptation to publish the discussions because of their volume, both in decibels and length, but they remain for many participants a part of a new ecology that will bear fruit in years to come.

We are very grateful to several people who made the conference and this book possible. Eugene Hughes, President, and Joseph Cox, Vice President for Academic Affairs, of Northern Arizona University have most effectively fostered the development of research personnel and resources on this campus. We were supported in spirit and in substance by Henry Hooper, Dean of Graduate Studies, who allocated funds to cover all conference expenses.

The Director of the Bilby Research Center, Richard Foust, provided advice, organizational expertise, secretarial help, and many other indispensable services. Dr. James Wick, Chair, Department of Biological Sciences, has overseen the department's development for more than 20 years, and without his commitment to creative scholarship neither conference nor book would have materialized.

PETER W. PRICE
C. N. SLOBODCHIKOFF
WILLIAM S. GAUD

Flagstaff, Arizona
September 1983

CONTENTS

A NEW ECOLOGY

CHAPTER 1

Introduction: Is There a New Ecology?

PETER W. PRICE
WILLIAM S. GAUD
C. N. SLOBODCHIKOFF
Department of Biological Sciences
Northern Arizona University
Flagstaff, Arizona

CONTENTS

1   A NEW ECOLOGY

Much of the work discussed in this book illustrates developments in ecology that have gained impetus within the last 10 years. It is new to this extent. Some areas are much younger. Many of the approaches taken here and the conceptual bases for them are presently not widely used but deserve to influence a wider range of ecologists. In the future, we believe that the areas discussed herein will flourish and result in major developments in ecological thought and practice.

There is certainly a new realization that plant populations provide a very heterogeneous environment for herbivores, contrasting sharply with a large literature on population dynamics that regards plants as homogeneous resources (Price, 1983a). This new attention to the detailed understanding of patterns of resource availability was sharply focused by Edmunds and Alstad (1978), Whitham (1978), and Whitham and Slobodchikoff (1981), and is discussed in Chapters 2 and 4. Resources also act as central themes in other chapters (3, 8, 13, 15) reflecting a realization of the need in population and community studies to work from the bottom up rather than from the top down. Looking at patterns and trying to guess the mechanisms have been unprofitable, producing a literature that is rapidly being forgotten on such topics as species-per-genus ratios, dominance–diversity curves, and diversity–stability relationships. Resources form the starting points of food webs to which populations and communities respond, and the patterns of resource availability and use need to be understood in detail before ecological mechanisms, which are the basis of ecological organization, can be described adequately. Connell (1980, p. 133) asked the question: *Can ecologists judge availability [of resources] as the organisms do?* We must be able to answer in the positive to this question and the chapters in this book approach that answer. A new ecology must continue to give this question a central place.

The rich possibilities for mutualistic interactions among plants, animals, and microorganisms will eventually lead to a burgeoning literature in major ecological journals. Conventional "wisdom" presently holds that mutualism's "importance in populations in general is small" (Williamson, 1972, p. 95) and that such relationships are "relatively uncommon in many natural ecosystems" (May, 1973, p. 4, but see May, 1981, for signs of change in opinion). The growing emphasis on mutualism as an ecological phenomenon fully as important as the phenomena of competition and predation is illustrated by a recent review (Boucher et al., 1982) in which microbes figure prominently. Resources are also profoundly modified by microorganisms, a subject addressed in Chapter 3, needing much more attention in ecology. The importance of mutualism is discussed also in Chapter 16. As Boucher et al. observe (1982, p. 337), "the study of mutualism has made major advances in just the past decade."

The study of life history traits, the subject of Part II of this volume, is a rapidly growing field in ecology (Stearns, 1982). Variation in life-history traits shows the diverse ways in which organisms have solved a complex array of problems including resource availability and habitat variability. Of course the idea that environmental variation resulted in variation in populations and species was Darwin's (1859) empirical observation, but since then we have passed through a bottleneck of typological thinking in ecology, typified by r and K selection theory, in which there is one solution to a given selective regime. Since that time there has been a growing sophistication of experimental analysis in ecology and the realization that alternative solutions to any problem exist in many species. We see in Chapters 5, 6, and 7 some of the major new approaches to life history studies: careful experimental analysis of mechanisms; recognition of a diverse array of solutions for organisms living in variable, unpredictable environments; and a broad comparative basis, both taxonomic and geographical, revealing the full range of variation in life history mechanisms.

The ecology and evolution of one specific life history trait, social behavior, has recently generated considerable debate. The seminal papers of Hamilton (1964, 1972) suggested kin selection as an elegant answer to the question of why animals are social. This answer, however, has not proved to be sufficient in explaining the diversity of social behaviors seen in animal groups, and ecological mechanisms relating to resource availability (Jarman, 1974) and antipredator defense (Triesman, 1975) have been suggested. A new approach has emphasized cooperative behavior (Axelrod and Hamilton, 1981). This approach, inspired by game theory (Maynard Smith, 1976, 1982), suggests that cooperative behavior may arise in response to a number of factors, including kin selection, resource abundance, and antipredator defense, as long as there is a net benefit to the individual to cooperate in a social group. Chapters 8, 9, and 10 reflect this novel approach and a primer on game theory is provided in Chapter 10.

Any textbook on general ecology may be consulted to perceive the importance ascribed to interspecific competition in universally influencing populations and communities. The alternative view that interspecific competition may be unimportant, or only one of several equally important ecological processes, raised many times (e.g., Gleason, 1926; Ramensky, 1926; Andrewartha and Birch, 1954; Whittaker, 1967) was never given much credence until the last decade. Since the early 1970s a growing volume of literature has recognized the commonness of nonequilibrium conditions in which interspecific competition is unlikely to be influential (see Pickett, 1980; Price, 1980, 1983b, for summaries), and the importance of patchiness, disturbance, and other factors (e.g., Dayton, 1971; Wiens, 1973; Connell, 1978; Lawton and Strong, 1981; Strong et al., 1983). We regard this as a healthy, desirable development in which alternative hypotheses on the development and dynamics of community organization are gaining equal plausibility. Part IV in this volume is devoted to this change in perspective in community ecology.

The chapters of synthesis in Part V pick out many themes linking the major areas treated in this volume.

2   THE SCIENTIFIC METHOD AND COMMUNICATION IN ECOLOGY

There are few graduate programs in ecology that deal specifically with the scientific method. This is unfortunate at a time when there is growing dissatisfaction in some sectors of the scientific community about evolutionary biology and ecology in relation to what is and what is not proper science (e.g., Peters, 1976; Brady, 1979, 1982; Bondi, 1980; Halstead, 1980; Popper, 1980). An additional problem is communication. Brady (1982, p. 79) has recognized a deficiency in communication between critics and defenders of evolutionary and ecological theory: "The critics . . . have not been exacting enough with their formulations. The defenders . . . have spent little or no effort finding out what the critics might actually have in mind." This pinpoints a second major failing in the training of ecologists today, that is, the meager emphasis on communication in science as a form of art. The two problems, the application of the scientific method and communication, are coupled and their correction represents a major challenge to ecologists in the future.

In their writing, scientists would profit from Joseph Pulitzer's admonition to writers: "Put it before them briefly so they will read it, clearly so they will appreciate it, picturesquely so they will remember it and, above all, accurately so they will be guided by its light." Few ecologists can claim these attributes for each or any of their scientific communications.

And yet the scientific method incorporates the clarity and precision Pulitzer demanded, if only ecologists would use it. Clarity comes from an explicit hypothesis posed about a certain question, stated early in a communication. Accuracy comes from the fact that the hypothesis is falsifiable: it is accepted or rejected on the basis of observations and experiments (type I and type II errors considered).

Ecology has moved from a strongly descriptive discipline to an experimental science. Descriptive studies of vegetation types, animal distributions and communities, food webs, and types of interaction (e.g., Elton, 1927; Tansley, 1939; Shelford, 1963) have been replaced by studies of mechanisms. The transition from description to mechanisms, however, has not been accompanied by a wholesale move to a more rigorous application of the scientific method. Answers have been largely speculative at best. Had the scientific method come into play adequately as the interpretive phase in ecology developed, description would have provided the first link in erecting hypotheses about ecological mechanisms. Tests of these hypotheses would have provided us with some concrete information on the validity of some mechanisms and interpretations. Such an approach applied throughout the field would undoubtedly have carried us beyond the condition we find ourselves in today, where we still debate the roles and influences of even the most obvious kinds of interactions. Even in 1980 Strong asserted that most ecological research "is either phenomenological on the one hand or corroborative on the other" (p. 273), with explicit null hypotheses infrequently used.

Since we believe that the scientific method is integral to ecological research, we would like to summarize the steps of the method, in the sequence appropriate for ecological questions. The elements of the scientific method are

1. Observation and description of a condition in nature.
2. Formulation of an interesting question relating to this condition.
3. Erection of hypotheses.
   (a) Erection of an hypothesis to explain the condition:
      (i) The hypothesis suggests the simplest possible explanation, separating the condition from its hypothetical cause(s); this null hypothesis may invoke nothing but random processes as leading to the observed condition (see Strong, 1980).
      (ii) The hypothesis must be verifiable or falsifiable using observations and experiments.
      (iii) An explicit statement should accompany the hypothesis on how it could be falsified.
   (b) Erection of alternative hypotheses that may account for the condition under study.
4. Tests among hypotheses.
   (a) Observations and experiments are designed to test as directly as possible the hypotheses.
   (b) An explicit method must be devised and stated in order to distinguish among hypotheses.
   (c) Tests among hypotheses must be objective and not biased towards a particular hypothesis.
   (d) Tests must incorporate the potential for falsifying hypotheses.
   (e) The extent to which the real world is modified by such experimental practices must be explicitly addressed and evaluated, and should be minimized with appropriate controls. Testing in the real world changes it, and testing in modified environments may alter reality.
5. Revision of hypotheses and asking new questions.
   (a) Once the original hypotheses have been tested and the results ascertained, they may require revision. Retesting with new data and/or the formulation of deeper, more searching questions may be necessary.
6. Communication of results.
   (a) Specialized terms are defined clearly and briefly.
   (b) Definitions, questions, and hypotheses are explicitly stated and distinguished.

(c) Distinction is made among established fact, extrapolations from fact, possibilities, speculations, guesses, and fresh working hypotheses to be examined later.

Testing hypotheses by the scientific method provides two possible outcomes: (1) accepting the null hypothesis (because it was true, or because the evidence was not sufficient to reject it even though it was false); and (2) rejecting the null hypothesis (because the evidence was sufficient to prove it false, or because the evidence was misleading). The former of these two conclusions constitutes a negative result (see below).

We see in the literature a growing awareness that the scientific method must be applied to ecological questions. As more ecologists apply this method, there will be increased improvement in the scientific rigor with which ecological mechanisms are developed. Much of this rigor will come from the new ecologists being trained today, and from those faculty members and journal editors who are deeply concerned with the proper application of the scientific method and communication in science.

3 THE IMPORTANCE OF NEGATIVE DATA

The scientific method necessarily includes the reporting of negative results. Some biological disciplines, such as biogeography, have taken this for granted. Ecology has not. Researchers and editors have considered it meritorious to publish evidence of interactions such as competition, however poor or misleading the evidence is, whereas a scientifically rigorous demonstration of competition's absence in a certain community is regarded as uninteresting. Whereas a biogeographer can say that a certain animal occurs in one area and not in another, the ecologist cannot say that a certain type of interaction occurs in one type of community and not in another. We have little idea of the relative commonness of interspecific competition, mutualism, parasitism, amensalism, and so on, in different kinds of communities, because we have not paid enough attention to negative evidence. This has allowed our science to become dominated by simple themes with apparently universal application, because the evidence establishing both the presence and absence of a certain phenomenon has not been treated equitably. Part of a new ecology must involve a new appreciation for negative data, as a step in the development of appropriate hypotheses. This requires a new ethic among investigators, authors, reviewers of research proposals and articles, research granting agency personnel, and editors; that is, an ethic that establishes the morality of reporting negative results.

Part of the difficulty in accepting negative data can be ascribed to an incorrect application of the scientific method. As statistics have shown, hypotheses can be accepted or rejected with different levels of confidence. A biogeographer who spends several years trapping and studying the mammal fauna of several mountaintops can say with nearly 100% confidence that a particular mammal is not present. Ecological experiments can rarely approach such confidence levels. The complexity of variation between organisms and their environment leads to much higher levels of uncertainty than a biogeographer would normally face. Such uncertainty allows detractors of negative data to say that the hypotheses producing the negative data were incorrectly stated, and a set of reformulated hypotheses, usually untestable under the particular circumstances, would have produced a positive result that had less associated uncertainty. Without question, the goal of the scientific method is to reduce the uncertainty in accepting or rejecting hypotheses. However, all the hypotheses must be testable. If a testable hypothesis produces negative results, with some degree of uncertainty, then that hypothesis is preferable to an untestable one that fits more closely a current or popular paradigm.

4 SYNTHESIS OF DISCIPLINES

With the increasing interest in mechanisms and processes in ecology, accompanied by a sincere effort to analyze them in detail, there is a new need for collaboration between scientists of divergent backgrounds. Chapter 3 emphasizes the need for understanding the role of microorganisms in plant–microorganism–herbivore interactions. Since microbial techniques and understanding are specialized, adequate study of such interactions requires the collaboration of microbial ecologists with plant–herbivore ecologists. In addition, plant quality and variation within plants need to be understood in more detail (see Chapter 2), not only in a descriptive way but also in terms of the genetic and physiological mechanisms involved. It now seems impossible to understand one trophic level unless the details of the resources that trophic level depends on are understood. Thus, instead of having one investigator studying plant–herbivore interactions it is much more realistic, and more productive, to have an association among a plant physiologist, a phytochemist, a microbiologist, and a person working on the dynamics of the herbivores and their enemies.

Other areas in ecology are also in need of more collaboration than is now apparent. As mutualistic relationships receive the attention they deserve there will be a growing need for understanding both parties in the system (see also Chapters 14 and 16), many of which are microbes. For example, the development of galls by herbivores is not well understood except in the crown gall system (reviewed by Ream and Gordon, 1982), but no doubt similarly complex interactions between vectored plasmids or viruses and the host plant will be involved.

The same broadening of perspective could be justified in any other area of ecology. Unfortunately, the training of ecologists does not seem to be adequate for the task. Ecologists trained in the broad conceptual arena usu-

ally do not get an adequate exposure to natural products chemistry, microbiology, plant physiology, or other pertinent fields. Perhaps this is expecting
too much, and the answer may lie in more collaboration among researchers.
However, there is a challenge in training ecologists to make them fully aware
of the need for integrating several disciplines in any realistic study of a
natural system.

The other stricture on collaboration is the funding available. Funding
agencies have tended to keep grants relatively small, as a way of spreading
limited dollars widely. This contributes to a fractionation of effort and is
divisive at a time when growing understanding seems to demand a broadening of the ecological base to include other disciplines more intimately. An
alternative is to foster more collaborative efforts with some larger grants
supporting three or four workers from different disciplines in a tightly integrated research effort. We predict that this approach will advance our
understanding of ecology more rapidly. The emphasis should remain on a
detailed understanding of mechanisms, and research teams should be small,
with each member working on an essential component of the system. The
initiative for this development should come from the research community.
Graduate students sensitized to the need for a broader yet more detailed
approach may well be the ones that establish the more integrated research
perspective needed in many areas of ecology. In this way the synthetic and
reductionist approaches so necessary in ecology may best be served (see
also Bartholomew, 1982).

5 FADS IN ECOLOGY

The strongly reinforcing system in ecology for publishing positive results
has delayed progress perhaps by decades. Many scientists feel compelled
to fit data into some existing body of theory, and do not feel equally compelled to falsify theory. This bias predisposes ecology to faddishness. Once
a theory is established, researchers struggle to fuel it, and confirm its validity. A healthier science would emerge should equal time be devoted to
attempts at falsifying the theory.

The most notable case of a fad in ecology is interspecific competition. So
dominant has this concept been that its status has reached that of a paradigm,
in Kuhn's (1970) sense (Strong, 1980). The body of theory is vast, and yet
very little has been tested objectively. Assuming that its construction started
with Gause, only after 50 years of building an edifice to competition is serious
doubt being cast on the evidence for its foundation (e.g., Connell, 1980;
Simberloff, 1980; Strong, 1980; Arthur, 1982; Strong et al., 1983). Connell
(1980) could find only one study in which adequate experimentation had
been performed to establish that competition played a significant role in the
interaction between species. After an extensive review of interspecific competition, Arthur (1982, p. 181) concluded: "However, as regards the relative

commonness of character displacement, character release and the evolution of competitive ability, and indeed the extent to which these processes occur separately to each other, almost nothing is known." Had more objective tests and more negative data filtered into the literature at an early date the edifice to competition would still exist but would be smaller and sounder, leaving room for early construction of alternative interactive mechanisms which are only now beginning to be built.

This is not a plea for less theory, but for undogmatic theory (see Bartholomew, 1982), a dispassionate approach to theory. Theory in ecology has been creative and has had heuristic value. The answer lies in the development of testable and falsifiable theory, its rapid testing, and the publication of negative results, if warranted. This process should rapidly select productive lines of research and cause early avoidance of misleading trends.

In conclusion, there are many challenges for ecologists entering the field. Many others are discussed in the chapters that follow. Young ecologists have much more potential for changing ecology for the better than anyone established in the field. Graduate students in ecology are better trained now than they have ever been, so the future holds great promise.

ACKNOWLEDGMENTS

We thank Hugh Dingle for discussions on some of these topics and William Boecklen and Susan Mopper for critical comments. Financial support was provided to P.W.P. through NSF grant DEB 80-21754.

LITERATURE CITED

Andrewartha, H. G., and L. C. Birch. 1954. *The distribution and abundance of animals.* University of Chicago Press, Chicago.

Arthur, W. 1982. The evolutionary consequences of interspecific competition. *Adv. Ecol. Res.* **12**:127–187.

Axelrod, R., and W. D. Hamilton. 1981. The evolution of cooperation. *Science* **211**:1390–1396.

Bartholomew, G. A. 1982. Scientific innovation and creativity: A zoologist's point of view. *Am. Zool.* **22**:227–235.

Bondi, M. 1980. Evolution. *New Sci.,* **87**:611.

Boucher, D. H., S. James, and K. H. Keeler. 1982. The ecology of mutualism. *Annu. Rev. Ecol. Syst.* **13**:315–347.

Brady, R. H. 1979. Natural selection and the criteria by which a theory is judged. *Syst. Zool.* **28**:600–621.

Brady, R. H. 1982. Dogma and doubt. *Biol. J. Linn. Soc.* **17**:79–96.

Connell, J. H. 1978. Diversity in tropical rain forests and coral reefs. *Science* **199**:1302–1310.

Connell, J. H. 1980. Diversity and the coevolution of competitors, or the ghost of competition past. *Oikos* **35**:131–138.

Darwin, C. 1859. *On the origin of species by means of natural selection.* Murray, London.

Dayton, P. K. 1971. Competition, disturbance, and community organization: The provision and subsequent utilization of space in a rocky intertidal community. *Ecol. Monogr.* **41**:351–389.

Edmunds, G. F., Jr., and D. N. Alstad. 1978. Coevolution in insect herbivores and confiers. *Science* **199**:941–945.

Elton, C. 1927. *Animal ecology.* Macmillan, New York.

Gleason, H. A. 1926. The individualistic concept of the plant association. *Bull. Torrey Bot. Club* **53**:7–26.

Halstead, B. 1980. Popper: Good philosophy, bad science? *New Sci.* **87**:215–217.

Hamilton, W. D. 1964. The genetical evolution of social behaviour. Parts I and II. *J. Theor. Biol.* **7**:1–52.

Hamilton, W. D. 1972. Altruism and related phenomena, mainly in social insects. *Annu. Rev. Ecol. Syst.* **3**:193–232.

Jarman, P. T. 1974. The social organization of antelope in relation to their ecology. *Behaviour* **48**:215–267.

Kuhn, T. S. 1970. *The structure of scientific revolutions,* 2nd ed. University of Chicago Press, Chicago.

Lawton, J. H., and D. R. Strong, Jr. 1981. Community patterns and competition in folivorous insects. *Am. Nat.* **118**:317–338.

May, R. M. 1973. *Stability and complexity in model ecosystems.* Princeton University Press, Princeton, New Jersey.

May, R. M. 1981. Models for two interacting populations. In R. M. May (ed.), *Theoretical ecology, principles and applications,* pp. 78–104. Sinauer, Sunderland, Massachusetts.

Maynard Smith, J. 1976. Evolution and the theory of games. *Am. Sci.* **64**:41–45.

Maynard Smith, J. 1982. *Game theory and evolution.* Cambridge University Press, Cambridge.

Peters, R. H. 1976. Tautology in evolution and ecology. *Am. Nat.* **110**:1–12.

Pickett, S. T. A. 1980. Non-equilibrium coexistence of plants. *Bull. Torrey Bot. Club* **107**:238–248.

Popper, K. 1980. Evolution. *New Sci.* **87**:611.

Price, P. W. 1980. *Evolutionary biology of parasites.* Princeton University Press, Princeton, New Jersey.

Price, P. W. 1983a. Hypotheses on organization and evolution in herbivorous insect communities. In R. F. Dunno and M. S. McClure (eds.), *Variable plants and herbivores in natural and managed systems,* pp. 559–596. Academic Press, New York.

Price, P. W. 1983b. Communities of specialists: Vacant niches in ecological and evolutionary time. In D. Strong, D. Simberloff, and L. Abele (eds.), *Ecological communities: Conceptual issues and the evidence.* Princeton University Press, Princeton, New Jersey.

Ramensky, L. G. 1926. Die Grundgesetzmässigkeiten in Aufbau der Vegetationsdecke. *Bot. Centralbl. N.F.* **7**:453–455.

Ream, L. W., and M. P. Gordon. 1982. Crown gall disease and prospects for genetic manipulation of plants. *Science* **218**:854–859.

Shelford, V. E. 1963. *The ecology of North America.* University of Illinois Press, Urbana.

Simberloff, D. 1980. A succession of paradigms in ecology: Essentialism to materialism and probabilism. *Synthese* **43**:3–39.

Stearns, S. C. 1982. Components of fitness. *Science* **218**:463–464.

Strong, D. R. 1980. Null hypotheses in ecology. *Synthese* **43**:271–285.

Strong, D., D. Simberloff, and L. Abele (eds.). 1983. *Ecological communities: Conceptual issues and the evidence.* Princeton University Press, Princeton, New Jersey.

Tansley, A. G. 1939. *The British Islands and their vegetation.* Cambridge University Press, London.

Triesman, M. 1975. Predation and the evolution of gregariousness. I. Models for concealment and evasion. *Anim. Behav.* **23**:779–800.

Whitham, T. G. 1978. Habitat selection by *Pemphigus* aphids in response to resource limitation and competition. *Ecology* **59**:1164–1176.

Whitham, T. G., and C. N. Slobodchikoff. 1981. Evolution by individuals, plant-herbivore interactions, and mosaics of genetic variability: The adaptive significance of somatic mutations in plants. *Oecologia* **49**:287–292.

Whittaker, R. H. 1967. Gradient analysis of vegetation. *Biol. Rev.* **42**:207–264.

Wiens, J. A. 1973. Pattern and process in grassland bird communities. *Ecol. Monogr.* **43**:237–270.

Williamson, M. 1972. *The analysis of biological populations.* Arnold, London.

PART I

Resources and Populations

The Variation Principle: Individual Plants as Temporal and Spatial Mosaics of Resistance to Rapidly Evolving Pests

THOMAS G. WHITHAM
Department of Biological Sciences
Northern Arizona University
and
Biology Department
Museum of Northern Arizona
Flagstaff, Arizona

ALAN G. WILLIAMS
ANNE M. ROBINSON
Department of Biological Sciences
Northern Arizona University
Flagstaff, Arizona

CONTENTS

1   INTRODUCTION

With the development of so-called "superior" strains of wheat, oats, corn, and other important crops, agronomists learned at often devastating cost the susceptibility of these strains to rapidly evolving pathogens and parasites. For example, after 1–3 years of their widespread use, cultivars of oats employing single genes for resistance succumbed to the rapidly evolving crown-rust and were removed from use when up to 30% of the oat crop was lost (Knott, 1972; Frey et al., 1973). Similarly, the southern corn leaf blight overcame the resistance of corn in 1970 because breeding practices had reduced 85% of the corn in the United States to near genetic homogeneity (Roane, 1973). The resulting epidemic was severe and emphasized the importance of diversity in the host population as a deterrent against rapidly evolving pathogens. Because of these and other examples, the importance of host variation in the population has long been recognized (Rosen, 1949; Vanderplank, 1963, 1968; Browning and Frey, 1969; Pathak, 1970; Thurston, 1971; Knott, 1972; Fry et al., 1973, 1979; Browning, 1974; Gallun et al., 1975; Browning et al., 1977, 1979; Cowling, 1978; Day, 1978; Schmidt, 1978; Leonard and Czochor, 1980; Robinson, 1980; Segal et al., 1980).

An excellent example of how diverse defenses may deter rapidly evolving pathogens, parasites, or herbivores from becoming virulent and causing serious economic loss was demonstrated by Pimentel and Bellotti (1976). Under laboratory conditions the housefly, *Musca domestica,* was exposed to a single toxicant at a concentration that resulted in 80% mortality (Fig. 1). The survivors were then reared and bred, and the experiment was then repeated on the next generation. With as few as seven generations of repeated exposure to the same toxicant, mortality had dropped from 80 to 30% and similar results were achieved with five other replicates, each of which used different toxicants. In comparison, another experiment using the same toxicants was designed to examine the effects of diverse plant defenses. The fly population was divided into subpopulations, each of which was exposed to a single but different toxicant and the survivors of all subpopulations allowed to interbreed. Since the toxicants were deployed spatially in combination, this experimental design closely approximates a diversity of de-

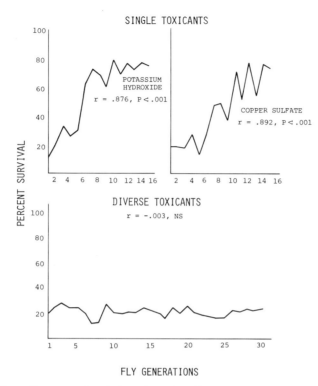

Figure 1. Houseflies exposed to single toxicants in which the survivors then interbred, rapidly evolved resistance as evidenced by increased survival over a few fly generations. When flies were exposed to diverse toxicants used in spatial combination (see text), no increase in survival was observed over 32 generations. A diversity of defenses stabilized the fly population and prevented the evolution of specific resistance traits. Adapted from Pimentel and Bellotti (1976).

fenses in which different hosts employ different defenses. After 32 generations of exposure to diverse toxicants, *no* detectable change had occurred in the fly population. Thus, when exposed to a single toxicant, flies rapidly evolved mechanisms to cope with the toxicant, but when the number of toxicants increased, they were unable to evolve the appropriate detoxification pathways.

As Pimentel and Bellotti (1976) point out, multiple defenses in a host population are very difficult to break because, to overcome host resistance, the parasite must acquire all the genes necessary to overcome the host's multiple defenses. It is even argued that such variation can be used to stabilize the pathogen population and select for simple races that lack the ability to reach epidemic proportions (Vanderplank, 1963; Browning and Frey, 1969; Knott, 1972; Frey et al., 1973, 1979; Pimentel and Bellotti, 1976; Leonard and Czochor, 1980). If true, this may explain why outbreaks in native systems are uncommon and why the general levels of insect damage to native plants are kept at relatively low levels (Mattson and Addy, 1975).

Whitham (1981, 1983) expanded on the above logic based on the importance of variation and multiple defenses in the plant *population* to argue that the same ideas apply to *individuals* in the host population as well. Thus, variation within an individual plant (particularly a long-lived host or clonally reproducing species) may be an evolved trait that negates the evolutionary advantages of pests with shorter generation times and greater recombination potential than their hosts (Whitham, 1981). Within-plant variation in time and space employing qualitative and/or quantitative variation in defenses, nutrition, or any other factor affecting parasite fitness may make the plant appear as a chameleon of different or changing resistances to herbivore attack. Such variation may have the same effects on the pest population as those quantified by Pimentel and Bellotti (1976).

The analogy between an individual plant and the plant population is appropriate for several reasons. First, a plant is at the very least a metapopulation of genetically identical individuals because of a plant's modular construction and the capacity of all modules to reproduce (Harper, 1977; White, 1979, 1980). A plant may even represent a true population exhibiting genetic variation due to chimeras and somatic mutations (Whitham and Slobodchikoff, 1981). Furthermore, due to developmental patterns of plant growth and induced defensive responses, different modules of the same plant may behave differently in response to parasite attack.

Second, just as some parasites specialize on a single host species or population, even more specialized parasites may evolve to attack a specific host genotype or individual plant. For example, Flor (1971) demonstrated a gene-for-gene interaction between some fungal parasites and their hosts. Similarly, there exists a genetic interaction between the Hessian fly, *Mayetiola destructor,* and wheat in which the genotype of the fly determines which wheat genotypes will be susceptible or resistant (Hatchett and Gallun, 1968, 1970). This has been carried a step further by the black pineleaf scale, *Nuculaspis californica,* which may become genetically adapted to an individual ponderosa pine tree, *Pinus ponderosa* (Edmunds and Alstad, 1978, 1981), in apparent response to the great intertree variation in terpenes and other plant defenses (Sturgeon, 1979). If an individual host tree represents a sufficiently large resource to favor the evolution of a genetically distinct population of parasites, then an individual tree may be considered as a monoculture.

Third, just as large populations risk increased exposure to rapidly evolving pathogens, long-lived individual plants also risk increased exposure to rapidly evolving pests. For example, agroecosystems are highly exposed to rapidly evolving pests because they form genetically uniform *monocultures in space.* In comparison, long-lived individual plants or clones may be exposed to similar pressures because they form *monocultures in time* (Whitham, 1981, 1983; Whitham and Slobodchikoff, 1981). Both herbaceous and woody perennials may be long-lived and form large clones. Clones of creosote bush, *Larrea tridentata,* can be as old as 11,700 years (Vasek, 1980).

Individual clones of aspen, *Populus tremuloides,* can cover 200 acres, be composed of 47,000 trees and may date back to the Pleistocene (Kemperman and Barnes, 1976). Clones of goldenrod may be 1000 years old and composed of 10,000 stalks (William J. Platt, personal communication). Bracken fern, *Pteridium aquilinum,* forms clones covering 138,000 m^2 and may be as old as 1400 years (Oinonen, 1967a,b). Clones of the grass, red fescue, *Festuca rubra,* may occupy 45,240 m^2 and are estimated to be hundreds if not thousands of years old (Harberd, 1961). When the dioecious Canadian pondweed, *Elodea canadensis,* was introduced in Europe about 1840 no male plants arrived. It nevertheless spread by cloning over the next 40 years to occupy the waterways of the entire region (Gustafsson, 1946). Thus, even though the clone is relatively young, it approaches a traditional monoculture in size.

Due to the genetic exposure of long-lived plants or clones, the importance of variation which has been demonstrated at the population level may well extend to the individual host. The following sections of this chapter will examine (1) the various mechanisms whereby an individual plant can become a mosaic of resistance in time and space, (2) our current knowledge of within-plant variation, and (3) the ecological implications of variation and their associated testable hypotheses.

2 GENETIC VARIATION WITHIN THE INDIVIDUAL PLANT

Genetic variation within individual plants has not generally been recognized or appreciated by ecologists. Cytologists, botanists, and horticulturists, however, have long been aware of such variation. Any genetic variation that arises (point mutations, deletions, duplications, inversions, aneuploidy, polyploidy, and extrachromosomal mechanisms of inheritance) can be perpetuated and is heritable (Whitham and Slobodchikoff, 1981). If a somatic mutation arises in a meristematic cell of a bud, the derivative cells will also carry the mutation and it can spread with the annual growth of the plant. The mutation can be expressed in the gametes of sexual reproduction (Stewart and Dermen, 1979) and/or develop into an independent plant through various naturally occurring modes of asexual reproduction (tubers, tubercles, tuberous roots, bulbs, pseudobulbs, cormels, stolons, runners, rhizomes, offsets, root crowns, stem joints, layering of lower branches, inflorescence bulbils, apomictic seeds, and adventitious embryony).

Long life-span, large clone size, and the complete regeneration of the meristematic tissues each year (i.e., buds) make it highly probable that variation will arise within the individual plant or clone. Since heritable mutations can arise in any bud, the reproductive population is far greater than the number of plants in the population and is directly proportional to the number of buds or repetitive modules. The following example demonstrates the size of the bud population from which heritable variation may arise. A 20 m narrowleaf cottonwood tree, *Populus angustifolia,* has approximately 30,000 buds (Whitham, unpublished data), each of which may mutate to form a bud sport. Assuming that buds are added at a geometric rate from the seedling stage and remain constant at 30,000 buds per year from age 16 to 100 years when the tree dies (see White, 1980, for a similar example), the aboveground bud population over the life-span of the tree is approximately 2,600,000. Considering that narrowleaf cottonwood also reproduces vegetatively and that an individual clone may be composed of 60 trees, over the life-span of these trees the bud population would be 156,000,000. If the clones date back to the Pleistocene as with the closely related species *Populus tremuloides* (Kemperman and Barnes, 1976), the vegetatively derived bud population from a single seedling would be approximately 18 billion buds.

The above estimates of the meristematic population are conservative because they have excluded the cambium, below ground meristematic tissues, and the number of cells per bud primordium. Depending on the plant group, the number of meristematic cells per bud primordium varies (Esau, 1975), and as the number of cells increases, the greater is the probability a mutation will arise. *Equisetum* (horsetail) possesses a single mother cell per primordium, whereas angiosperms have many initiating cells. By examining the apical meristems of the haploid gametophyte of the fern, *Onoclea sensibilis,* Klekowski (1983) found an average of 0.0242 mutations per apical cell per shoot apex. Considering the number of cells per bud primordium of higher plants and the size of the bud population, many deviant buds should arise simultaneously on the same plant.

Because of the heritability of genetic variation within an individual plant, differential selection coefficients may operate on different parts of the same plant. Just as variation within a population can be selected on, genetic variation within a single plant is also the ''stuff'' of evolution and may allow the plant as an ''archipelago of similar but distinct genetic islands'' to change in gene frequency through time and evolve (Whitham and Slobodchikoff, 1981, p. 287). At present we do not know how common genetic variation is within a single plant. Most researchers simply do not look for it; the variation may not produce visible phenotypes; and within a single plant it is often dismissed as a freak occurrence, environmentally caused or as a mistake in sampling because plants are supposed to be genetically uniform throughout. The following examples clearly demonstrate that genetic variation within a single plant can arise, may be quite common, and be evolutionarily significant.

Lewis et al. (1971, p. 564) state ''Constancy of chromosome number within all individuals is merely convenient fiction . . . '' and this variation needs critical evaluation. In their examination of the herbaceous perennial, spring beauty, *Claytonia virginica,* individual plants were found to have multiple genotypes in which different parts of the same plant could have different chromosome numbers. Table 1 shows that different roots of the same plant had as many as four different chromosome numbers ranging from 28 to 52. Had such multiple genotypes been found in only a few plants, the

Table 1. Multiple Genotypes in Roots from Corms and Floral
Stems and from Microsporocytes of Individual Plants of
Claytonia virginica[a]

Diploid Chromosome Number		
Roots from Corms	Roots from Flora Stems	Microsporocytes[b]
28(2)	28(1)	29(2)
28(4)	30(2)	30(5)
28(2), 29(2)	28(1)	28(5)
28(3), 29(1)	28(2), 28 and 29(1)[c]	28(2), 29(1)
28(2), 29(2)	28(1), 32(1)	28(2)
28(1), 29(3)	28(2)	29(3), 31(1), 33(1)
28(4), 29(1)	31(1)	29(8), 31(3)
28(3), 30(4)	28(1), 30(1), 31(1)	30(1)
	30 and 52(1)[c,d]	

[a] Numbers in parentheses indicate number of roots and floral stem apices studied. Each line is for one plant. (From Lewis et al., 1971. Copyright 1971 by the American Association for the Advancement of Science.)

[b] Chromosome numbers expressed as $2n$.

[c] Two numbers in a single adventitious root (aneusomaty).

[d] Plus plantlet number 1 having $2n = 28$ in floral stem roots.

variability might have been dismissed as an anomaly. However, 68% of the population studied exhibited multiple genotypes *within* the same plant. Lewis et al. (1971) note that plants collected in the field are just as likely to exhibit within-plant variability as are plants grown in the greenhouse. This suggests that variability is not induced by the rigors of field growth but is "locked genetically." Furthermore, Lewis (1970) also reports that over a several year period the mean chromosome number of the population has shifted, indicating that selection is acting on the population and the variability within the individual.

There is considerable evidence that the above example of *Claytonia virginica* (family Portulacaceae) is not an isolated case, and variable chromosome numbers within an individual are found in diverse plant families. Berger and Witkus (1946) describe populations of *Xanthisma texanum* (Compositae) that are composed of two cytologically different types. One type has eight chromosomes in all cells of the plant, whereas the other type has roots with eight chromosomes and shoots (also germ line cells) with 10 chromosomes. Since the two types are morphologically indistinguishable, variation of this kind is probably largely undetected. Similar observations have been made with *Poa alpina* (Gramineae) (Müntzing, 1946, 1948), *Haplopappus spinulosis* spp. *cotula* (Compositae) (Li and Jackson, 1961), *Allium cepa* (Liliaceae) (Berger and Witkus, 1946), *Rubus* sp. (Rosaceae) (Hull and Britton, 1956), and *Sorghum purpureosericeum* (Gramineae) (Darlington and Thomas, 1941).

What is the effect of chromosomal variation within a single plant? Conservatively, with aneuploidy the same genetic material is present but spread over different numbers of chromosomes. With polyploidy, the dosage of individual genes increases and, when the plant flowers, the sexually derived progeny from polyploid tissues will be more heterozygous than those derived from diploid tissues (Solbrig and Solbrig, 1979, p. 113). In both aneuploids and polyploids the linkages, crossing-over potential, and the chance for inversions to occur will be affected and vary between different parts of the same plant. Consequently, these examples of chromosomal variation within a plant permit natural selection to occur within the individual.

Hybridization between different species and subsequent somatic doubling resulting in allopolyploidy represent a common form of speciation in the plant kingdom. It is estimated that 47% of the angiosperms are polyploid and in some groups (e.g., the pteridophytes or ferns and fern allies) 95% of the species are polyploids (Grant, 1971, p. 234). An example of incipient speciation resulting from chromosomal variation within an individual plant was observed by Butters and Tryon (1948). The fern *Woodsia abbeae* is a sterile hybrid of *W. cathcartiana* and *W. ilvensis*. In scrutinizing the fronds on a field trip they observed that, whereas the base of one frond produced the typical sterile sporangia, the tip of the same frond produced what appeared to be fertile sporangia (Fig. 2). The spores of the latter were collected and found to be viable. They also found the individual cells of the fertile frond tip were 1.28 times larger than the cells from the sterile base, indicating that the base was diploid and the tip tetraploid.

This example of incipient speciation demonstrates the ecological impor-

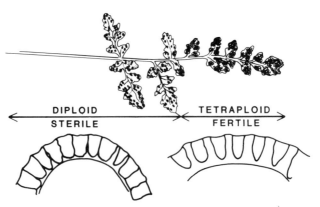

Figure 2. Drawing of a single frond of a natural diploid hybrid of *Woodsia cathcartiana* and *W. ilvensis*. Although this hybrid is sterile and reproduces only vegetatively, a somatic mutation occurred in the meristematic region of one frond to produce tetraploid tissues. Although the diploid base of the frond produced only sterile sporangia, the tetraploid tissues of the tip produced fertile sporangia. Portions of annulus of both sterile and fertile sporangia are shown in cross section. This represents an example of incipient speciation within a single plant. Redrawn by Pam Lungé from Butters and Tryon (1948).

tance of a somatic mutation occurring in the meristem and is not an isolated example. The diploid perennial herb *Primula kewensis* is a sterile hybrid of *P. floribunda* and *P. verticillata*. On *three* different occasions the otherwise sterile hybrid plant spontaneously gave rise to fertile tetraploid branches that produced viable seed (Grant, 1971, p. 192). Thus, repeated somatic mutations occurring in the bud population were responsible for the restoration of fertility. Although such mutations are thought to be rare events, considering the life-span of plants, their cloning behavior, and the size of the bud populations, genetic variation within a single plant is to be expected.

Some of the best evidence of genetic variation within an individual is observed in the fungi in which somatic variation in the vegetative phase of the life cycle is common. For example, even though no known sexual reproduction occurs in the deuteromycotina, significant genetic variation exists due to heterokaryosis and parasexuality (Moore-Landecker, 1982, p. 344). Heterokaryosis is the condition of the same cell containing genetically different nuclei, and parasexuality is the process whereby mitotic divisions produce nuclei that are genetically different from the original nuclei. Although crossing-over is generally associated with meiosis, as part of the parasexual cycle *mitotic crossing-over* may occur, which results in recombination (Pontecorvo, 1956, 1958; Papa, 1978). Mitotic crossing-over also occurs in higher plants to produce chimeras (Stewart, 1978). In the fungus *Macrophomina phaseoli,* mitotically derived variation is very common and the same mycelium may possess haploid, diploid, aneuploid, and possibly polyploid nuclei (Knox-Davies, 1967). It is thought that these nonsexual (no gametic fusion or meiosis) mechanisms of vegetative reproduction have provided an important source of heritable variation in the filamentous ascomycetes and basidiomycetes (Moore-Landecker, 1982, p. 344).

Genetic variation may also arise within an individual plant in which two or more genetically distinct tissues may grow adjacent to one another. Because the apical meristems of the stems of most angiosperms are thought to be composed of three distinct layers in which each layer gives rise to relatively independent cell lineages, the derivative tissues may express different genotypes and compete with one another for expression (Stewart, 1978). If a somatic mutation occurs in only one layer of the meristem, the derivative tissues of the affected portion will be one genotype and the derivative tissues of the rest of the meristem will be of another genotype. Such plants are referred to as chimeras and may be quite common. For example, Stewart and Dermen (1979) have recorded stable plastid chimeras (variable leaf color) in 60 genera in 10 plant families of monocotyledons.

If the derivative tissues of the affected meristematic layer include the floral parts, it is quite possible for different branches of the same plant to produce different types of gametes. Baur (1930) found that self-pollinated flowers arising on totally green branches of *Pelargonium* gave rise to only green seedlings, and self-pollinated flowers arising on totally colorless branches produced only colorless seedlings. However, if the two branch types were crossed, some of the seedlings produced green, colorless, and chimeral branches. Thus, once the variation arises, it may be maintained by both vegetative and sexual reproduction.

Figure 3 shows an example of a plastid chimera in the dicotyledonous evergreen shrub *Euonymus japonica,* which grows to a height of about 2.5 m. As demonstrated in this photograph, there is great variation both within and between plants in the chimeral pattern. Some branches produced completely green leaves; other branches produced completely yellow leaves; and most interestingly, other branches produced leaves with one half green and the other half yellow, or the leaves were a blend of yellow and green in various dosages. Although the genetic variation shown here produces a visible effect, other somatic mutations may not be visible, and affect the defensive chemistry of the plant. For example, Soost et al. (1961) described a somatic mutation that arose in citrus in which the acid concentration of the fruit was so high that the fruit was unpalatable. When the mutant branch was unknowingly used as propagation stock for other orchards, economic loss resulted. Additionally, Shamel (1943) estimated that in the average orchard 25% of the trees had produced visible deviations from the original parent stock demonstrating that citrus are very difficult to maintain as pure clones.

An example of the genetic variation which may exist within a single clone is described by Shepard et al. (1980). Although genetic explanations are lacking, a phenomenal degree of variation was observed in protoplast-derived clones of 'Russet Burbank' potatoes, *Solanum tuberosum* spp. *tuberosum.* Russet Burbank is itself a somatic mutation of 'Burbank.' The clones were derived from single leaf cell protoplasts from a single clone. Since no known chemical or physical mutagens were used, it was anticipated

Figure 3. Photograph of a plastid chimera of the evergreen shrub *Euonymus japonica,* demonstrating extensive genetic variation within an individual plant. Different parts of individual leaves, whole leaves, small and large branches differ genetically in chlorophyll production and the pattern is highly variable from shrub to shrub. Photo by T. G. Whitham.

that the regenerated plant population would be composed of carbon-copy replicas of the parent stock. To their surprise, however, extensive variation was observed and found to be stable over a period of three tuber generations. Most importantly, the variation was not trivial. Clonal variation included resistance to the fungus *Alternaria solani,* and to late blight, *Phytophthora infestans,* which caused the Irish potato famine in the 1840s. Although Russet Burbank is highly susceptible to late blight, several of the protoplast-derived clones were found to be highly resistant, and the resistance most closely resembled polygenic rather than single major gene resistance. Other important characters that varied among clones were tuber shape, yield, maturity date, photoperiod requirements for flowering, and plant morphology. Similar results have been obtained with cultivars of sugarcane (Heinz et al., 1977). Although it is well known that the culturing process itself can induce genetic variability (Sunderland, 1977), D'Amato (1975) states that as much as 90% of the genetic variation preexists *in vivo.* It would appear that when individual cells are liberated from the constraining influence of adjacent cells in their native environment, they are then free to express their true genetic makeup.

Shepard et al. (1980, p. 23) suggest "that a somatic cell population directly isolated from the plant represents a vast source of genetic variation of interesting practical and genetic consequence." The implications of such variability within an individual plant or clone are great, and likely to be important in plant evolution. They also note that many vegetatively propagated plants frequently express spontaneous mutations. For example, with the 'Centennial' cultivar of sweet potato, *Ipomoea batatas,* two plants out of every 100 differed morphologically (Hammett, 1979). Viewed in this light it is not surprising that some horticulturists have stated that the progeny of asexually propagated crops, such as cherry, lemon, grape, sweet potato, chrysanthemum, dahlia, coleus, and others, are no more uniform than that of seed propagated crops (Edmond et al., 1975). Such intraclonal variation is further supported by Cramer (1907, cited in Shamel and Pomeroy, 1936), who estimated that of the 8000 plant varieties cultivated in Europe in 1899, 5000 originated as somatic mutants. Numerous other examples of intraplant variation and the adaptive significance of somatic mutations in plants are developed by Whitham and Slobodchikoff (1981).

If both individual plants and the host population varied as much as the above examples suggest, a specialist parasite might encounter severe difficulties trying to locate suitable hosts or parts of hosts. In a highly heterogeneous host environment, host selection may become a very crucial and limiting window in the parasite life cycle. If different parts of the same plant varied qualitatively and/or quantitatively in their defenses, such within-plant diversity may be highly adaptive in preventing parasites from evolving specialized adaptations. These and other implications will be developed in the discussion.

3   INDUCED VARIATION WITHIN THE INDIVIDUAL PLANT

Pathogen attack can induce the production of temporal and spatial host defenses that affect pathogen success. The active induced defense of a plant can create variation in time if the reaction is systemic and spreads throughout the entire plant, or variation in both time and space if the host response is localized. The invading pathogen will then be forced to either (1) adapt to rapidly changing host quality; (2) migrate to unaffected host tissues; or (3) migrate to an unaffected host. Since the mosaic pattern of host quality is an induced response to parasite attack, the success of subsequent attacks is affected by the parasite's ability to discriminate between patches that have become unpredictable in time and space. Those pathogens unable to do so should suffer a reduction in growth, survival, and reproduction.

Green and Ryan (1972) discovered that damage to the leaves of potato by the Colorado potato beetle induces a rapid and systemic defensive reaction throughout the plant that could provide protection against persistent or subsequent attack. Apparently, feeding by the insect released pectic polysaccharides from the cell wall of the leaf (Ryan et al., 1981) which are then transported through the phloem, primarily to younger leaves (Makus et al., 1980). The magnitude of the chemical signal that is released depends on both the location and amount of insect feeding (Green and Ryan, 1972). Within hours, the levels of four different proteinase inhibitors rise (Graham and Ryan, 1981), and within 48 hr, up to 10% of all soluble proteins in the undamaged leaves are inhibitor proteins (Gustafson and Ryan, 1976). These proteins, which exhibit both specificity (Applebaum, 1964) and concentration effects (Gatehouse et al., 1979), bind with the digestive proteinases of the herbivores to significantly retard growth rate and weight at pupation (Lipke et al., 1954). Far from being an isolated mechanism, this plant defense has been identified in 20 diverse families from four major divisions (McFarland and Ryan, 1974). In addition to systemic induction, pheromone contact between trees has been recently proposed which opens up yet another avenue of temporal and spatial variation (Rhoades, 1983; Baldwin and Schultz, 1983).

Inducible defenses in mountain birch, *Betula pubescens,* created a patchwork of highly defended areas near sites of previous herbivory (Haukioja and Niemelä, 1976, 1977, 1979). To simulate herbivory by the moth *Oporinia autumnata,* leaves of birch were mechanically damaged. Two days later, adjacent undamaged leaves were fed to larvae, and growth was compared to a control group that was fed leaves from the same tree but distant from the damaged leaves. The larval growth was reduced on leaves near the damaged site but was unaffected on the distant leaves, indicating that the plant was now composed of leaves with differential defensive capabilities. This mosaic pattern of defense would reduce the fitness of the inducing insect, as well as any subsequent invading larvae foraging near the induced zone.

Furthermore, the larvae would have to forage longer to find induction-free zones.

The production of phytoalexins in response to fungal invasion is similar to the proteinase–inhibitor system except the response is very localized instead of systemic, which emphasizes spatial rather than temporal variation. The result is a rapid change in host substrate quality. In response to cell wall fragments released from either the invading fungus or the damaged plant (Ryan et al., 1981), a wide diversity of phytoalexins can be produced (West, 1981) in high concentrations in or near damaged cells (Deverall, 1972). Gaumann and Hohl (1960, cited in Deverall, 1972) were able to extract orchinol, a specific phytoalexin, in high concentrations after infection by the pathogenic fungus *Rhizoctonia repens,* but only in lower concentrations 1 cm away. This localization of the defense may be an economical way for the plant to combat a sedentary pathogen, whereas a systemic defense may be more effective against mobile insects. As the plant changes, the fungus is faced with a rapidly changing host defense that must be countered by a changing offense. To compound the complexity, the magnitude of response is also affected by age and intensity of attack (Cruickshank and Perrin, 1963).

The production of immune bodies to combat bacterial invasion also creates a temporal mosaic of varying defensive capabilities. There is a growing body of literature showing that the inducers of the immune response are lipopolysaccharides of the bacterial cell wall (Goodman, 1980). They induce production of granules, fibers, and membrane fragments that cause bacteria to agglutinate and lyse. The reaction varies through time (Lippincott et al., 1977) and with inoculum dose (Averre and Kelman, 1964) so the response is unpredictable to the invading bacterium. The response can negatively affect subsequent infection of the same or different species of bacteria (Goodman, 1980). Thus, to an invading bacterium, the plant will exhibit a variable pattern of defenses in time and space, dependent on its past associations with the same or other bacterial species.

Viruses also induce a similar defense that can protect the plant from subsequent invasion of the same or different viruses. The reaction can be both local or systemic and can decrease lesion numbers, size, and/or final concentration of the virus. The mechanism of production is poorly understood, but possibly works at the molecular level by interfering with RNA synthesis and replication (Hamilton, 1980). Whatever the mechanism, the effect on the virus will be the same. Its potential success on its host is dependent on the host's past associations with other viruses and changes through time and space.

Premature abscission of plant parts may also be a potent induced defense against such sedentary organisms as leaf miners, galling insects, parasitic plants, fungi, or bacteria. Faeth et al. (1981) found that mined leaves from three species of oaks were more likely to be abscised by the tree than mine-free leaves. Early leaf fall surpassed predation and parasitism as the highest cause of mortality for both leaf miners, the coleopteran *Bracys ovatus* (34.2%) and the lepidopteran *Tischeria citrinipennalla* (38.6%). Similarly, with the gall aphid, *Pemphigus betae,* Williams and Whitham (in preparation) have shown that compared to ungalled leaves of narrowleaf cottonwood, *Populus angustifolia,* leaves with one aphid gall have a five-fold increase in the probability of premature abscission and leaves with three or more galls have a 20-fold increase in the probability of premature abscission. Thus, there is a dosage response that varies with the intensity of attack. The host response has reduced parasite loads on some narrowleaf cottonwoods by 30% and on some Fremont cottonwoods, *P. fremontii,* by 70%. The abscission process is complicated and involves many different hormones (Milborrow, 1974) but one step is the reabsorption of nutrients from the leaf to the tree. To the aphid, the leaf is no longer a stable resource but becomes a rapidly diminishing one through time. The abscission of leaves, leaflets, stipules, stems, buds, bark, flowers, flower parts, fruits, seeds, and even branches (Addicott, 1978, 1982) may all potentially provide the same defense.

Hydrogen cyanide (HCN) can be released when a plant is attacked by chewing insects with the intensity of response varying between plant organs so the plant's defensive system is a mosaic in space. The production of HCN is immediate and localized. Since it affects the respiratory chain of electron transport and can be fatal in small doses, the release of HCN is probably to arrest the attack of the herbivore during initial attack. This defense in plants has been recognized for centuries in such important food crops as apricots, peaches, and almonds. Now, over 800 species in 80 families of fungi, ferns, gymnosperms, dicotelydons, and monocotyledons have been shown to produce HCN (Seigler, 1977). Leaves have the highest concentrations of cyanogenic compounds although they can be found in the roots, seeds, and other plant parts (Seigler, 1977). Intraorgan variation has never been specifically investigated but would add another layer of variation to the plant's defenses.

Pathogens themselves can produce mosaics in a plant that can alter the fitness of subsequent invading pathogens by creating a resource sink at the site of infection. A resource sink is an area that has a greater translocation rate of nutrients into the site than unaffected sites and that causes depletion of nutrients in surrounding areas. Pathogens and parasites creating sinks may be insects (Way and Cammell, 1970; Norris, 1979), parasitic plants (McDowell, 1964; Wood, 1967; Grovier et al., 1967; Whitney, 1973; Knutson, 1979), bacteria (Kelman, 1979), or fungi (Williams, 1979). For example, Way and Cammell (1970) experimentally established aphids on one leaf of Brussels sprouts. The following day, an uncolonized leaf next to the infected one was treated with $^{14}CO_2$ for 1 hr and the distribution of ^{14}C-labeled sucrose in the plant was measured after 3 hr. Compared to the control, the aphids increased the rate of movement of sucrose out of the uninfected leaf by 2.8 times and caused the rate of movement of sucrose into the infected leaf to

Table 2.   Effect of Aphid Feeding on the Translocation Rate of
^{14}C-Labeled Sucrose between Leaves of the Same Plant[a]

	Translocation Rate (mg labeled Sucrose/hr)		
	Out of Labeled Leaf	Into Adjacent Leaf	Into Meristem and Young Leaves
Aphids on adjacent leaf	3.35	0.81	0.19
Aphid free control	1.16	0.01	0.26

[a] Adapted from Way and Cammell (1970).

increase by 80 times. Furthermore, they reduced the amount of sucrose going to the meristem and the growing leaves (Table 2). Thus, even though the plant is not actively involved in the creation of this mosaic, and may even have lowered fitness because of the attack by the first pest (but, see Lovett and Duffield, 1981), the impact of the second invading pest may well be lowered because of its inability to identify suitable habitat altered by the first pest.

The fact that induced plant defenses vary both temporally and spatially creates another level of variation with which plant pests must cope. The required pest traits of overcoming often rapidly changing defense levels; locating narrow windows of a plant's temporal and spatial vulnerability; or completing a life cycle before the host responds should greatly diminish the suite of pests capable of utilizing a host. In comparison with an unchanging host (either susceptible or resistant) within-plant variation should greatly increase the uncertainty of the pest's continued existence.

4   DEVELOPMENTAL PATTERNS WITHIN THE INDIVIDUAL PLANT

Developmental patterns of plant growth are produced by the turning on or off of genes and can produce extensive temporal and spatial variation within a single plant. These growth patterns are important to plant pests because they involve dynamic changes in leaf toughness, nutrient content, leaf moisture, quantitative and qualitative chemical defenses, and mechanical defenses. Since these characteristics may vary spatially (within a leaf, between leaves, and between branches) and temporally (diurnally, seasonally, and over the life-span of the plant) plant pests are confronted with developmental mosaics of variability through which plant tissues may escape in time and space.

Plant maturation can produce extensive within-plant variation because the nutritional quality and level of plant defenses of juvenile and mature tissues can vary greatly. The honeylocust, *Gleditsia triacanthos,* has juvenile tissues at the base of the tree which produce large thorns that protect these thin-barked trees from ground dwelling herbivores. The mature tissues at the top of the tree, however, do not produce thorns. O'Rourke (1949) found that rooted cuttings collected from the top of the tree failed to develop thorns whereas rooted cuttings collected from the base of the tree developed thorns. This simple example demonstrates that a single plant possesses the regulatory machinery to turn defenses on or off and that the plant is capable of making itself a temporal and spatial mosaic of defenses.

Spatial variation can also occur on a very localized scale within a single leaf. Phenolics, which have been strongly implicated in the defensive chemistry of many plants (Levin, 1971), are not uniformly distributed within or between leaves of narrowleaf cottonwood, *Populus angustifolia.* Zucker (1982) found that the concentration of phenolics increased from the base of the leaf to the tip and that large leaves had a lower concentration than small leaves. Leaf choice by the gall-forming aphid, *Pemphigus betae,* showed a strong preference for galling the base of large leaves where the concentration of phenolics was lowest. Failure to settle at these relatively rare sites resulted in greater aphid mortality and lower reproductive success (Whitham, 1978, 1980).

Pillemer and Tingey (1976) presented an example of both temporal and spatial variation within a single leaf. Mortality of nymphal leafhoppers was found to be closely correlated with the density of hooked trichomes on the leaves of field beans, *Phaseolus vulgaris.* Because trichome number was determined early in leaf development, as leaves expanded the density of trichomes declined. Thus, when the tissues were immature and most desirable to herbivores, the density of trichomes was highest. Trichomes were also most dense on the lower surface of the leaf. Since the leafhoppers fed almost exclusively on the underside of the leaf, the trichomes were ideally positioned for defensive purposes.

Developmental changes in epidermal features may also determine resistance to attack by pathogens. Aust et al. (1980) found that as new leaves of barley were produced over a 6-week period, their resistance to powdery mildew, *Erysiphe graminis hordei,* increased (Fig. 4). By exposing leaves to the same number of fungal spores at the onset of the experiment 21.7% of the conidia produced lesions on new leaves, whereas at the end of the experiment only 0.4% produced lesions on new leaves. Increased resistance was apparently the result of changes in leaf morphology in which leaves produced later in the season had greater numbers of silicified cells and a thicker epidermis.

During plant development, individual plant parts change in their value to the plant and in their vulnerability to attack. Because damage to young leaves and buds results in greater loss of fitness than equal damage to mature tissues, young plant parts are considered to be more valuable than mature tissues (McKey, 1974, 1979; Rhoades, 1979). High nitrogen content and reduced leaf toughness also make young tissues more vulnerable to attack

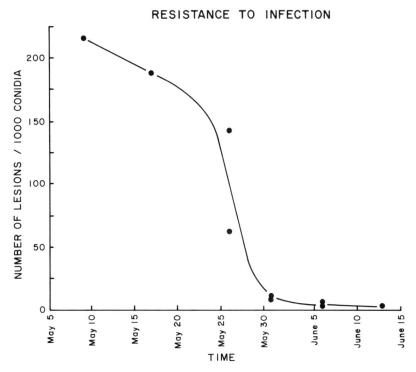

Figure 4. Resistance of newly produced barley leaves to powdery mildew changes over a short period of time. New leaves produced early in the season are far more susceptible to attack than new leaves produced later in the season. Developed from data of Aust et al. (1980).

(Mattson, 1980). In response to the changing value of each plant part, secondary compounds vary throughout plant development and are often highest in the immature tissues (McKcy, 1974, 1979). In 33 of 42 tests, young leaves had higher concentrations of phenolics, alkaloids, steroidal compounds, cyanogenic glycosides, proteinase inhibitors, and other defense compounds than did mature tissues (McKey, 1979).

As the growth season begins, toxins should be transferred from storage organs to new, nutrient-rich foliage. For example, throughout the growth season, solanidine concentrations in the potato decrease in the tubers and increase in the foliage (Street et al., 1946). In the species *Saponaria officinalis* and *Polemonium coeruleum,* saponin, implicated as a growth inhibitor and toxin, decreased in concentration in the underground organs as concentrations increased in the reproductive organs (Drozdz, 1962, cited in Applebaum and Birk, 1979). As flowering and seed set occur, distribution of defense compounds may then shift from foliage to flowers and seeds, especially in annual plants and in those plants whose vegetation dies back each year (McKey, 1974). Sander (1956, cited in McKey, 1974) found that the growth inhibitor tomatine is lost from tomato roots through partial translocation to flowers. Sander and Angermann (1961, cited in McKey, 1974) found that after pollination, tomatine concentrations within the ovary increased to 1.5% whereas the remainder of the tomato plant maintained concentrations of 0.15% tomatine. Green tomato fruits have been found to contain 0.087% tomatine and to steadily decrease in concentration until only trace amounts were found in very ripe fruits (Roddick, 1974). Consequently, tomatine is most highly concentrated in those tissues that are most valuable to the plant at that time and may decrease in concentration as the tissues' value decreases.

In addition to the redistribution of a single defensive compound, plants may increase one defense while decreasing another. For example, Evans and Cowley (1972a) found that toxic compounds in *Digitalis purpurea* varied in concentration and composition during seed germination and growth. As the seeds germinated, cardenolide concentrations decreased but then rapidly increased during primary leaf formation, whereas sapogenins were most abundant during cotyledon development and primary leaf formation. Cardenolides then became most concentrated in the young leaves, whereas spirostanols accumulated in the mature second year leaves (Evans and Cowley, 1972b). Consequently, herbivores must simultaneously face the heart interference activity of cardenolides, the growth inhibition and toxic effects of sapogenins and spirostanols, as well as rapid changes in the concentrations of all three compounds.

Temporal variations in foliage quality which affect herbivore growth rates may occur in as little as 24 hr. Diurnal cycles have been observed in the concentrations of terpenoid compounds of three species of *Juniperus* (Tatro et al., 1973). Similarly, Fairbairn and Wassel (1964) found that morphine was more concentrated in the leaves of poppy, *Papaver somniferum,* in the morning whereas thebaine and codeine reached their highest levels in the afternoon (see Raupp and Denno, 1983, for other examples). Since some allelochemicals are known to both attract and repel insects (Rudinsky, 1966; Sturgeon, 1979) diurnal cycling may significantly affect both plant pests and their predators. Although diurnal changes in the plant's water relations do not represent a developmental pattern of plant growth, the effect of this environmentally induced variation can affect herbivores. Scriber (1977) found changes in relative humidity caused diurnal fluctuations of more than 10% in the water content of wild cherry, *Prunus serotina,* leaves. In the laboratory, *Hyalophora cecropia* larvae fed wild cherry leaves low in water were 48% less efficient in converting leaf dry matter to larval biomass, 27% less efficient in nitrogen utilization, and weighed 38% less at pupation than did larvae fed leaves higher in water.

In addition to diurnal variations, plants commonly exhibit seasonal trends in foliage quality characteristics which may limit the plant's vulnerability to a narrow window of time. For example, Milton (1979) found in 11 tree species tested, fiber content was 36% less and protein content 33% greater in young

foliage, the preferred food of the howler monkey, *Alouatta palliata*. Similarly, Rausher (1981) found that the larvae of the pipevine swallowtail, *Battus
philenor*, readily consumed the early season foliage of *Aristolochia reticulata*, but seldom fed on the mature foliage. Corresponding with a seasonal
decrease in nitrogen and an increase in leaf toughness, the growth rate and
size at dispersal of larvae fed mature leaves were much lower than those
fed young leaves. In another example, seasonal variations in protein, nitrogen, leaf toughness, and tannin concentrations also affected herbivore
growth rates and survival on oak trees. Feeny (1970) found that with the
winter moth, *Operophtera brumata*, feeding on oak leaves was concentrated
in the spring, when nitrogen and protein levels were twice as high as late
summer leaves and tannin concentrations were eight times lower. Oak leaves
also increased in toughness sevenfold in 22 days and larvae fed mature leaves
had significantly lower weights than larvae fed young leaves, and produced
no adult survivors. Similarly, within bracken fern, seasonal and spatial variations combine to affect herbivore levels. Mammalian feeding on young
fronds was completely inhibited by cyanide production, and both sheep and
deer browsed only on the young acyanogenic fronds interspersed with cyanogenic fronds (Cooper-Driver et al., 1977). The locust *Schistocerca gregaria*
did not appear to spatially select fronds but was strongly inhibited from
feeding during seasonal peaks in cyanide production. Seasonal variations in
foliage quality have also been determined for numerous other species of
plants (Tamm, 1951; Schweitzer, 1979; Haukioja et al., 1978; Coley, 1980;
Schultz et al., 1982; Raupp and Denno, 1983).

Seasonal changes in host quality may also be responsible for pest migration patterns. Watt (1979) found that as oats matured, the reproductive rate
of the English grain aphid, *Sitobion avenae*, decreased and mortality rates
increased. From additional work, Watt and Dixon (1981) hypothesized that
the development of alatae, or winged aphids, was in part due to the decline
in host quality on late stages of wheat. More alatiform nymphs were found
on advanced wheat stages than on earlier stages, and subsequent emigration
from the wheat was largely responsible for the decline in aphid numbers.
Thus, plants that rapidly change in quality may limit their susceptibility to
pests that can change equally fast.

Variations in foliage quality may extend through more than a single growth
season, and herbivores are known to distinguish between foliage of different
years. For example, diprionid sawflies, with the exception of a single species,
oviposited only on 1-year-old conifer needles (Knerer and Atwood, 1973).
Blais (1952) found that spruce budworm mortality was higher, development
retarded, and fecundity reduced in insects fed on previous year growth as
compared to those fed on current year growth. Early instars of the Douglas
fir tussock moth preferentially fed on new needle growth and would disperse
rather than feed on 1-year-old growth (Mason and Baxter, 1970). Not all
herbivores prefer current year growth, however. Parry (1976) compared

feeding behavior of the aphid *Elatobium abietinum* on previous and current year needles of the Sitka spruce. In the summer prior to new needle maturation, the aphids preferentially settled on previous year growth. When forced to feed on new growth, the aphids delayed time before sap uptake and ceased more quickly than when using old growth. Nitrogen and amino acid analysis revealed that before maturation, needles contained amino acid imbalances apparently unfavorable to aphid survival; the survival rates of aphids fed on current year branches supplemented with the essential amino acids were approximately twice those fed on untreated branches.

As plants mature, reproductive condition may determine the allocation of defenses or nutrient resources in the plant. Blais (1958) found that with balsam fir, *Abies balsamea,* flowering condition affected resistance of the tree to spruce budworm attack. Mature flowering trees suffered greater defoliation and died before immature, nonflowering trees. Amino acid concentrations may in part explain flowering trees' vulnerability to attack. Kimmins (1971) compared amino acid concentrations of new and old foliage in nonflowering and flowering trees. New foliage of both classes contained greater concentrations of amino acids than old foliage and new foliage from flowering trees contained greater concentrations of amino acids than any other foliage. Similarly, Niemelä et al. (1980) found the level of defoliation of tamaracks, *Larix laricina,* during larch bud moth outbreaks to be positively correlated with age and reproductive index of the trees. During the outbreak, old trees and trees with large numbers of cones suffered significantly more damage than young trees and trees with few cones. It was suggested that young trees with greater reproductive value invested more resources in defense than older trees.

Plants may also change over their life-span in their quality as overwintering sites for both herbivores and the predators of their herbivores. With insect pests that require the deeply fissured bark of mature host trees for overwintering sites, young smooth-barked trees may escape attack simply because they do not provide the prerequisite overwintering sites. In a similar vein, litter accumulation beneath older plants may provide an important habitat for many of the predators of the plant's herbivores. Waloff (1968) found that as Scotch broom, *Sarothamnus scoparius,* grew, increasing amounts of leaf litter accumulated beneath the shrubs, providing shelter for a number of insect predators. These predators were effective at controlling all but a few species of herbivores and, by providing shelter for the predators (particularly in winter), the Scotch broom maintained a capable defense against herbivory.

When the above developmental patterns of plant growth and the previously described genetic and induced sources of within-plant variation act in concert, plant pests should encounter an everchanging host environment. Such variability should place severe restrictions on pest survival, reproduction, and their ability to evolve specialized counteradaptations.

5 DISCUSSION

This chapter has addressed what we believe is a fundamental problem in plant–parasite interactions, that is, the apparent advantage rapidly evolving pathogens, parasites, and herbivores have over their sedentary, long-lived host plants. The work of Parlevliet and Zadoks (1977) with the pathogen *Puccinia recondita* emphasizes the problem. They estimated, using known mutation and uredospore production rates, that when 1% of the leaf area of wheat is infected with mature uredosori, *in a single day* 100,000 mutants would be produced per hectare of wheat.

In the face of such rapidly evolving pests, it is inconceivable that any single fixed plant defense could remain effective over long periods of time. For example, in a small scale experiment, Howatt and Grainger (1955) showed how easily resistant potatoes (*Solanum demissum* × *S. tuberosum*) could be overcome by the rapidly evolving fungus *Phytophthora infestans*. Under greenhouse conditions 10,000 potato seedlings were exposed to cultures derived from single spores of the avirulent race (O). Although race (O) quickly infected plants without resistance genes (R) as was expected, within weeks plants with single genes for resistance were also becoming infected and eventually even plants with multiple genes for resistance were infected. Thus, within a few weeks several mutations occurred (some in combination) which permitted the formerly avirulent pathogen to become virulent and successfully attack the formerly resistant host population. Such phenomenal loss of effective resistance is particularly interesting when considering the small scale of the experiment. This was not a monoculture occupying millions of hectares as we usually consider; 10,000 potato seedlings would probably be comparable in leaf number and area to a single 20 m cottonwood tree with approximately 30,000 shoots or 180,000 leaves (Whitham, unpublished data).

With the above examples and others cited earlier in mind, the main goal of this chapter has been to point out the potential importance of within-plant variation as a defense against rapidly evolving plant pests. If any lessons have been learned from agricultural systems, the most important one is that "vulnerability stems from genetic uniformity" (National Academy of Sciences, 1972, p. 286). Such realizations probably account for more recent attempts of agriculturists to mimic the diversity of natural ecosystems by temporally and spatially deploying diverse resistance genes. Since it is known that variation in the host population can be highly adaptive, even crucial in agroecosystems, it is also likely to be important within individual native plants (particularly long-lived or clonally reproducing species) that, except for within-plant variation, would otherwise be monocultures in time. Because within-plant variation has been largely ignored as "noise in the system" (Whitham, 1983), it may now be appropriate to consider the potential of within-plant variation as an adaptation.

At least three genetically based mechanisms may produce quantitative and/or qualitative variation within an individual host plant or clone. Plants may develop as genetic chimeras and mosaics in which different tissues, branches, etc. are differentially defended. As shown in *Euonymus japonica* (see Fig. 3), different branches of the same plant can be genetically different and present a continuum of dosage effects. Developmental patterns of plant growth can also produce temporal and/or spatial mosaics of genetic expression such as the thorny and thornless condition of honeylocust in which developmental traits are maintained in rooted cuttings. Induced plant defenses are probably also under genetic control in which herbivore or parasite attack can trigger the production of specific digestibility inhibitors either on a very localized or systemic scale.

Since the above genetic mechanisms of within-plant variation are very likely to be interactive, as the number of variables increases the complexity of the plant–parasite interaction must also increase. Mattson et al. (1982) suggest that the individual characteristics of a plant make up an "*n*-dimensional trait vector" and the success of a herbivore depends on the closeness of fit between its corresponding trait vector and that of the plant. In other words, herbivores must have traits that are well adapted to plant traits. By creating a heterogeneous environment, the plant may make itself appear as a mosaic or chameleon of changing resistances (Whitham, 1981). This reduces the probability that the right plant pest will match up with the appropriate plant resource, and results in inappropriate pest settling decisions and reduced pest fitness. Compared with a uniform host substrate, individual plants as temporal and spatial mosaics of resistance should have several important effects on their pests that reduce their impacts on the host. The remainder of this chapter will develop the various implications and predictions of within-plant variation as a plant defense.

Because decreasing patch size is often associated with reduced herbivore loads, a single plant may reduce the size of its herbivore load by increasing variation, which reduces the size of patches available for herbivores. Although there are counterexamples, particularly with *generalist* herbivores in which small or isolated patches suffer increased damage (Futuyma and Wasserman, 1980), numerous other studies show that reduced patch size or increased plant species diversity can effectively reduce the *specialist* herbivore load (van Emden, 1965; Tahvanainen and Root, 1972; Root, 1973; Dempster and Coaker, 1974; Altieri et al., 1977; Brown and Kodric-Brown, 1977; Bach, 1980a,b; Risch, 1981). Since increased *plant species diversity* can reduce the specialist herbivore load, spatial *diversity within a single plant* or clone may have the same effect on the pest population.

In association with patch size, temporal aspects of the "window of vulnerability" may have similar effects on plant pests. The width of the window can greatly reduce the susceptibility of the host plant to attack by requiring successful pests to be precisely synchronized with their hosts. For example, over a 35 day period the resistance of newly produced barley leaves to the same number of fungal spores increased 54-fold (see Fig. 4). Plants may so narrow their period of acceptability as to escape herbivory entirely. Waller (1982) found that young *Berberis trifoliata* and *Celtis reticulata* foliage was readily accepted by leaf cutting ants under experimental conditions, but were seldom taken by ants in natural conditions. Leaves of both species were found to rapidly become tough and decrease in palatability between late March and early June. Waller suggested that young leaves became unacceptable so rapidly that the probability of an ant finding a suitable leaf was small and the species were avoided in favor of plant species that retain palatability for longer periods.

Host heterogeneity concentrates plant parasites at specific sites where they suffer greater mortality due to intraspecific competition. Whitham (1978) showed that within-plant variation in narrowleaf cottonwood, *Populus angustifolia,* caused gall forming aphids, *Pemphigus betae,* to compete for superior gall sites, exhibit territorial behavior (Whitham, 1979, 1980), and suffer a 34% reduction in the number of progeny (Whitham, 1983). It is also likely that host heterogeneity encourages cannibalism among plant pests by concentrating them at specific sites where the possibility of feeding on an even better resource is increased. Fox (1975, p. 88) notes that "surprisingly, a large proportion of observations of cannibalism among terrestrial animals is for species that are usually 'herbivores'. . . ."

Concentration of plant pests also increases the probability of their detection and/or increases the foraging efficiency of predators which may then reduce the herbivore load of the plant. In comparisons of adjacent branches of the same tree which differed in the density of gall aphids, Whitham (1981) observed that predators preferentially foraged on branches where gall densities were highest and removed nearly twice the percentage of these plant parasites. Similarly, Southwood (1979) observed that when the foliage of broom, *Sarothamnus scoparius,* became less palatable, several species of mirids concentrated at rare high quality sites where they then preyed on one another and competed for the remaining resources of the host plant.

Rapid changes in host quality place many herbivores and parasites under severe selection pressures to evolve a rapid development rate to complete their life cycle before their host becomes unsuitable. Faeth et al. (1981) found that some leaf miners could not complete development before their host trees succeeded in dropping the parasitized leaves and killing the leaf miners. As Mattson et al. (1982) point out, the duration of plant suitability and herbivore life cycle should be approximately equal if the parasite is to successfully utilize the host.

In addition to the constraints placed on the development rates of plant pests, rapid changes in host quality may be responsible for host shifts, costly migrations, and special adaptations to cope with more than one host. Work by Dixon (1971) suggests that seasonal variation in plant quality may force host shifts and require specialized adaptations by the bird cherry oat aphid, *Rhopalosiphum padi.* The aphid leaves its primary host, the bird cherry,

Prunus padus, at the end of June as soluble nitrogen levels fall below 4% and returns in September as nitrogen levels rise above the 4% level in senescent leaves. In order to determine if the different life stages of the aphids have different nutrient requirements, adult wingless aphids normally inhabiting the secondary host grasses were caged on bird cherry foliage; all aphids died within 5 days and produced no viable young. Similarly, adult aphids, which normally live on bird cherry in September, were caged on oats and survived but produced far fewer nymphs, demonstrating that in both instances adaptations to one host came at the expense of adaptations to the other host. In another example, the gall-producing aphid, *Pemphigus betae,* is literally forced by some host trees to migrate in midsummer. Although these aphids have the advantage of living in the relative safety of a hollow gall, a trade-off occurs when the leaf they gall may be selectively abscised by the host tree before the aphids complete development. Since continued existence on dropped leaves is not possible, it would appear that these aphids have evolved a complex life cycle in which they migrate to a secondary host to avoid their primary host's induced defenses. This may explain why galling aphids are 3.5 times more likely to have a complex life cycle than their free-living relatives (Williams and Whitham, unpublished data).

Induced plant defenses that deter further feeding and produce a mosaic of suitable and unsuitable host tissues may be largely responsible for the dispersed feeding patterns of herbivores in which much of the host is left uneaten. Carroll and Hoffman (1980) demonstrated that when leaves of squash, *Cucurbita moschata,* were damaged, feeding deterrents accumulated within 20–40 min. Associated with these induced defenses, adult coccinelid beetles, *Epilachna tredecimnotata,* traveled an average of 6 m after a single feeding period whereas less mobile larvae were never observed feeding within 2 m of a fresh feeding scar. The observation that some herbivores move before the food supply is consumed may well reflect the avoidance of plant parts recently supplied with induced defenses rather than unutilized and available resources. Furthermore, since movement makes insects more conspicuous to predators and parasitoids (Richerson and DeLoach, 1972), within-plant variation increases the exposure of plant pests that must move to feed (Schultz, 1983).

Temporal host variation may also be responsible for boom and bust cycles in herbivore populations. Snowshoe hare and mountain hare feeding on willow, birch, and aspen induce the production of adventitious shoots that are much higher in resins and other secondary compounds than the preferred mature tissues (Bryant, 1981). Feeding experiments showed that the hare could not maintain weight or survive on these resinous tissues, but could maintain weight and thrive on mature tissues (Bryant and Kuropat, 1980). Because heavy browsing greatly reduces the availability of mature tissues, only the juvenile tissues which remain unpalatable for three years are available as hare forage (Bryant, 1980). Consequently, the famous 10-year cycle of the snowshoe hare, rather than representing a predator–prey cycle as is commonly assumed, may represent a plant–herbivore cycle caused by temporal changes in the plant's defensive status.

Since temporal and spatial variation place severe constraints on plant pests, it must place strong selection pressures on them to circumvent the host's defenses and prevent induced changes in resource quality. For example, even though it takes squash only 40 min to systemically translocate feeding deterrents to the site of injury, it takes the beetle, *Epilachna tredecimnotata,* as long as 1–2 hr to consume a leaf. By using its enlarged mandibular teeth to cut the vascular tissues of the leaf, the beetle can avoid the induced defenses of the plant and eat the leaf as its leisure (Carroll and Hoffman, 1980). Similarly, cyanogenesis may also be circumvented by specialized feeding behavior. In many cases, the tissue must be crushed to combine components of the epidermal layer with components of the mesophyll layer for the liberation of HCN to occur (Kojima et al., 1979). If an insect such as a leaf miner specialized on one of these two layers, it could avoid this generalized defense. Because of body size limitations, however, a similar foraging behavior would not be possible for most other herbivores. From the plant's perspective the problem becomes one of scaling the appropriate defenses to the size of its pest.

Although the ecological implications for the plant are unclear, if we view an individual host plant as an *n*-dimensional mosaic of temporal and spatial variation, as herbivore populations respond to changes in host quality, the net effect is a shift in the patterns of herbivore abundance and species richness on the plant. Futuyma and Gould (1979) found that the number of phytophagous larvae species on 18 woody plant species dropped by 54% between early and late June and larval densities on the plants dropped by more than 32% during the same period. In contrast, Lawton (1976) found that the species diversity and herbivore density on bracken fern to be lowest on early season fronds because they are highest in allelochemical defenses and are small sized (Lawton, 1978; Jones, 1983). Although the above examples are concerned with temporal effects on herbivore abundance and diversity, spatial patterns are also apparent. Lawton and Schroder (1977) and Strong and Levin (1979) found that species richness was correlated with plant architectural complexity and growth form indicating that as spatial variation increases additional species may be added. Even though these patterns are real, whether or not they have any evolutionary significance for the host plant is presently unknown.

Perhaps the most important and controversial aspect of within-plant variation is the potential for an individual plant or clone to manipulate the evolution of its pests along less harmful pathways. Agriculturists have argued that it is possible to direct the evolution of plant pests and to select for simple races that lack the ability to reach epidemic proportions (Vanderplank, 1963; Browning and Frey, 1969; Knott, 1972; Frey et al., 1973, 1979; Pimentel and Bellotti, 1976; Leonard and Czochor, 1980). Where uniformity or single defenses only result in the evolution of specialized races capable of coping with the defense, variation or a diversity of defenses may be exceptionally difficult to circumvent and evolutionarily surmount. It is interesting to point out that the evolution of specialized pest races and gene-for-gene interactions with their hosts are often associated with a parthenogenetic reproduction or an asexual phase during the pest life cycle (see Gallun and Khush, 1980; Vanderplank, 1982, pp. 47 and 84, for lists of pest species). Presumably asexual reproduction permits a pest with the appropriate genetic match with its host to rapidly spread and avoid the breakup of an adaptive gene complex that might be lost with sexual reproduction. Whereas variation at the population level does not prevent pest race formation (see Edmunds and Alstad, 1978), an additional level of variation within the individual host would greatly diminish the resource base on which a specific pest race could spread. Consequently, if individual plants (particularly long-lived or clonal plants) are temporal and spatial mosaics as we suggest, this variation may provide the answer to the dilemma posed at the beginning of this chapter: Why don't rapidly evolving plant pests break the defenses of their long-lived host plants which according to current dogma remain unchanged over a life that may span thousands of years?

In an attempt to examine the impact of within-plant variation on a parasite population, Whitham (1981, 1983, unpublished data) found that individual trees of *Populus angustifolia* were spatial mosaics of resistance to gall aphid attack. The variation in resistance between adjacent branches of the same tree to the gall aphid, *Pemphigus betae,* can be as great as the variation in resistance between trees in the population (0–72% aphid mortality both within and between trees) and the variation in resistance remains stable from year to year. Regardless of the mechanisms involved the host tree is a mosaic of resistance in which the various parts of the mosaic have very different and pronounced effects on aphid fitness. Since these aphids realize high reproductive success only on a small portion of the tree while suffering high mortality and low reproductive success on the rest of the tree, it would appear that adaptation to part of the mosaic occurs at the expense of adaptation to other parts of the mosaic. This strongly suggests that there are trade-offs involved. These trade-offs may allow us to address the question: Why have these aphids not evolved adaptations that allow them to effectively utilize the whole plant rather than being restricted to a rather small subset of resources? The answer may well lie in the pattern of variation and the potential of the host to manipulate the aphids.

When considering the abilities of pathogens and parasites to change faster than their hosts, by employing variation the potential exists for host plants to direct the evolution of their pests or act as a stabilizing factor in their evolution. For instance, if there are trade-offs associated with survival on different parts of the mosaic, adaptations to the poorer quality sites make those individuals poorer competitors on good quality sites and vice versa. Although superior host leaves and branches are relatively rare, the reproductive success of the few aphids that successfully sequester the superior resources may provide the bulk of the next aphid generation. By manipulating the ratio of good to poor quality sites the host may direct the evolution of its parasites and restrict them to a relatively small portion of the host where their impact is minimized.

With the above logic the central issue becomes "What maintains variation in the host or why does the entire plant not employ the resistance of the most resistant branches?" If no one single "ultimate weapon" or defense can remain effective for long periods of time as agricultural examples suggest, then a diversity of defenses will be evolutionarily favored. The work of Pimentel and Bellotti (1976) showed that while flies rapidly evolved resistance to single toxicants, when the same toxicants were used *spatially* in combination, no resistance evolved. Similarly, quantitative and/or qualitative spatial variation within a single tree or clone may prevent the evolution of specialized aphid races and restrict them to a small subset of the host plant. If it is not possible to "win" and completely exclude a pest, isolation and severe restrictions on their biology (increased competition and predation, induced defenses such as premature leaf fall, and complex life cycles) may evolutionarily represent the best strategy. Consequently, if parasites are limited by their ability to cope with variation, then the concept of generalist and specialist pests may become less meaningful. Independent of the number of host species utilized, most pests may simply evolve to cope with the same level of temporal and spatial variation in their host(s) defenses.

If temporal and spatial variation are important plant defenses against rapidly evolving pests as we suggest, two additional predictions can be made. First, individual plants and plant populations historically isolated from pathogens, parasites, and/or herbivores should exhibit less internal variation than individual plants and populations with a long historical association with pests. Second, in the presence of pests, plants with variation should achieve greater inclusive fitness than plants that are more temporally and spatially uniform.

6 SUMMARY

Individual plants, because of their often long life-span and large clone size, are exposed to rapidly evolving pests. Here we explore the hypothesis that variation within an individual plant is a potent defense against pests that would otherwise hold an evolutionary edge over their sessile hosts. After developing the logic of variation as a plant defense in the introduction, the main body of the chapter has been devoted to a review of the literature identifying what is currently known about variation within individual plants that is ecologically relevant to plant pathogens, parasites, and herbivores. The mechanisms of within-plant variation discussed include (1) genetic variation in which different parts of the same plant are genetically different such that the host is a mosaic of defenses; (2) inducible defenses in which the responses of the host to attack are highly variable both temporally and spatially; and (3) developmental patterns in which genes are turned on or off over the life-span of the plant and vary with different parts of the same plant. When these three major sources of within-plant variation are superimposed on one another, the attacking pest should encounter a degree of complexity that effectively prevents pests from breaking the defenses of their hosts.

The ecological implications of variation within an individual plant are developed in the discussion. Host complexity may (1) cause pests to make inappropriate settling decisions; (2) reduce effective patch size and allow plants to escape in time and space; (3) narrow the window of host vulnerability; (4) concentrate pests at sites where they suffer increased competition and predation; (5) affect pest development rates by reducing time available for completion of their life cycle; (6) be a driving force in the evolution of complex life cycles; (7) cause the dispersed feeding patterns of herbivores due to rapidly changing host defenses; (8) cause boom and bust cycles of pest populations; (9) direct the evolution of specific counteradaptations by pests; (10) affect herbivore abundance and species richness; and (11) manipulate pest evolution along less harmful pathways. Thus, the ramifications of variation within an individual plant (particularly long-lived plants or large clones) are very broad. The importance of variation as a defense within a plant may be just as important as variation at the population level which has been well demonstrated in agricultural systems.

When viewed in this light the extensive variation that is observed both within and between individual plants may be adaptive rather than environmental noise in the system. The implications generate testable hypotheses that we hope this chapter has emphasized, and will encourage others to critically examine variation as a plant defense that is evolutionarily very difficult for pests to surmount or circumvent.

ACKNOWLEDGMENTS

We gratefully acknowledge the support of NSF Grants DEB7905848 and DEB8005602. We thank Michael Kearsley, H. Lloyd Mogensen, and Peter W. Price for their constructive comments on the manuscript.

LITERATURE CITED

Addicott, F. T. 1978. Abscission strategies in the behavior of tropical trees. In P. B. Tomlinson and M. H. Zimmermann (eds.), *Tropical trees as living systems*, pp. 381–398. Cambridge University Press, Cambridge.

Addicott, F. T. 1982. *Abscission*. University of California Press, Berkeley.

Altieri, M. A., A. van Schoonhoven, and J. Doll. 1977. The ecological role of weeds in insect pest management systems: A review illustrated by bean (*Phasealus vulgaris*) cropping systems. *PANS* **23**:195–205.

Applebaum, S. 1964. Physiological aspects of host specificity in the Bruchidae—I. General considerations of developmental compatibility. *J. Inst. Physiol.* **10**:783–788.

Applebaum, S. W., and Y. Birk. 1979. Saponins. In G. A. Rosenthal and D. H. Janzen (eds.), *Herbivores: Their interaction with secondary plant metabolites*, pp. 55–133. Academic Press, New York.

Aust, H. J., E. Bashi, and J. Rotem. 1980. Flexibility of plant pathogens in exploiting ecological and biotic conditions in the development of epidemics. In J. Palti and J. Kranz (eds.), *Comparative epidemiology*, pp. 46–56. Centre for Agricultural Pub. and Documentation, Wageningen, Netherlands.

Averre, C. W., III, and A. Kelman. 1964. Severity of bacterial wilt as influenced by ratio of virulent to avirulent cells of *Pseudomonas salanaciarum* in inoculum. *Phytopathology* **54**:779–783.

Bach, C. E. 1980a. Effects of plant diversity and time of colonization on an herbivore–plant interaction. *Oecologia* **44**:319–326.

Bach, C. E. 1980b. Effects of plant density and diversity on the population dynamics of a specialist herbivore, the striped cucumber beetle, *Acalymma vittata* (Fab.). *Ecology* **61**:1515–1530.

Baldwin, I. T., and J. C. Schultz. 1983. Rapid damage-induced changes in leaf chemistry and evidence for inter-tree communication (in review).

Baur, E. 1930. *Einführung in die vererbungslehre*. Verlag von Gebrüder Bontraeger, Berlin.

Berger, C. A., and E. R. Witkus. 1946. Polyploid mitosis as a normally occurring factor in the development of *Allium cepa* L. *Am. J. Bot.* **33**:785–787.

Blais, J. R. 1952. The relationship of the Spruce budworm (*Choristoneura fumiferana*, Clem) to the flowering condition of balsam fir *Abies balsamea* (L.) Mill. *Can. J. Zool.* **30**:1–29.

Blais, J. R. 1958. The vulnerability of balsam fir to Spruce budworm attack in Northwestern Ontario with special reference to the physiological age of the tree. *For. Chron.* **34**:405–422.

Brown, J. H., and A. Kodric-Brown. 1977. Turnover rates in insular biogeography: Effect of immigration on extinction. *Ecology* **58**:445–449.

Browning, J. A. 1974. Relevance of knowledge about natural ecosystems to development of pest management programs for agro-ecosystems. *Proc. Am. Phytopathol. Soc.* **1**:191–199.

Browning, J. A., and K. J. Frey. 1969. Multiline cultivars as a means of disease control. *Annu. Rev. Phytopathol.* **7**:355–382.

Browning, J. A., M. D. Simons, and E. Torres. 1977. Managing host genes: Epidemiologic and genetic concepts. In J. G. Horsfall and E. B. Cowling (eds.), *Plant disease*, Vol. 1, pp. 191–212. Academic Press, New York.

Browning, J. A., K. J. Frey, M. E. McDaniel, M. D. Simons, and I. Wahl. 1979. The bio-logic of using multilines to buffer pathogen populations and prevent disease loss. *Indian J. Genet. Plant Breed.* **39**:3–9.

Bryant, J. P. 1980. The regulation of snowshoe hare feeding behavior during winter plant antiherbivore chemistry. In K. Meyer (ed.), *Proceedings of the International Lagomorph Conference, 1st.* Guelph University, Canada.

Bryant, J. P. 1981. Phytochemical deterrence of snowshoe hare browsing by adventitious shoots of four Alaskan trees. *Science* **213**:889–890.

Bryant, J. P., and P. J. Kuropat. 1980. Subartic browsing vertebrate winter forage selection: The role of plant chemistry. *Annu. Rev. Ecol. Syst.* **11**:261–285.

Butters, F. K., and R. M. Tryon, Jr. 1948. A fertile mutant of a *Woodsia* hybrid. *Am. J. Bot.* **35**:132–133.

Carroll, C. R., and C. A. Hoffman. 1980. Chemical feeding deterrent mobilized in response to insect herbivory and counteradaptation by *Epilachna tredecimnotata*. *Science* **209**:414–416.

Cates, R. G. 1980. Feeding patterns of monophagous, oligophagous and polyphagous insect herbivores: the effect of resource abundance and plant chemistry. *Oecologia* **46**:22–31.

Coley, P. 1980. Effects of leaf age and plant life history patterns on herbivory. *Nature (London)* **284**:545–546.

Cooper-Driver, G., S. Finch, and T. Swain. 1977. Seasonal variation in secondary plant compounds in relation to the palatability of *Pteridium aquilinum*. *Biochem. Syst. Ecol.* **5**:177–183.

Cowling, E. B. 1978. Agricultural and forest practices that favor epidemics. In J. G. Horsfall and E. B. Cowling (eds.), *Plant disease*, Vol. 2, pp. 361–381. Academic Press, New York.

Cramer, P. J. S. 1907. *Nature*, 3 vers. Vol. 6, No. 5. Verh. Holland Maatsch Wetenschappelijke, Haarlem, Netherlands.

Cruickshank, I. A. M., and D. R. Perrin. 1963. (5) Studies on phytoalexins. VI. Pisatin formation by cultivars of *Pisum sativum* L. and several other *Pisum* species. *Aust. J. Biol. Sci.* **16**:111–128.

D'Amato, F. 1975. The problem of genetic stability in plant tissue and cell cultures. In O. H. Frankel and J. G. Hawkes (eds.), *Crop genetic resources for today and tomorrow*, pp. 333–348. Cambridge University Press, London.

Darlington, C. D., and P. T. Thomas. 1941. Morbid mitosis and activity of inert chromosomes in Sorghum. *Proc. R. Soc. London, Ser. B* **130**:127–150.

Day, P. R. 1978. The genetic basis of epidemics. In J. G. Horsfall and E. B. Cowling (eds.), *Plant disease*, Vol. 2, pp. 263–285. Academic Press, New York.

Dempster, J. P., and T. H. Coaker. 1974. Diversification of crop ecosystems as a means of controlling pests. In D. P. Jones and M. E. Solomon (eds.), *Biology in pest and disease control*, pp. 106–114. Blackwell, London.

Deverall, B. J. 1972. Phytoalexins. In J. B. Harborne (ed.), *Phytochemical ecology*, pp. 217–233. Academic Press, New York.

Dixon, A. F. G. 1971. The life cycle and host preferences of the bird cherry-oat aphid, *Rhopalosiphum padi* L., and their bearing on the theories of host alternation in aphids. *Ann. Appl. Biol.* **68**:135–147.

Edmond, J. B., T. L. Sonn, F. S. Andrews, and R. G. Halfacre. 1975. *Fundamentals of horticulture*. McGraw-Hill, New York.

Edmunds, G. F., Jr., and D. N. Alstad. 1978. Coevolution in insect herbivores and conifers. *Science* **199**:941–945.

Edmunds, G. F., Jr., and D. N. Alstad. 1981. Responses of black pineleaf scale to host plant variability. In R. F. Denno and H. Dingle (eds.), *Insect and life history patterns: Habitat and geographic variation*, pp. 29–38. Springer-Verlag, New York.

Esau, K. 1975. *Plant anatomy*. Wiley, New York.

Evans, F. J., and Cowley, P. S. 1972a. Variation in cardenolides and sapogenins in *Digitalis purpurea* during germination. *Phytochemistry* **11**:2729–2733.

Evans, F. J., and Cowley, P. S. 1972b. Cardenolides and spirostanols in *Digitalis purpurea* at various stages of development. *Phytochemistry* **11**:2971–2975.

Faeth, S. H., E. F. Connor, and D. Simberloff. 1981. Early leaf abscission: A neglected source of mortality for folivores. *Am. Nat.* **117**:409–415.

Fairbairn, J. W., and G. M. Wassel. 1964. The alkaloids of *Papaver somniferum* L.: Evidence for rapid turnover of the major alkaloids. *Phytochemistry* **3**:253–258.

Feeny, P. 1970. Seasonal changes in oak leaf tannins and nutrients as a cause of spring feeding by winter moth caterpillars. *Ecology* **51**:565–581.

Flor, H. H. 1971. Current status of the gene-for-gene concept. *Annu. Rev. Phytopathol.* **9**:275–296.

Fox, L. R. 1975. Cannibalism in natural populations. *Annu. Rev. Ecol. Syst.* **6**:87–106.

Frey, K. J., J. A. Browning, and M. D. Simons. 1973. Management of host resistance genes to control diseases. *Z. Pflanzenkrankh. Pflanzenschutz* **80**:160–180.

Frey, K. J., J. A. Browning, and M. D. Simons. 1979. Management systems for host genes to control disease loss. *Indian J. Genet. Plant Breed.* **39**:10–21.

Futuyma, D. J., and F. Gould. 1979. Associations of plants and insects in a deciduous forest. *Ecol. Monogr.* **49**(1):33–50.

Futuyma, D. J., and S. S. Wasserman. 1980. Resource concentration and herbivory in oak forests. *Science* **210**:920–922.

Gallun, R. L., and G. S. Khush. 1980. Genetic factors affecting expression and stability of resistance. In F. G. Maxwell and P. R. Jennings (eds.), *Breeding plants resistant to insects,* pp. 63–85. J. Wiley, New York.

Gallun, R. L., K. J. Starks, and W. D. Guthrie. 1975. Plant resistance to insects attacking cereals. *Annu. Rev. Entomol.* **20**:337–357.

Gatehouse, A., J. Gatehouse, P. Dobie, A. Kilminster, and D. Boulter. 1979. Biochemical basis of insect resistance in *Vigna unguicuata. J. Sci. Food Agric.* **30**:948–958.

Goodman, R. N. 1980. Defenses triggered by previous invaders: Bacteria. In J. G. Horsfall and E. B. Cowling (eds.), *Plant disease,* Vol. 5, pp. 305–317. Academic Press, New York.

Graham, J. S., and C. A. Ryan. 1981. Accumulation of a metallo-carboxypeptidase inhibitor in leaves of wounded potato plants. *Biochem. Biophys. Res. Commun.* **101**:1164–1169.

Grant, V. 1971. *Plant Speciation.* Columbia University Press, New York.

Green, T. R., and C. A. Ryan. 1972. Wound-induced proteinase inhibitor in plant leaves: A possible defense mechanism against insects. *Science* **175**:776–777.

Grovier, R. N., M. D. Nelson, and J. S. Pate. 1967. Hemiparasitic nutrition in angiosperms. I. The transfer of organic compounds from host to *Odontites verna* (bell.) Dum. (Scrophulariaceae). *New Phytol.* **66**:285–297.

Gustafsson, A. 1946–1947. *Apomixis in higher plants.* Lunds Universitets Arsskrift, Sweden.

Gustafson, G., and C. A. Ryan. 1976. Specificity of protein turnover in tomato leaves. *J. Biol. Chem.* **251**:7004–7010.

Hamilton, R. I. 1980. Defenses triggered by previous invaders: Viruses. In J. G. Horsfall and E. B. Cowling (eds.), *Plant disease,* Vol. 5, pp. 279–303, Academic Press, New York.

Hammett, H. L. 1979. Sweet-potato mutations as influenced by cultivar. *HortScience* **14**:123.

Harberd, D. J. 1961. Observations on population structure and longevity of *Festuca rubra* L. *New Phytol.* **60**:184–206.

Harper, J. L. 1977. *Population biology of plants.* Academic Press, London.

Hatchett, J. H., and R. L. Gallun. 1968. Frequency of Hessian fly, *Mayetiola destructor,* races in field populations. *Ann. Entomol. Soc. Am.* **61**:1446–1449.

Hatchett, J. H., and R. L. Gallun. 1970. Genetics of the ability of the Hessian fly, *Mayetiola destructor,* to survive on wheats having different genes for resistance. *Ann. Entomol. Soc. Am.* **63**:1400–1407.

Haukioja, E., and P. Niemelä. 1976. Does birch defend itself actively against herbivores? *Rep. Kevo Subartic Res. Stat.* **13**:44–47.

Haukioja, E., and P. Niemelä. 1977. Retarded growth of a geometrid larva after mechanical damage to leaves of its host tree. *Ann. Zool. Fenn.* **14**:48–52.

Haukioja, E., and P. Niemelä. 1979. Birch leaves as a resource for herbivores: seasonal occurrence of increased resistance in foliage after mechanical damage of adjacent leaves. *Oecologia* **39**:151–159.

Haukioja, E., P. Niemelä, L. Iso-Iivari, H. Ojala, and E. Aro. 1978. Birch leaves as a resource for herbivores. I. Variation in the suitability of leaves. *Rep. Kevo. Subarctic Res. Stat.* **14**:5–12.

Heinz, D. J., M. Krishnamurthi, L. G. Nickell, and A. Maretzki. 1977. In J. Reinert and Y. P. S. Bajaj (eds.), *Plant cell tissue and organ culture,* pp. 3–17. Springer-Verlag, New York.

Howatt, J. L., and P. N. Grainger. 1955. Some new findings concerning *Phytophthora infestans* (Mont.) de By. *Am. Potato J.* **32**:180–188.

Hull, J. W., and D. M. Britton. 1956. Early detection of induced internal polyploidy in *Rubus. Am. Soc. Hortic. Sci.* **68**:171–177.

Jones, C. G. 1983. Phytochemical variation, colonization and insect communities: The case of bracken fern, *Pteridium aquilinum* L. (Kuhn). In R. F. Denno and M. S. McClure (eds.), *Variable plants and herbivores in natural and managed systems,* pp. 513–558. Academic Press, New York.

Kelman, A. 1979. How bacteria induce disease. In J. G. Horsfall and E. B. Cowling (eds.), *Plant disease,* Vol. 4, pp. 181–202. Academic Press, New York.

Kemperman, J. A., and B. V. Barnes. 1976. Clone size in American aspens. *Can. J. Bot.* **54**:2603–2607.

Kimmins, J. P. 1971. Variation in the foliar amino acid composition of flowering and non-flowering balsam fir (*Abies balsamea* L. (Mill.)) and white spruce (*Picea glauca* Moench. (Voss)) in relation to outbreaks of the spruce budworm (*Choristoneura fumiferna* Clem.). *Can. J. Zool.* **49**:1005–1011.

Klekowski, E. J., Jr. 1983. Mutational load in clonal plants: A study of two fern species. Evolution (in press).

Knerer, G., and C. E. Atwood. 1973. Diprionid sawflies: Polymorphism and speciation. *Science* **179**:1090–1099.

Knott, D. R. 1972. Using race specific resistance to manage the evaluation of plant pathogens. *J. Environ. Qual.* **1**:227–231.

Knox-Davies, P. S. 1967. Mitosis and aneuploidy in the vegetative hyphae of *Macrophomina phaseoli. Am. J. Bot.* **54**:1290–1295.

Knutson, D. M. 1979. How parasitic seed plants induce disease in other plants. In J. G. Horsfall and E. B. Cowling (eds.), *Plant disease,* Vol. 4, pp. 293–312. Academic Press, New York.

Kojima, M., J. E. Poulton, S. S. Thayer, and E. E. Conn. 1979. Tissue distributions of Dhurrin and of enzymes involved in its metabolism in leaves of *Sorghum bicolor. Plant Physiol.* **63**:1022–1028.

Lawton, J. H. 1976. The structure of the arthropod community on bracken. *Bot. J. Linn. Soc.* **73**:187–216.

Lawton, J. H. 1978. Host plant influences on insect diversity: The effects of space and time. In L. A. Mound and N. Waloff (eds.), *Diversity of insect faunas,* pp. 105–125, Symp. Roy. Ent. Soc. 9. Blackwell, London.

Lawton, J. H., and D. Schroder. 1977. Effects of plant type, size of geographical range and taxonomic isolation on number of insect species associated with British plants. *Nature (London)* **265**:137–140.

Leonard, K. J., and R. J. Czochor. 1980. Theory of genetic interactions among populations of plants and their pathogens. *Annu. Rev. Phytopathol.* **18**:237–258.

Levin, D. A. 1971. Plant phenolics: An ecological perspective. *Am. Nat.* **105**:157–181.

Lewis, W. H. 1970. Chromosomal drift, a new phenomenon in plants. *Science* **168**:1115–1116.

Lewis, W. H., R. L. Oliver, and T. K. Luikart. 1971. Multiple genotypes in individuals of *Claytonia virginica. Science* **172**:564–565.

Li, N., and R. C. Jackson. 1961. Cytology of supernumerary chromosomes in *Haplopappus spinulosus* ssp. *cotula. Am. J. Bot.* **48**:419–426.

Lipke, H., G. S. Fraenkel, and I. E. Liener. 1954. Effect of soybean inhibitors on growth of *Tribolium confusum. J. Agric. Food Chem.* **2**:410–414.

Lippincott, B. B., M. H. Whatley, and J. A. Lippincott. 1977. Tumor induction by *Agrobacterium* involves attachment of the bacterium to a site on the host plant cell wall. *Plant Physiol.* **59**:388–390.

Lovett, J. V., and A. M. Duffield. 1981. Allelochemicals of *Camelina sativa*. *J. Appl. Ecol.* **18**:283–290.

Makus, D., G. Zuroske, and C. A. Ryan. 1980. The direction and rate of transport of the proteinase inhibitor inducing factor out of wounded tomato leaves. *Plant Physiol. Suppl.* **65**:150.

Mason, R. R., and J. R. Baxter. 1970. Food preference in a natural population of the Douglas fir tussock moth. *J. Econ. Entomol.* **63**:1257–1259.

Mattson, W. J. 1980. Herbivory in relation to plant nitrogen content. *Annu. Rev. Ecol. Syst.* **11**:119–161.

Mattson, W. J., and N. D. Addy. 1975. Phytophagous insects as regulators of forest primary production. *Science* **190**:515–522.

Mattson, W. J., N. Lorimer, and R. A. Leary. 1982. Role of plant variability (trait vector dynamics and diversity) in plant/herbivore interactions. In *Proc. IUFRO Conf. genetics of host/parasite interactions*. Centre for Agric. Publ. and Documentation. Wageningen, Netherlands.

McDowell, L. L. 1964. Physiological relationships between dwarf mistletoe and ponderosa pine. *Diss. Abstr.* **25**(1):53.

McFarland, D., and C. A. Ryan. 1974. Proteinase inhibitor-inducing factor in plant leaves. *Plant Physiol.* **54**:706–708.

KcKey, D. 1974. Adaptive patterns in alkaloid physiology. *Am. Nat.* **108**:305–320.

McKey, D. 1979. The distribution of secondary compounds within plants. In G. A. Rosenthal and D. H. Janzen (eds.), *Herbivores: Their interaction with secondary plant metabolites,* pp. 56–133 Academic Press, New York.

Milborrow, B. V. 1974. The chemistry and physiology of abscisic acid. *Annu. Rev. Plant Physiol.* **25**:259–307.

Milton, K. 1979. Factors influencing leaf choice by howler monkeys: A test of some hypotheses of food selection by generalist herbivores. *Am. Nat.* **114**:362–378.

Moore-Landecker, E. 1982. *Fundamentals of the fungi.* Prentice-Hall, Englewood Cliffs, New Jersey.

Müntzing, A. 1946. Different chromosome numbers in root tips and pollen mother cells in a sexual strain of *Poa alpina. Hereditas* **32**:127–129.

Müntzing, A. 1948. Accessory chromosomes in *Poa alpina. Heredity* **2**:49–61.

National Academy of Sciences. 1972. *Genetic vulnerability of major crops.* National Academy of Sciences, Washington, D.C. 307 pp.

Niemelä, P., J. Tuomi, and E. Haukioja. 1980. Age-specific resistance in trees: defoliation of Tamaracks (*Larix laricina*) by larch bud moth (*Zeiraphera improbana* (Lep. Tortricidae)). *Rep. Kevo. Subarctic Res. Stat.* **16**:49–57.

Norris, D. M. 1979. How insects induce disease. In J. G. Horsfall and E. B. Cowling (eds.), *Plant disease,* Vol. 4, pp. 239–256. Academic Press, New York.

Oinonen, E. 1967a. Sporal regeneration of bracken in Finland in the light of the dimensions and age of its clones. *Acta For. Fenn.* **83**:3–96.

Oinonen, E. 1967b. The correlation between the size of Finnish bracken (*Pteridium aquilinum* (L.) Kuhn) clones and certain periods of site history. *Acta For. Fenn.* **83**:1–51.

O'Rourke, F. L. 1949. Honeylocust as a shade and lawn tree. *Am. Nurseryman* **90**:24–29.

Papa, K. E. 1978. The parasexual cycle in *Aspergillus parasiticus. Mycologia* **70**:766–773.

Parlevliet, J. E., and J. C. Zadoks. 1977. The integrated concept of disease resistance: A new view including horizontal and verticle resistance in plants. *Euphytica* **26**:5–21.

Parry, W. H. 1976. The effect of needle age on the acceptability of Sitka spruce needles to the aphid, *Elatobium abietinum* Walker. *Oecologia* **23**:297–313.

Pathak, M. D. 1970. Genetics of plants in pest management. Conference on Principles of Pest Management. March 25–27, Raleigh, North Carolina.

Pillemer, E. A., and W. M. Tingey. 1976. Hooked trichomes: A physical plant barrier to a major agricultural pest. *Science* **193**:482–484.

Pimentel, D., and A. C. Bellotti. 1976. Parasite-host population systems and genetic stability. *Am. Nat.* **110**:877–888.

Pontecorvo, G. 1956. The parasexual cycle in fungi. *Annu. Rev. Microbiol.* **10**:393–400.

Pontecorvo, G. 1958. *Trends in genetic analysis.* Columbia University Press, New York.

Raupp, M. J., and R. F. Denno. 1983. Leaf age as a predictor of herbivore distribution and abundance. In R. F. Denno and M. S. McClure (eds.), *Variable plants and herbivores in natural and managed systems,* pp. 91–124. Academic Press, New York.

Rausher, M. D. 1981. Host plant selection by *Battus philenor* butterflies: the roles of predation, nutrition and plant chemistry. *Ecol. Monogr.* **51**(1):1–20.

Rhoades, D. F. 1979. Evolution of plant chemical defense against herbivores. In G. A. Rosenthal and D. H. Janzen (eds.), *Herbivores: Their interaction with secondary plant metabolites,* pp. 3–54. Academic Press, New York.

Rhoades, D. F. 1983. Herbivore population dynamics and plant chemistry. In R. Denno and M. McClure (eds.), *Variable plants and herbivores in natural and managed systems,* pp. 155–220. Academic Press, New York.

Rhoades, D. F. 1983. Responses in alder and willow to attack by tent caterpillars and webworms: Evidence for phenomenol sensitivity of willow. In P. Hedin (ed.), *Mechanisms of plant resistance to insects.* American Chemical Society Symposium (in press).

Rhoades, D., and R. Cates. 1976. Toward a general theory of plant antiherbivore chemistry. *Recent Adv. Phytochem.* **10**:168–213.

Richerson, J. V., and C. J. DeLoach. 1972. Some aspects of host selection by *Perilitus coccinellae. Ann. Entomol. Soc. Am.* **65**:834–839.

Risch, S. J. 1981. Insect herbivore abundance in tropical monocultures and polycultures: An experimental test of two hypotheses. *Ecology* **62**:1325–1340.

Roane, C. W. 1973. Trends in breeding for disease resistance in crops. *Annu. Rev. Phytopathol.* **11**:463–486.

Robinson, R. A. 1980. New concepts in breeding for disease resistance. *Annu. Rev. Phytopathol.* **18**:189–210.

Roddick, J. G. 1974. The steroidal glycoalkaloid α-tomatine. *Phytochemistry* **13**:9–25.

Root, R. B. 1973. Organization of a plant-arthropod association in simple and diverse habitats: The fauna of collards (*Brassica oleracea*). *Ecol. Monogr.* **43**:95–124.

Rosen, H. R. 1949. Oat percentage and procedure for combining resistance to crown rust, including race 45, and Helminthosporium blight. *Phytopathology* **39**:20.

Rudinsky, J. A. 1966. Scolytid beetles associated with Douglas fir: Responses to terpenes. *Science* **152**:218–219.

Ryan, C. A. 1983. Insect-induced chemical signals regulating natural plant protection responses. In R. F. Denno and M. S. McClure (eds.), *Variable plants and herbivores in natural and managed systems.* Academic Press, New York (in press).

Ryan, C. A., P. Bishop, G. Pearce, A. G. Darvill, M. McNeil, and P. Albersheim. 1981. A sycamore cell wall polysaccharide and a chemically related tomato leaf polysaccharide possess similar proteinase inhibitor-inducing activities. *Plant Physiol.* **68**:616–618.

Sander, H. 1956. Studien über bildung und abbau von Tomatin in der tomatenpflanze. *Planta* **47**:374–400.

Sander, H., and B. Angermann. 1961. Uberden biologischen abbau des tomatins. *Tagungsber., Deut. Akad. Landwirt. Berlin* **27**:163–170.

Schmidt, R. A. 1978. Diseases in forest ecosystems: the importance of functional diversity. In J. G. Horsfall and E. B. Cowling (eds.), *Plant disease,* Vol. 2, pp. 287–315. Academic Press, New York.

Schultz, J. C. 1983. Impact of variable plant defensive chemistry on susceptibility of insects to natural enemies. In P. Hedin (ed.), *Mechanisms of plant resistance to insects.* American Chemical Society Symposium (in press).

Schultz, J. C., P. S. Nothnagle, and I. T. Baldwin. 1982. Seasonal and individual variation in leaf quality of two northern hardwood tree species. *Am. J. Bot.* **69**:753–759.

Schweitzer, D. F. 1979. Effects of foliage age on body weight and survival in larvae of the tribe Lithophanini (Lepidoptera: Noctuidae). *Oikos* **32**:403–408.

Scriber, J. M. 1977. Limiting effects of low leaf water content on the nitrogen utilization, energy budget and larval growth of Hyalophora cecropia (Lepidoptera: Saturniidae). *Oecologia* **28**:269–287.

Segal, A., J. Manisterski, G. Fischbeck, and I. Wahl. 1980. In J. G. Horsfall and E. B. Cowling (eds.), *Plant disease,* Vol. 5, pp. 75–102. Academic Press, New York.

Seigler, D. S. 1977. The naturally occurring cyanogenic glycosides. In L. Reinhold, J. B. Harborne and T. Swain (eds.), *Progress in phytochemistry,* Vol. 4, pp. 83–120. Pergamon Press, New York.

Shamel, A. D. 1943. Bud variation and bud selection. In H. J. Weber and L. D. Batchelor (eds.), *The citrus industry,* Vol. 1, pp. 915–955.

Shamel, A. D., and C. S. Pomeroy. 1936. Bud mutations in horticultural crops. *J. Hered.* **27**:486–494.

Shepard, J. F., D. Bidney, and E. Shahin. 1980. Potato protoplasts in crop improvement. *Science* **208**:17–24.

Solbrig, O. T., and D. J. Solbrig. 1979. *Population biology and evolution.* Addison-Wesley, Reading, Massachusetts.

Soost, R. K., J. W. Cameron, W. P. Bitters, and R. G. Platt. 1961. Citrus bud variation, old and new. *Calif. Citrogr.* **46**(176):188–193.

Southwood, T. R. E. 1979. The components of diversity. In L. A. Mound and N. Waloff (eds.), *Diversity of insect faunas,* pp. 19–40. Symp. Roy. Ent. Soc. 9, Blackwell, London.

Stewart, R. N. 1978. Ontogeny of the primary body in chimeral forms of higher plants. In S. Subtelny and I. M. Sussex (eds.), *The clonal basis of development,* pp. 131–160. Academic Press, New York.

Stewart, R. N., and H. Dermen. 1979. Ontogeny in monocotyledons as revealed by studies of the developmental anatomy of periclinal chloroplast chimeras. *Am. J. Bot.* **66**:47–58.

Street, H. E., A. E. Kenyon, and G. N. Watson. 1946. The nature and distribution of various forms of nitrogen in the potato. *Ann. Appl. Biol.* **33**:1–12.

Strong, D. R., Jr., and D. A. Levin. 1979. Species richness of plant parasites and growth form of their hosts. *Am. Nat.* **114**:1–22.

Sturgeon, K. B. 1979. Monoterpene variation in ponderosa pine xylem resin related to western pine beetle predation. *Evolution* **33**:803–814.

Sunderland, N. 1977. Nuclear cytology. In H. E. Street (ed.), *Plant tissue and cell culture,* pp. 177–205. University of California Press, Berkeley.

Tahvanainen, J. O., and R. B. Root. 1972. The influence of vegetational diversity on the population ecology of a specialized herbivore, *Phyllotreta cruciferae* (Coleoptera: Chrysomelidae). *Oecologia* **10**:321–346.

Tamm, C. O. 1951. Seasonal variation in composition of birch leaves. *Phys. Plan.* **4**:461–469.

Tatro, V. E., R. W. Scora, F. C. Vasek, and J. Kumamota. 1973. Variations in the leaf oils of three species of *Juniperus*. *Am. J. Bot.* **60**:263–271.

Thurston, H. D. 1971. Relationship of general resistance: Late blight of potato. *Phytopathology* **61**:620–626.

Vanderplank, J. E. 1963. *Plant diseases: epidemics and control*. Academic Press, New York, 349 pp.

Vanderplank, J. E. 1968. *Disease resistance in plants*. Academic Press, New York.

Vanderplank, J. E. 1982. *Host-pathogen interactions in plant disease*. Academic Press, New York.

van Emden, H. F. 1965. The effect of uncultivated land on the distribution of cabbage aphid on an adjacent crop. *J. Appl. Ecol.* **2**:171–196.

Vasek, F. C. 1980. Creosote bush: Long-lived clones in the Mojave desert. *Am. J. Bot.* **67**:246–255.

Waller, D. A. 1982. Leaf-cutting ants and avoided plants: Defenses against *Atta texana* attack. *Oecologia* **52**:400–403.

Waloff, N. 1968. Studies on the insect fauna on Scotch broom (*Sarothamnus scoparius* (L.) Wimmer). *Adv. Ecol. Res.* **5**:88–108.

Watt, A. D. 1979. The effect of cereal growth stages on the reproductive activity of *Sitobion avenae* and *Metopolophium dirhodum*. *Ann. Appl. Biol.* **91**:147–157.

Watt, A. D., and A. F. G. Dixon. 1981. The role of cereal growth stages and crowding in the induction of alatae in *Sitobion avenae* and its consequences for population growth. *Ecol. Entomol.* **6**:441–447.

Way, M. J., and M. Cammell. 1970. Aggregation behavior in relation to food utilization by aphids. In A. Watson (ed.), *Animal populations in relation to their food resources*, pp. 229–247. Blackwell, London.

West, R. A. 1981. Fungal elicitors of the phytoalexin responses in higher plants. *Naturwissenschaften* **68**:447–457.

White, J. 1979. The plant as a metapopulation. *Annu. Rev. Ecol. Syst.* **10**:109–145.

White, J. 1980. Demographic factors in populations of plants. In O. T. Solbrig (ed.), *Demography and evolution in plant populations*, pp. 21–48, University of California Press, Berkeley.

Whitham, T. G. 1978. Habitat selection by *Pemphigus* aphids in response to resource limitation and competition. *Ecology* **59**:1164–1176.

Whitham, T. G. 1979. Territorial behavior of *Pemphigus* gall aphids. *Nature* (*London*) **279**:324–325.

Whitham, T. G. 1980. The theory of habitat selection: Examined and extended using *Pemphigus* aphids. *Am. Nat.* **115**:449–466.

Whitham, T. G. 1981. Individual trees as heterogeneous environments: Adaptation to herbivory or epigenetic noise? In R. F. Denno and H. Dingle (eds.), *Insect and life history patterns: Habitat and geographic variation*, pp. 9–27. Springer-Verlag, New York.

Whitham, T. G. 1983. Host manipulation of parasites: Within-plant variation as a defense against rapidly evolving pests. In R. F. Denno and S. McClure (eds.), *Variable plants and herbivores in natural and managed systems*, pp. 15–41. Academic Press, New York.

Whitham, T. G., and C. N. Slobodchikoff. 1981. Evolution by individuals, plant-herbivore interactions, and mosaics of genetic variability: The adaptive significance of somatic mutations in plants. *Oecologia* **49**:287–292.

Whitney, P. J. 1973. Transport across the region of fusion between beans (*Vicia faba*) and broomrape (*Orobanche crenata*). *Proc. Eur. Weed Res. Counc. Symp. Parasit. Weeds*, 154–166.

Williams, P. H. 1979. How fungi induce disease. In J. G. Horsfall and E. B. Cowling (eds.), *Plant disease,* Vol. 4, pp. 163–180. Academic Press, New York.

Wood, R. K. S. 1967. *Physiological plant pathology.* Blackwell, Oxford.

Zucker, W. V. 1982. How aphids choose leaves: The role of phenolics in host selection by a galling aphid. *Ecology* **63**:972–981.

Microorganisms as Mediators of Plant Resource Exploitation by Insect Herbivores

CLIVE G. JONES
Department of Chemical Ecology
New York Botanical Garden
Cary Arboretum
Millbrook, New York

CONTENTS

1 INTRODUCTION

Plants are exposed to a variety of biotic and abiotic stresses. Theories attributing resource defenses against, and exploiter adaptations for, utilizing the resource, such as plant defenses against insects (Feeny, 1975, 1976; Rhoades and Cates, 1976; Gilbert, 1979; Rosenthal and Janzen, 1979; Fox, 1981) and insect adaptations for herbivory (Schoonhoven, 1972; Futuyma, 1976; Wallace and Mansell, 1976; Gilbert, 1979; Rosenthal and Janzen, 1979; Cates, 1980; Dethier, 1980; Fox and Morrow, 1981), often assume that one exploiter has been the primary selection pressure on the resource, and that exploiter adaptation is primarily constrained by the nature of the resource. This ignores the effects of multiple exploitation on resource defenses, and how combinations of exploiters influence resource utilization. A more realistic model should incorporate these inter- and intratrophic level interactions (Whittaker and Feeny, 1971; Levin, 1976; Swain, 1977; Campbell and Duffey, 1979; Price et al., 1980).

I shall examine microbial influences on trophic interactions between plant resource and insect herbivore. The underlying assumptions are that plants are heterogenous resources varying in suitability to insect herbivores (Denno and Dingle, 1981; Denno and McClure, 1983). This heterogeneity places constraints on herbivore adaptation (Mattson et al., 1982; Denno and McClure, 1983). Microbial heterotrophs are ubiquitous, rapidly reproduce and adapt, and are capable of unique metabolic transformations (Brock,

1966). As such, they have properties that distinguish them from other consumers and give them a greater capacity for adjusting to resource heterogeneity. Direct and indirect interaction of plants, insects, and microorganisms is inevitable. Coupling of microorganism and insect changes the constraints on the herbivore, and interaction of microorganism and plant changes the nature of the resource. Both microbial–insect and microbial–plant interactions thus modify the plant–insect herbivore relationship. The major concerns will be the ways in which this modification occurs and the ecological and evolutionary implications. The characteristics of plants, insect herbivores, and microorganisms, and the types of trophic interactions involved in microbial mediation will be considered first.

1.1 Plant Resource Characteristics

There is strong evidence that plants show extensive qualitative and quantitative spatial and temporal variation in physical and chemical character. This includes nutritional and allelochemical variation down to the tissue and cellular levels (Southwood, 1973; Langenheim et al., 1978; McKey, 1979; Whitham, 1981, 1983; Whitham and Slobodchikoff, 1981; Alstad and Edmunds, 1983; Denno, 1983; Jones, 1983; Kareiva, 1983; Krischik and Denno, 1983). The net result of this variation is a heterogenous resource whose suitability to insects may vary extensively (Whitham et al., Chapter 2). Plant resources may be unsuitable for nonadapted herbivores and suboptimal for adapted herbivores (Gordon, 1961; Southwood, 1973; Alstad and Edmunds, 1983; Jones, 1983; Rauscher, 1983; Schultz, 1983; Scriber, 1983).

A broad classification of resource status has four categories of relevance to microbial–insect interactions. (1) Some nutritional components may be at optimal concentrations and available; for example, water in conducting tissues (Baker, 1978) and leaves (Southwood, 1973), protein, lipid, and carbohydrate in many seeds (Bewley and Black, 1978), carbohydrate in nectar (Bolten et al., 1979), and protein in pollen (Southwood, 1973). (2) One or more components may be at suboptimal concentrations and available; for example, most components in phloem other than pentoses (Baker, 1978), lipids or carbohydrates in some seeds (Bewley and Black, 1978), nitrogen in most tissues not actively growing or used as storage [e.g., phloem, xylem, cortex, epidermis, woody tissues (Mattson, 1980)]. (3) One or more components may be optimal or suboptimal but unavailable because (a) they occur in modified, nonutilizable form, such as nonprotein amino acids (Rosenthal and Bell, 1979) and toxic lipids in seeds (Seigler, 1979); (b) they are nutrient conjugates that require conversion to metabolites (e.g., polysaccharides and glycoproteins); (c) they are conjugated to nonnutrients (Reese, 1979), such as the glycosides of phenolics (Harborne, 1979) and cyanide (Conn, 1979), and protein–phenol (Feeny, 1976) and protein–resin complexes (Rhoades and Cates, 1976); (d) nonnutritional components are present that directly or indirectly interfere with utilization by their action on physiological and biochemical processes [e.g., cardiac glycosides (Vaughan and Jungreis, 1977); cyanide (Conn, 1979); saponins (Applebaum and Birk, 1979); phenolics (Beck and Reese, 1976)], and behavioral processes [e.g., deterrents and repellents (Hedin et al., 1974)]; (e) they are not utilizable, such as cellulose and many allelochemicals (Rosenthal and Janzen, 1979). (4) Nutritional components are effectively absent, such as carbohydrate in some seeds (Bewley and Black, 1978), many components other than carbohydrate in nectar (Baker and Baker, 1979), and many components in xylem of quiescent plant parts (Baker, 1978).

This classification distinguishes energetically different forms of the same metabolite (Gordon, 1972) (e.g., free glucose vs. glucose 6-phosphate vs. oligosaccharides vs. phenolic glycosides vs. cellulose), and links nutritional and allelochemical components through their effects on availability; hence, the nutritive potential of allelochemics [e.g., L-canavanine (Rosenthal et al., 1978); phenols (Bernays and Woodhead, 1982)] are included. The scaling of heterogeneity at the tissue level means that comparison of different resources is based on nutritional characteristics of the niche(s) utilized by the herbivore. This provides a mechanism for linking insect, plant, and microorganism that is less feasible at the more gross level of the whole plant.

1.2 Insect Herbivore Adaptations and Constraints

Insect herbivore adaptations are therefore mechanisms for adjusting to resource heterogeneity. Again, four nonexclusive options exist. (1) The herbivore utilizes alternative niche(s); for example, the winter moth on early season oak leaves due to later season tannin–protein complexes (Feeny, 1968, 1970), and timing of pine sawfly feeding due to resins (Ikeda et al., 1977). (2) The herbivore utilizes niche(s) for available, optimal, or adequate components and utilizes mechanisms for (a) concentrating suboptimal components [e.g., aphids (Dixon, 1975)]; (b) converting otherwise unavailable components [e.g., L-canavinine in bruchids (Rosenthal et al., 1978, 1982)]; (c) deconjugating otherwise unavailable components, such as β-glucosidase, lipase, and protease activity (i.e., normal digestive processing of macromolecules); (d) detoxifying interfering components, in the broadest sense [e.g., conversion metabolism of L-canaline (Rosenthal et al., 1978); prevention of protein–tannin complexes (Berenbaum, 1979; Bernays, 1981; Bernays et al., 1981); mixed function oxidase activity (Brattsten, 1979a,b); specific detoxification activity, such as rhodanase (Parsons and Rothschild, 1964); sequestration of cardenolides (Roeske et al., 1976; Duffey et al., 1978); alkaloids, (Bernays et al., 1977); phenolics (Eisner et al., 1971; C. Jones, unpublished data); and other compounds (Duffey, 1976)]. (3) The herbivore compromises by utilizing alternative niche(s) for unavailable components. For example, generalist grasshoppers surviving best on multiple host plant diets (Hodge, 1933; Mulkern, 1967; Otte, 1975; C. Jones and D. Whitman, unpublished data). (4) The herbivore modifies the host plant tissues and their nutritional status. For example, aphids (Dixon, 1975) and gall insects (Mani, 1964; Kirst and Rapp, 1974) acting as "sinks," and gall insects modifying phenolic distribution (C. Jones, D. Philips, and M. Blum, unpublished data; H. Larew, personal communication).

Insects utilize these options (Schoonhoven, 1972; Hedin et al., 1974; Beck and Reese, 1976; Rosenthal and Janzen, 1979; Dethier, 1980; Denno, 1983; Raupp and Denno, 1983; Rausher, 1983; Schultz, 1983; Scriber, 1983), but not all options are available to all herbivores. Examples of constraints are as follows: (1) Insects with specialized feeding mechanisms (e.g., sucking) may become restricted to certain types of tissue (i.e., those under pressure) and their nutritional inadequacies (i.e., phloem is dilute). The advantages of specialism for phenological and biochemical synchrony with the host plant (Denno and Dingle, 1981) may be offset by nutritional problems of an inadequate niche. (2) Generalists utilizing a mixture of tissues (e.g., leaf chewers) may not be able to exploit tissue-localized nutrient concentrations [e.g., nitrogen pulses in the phloem of senescing plants (Mattson, 1980)]. These local concentrations, even if ingested, may be diluted with co-ingested suboptimal tissues. Generalists, even if locally specialized (Fox and Morrow, 1981) may miss out on nutrient-rich tissues that are spatially or temporally asynchronous. (3) Insects processing large amounts of diluted nutrients through the gut (e.g., leaf feeding Macrolepidoptera and sucking insects) may not be able to sufficiently concentrate certain nutrients, particularly micronutrients and nitrogen (Mattson, 1980). For example, the efficiency of conversion of ingested food (ECI; Waldbauer, 1968) is related to nitrogen content (Iversen, 1974; Slansky and Feeny, 1977), and most herbivores must concentrate nitrogen by about 3- to 10-fold (ECI 10–30%; Mattson, 1980). Xylem insects must consume 10^2–10^3 body weight per day (Horsfield, 1977), and ECI values on wood are as low as 1% (Mattson, 1980). Increasing flow rate or enlarging gut volume may be constrained by feeding mechanism and body size. (4) Many insect herbivores have specific requirements in terms of quantitative nutritional balance, specific feeding cues, and, in some cases, qualitatively specific nutrients. However, most qualitative requirements are similar to other heterotrophs, with the notable exception of their inability to synthesize sterols (Fraenkel, 1959,1969; Dadd, 1973). So essential amino acids, B vitamins, minerals, and sterols must be obtained from the resource, together with nitrogen, lipids, water, and an energy source (Dadd, 1973). Metabolic limitations exist. (a) Refractile compounds (e.g., cellulose) cannot be used. (b) Complex conjugates may not be utilizable or energetically worthwhile (e.g., lignin). (c) Synthetic ability is limited (i.e., no sterol, organic nitrogen, or B vitamin synthesis). (d) Detoxification capacity is usually restricted to "unique" metabolism of host plant defenses [e.g., rhodanases (Parsons and Rothschild, 1964)] and oxidative metabolism [e.g., mixed function oxidases (Brattsten, 1979a)]. (e) Certain metabolic activities (e.g., reduction, lipophilic conversion, anaerobic digestion) are limited or absent. Given the adaptations and constraints due to different types of plant heterogeneity, it will be apparent that a number of microbial characteristics are relevant.

1.3 Microbial Characteristics

In comparison to other consumers (and here I exclude viruses and microbial autotrophs and include protozoa on functional considerations), microorganisms have a number of special properties. By virtue of their small size (hence large surface to volume ratio), rapid growth and reproductive rates, high mutability, tendency to occur in complex microbial communities, and unique metabolic character (Brock, 1966; Carlile and Skehel, 1974; Lynch and Poole, 1979), they can (1) concentrate nutrients from a dilute medium due to high osmotic potential (Koch, 1979; Martin, 1979); (2) synthesize nutrients from simple, ubiquitous precursors (Barker, 1956; Annison and Lewis, 1959; Alexander, 1964; Stewart, 1966; Turner, 1971); (3) convert refractile components and nonnutrients to nutrients, and (4) deconjugate metabolites from nutrient–nutrient complexes and nutrient–nonnutrient complexes (Barker, 1956; Annison and Lewis, 1959; Archer, 1973; Clarke, 1974; Mortlock, 1976; Martin, 1979; Kaplan and Hartenstein, 1980); (5) detoxify free or conjugated nonnutrients by oxidative and reductive metabolism (Barker, 1956; Anderson, 1968; Turner, 1971; Archer, 1973; Hill, 1978; Rosazza, 1978; Bourquin and Pritchard, 1979; Suflita et al., 1982); (6) operate at high metabolic efficiencies, and (7) occupy physically small niches at high densities, thereby utilizing resources efficiently (Haddock and Hamilton, 1977; Jones, 1980); (8) rapidly increase in density in response to resource quantity, and (9) adjust to changes in resource quality and quantity by substrate induction of a diversity of enzymes (Konigs and Veldkamp, 1980); (10) adapt metabolically by evolving new enzyme systems in response to qualitative resource selection pressures, and (11) preadapt by fixing mutant inducible and noninducible enzyme systems (Konigs and Veldkamp, 1980; Slater and Godwin, 1980); (12) rapidly adjust to changes in resource quality, and maintain an overall, high utilization efficiency in the microbial community via rapid inter- and intraspecific competition (Baker, 1980).

Although all consumers possess similar attributes, with the exception of many enzyme systems, it is the small size, rapidity, and combination of all properties that should give them a greater capacity for adjusting and adapting to resource heterogeneity than insects. On the other hand, microbial tracking of resources is rooted in a phenomenal reproductive capacity and production of resistant spores (Sussman, 1965). This is inefficient compared to active host plant location mechanisms of, say, insects. So the likely advantages of microbial–plant and microbial–insect associations are metabolic enhancement of plant or insect, increased microbial dispersion efficiency, and increased niche availability for all three parties.

1.4 Trophic Interactions

There are four kinds of interaction between microorganisms, insects, and plants: (1) mutualism (van Beneden 1875), in which both parties benefit; (2) commensalism, in which one party benefits and the other is unaffected; (3) competition, in which both parties utilize the same resource, and (4) predation, in which one party utilizes the other (Boucher et al., 1982). Mutualism may be symbiotic (de Bary, 1879), where there is continuous physical and physiological linkage between both parties, and nonsymbiotic, in which association involves contact but no permanent physical or physiological connection (Boucher et al., 1982). Symbiosis and nonsymbiosis lie on a gradient, and all four types of interaction, particularly mutualism, cover a spectrum from obligate (essential) to facultative (beneficial but nonessential). Mixtures of these interactions are also common. These definitions will be used throughout to avoid confusion (Boucher et al., 1982). I shall focus first on the utilization of microorganisms by insect herbivores involving mutualism and predation, then consider other interactions that may alter the distribution, abundance, and character of the plant and insect herbivore. These include plant pathogens, plant mutualists, and insect pathogens. The occurrence, function, patterns, adaptations, and constraints will be examined from the viewpoint of their impact on the evolution and ecology of the insect–plant relationship.

2 INSECT UTILIZATION OF MICROORGANISMS.

The combination of microbial characteristics and insect herbivore options suggest that microorganisms could have a significant impact on insect herbivore resource exploitation, either by augmenting existing adaptations, or by contributing missing characteristics. So herbivores should utilize microorganisms to improve the nutritional status of the plant (Buchner, 1953). If plants are frequently nutritionally problematical, these associations should be prevalent, perhaps more so than is currently realized. Since microorganisms could modify the resource prior to ingestion by the insect, nonsymbiotic obligate and facultative mutualism and insect predation on microorganisms and plant–microbial substrates are as likely as symbiotic mutualism. Since the types of nutritional problems are varied (see Section 1.1), the functions of microorganisms are also likely to be diverse. Do these associations occur, how prevalent are they, what types exist, and what functions do they fulfill?

2.1 Diversity of Associations

Table 1 exemplifies taxonomic, functional, and resource-related diversity of microbial interactions of proved or presumptive benefit to insects. Some

Table 1. Taxonomic, Resource-Related, and Functional Diversity in Insect Utilization of Microorganisms[a]

	Order	Family	Specific Example	Food Source	Insect Stage	Microorganism Involved	Location of Microorganism	Type of Interaction	Microbial Function	References
1	Orthoptera	Blattidae	*Blattella germanica*	omnivory	L/A	*Corynebacterium blattellae*; flagel. protozoa	int. myc. abd. fat. gonad; ext. alim.	fac.[a,b] sym. mut. spec. specif.; obl? sym. mut. pluri.	peptides, sulfur amino acids, fatty acids? N_2 fixation, cellulase, physical breakdown	Glaser (1930); Henry and Cook (1964); Brooks and Kringen (1972)[b]; Bracke and Markovetz (1980)
2		Cryptocercidae	*Cryptocercus punctulatus*	rotting wood	L/A	bacteria; flagel. protozoa	ext. alim.	obl. sym. mut. extensive pluri.	cellulase, physical breakdown, N_2 fixation	Cleveland et al. (1934); Steinhaus (1967); Breznak et al. (1974)
3		Acrididae	*Melanoplus differentialis*	sunflower leaves	L/A	wilt, rust, and other fungi	leaf surface	fac. nonsym. pred. pluri.	preference, increased growth/survival	Lewis (1979)
4	Isoptera	Mastotermitidae Kalotermitidae	*Pterotermis occidentalis* *Kalotermes* spp. *Cryptotermes brevis*	wood, rotting wood	L/A	flagel. protozoa and bacteria (micrococci)	ext. alim	obl. sym. mut. extensive pluri.	*protozoa:* physical breakdown, cellulase, cellulose to lipid; *bacteria:* N_2 fixation, amino acid synthesis, lactose fermentation, uric acid degradation	*protozoa:* Kirby (1942, 1944, 1946); Koch (1967); Honigberg (1970); Mauldin et al. (1972); Mauldin and Smythe (1973); Mauldin (1977); To et al. (1980) *bacteria:* Breznak et al. (1973); Breznak (1975); Potrikus and Breznak (1977, 1980a,b,c); Mauldin et al. (1978); Schultz and Breznak (1978) *fungi:* Carter et al. (1972); Carter and Smythe (1973)
		Hodotermitidae	*Zootermopsis brevis*	(+ fungus?)		(+ fungus?)	(wood surface)	(fac. sym? pred. pluri.)		
5		Rhinotermitidae	*Reticulotermes flavipes* *Coptotermes formosanus*	as above + fungus	L/A	as above + wood rot fungi	as above + wood surface	as above + fac? nonsym. pred. pluri.	as above + free and bound amino acids, fatty acid composition	
6		Termitidae		wood, grass, herbs + fungus	L/A	fungi	fungal garden	obl. sym. mut. spec. specif.	sole/major/supplement food, fungal cellulase, lignin/phenolic degrading enzymes	Honigberg (1970); Martin and Martin (1978, 1979); Batra and Batra (1979); Martin (1979)
7			*Nasutitermes* *Rhyncotermes* spp.			as above + bacteria	as above + ext. alim.	obl? sym. mut.	as above + N_2 fixation	Prestwich et al. (1980); Prestwich and Bentley (1981)
8	Psocoptera	Trogiidae	*Dorypteryx pallida*	cellulose, starch	L/A	rickettsia-like	int. malpig. ovaries	obl? sym. mut?		Hertig and Wolbach (1924)
9	Mallophaga and Anoplura			vertebrates	L/A	*Rickettsia* and diplococci	int. myc. mid. hind. malpig.	obl. sym. mut.		Steinhaus (1967)

Order	Family	Genus/species	Food	Stage	Symbiont	Location	Symbiosis	Function	Reference
10 Thysanoptera	Thripidae	*Caudothrips buffai*	fungi	L?A	fungus	int. myc. ovocyte / fungal surface	obl? sym. mut.		Bournier (1961)
11	other genera		plants	A	NO bacteria		obl. nonsym. pred. spec. specif.		
12 Hemiptera	Reduviidae	*Rhodnius prolixus*	vertebrate blood	L/A	*Actinomyces* sp.	int. fore. mid. lumen	obl. sym. mut. spec. specif?	growth, molting, egg production, B vitamins	Wigglesworth (1936)
13	Lygaeidae	*Oncopeltus fasciatus*	milkweed seeds/pods	L/A	protozoa: *Proteus recticolens*	ext. 3 of 4 gastcaec.	fac? sym. mut? pluri. spec. specif?		Steinhaus (1941)
					P. insecticolens bacteria: *Erbethella* sp. *Streptococcus* sp.	ext. all 4 gastcaec.			
14	other genera		seeds	A	NO bacteria				Forbes (1892)
15	Scuterellidae	some genera	sap	A	bacilli	ext. gastcaec.	fac? sym. mut?	inhibit pathogens?	Glasgow (1914)
16	other genera		sap	A	NO bacteria				Forbes (1892)
17	Pentatomidae	*Nezara viridula*	seeds	L/A	bacteria and fungi	ext. lab. sty. sal. gastcaec.	fac? sym. mut? pluri.		Forbes (1892); Ragsdale et al. (1979)
18	Coreidae	some genera	sap	A	bacteria	ext. gastcaec.	fac? sym. mut?		Forbes (1892)
		other genera	insects	A	NO bacteria				
19	Phyrrhocoridae		seeds	females	bacteria	ext. gastcaec.	obl. sym. mut.	sex maturation, eggs?	Glasgow (1914)
20 Homoptera	Cicadellidae	*Eutettix tenellus*	leaf sap	L/A	eubacteria	int. cytopl. mid.	obl. sym. mut.		Swezy and Severin (1930)
21	Chermidae		conifer needles, twigs, galls, sap	L/A	bacteria	int. mononucleate myc.	obl. sym. mut.		Steinhaus (1967)
22	Psyllidae	*Psylla* sp.	fruit, twig gall, lvs, sap	L/A	bacteria	int. myc. fat.	obl. sym. mut.		Steinhaus (1967)
23	Orthezidae	*Orthezia insignis*	root epid. cortex	L/A	bacteria	int. myc. fat.	obl. sym. mut.		Buchner (1921)
24	Eriococcidae	*Pseudococcus citri*	sap	L/A	bacteria	int. tracheated myc.	obl. sym. mut. pluri. spec. specif.	B vitamins	Koch (1967); Steinhaus (1967)
					Coccidomyces yeast	int. myc.			
25	Cicadidae	*Magicicada septemdecim*	xylem root	L/A?	rickettsia-like	int. lumen alim. malpig.	obl. sym. mut.	oocyte development	Cowdry (1925)
26	Coccidae	*Lecanium* sp.	sap	L/A	*Lecaniola* sp. yeast	ext. conntiss. body cavity	obl. sym. mut. spec. specif.		Buchner (1958)

61

Table 1. (*Continued*)

Order	Family	Specific Example	Food Source	Insect Stage	Microorganism Involved	Location of Microorganism	Type of Interaction	Microbial Function	References
27		Hippeococcus montana	sap + ant tended	females	NO yeasts	sterile myc.	lost assoc.	ant tending necessary for oocyte development	Buchner (1957)
28	Kermidae	Kermes quercus	twigs oak	L/A	Kerminicola sp. yeast	ext. alim.	obl. sym. mut. spec. specif.		Steinhaus (1967)
29	Membracidae	Oxyrachis trandus	Acacia sap	L/A	bacteria	int. myc.	obl. sym. mut.		Paillot (1933); Thakur and Rao (1976)
30	Aphididae	Shizaphis graminum Neomyzus circumflexus Tuberolachnus salignus Acyrthosiphon pisum	phloem	L/A	eubacteria	int. myc. ext. hemolymph	obl. sym. mut. variable specificity? pluri?	acetate to lipid, sterol metabolism/synthesis? sulfur amino acid and tryptophan synthesis, vitamin precursor conversion	Mittler (1953, 1958); Erharhdt (1968a,b); Houk et al. (1976); Campbell and Nes (1983)
31 Neuroptera	Chrysopidae	Chrysopa carnea	insect honeydew and pollen	A	Torulopsis sp. yeast	ext. crop	fac? sym. mut.	B vitamins, amino acids, female fecundity	Hagan and Tassan (1972)
			insects	L	NO yeasts				
32		C. oculata	as above	A	rickettsia-like		as above?	as above?	Cowdry (1923)
33 Coleoptera	Bostrichidae	Rhizopertha dominica	grain	L/A	bacteria	int. myc.	obl. sym. mut.		Mansour (1934a,b); Huger (1956)
34		Sinoxylon certoniae Bostrycholites zichlei	wood			ext. alim.			
35	Lyctidae	Lyctus linearis	dry wood	L/A	bacteria (>2 spp.)	int. myc. fat.	obl. sym. mut. pluri.		Gambetta (1926); Koch (1936a)
36	Curculionidae	some genera	all plant parts	L/A	bacteria	int. myc. fore mid. hind.	fac? sym. mut.		Koch (1967); Steinhaus (1967)
37		Sitophilus oryzae S. granaria	grain			int. tracheated myc. ext. alim. gastcaec.	obl. sym. mut. pluri.	N$_2$ fixation? B vitamins	Pant and Dang (1972)
38		S. granaria Egyptian variety		A	NO bacteria	sterile myc.	lost assoc.	due to temperature?	Koch (1967)
39	Cucujidae	Oryzaephilus suranimensis	grain	L/A	bacteria-like	int. myc. alim. fat.	obl. sym. mut.	B vitamins	Koch (1936b, 1967); Steinhaus (1967)

No.	Family	Species	Substrate	L/A	Symbiont	Location	Type	Function	Reference
40	Chrysomelidae	*Bromius obscurus*	roots	L only	bacteria (3 spp.)	int. gut. ext. gastceac. int. lumen malpig.	fac? sym. mut. pluri. spec. specif.		Stammer (1935, 1936)
	(many genera examined have no bacteria)								
41		*Donacia* spp.	root sap aquatics	L only	bacteria	int. malpig.	fac? sym. mut.		
42		*Cassida* spp.	leaves	L/A	bacteria	int. gastcaec.	fac? sym. mut?		
43		*C. nebulosa* *C. flaveola*		L/A	NO bacteria				
44	Tenebrionidae	*Tribolium* spp.	grains, etc.	L/A	bacteria	int. myc. fat.	obl? sym. mut.		Steinhaus (1967)
45		*Beliothernus cornutus*	fungal sporocarp	L/A	*Gannoderma applanatum*	fungal surface	obl. nonsym. pred. spec. specif.	sole food source	Martin et al. (1981)
		Neomida bicornis		A	*Coriolis versicolor*				
46	Bruchidae		seeds/pods		NO symbionts				Koch (1967)
47	Lagriidae		fresh/decaying lvs. detritus	L/A	bacteria	ext. alim.	fac? sym. mut?		Stammer (1929)
48	Scarabeidae	*Popillia japonica*	roots lvs/fruit	L / A	flagel. protozoa	ext. alim.	obl. sym. mut.		Steinhaus (1967)
49		*Potosia cuprea*	decaying conifer lvs.	L	bacteria	ext. alim. fermentation chamber	obl. sym. mut?	cellulase + ?	Werner (1926a,b)
					fungi	leaf surface	fac? nonsym? pred.		
50	Lymexylidae		sap. rot. wood, fungus	L	ambrosia fungi contam. fungi	in wood	obl/fac. sym/nonsym. spec. specif/non specif. pluri.	concentrate xylem, supplementary food source	Francke-Grosmann (1967)
51	Anobiidae	*Stegobium paniceum* *Lasioderma serricorne*	grains, etc.	L/A	*Saccharomyces* spp. yeasts	ext. mid.	obl. sym. mut. spec. specif?	B-vitamins, choline, other vitamins sterols, uric acid N$_2$, essential and nonessential amino acids	Koch (1933); Blewett and Fraenkel (1944); Pant and Dang (1972); Jurzitza (1979)
52	Ipidae	*Coccotrypes dactyliperda*	data endosperm	L/A?	*Bacillus*	int. myc. malpig.	obl. sym. mut.		Buchner (1961); Koch (1967)
53		other genera	wood	L/A	bacteria and yeasts	ext. alim.	fac? sym. mut?		

Table 1. (*Continued*)

Order	Family	Specific Example	Food Source	Insect Stage	Microorganism Involved	Location of Microorganism	Type of Interaction	Microbial Function	References
54	Cerambycidae	some genera	decaying/live wood conifer	L	yeasts	ext. alim. mid.	obl. sym. mut. spec. specif. pluri.	N₂ utilization and metabolism	Steinhaus (1967)
					other fungi	in wood			
		other genera	fresh wood deciduous	L	NO symbionts				Koch (1967); Steinhaus (1967)
55	Nitidulae Erotylidae Mycetophagidae Ciidae	various spp.	fungus	A	Polporaceae Tricholomataceae Clavariaceae (various spp.)	on wood	obl. nonsym. pred. spec. specif.	sole food source + enzymes	Martin et al. (1980)
56	Scolytidae	Xyloterini Corythylini Xyleborini Webbini	sapwood diseased trees + fungi	L/A	blue stain fungi (e.g. Hyphomycetes)	in wood	obl/fac. sym/nonsym. spec. specif/non specif. pluri.	major and supplement food, concentrate sap, digestive enzymes	Rumbold (1941); Francke-Grosmann (1967); Kok (1979); Martin (1979)
57		*Dendroctonus* sp.	bark and fungus	L/A	*Candida* sp.	ext. alim.	obl. sym. mut. pluri.	ergosterol, oocyte maturation	
					fungi	in wood	obl. sym. pred.	main food	
58	Platypodidae	*Xyloborus ferrigineus*	heartwood + fungus	L/A	*Staphylococcus*	int. lumen gut	obl. sym. mut. pluri.	oocyte formation	Francke-Grosmann (1967); Norris (1972); Martin (1979)
					blue stain fungi	in wood	fac? sym. mut?	acquired enzymes	
59	Lepidoptera Sphingidae Geometridae Noctuidae	*Smerinthus ocellatus Ennomos autumnaria Agrotis ashworthii*	lvs.	L	bacteria (nonpathogenic)	ext. hemolymph	fac? sym. mut? frequent occurrence		Cameron (1934)
60	Gallerinae	*Galleria mellonella*	beeswax	L	bacteria	ext. alim.	fac. sym. mut.	larvae can survive aseptically	Wollman (1921); Dickman (1933); Niemierko (1959)
61	Nepticulidae Gracilariidae	*Nepticula malella Gracilaria syringella*	leafminers	L	NO symbionts				Portier (1911); Hering (1926)

		Species	Food	L	bacteria & fungi	location	type	function	References
62	Bombycidae	*Bombyx mori*	mulberry leaves	L		ext. alim & leaf surface	fac. nonsym. pred?	cellulase? sterols? B vitamins? protease? physical breakdown?	Portier (1911); Glaser (1925); Ito and Tanaka (1962); Matsubara et al. (1967); Ito (1969); Khokhlacheva and Azimdzhavov (1977); Jones et al. (1981a,b)
64	Diptera Ceratopogonidae	*Dasyhelea obscura*	sap / blood	L only / A	rickettsia-like / NO bacteria	int. myc. sal.	obl? sym. mut.		Buchner (1965)
65	Cecidomiidae	*Asteromyia carbonifera*	stem gall	L	*Sclerotium asteris*	gall wall	obl. sym. mut. spec. specif.	protects from parasitoid of insect	Weis (1982)
66	Anthomyidae	*Hylema antiqua*	onion	L/A	bacteria and fungi	surface	fac. sym/nonsym. mut/pred. pluri.	oviposition stimulant, larval food	Friend et al. (1959); Hough et al. (1981)
67	Tipulidae	*Tipula abdominalis*	aquatic leaf detritus	L	flagel. protozoa, protomonad protozoa / bacteria	ext. alim. / ext. alim. gast-caec.	obl. sym. mut. pluri.	proteinases, physical breakdown	Geiman (1932); Meitz (1975); Martin et al. (1980)
68	Drosophilidae	*Drosophila* sp.	microorg.	L	fungal sporophores, yeasts, bacteria	surface	obl. nonsym. pred. pluri.	sole food source	Baumberger (1919); Kimura (1980)
69	Tephritidae	*Dacus oleae*	olive fruit	L	*Pseudomonas savastonoi*	ext. alim.	obl. sym. mut. spec. specif.	lipase? free fatty acids?	Petri (1910)
70		*Dacus dorsalis*	honeydew	A	bacteria	ext. alim.	obl. sym. mut. spec. specif?	amino acids, growth factors	Hagan and Tassan (1972)
71	Hymenoptera Formicidae	*Camponotus ligniperda*	wood + seeds, insects	L/A	bacteria	int. basal gut membrane	obl? sym. mut.		Koch (1967)
72		*Formica fusca* / *F. rufa*	honeydew nectar, insects	L/A	bacteria	int. gut	obl? sym. mut?		
		F. sanguinea			NO bacteria	embryogeny shows development	lost assoc. then loss of myc.		
73		*Attini*	fungus and leaf sap	L/A	Basidiomycetes	fungal garden	obl. sym. mut. spec. specif.	sole/major food, digestive and detoxifying? enzymes. α-amylase, chitinase, proteases, cellulases, phenoloxidases	Weber (1972, 1979); Martin et al. (1973); Boyd and Martin (1975); Martin (1979)

Table 1. (*Continued*)

Order	Family	Specific Example	Food Source	Insect Stage	Microorganism Involved	Location of Microorganism	Type of Interaction	Microbial Function	References
74	Ichneumonidae	*Cassinaria infesta*	insects	L	rickettsia-like	int. myc. gut	obl? sym. mut.		Cowdry (1923)
75	Apidae	*Apis mellifera*	nectar pollen	L	bacteria NO bacteria	ext. alim. hind. fore. mid.	fac. sym. pred.	contaminant vitamins	Haydak and Dietz (1972)
76	Siricidae	*Tremex* sp.	hardwood dead + fungus	L	Basidiomycetes	in wood	fac? nonsym? pred. spec. specif.	food supplement, host selection, habitat for development	Francke-Grosmann (1967); Madden and Coutts (1979)
77		*Sirex* sp.	fungus	L	as above	in softwood	obl. sym. pred. spec. specif.	sole food, cellulose, lignin degradation	

[a] Insect taxonomy is based on Borror et al. (1976). Key and review references are included, in addition to those in the text (see Section 2.1). Food source refers to that consumed by feeding stages. Insect stage refers to the feeding stages utilizing microorganisms, and does not include transmission or transport of microorganisms.

[b] Abacterial insects will survive on nonminimal diets. *Abbreviations*: A, adult; abd., abdominal tissues; alim., alimentary canal; assoc., association; conntiss., connective tissues; contam., contaminant or casual microorganisms; cytopl., cytoplasm; epid., epidermis; ext., extracellular; fac., facultative (nonessential); fat., fat bodies; flagel., flagellate; fore., foregut; gastcaec., gastric caecae; hind., hindgut; int., intracellular; L, larva; lab., labium; lvs., leaves; malpig., Malpighian tubules; mid., midgut; mut., mutualism; myc., mycetome or mycetosome; nonspec. specif., not species specific; nonsym., nonsymbiotic; obl., obligate (essential); only, other stages examined lack these organisms, in other cases the other stages have not been examined or reported; pluri., plurisymbiotic and/or multiple predation on more than one species; pred., predation; sal., salivary glands; spec. specif., species specific; sty., stylets; sym., symbiotic; ?, status or role or function uncertain.
Note, all entries not separated by a horizontal line from the entry above have the same information applicable.

obligate microphagous species have been included because of their association with plants, some nonherbivores, and insects apparently lacking associations are included for comparative purposes. The literature documenting this diversity is extensive, and no attempt will be made to review it (see Brooks, 1963; Buchner, 1965; Francke-Grosmann, 1967; Hartzell, 1967; Koch, 1967; Steffan, 1967; Steinhaus, 1967; Krishna and Weesner, 1969, 1970; Trager, 1970; Weber, 1972; Breznak, 1975; Batra, 1979; Martin, 1979; Houk and Griffiths, 1980). However, there are extensive gaps in our knowledge, and a strong bias toward obvious, obligate mutualism, particularly of insects on unusual or exceptionally poor substrates (e.g., wood). Three areas in particular are underrepresented in the literature: (a) studies on the roles of nonpathogenic, facultative mutualist or commensal bacteria that are ubiquitous on the insect body surface and alimentary canal (Steinhaus, 1967); (b) documentation and functional studies on facultative or consequential microphagy on the phylloplane (especially leaf surfaces); (c) studies on the occurrence and role of microbial associations in typical phytophagous orders (Steinhaus, 1967), particularly Lepidoptera, foliage-feeding Coleoptera, Hymenoptera, and Orthoptera.

Given the above ommissions and biases, some obvious patterns, and lack of them do emerge from Table 1. (1) Mutualism with, and/or predation on microorganisms occurs in most orders that have phytophages. (2) Most phytophagous orders have many phytophagous representatives that utilize microorganisms. (3) Many microbial taxa are involved both between and within phytophagous orders. (4) There are taxonomic discontinuities within orders (e.g., 12–19, 46; Table 1). (5) There is some degree of microbial taxonomic correspondence within orders in some cases (e.g., 1–2, 4–5, 20–30; Table 1). (6) There are disjunctions by sex and stage within a species (e.g., 19, 31, 40, 41, 64; Table 1). (7) There are insects that do not apparently have associations (e.g., 14, 16, 18, 46, 54, 61; Table 1) and (8) insects that have lost associations (e.g., 27, 38, 72; Table 1). (9) There does not appear to be a single way in which microorganism and insect can be associated (e.g., symbiotic and nonsymbiotic mutualism both occur; both obligate and facultative predation and mutualism occur; intracellular and extracellular associations occur; intracellular locations are varied). (10) Plurisymbiotic relationships with an insect are common, both within a type of microorganism (e.g., protozoa; see 1, 2, 4, 13, 67; Table 1), and between types of microorganisms (e.g., protozoa, bacteria and fungi; see 1, 2, 4, 7, 13, 17, 67, 68; Table 1). (11) Both species specificity (e.g., 1, 6, 24, 26, 28, 30, 45, 50, 55, 56, 65, 69, 73, 76, 77; Table 1) and nonspecificity (e.g., 1–5, 30, 50, 56, 58, 67, 68; Table 1) occur. (12) There are certain types of relationship that do not occur or are infrequent, for example, nonsymbiotic mutualism with protozoa, and endosymbiotic mutualism with fungi other than yeasts (Steinhaus, 1967). Overall it is clear that insect utilization of microorganisms is prevalent and diverse in form. This diversity should extend to function.

2.2 Functional Diversity

Table 1 includes putative or confirmed functions for microorganisms. Again, a number of characteristics emerge. (1) There is no single function that microorganisms can fulfill. (2) No one species of microorganism has exclusive domain over any one function. (3) A given class of microorganisms does not dominate certain activities, unless they are the only ones capable of that activity (e.g., cellulase is not restricted; N_2 fixation is bacterial). (4) There is a good overall correspondence between function fulfilled and the nutritional status of the host plant (Buchner, 1953), although data on the exact nutritional status of the host plant are lacking, and the microbial functions inadequately explored in many cases. In particular, there is a relative paucity of data on concentration of macro- and micronutrients (but see 50, 56; Table 1; Martin, 1979), and a conspicuous dearth of data on detoxification (but see 6, 73; Table 1; Martin, 1979).

Microorganisms beneficially impact growth, development (molting, sexual maturation), survivorship, and fecundity (egg production). In addition, they can physically break down food, inhibit pathogens or parasitoids, produce oviposition stimulants, host attractants, and even synthesize insect defenses (Duffey, 1976; Brand et al., 1979). At the trophic, metabolic level these activities include (i) the microorganism as a sole food source in obligate predatory and mutualistic (insect provisions microorganism from plant) interactions; (ii) the microorganism as a supplemental food source in predatory and mutualistic, obligate and facultative associations; (iii) the provisioning of macro- and micronutrients by biochemical and physiochemical activities. These activities are (a) concentration; (b) synthesis of absent or deficient components (e.g., 1–7, 12, 24, 30, 31, 37, 39, 51, 54, 56, 57, 62, 73; Table 1); (c) enzymatic conversion of refractile or conjugated components (e.g., 1–7, 49, 62, 73, 77; Table 1); (d) detoxification of interfering components (e.g., 6, 73; Table 1; and see Section 2.6). These activities correspond closely to many of the categories of plant heterogeneity (see Section 1.1), insect adaptation and constraints (see Section 1.2), and the functions that microorganisms possess in general that distinguish them from other heterotrophs (see Section 1.3). The underlying assumptions made initially (see Section 1) appear to be valid.

2.3 Plant Nutritional Status as a Determinant of Associative Pattern

Buchner (1953) pointed out that the occurrence of microbial–insect "symbiosis" was very much a function of the food source, so much so that the main body of his book (1965) is a specialized section partitioning associations by type of resource. Many workers have concurred with this view, as does the overall associative and functional diversity examined earlier (see Section 2.2, characteristic 4). Since that time we have become aware of a number of factors that were unknown or not considered by Buchner, and these factors will comprise the remainder of this chapter. As a preface to these considerations, it is appropriate to expand Buchner's hypothesis based on the knowledge that plant heterogeneity, particularly at the within-plant level, is extensive, and that the ways in which heterogeneity occur are diverse. What determines whether or not an association occurs, how necessary it is, and the function(s) fulfilled by the microorganism?

If the nature of the plant resource determines the herbivore constraints, then the patterns of association will be determined by the nutritional status of the plant, at the smallest spatial and temporal niche occupied by the herbivore, i.e., the overall nutritional status of different plant tissues necessary for growth, development, and survivorship to recruitment. Such a statement must not exclude the nonnutritional components of trophic interactions. Thus: (1) Insect microbial associations will tend to occur for any insect that utilizes tissues that are consistently inadequate in any one or more nutritional components necessary for survival, growth, and reproduction *and/or* consistently contain nonnutritional components that interfere with these processes. (2) The degree to which associations are necessary (obligate) will be proportional to the degree to which one or more inadequate components are either essential for, or only differentially enhance growth, survivorship, and reproduction. Similarly, necessity will equate to the degree interfering components inhibit or only reduce growth, survivorship, and reproduction. (3) The function(s) fulfilled by the microorganism will correspond to the nutritional category of the resource (see Section 1.1). Thus, microorganisms will not provide or augment optimally concentrated components; will concentrate suboptimal components, convert unavailable components, detoxify interfering nonnutrients, and synthesize components that are absent, suboptimal, or unavailable. If this modified hypothesis is realistic, certain overall patterns of diversity and function are to be expected; particularly, prevalence in phytophages (see Section 2.1, characteristics 1–3), diversity in form of association (see Section 2.1, characteristic 9), and multiple functions (see Section 2.2, characteristic 1).

To demonstrate whether or not such a tight matching among resource character, herbivore adaptation, and microbial function occurs is not feasible at present. The evidence presented so far suggests that it is a reasonable basis for future studies. This will require an emphasis on accurate quantification of the nutritional and nonnutritional status of the resource utilized by the herbivore, at the ecologically appropriate spatial and temporal level, which complements the developing thrust on the metabolic basis of microbial function (e.g., Brooks and Kringen, 1972; Hagen and Tassan, 1972; Pant and Dang, 1972; Breznak, 1975; Hervey et al., 1977; Mauldin, 1977; Mauldin et al., 1978; Schultz and Breznak, 1978; Garling, 1979; Kok, 1979; Martin, 1979; Prestwich et al., 1980; Prestwich and Bentley, 1981; Campbell and Nes, 1983). The extension of this hypothesis to examine other factors that may determine or modify patterns of association, necessary adaptations, and influence on the plant–insect relationship will now be considered.

2.4 Insect and Microbial Factors Modifying Patterns of Association

What characteristics, other than the nutritional status of the host plant, determine or modify the patterns of association, and explain the patterns identified previously (see Sections 2.1 and 2.2)? Ecological considerations of insect specialism, evolutionary considerations of insect taxonomy, and taxonomic and functional characteristics of the microorganism will be examined.

2.4.1 Insect Specialism and Generalism.

The terms ecological specialist and generalist (e.g., Rhoades and Cates, 1976; Rhoades, 1979; Fox and Morrow, 1981) cover a wide gradient. Extreme specialists feed on specific tissues of one plant species [e.g., some seed endosperm bruchids (Janzen, 1971)], whereas an extreme generalist feeds across species and tissues [e.g., *Romalea guttata* (= *microptera*) feeds on stem, leaves, fruits, flowers, and seeds of grasses, forbs, and low-growing shrubs, utilizing young, old, senescing, and dead tissues, and is opportunistically carnivorous, cannibalistic, and mycophagous (C. Jones, D. Whitman, and M. Blum, unpublished data)]. Does specialism or generalism have any effect on patterns of microbial association (see Section 2.1, characteristics 6–8)?

Obligate Associations. Insects feeding on tissues that are intrinsically nutritionally inadequate should have obligate requirements for microorganisms irrespective of the number of plant sources at which food is obtained. Thus, many termites have a wide host plant range (Krishna and Weesner, 1970) but always have mutualists, as do generalist leaf-cutting ants (Fowler and Stiles, 1980), some generalist seed/stored product insects (e.g., 19, 33, 37, 39, 51; Table 1), and generalist aphids (e.g., *Myzus persicae*; van Emden, 1972). The same is true of host specific aphids (e.g., *Brevicoryne brassicae*; van Emden, 1972), and tephritids (e.g., 69; Table 1). Furthermore, some specialists, such as bruchids, apparently do not have any associations (see 46; Table 1). So at the obligate level there is no obvious effect of specialization.

Facultative Associations. If food is barely adequate and microorganisms may benefit the insect, then the occurrence of facultative associations is more likely to reflect (1) the degree of availability of alternate sources of complementary status (i.e., provide what is missing, or lack what is interfering) both at the within-plant or between-plant levels, and (2) the ability of the herbivore to select the best resources from the available mixture, rather than specialism per se.

Specificity increases the likelihood of the resource being suboptimal and decreases the likelihood of complex allelochemical encounter and nutritional heterogeneity. Conversely, generalism decreases the likelihood of overall resource suboptimality, and increases allelochemical encounter and nutritional heterogeneity (Gordon, 1961; Kreiger et al., 1971; Scriber and Feeny, 1979). So, if we assume, for the moment, that herbivores cannot choose the most optimal resources from among those available to them, then, on the basis of availability alone, both specialists and generalists should benefit from facultative microbial associations. Specialists should benefit from resource enhancement, while "solving" their own specific allelochemical problems; generalists should benefit from microbial detoxification, while the diversity of resources encountered compensates nutritionally.

However, we know that all insect herbivores, even extreme generalists like lubber grasshoppers, can be highly selective given the option (Fox and Morrow, 1981; C. Jones, D. Whitman, and M. Blum, unpublished data). Is the ability to choose more important than the degree of specialization? If so, both specialists occurring on, and generalists actively selecting, a near-optimal diet should have a more facultative requirement for microorganisms. This appears to be the case (for examples, see 3, 17, 59, 75; Table 1, for generalists. See 13, 42, 46, 59, 62, 66; Table 1, for specialists with facultative or lesser microbial associations). It is possible that many folivores have these relationships. Similarly, generalists that are not highly selective and specialists on suboptimal diets should both have more facultative relationships. Again, this appears to be so (see wood, sap, honeydew feeders, and omnivores in Table 1). Whether or not the same end result has arisen in both specialists and generalists via different mechanisms (such as competitive displacement onto poorer resources followed by microbial association, or expansion onto poorer resources because of microbial associations) must await further studies.

Location and Specificity of Association.   Does specialism or generalism influence where microbial associations take place (i.e., in insect or on/in plant), and the degree of species specificity between insect and microorganism (see Section 2.1, characteristics 9, 11, and 12)? Specialists are closely associated with their host plant. One would predict that they should show species specific relationships with variable microbial locations depending on the necessity of the association, the constancy of the resource, and the availability of microbial transmission mechanisms. Fluctuating resources (e.g., fruits, seeds) should select for endosymbiosis (microorganism in insect) if transmission is feasible, as it usually is (see Section 2.6), and an obligate requirement exists (e.g., 33, 37, 39, 51, 52, 69; Table 1). This would favor species specificity (e.g., 51, 69: Table 1). More constant resources, irrespective of quality (e.g., wood, sap), increase the likelihood of microorganism–insect co-occurrence. If sufficiently frequent to guarantee microbial function, exosymbiosis in obligate specialists and nonsymbiosis in facultative specialists could occur (see Keeler, 1981, for parallels in ant/plant mutualism). However, determining degrees of species specificity and distinguishing exosymbiotic from nonsymbiotic are problematical because at high frequencies of co-occurrence they are indistinguishable without detailed behavioral and nutritional studies. This may explain the confusing variety of associations found in wood/fungus feeders occurring on the resource (e.g., 50, 54, 56; Table 1).

The situation is clearer for generalists, where continuity with one host is not necessarily high. Facultative associations are likely to be non-species-specific with nonsymbiotic mutualism or predation (e.g., 3, some species in 50, 56, and 76; Table 1). Generalists with obligate requirements must either carry their symbionts (endosymbiosis, e.g., 1, generalists in 4 and 5; Table 1) or maintain microbial cultures away from the host plants (e.g., 6, 73; Table 1) and tend to be more species specific. Plurisymbiosis (see Section 2.1, characteristic 10) would be expected to decrease species specificity, provided that more than one species serves the same function (e.g., 1, 2, 4, 5; Table 1; see Section 2.6).

Overall, insect specialism and generalism at the plant species level may influence the location and specificity of microbial associations, but have no obvious effect on whether an association occurs, or how obligate it is, this being determined more by the nutritional status of the tissue(s) fed on than by the number of different plant sources of that type of tissue consumed. The apparent unimportance of specialism or generalism in this context is ecologically intriguing, and of potential relevance to theories of plant defense (Feeny, 1976; Rhoades and Cates, 1976; Fox, 1981) and plant resource allocation strategies (Chew and Rodman, 1979).

2.4.2 Insect Taxonomy and Evolution.   If certain taxa are less intrinsically capable of utilizing plant resources than others, they may have evolved microbial associations permitting successful competition with more efficient taxa. The arguments against this are twofold. First, it assumes that all plant resources are equally problematical or equally suitable, irrespective of the herbivore. This is not so. For example, lipid-deficient bean seeds (see Bewley and Black, 1978) present different problems from dilute nutrients in xylem (see Baker, 1978). Even though both are problematical, they are obviously not energetically, nutritionally, or allelochemically equal. Second, it predicts taxonomic disjunctions in the presence or absence of microbial associations at the ordinal level. Table 1 shows that this is not the case. All orders containing phytophages have some representatives that use microorganisms, even if these representatives are not phytophagous (see Section 2.1, characteristics 1–3). For example, extensive mutualistic associations do not appear to occur in phytophagous Orthoptera and Lepidoptera, and yet other nonphytophages do have associations. When extended to the familial, generic or specific level, obvious gaps do appear (see Section 2.1, characteristic 4; 11, 12–19, 40–43, 46, 54, 61, 72; Table 1). However, a purely taxonomic rationale supposes fundamental differences in efficiency at these levels that is not justifiable, that is, closely related insects should always have associations *or* always lack them. This is not so (e.g., 15, 18, 42, 43, 54; Table 1).

It seems more reasonable to assume that success in a given taxon that has associations would result in adaptive radiation and increases in species diversity (Garling, 1979), that is, the number of related species with similar microbial associations. In this sense taxonomy is a consequential and subsequent modifier (see Section 2.1, characteristic 4) rather than a determinant, because the initial "stimulus" for microbial associations would be the status of the resource (Garling, 1979; cf. Weber, 1972). The loss of associations (taxonomic disjunction [see Section 2.1, characteristics 7 and 8]) is perhaps more likely to reflect the selection of alternative niches or a change in the nutritional status of the resource, rather than a major evolutionary change in the metabolic capacity of the insect, although this is, of course, possible.

2.4.3   Microbial Taxonomy and Function.   Does the class of microorganism or the function fulfilled determine and/or modify pattern (see Section 2.1, characteristics 3, 5, 10–12, and Section 2.2, characteristics 1–3)?

Microbial Taxonomy.   This assumes that only certain microbial taxa are capable of forming associations, that these are restricted in distribution, and are specific to certain types of insect. This is unlikely because associations are formed across taxa, including a variety of species of bacteria, fungi, and protozoa (see Table 1), and multispecies complexes with a single insect within microbial taxa (e.g., 2, 4, 35, 40; Table 1) and across taxa (e.g., 5, 13, 49, 67; Table 1) occur. As with insect taxonomy, disjunct distributions (e.g., no protozoa in Homoptera, 20–30; Table 1), and taxonomically related associations (e.g., yeasts in Homoptera, 20–30; Table 1) are common. However, the underlying diversity suggests, again, that this modifies but does not determine pattern. That is, the pool of microorganisms potentially available for forming associations is large and not limiting.

Function Fulfilled by the Microorganism.   This supposes that specific and necessary biochemical activities (e.g., cellulase, B vitamin synthesis, N_2 fixation) are restricted microbial characteristics, and that only microorganisms with "special" biochemical and physiological characteristics are compatible with insects. This does not seem reasonable because there are so many insects with associations, extensive duplication of biochemical activities in unrelated microorganisms (e.g., 12, 24, 31, 54, 62; Table 1), and functionally similar but taxonomically different associations with one insect species (e.g., 1, 67; Table 1). Although the function fulfilled by the microorganism may be crucial to the insect, a lack of incompatibility and a large pool of microorganisms with similar biochemical activities may explain diversity of associations, but not the underlying reason for the association.

2.5   Plant Allelochemicals as Modifiers of Association

The extensions of Buchner's (1953) original hypothesis that I have considered are based on the overall nutritional status of the host plant tissues. The integration of nonnutritional (allelochemical) components within an overall nutritional framework (see Section 1.1) has distinct advantages for an overview of insect utilization of microorganisms. Since there is a conspicuous lack of information concerning the role of allelochemicals in determining and modifying patterns of association, and these compounds are important in insect–plant interactions, there is justification for consideration of these compounds separately.

2.5.1 Diffuse Defenses and Broad Spectrum Activity. The selection pressures for the evolution of plant defenses are considered to have arisen from interaction between plants and other organisms such as viruses, bacteria, fungi, other plants, invertebrates, and vertebrates (Cruickshank, 1963; Sondheimer and Simeone, 1970; Whittaker and Feeny, 1971; Bate-Smith, 1972; Deverall, 1972; Muller and Chou, 1972; Bell, 1974; Freeland and Janzen, 1974; Hedin et al., 1974; Gilbert and Raven, 1975; Atsatt and O'Dowd, 1976; Levin, 1976; Swain, 1976, 1977; Wallace and Mansell, 1976; Hedin, 1977; Kuc and Shain, 1977; Rosenthal and Janzen, 1979; Denno and McClure, 1983). Diffuse models of defense selection (e.g., Fox, 1981; Jones, 1983) assume that target specificity of allelochemicals can only arise when the target has been or is the primary selection pressure on the plant. Since the number of stress factors (abiotic and biotic) involved over evolutionary time is great, and the degree of stress due to any one factor is variable over evolutionary time, a diffuse model predicts that the evolutionary "strategy" of the plant is the production of compounds of broad spectrum biological activity. Specific allelochemical interaction between, say, an insect herbivore and a plant allelochemical is the result of adaptation and inherent susceptibility of the insect, evolving in response to the compound, rather than the compound evolving in response to the insect. Once this specificity has arisen, the insect may select for directional shifts in plant activity (e.g., select for more active variants of the compound, or higher concentrations), but this modifies the defensive patterns and is not the cause. Production of a diversity of cross-reactive compounds and the variability in their occurrence within and between plants in space and time are seen as an underlying mechanism of resistance to biotic stresses (Jones, 1983). No attempt will be made to review the extensive literature exemplifying broad spectrum activity of allelochemicals to insects and microorganisms (see refs. above). The few examples shown in Table 2 are sufficient to demonstrate the phenomenon, and it is likely that further studies will expand the list. Specificity [e.g., juvenile hormone mimics (Slama et al., 1974) and phytoecdysteroids (Jones and Firn, 1978)] appears to be the exception rather than the rule.

2.5.2 Effects on Insect–Microbial Associations. A precedent for nutritional benefit and allelochemical detoxification is well known in ruminant and nonruminant vertebrates (Nagy et al., 1964; Oh et al., 1967; Arnold and Hill, 1972; Bate-Smith, 1972; Freeland and Janzen, 1974; Hungate, 1975;

Table 2.   Examples of Classes of Compounds, or Individual Compounds, with Biological Activity to Both Insects and Microorganisms

Class of Compound	"Anti-insect" Reference	"Antimicrobial" Reference
Tannins	Feeny (1970)	Cook and Taubenhaus (1911); Bell (1974)
Simple phenolics	Beck and Reese (1976)	Bell (1974)
Isoflavanoids	Sutherland et al. (1980)	Russell et al. (1978)
Gossypol	Bottger and Patana (1966)	Bell (1974)
Benzoxazoles	Klun et al. (1967)	Virtanen (1965)
Isothiocyanates	Feeny (1977)	Virtanen (1965)
Sesquiterpene lactones	Rodriguez et al. (1976)	Rodriguez et al. (1976)
Pyrrolizidine alkaloids	McLean (1970)	McLean (1970)
Other alkaloids, e.g., tomatine, solanine, nicotine	Schmeltz (1971)	Bell (1974)
Cyanogenic glycosides	Jones (1972)	Jones (1972)
2-Furaldehyde	Jones et al. (1981a)	Jones et al. (1981b)

Swain, 1977; Schwartz et al., 1980a,b; Waterman et al., 1980; Belovsky, 1981; Wolin, 1981; Owen-Smith and Novellie, 1982; Troyer, 1982; Vaughan, 1982). The evidence for plant allelochemical inhibition of insect–microbial systems is scanty but tantalizing. An increasing number of cases of insect pathogen inhibition have been reported (see Section 3.4). Martin (1979) suggests that phenolic detoxification is due to fungal enzymes in many mycophagous species. Intracellular elimination of bacterial mutualists by synthetic and natural antibiotics (in the medical sense) is common (Behrenz and Technau, 1959; Koch, 1967; Brooks and Kringen, 1972; Pant and Dang, 1972), and is used as the major investigative tool. Fluctuating protozoan populations in lower termites are common (Honigberg, 1970) and considerable evidence exists to implicate plant allelochemicals in the elimination of protozoa from termites (Beal et al., 1974; Carter et al., 1975, 1981; Mauldin et al., 1981). Jones et al. (1981a,b) suggested that exacerbation of growth inhibition of *Bombyx mori* exposed to 2-furaldehyde was due to inhibition of nutritional contributions of leaf surface and enteric microorganisms by this compound. Milne (1961) reported that growth retardation of *Lasioderma serricorne* (see 51; Table 1) with nicotine treatments was due to inhibition of symbionts, and nicotine had no effect if vitamin supplements (the microbial contribution) were added to the diet.

If inhibition of microbial–insect systems by plant allelochemicals is common, then it may well have shaped the evolution of patterns of association and adaptation of microorganism and insect, and hence the insect–plant relationship. The advantages of microbial association to the insect may be offset by disadvantages of microbial susceptibility, and hence, indirectly, an adverse effect on the insect (a problem not faced by insects with no mutualistic microbial associations). On the other hand, the ability of microorganisms to detoxify allelochemicals may markedly benefit the insect (a potential solution not available to insects without microbial associations). What adaptations by both parties are of relevance to these allelochemical interactions and other, functionally similar stress factors?

2.6 Adaptation of Mutualism to Plant Allelochemicals

If insects are to benefit from facultative or obligate predation on microorganisms or plants, there must be no net negative effect on the insect only. However, mutualism requires that there be no net negative effect on fitness of both parties (see Keeler, 1981; Boucher et al., 1982). Therefore four requirements must be fulfilled. Adaptations (1) to prevent loss of association and (2) maintain continuity within and between generations must occur. (3) A habitat for microorganism or microbial–insect combination must be provided. (4) A suitable substrate for microorganism and insect must be provided. A variety of biotic and abiotic factors could affect these requirements [e.g., resource, insect, and microorganism distribution and abundance (see Sections 3.1 and 3.4); microbial competition for the resource (Janzen, 1977); plant mutualistic and pathogen effects on the resource (see Sections 3.1 and 3.2); physical factors such as oxygen tension (Cleveland, 1925b; Veivers et al., 1980); climatic and edaphic factors such as temperature and soil density (Noirot, 1970)] in addition to nutritional and allelochemical aspects of the resource–exploiter complex. I shall examine only adaptations of potential value to allelochemical aspects, although these morphological, biochemical, physiological, and behavioral features may serve multiple adaptive roles.

If plant allelochemicals have shaped insect–microbial associations, as is the case with insect–plant interactions, then I would predict that one or more of the following adaptations should be involved for insect–microbial mutualist systems:

1. Insects should utilize existing detoxification mechanisms or evolve new mechanisms to metabolize compounds that are active against their microorganisms, even if these compounds are not directly toxic to the insect. It is therefore feasible that some detoxification mechanisms in these insects may have a higher specific activity than is warranted by the dose–response curve against the insect, if the microorganisms are highly susceptible. Since increasing generalism will expose the mutualists to a broader array of allelochemicals, generalists should use the same broad spectrum detoxification mechanisms as their nonmutualist counterparts (Kreiger et al., 1971; Bratts-

ten, 1979b), and broad spectrum detoxification of the "antimicrobial" component should be more prevalent in generalists than specialists with obligate mutualism. Specific detoxification mechanisms should be more common in obligatory mutualistic specialists. These types of detoxification mechanisms, as a whole, are more likely to be found in obligate mutualism than in facultative, and this may well be the case for the majority of other adaptations.

2. Microorganisms should utilize or evolve detoxification mechanisms in the same manner as insects. However, the ability of microorganisms to metabolize natural products and xenobiotics is extensive (see Section 1.3). It seems likely that selection should favor microbial, rather than insect, detoxification in mutualists. It may be that microbial detoxification in mutualistic insects is more prevalent and less substrate-specific than insect detoxification in insects with no mutualists. This may contribute to the extensive generalism found in mutualistic Orthoptera, Attini, and lower termites. Evidence for microbial detoxification in mutualists is scanty and limited to metabolism of termiticidal wood extractives (Sands, 1969) and production of phenol-oxidizing enzymes in fungi (see Martin, 1979).

3. Insect and microbial detoxification mechanisms should produce metabolites nontoxic to both parties. This is important given the oxidative, lipophobic insect mechanisms vs. oxidative and/or reductive, lipophilic microbial mechanisms. Production of hypertoxic metabolites is known in some systems [e.g., L-canaline from L-canavanine (Rosenthal et al., 1978); n-methyl hydroxylation of parathion and desulfuration of Schadran (Nakatsugawa and Morelli, 1976); *Senecio* alkaloid metabolites (Mattocks, 1972); mammalian gut microbial degradation of dinitrotoluene (Mirsalis et al., 1982)] but the significance to insect–microbial mutualism is unknown.

4. Microbial mutualists should not synthesize toxins active against the insect, the conspecific, or other plurisymbiotic species in association, but may produce toxins active against pathogens or competitors. The same would be true for insect production of antibacterial or fungistatic compounds [e.g., higher termite production of contaminant fungistatic secretions on fungal gardens (Sands, 1969)]. Microbial biosynthesis of toxins is well known (Turner, 1971; Buchanan and Gibbons, 1974; Steyn, 1980) but their significance in mutualism has not been considered.

5. Microorganisms in mutualistic associations should be inherently resistant to plant allelochemicals. General microbial resistance is well known (Bell, 1974; Sijpesteijn et al., 1977). Studies on intracellular symbiont elimination have shown resistance in a number of cases, albeit to microbial and synthetic antibiotics [e.g., trypaflavin and sodium sulfathiazole in *Periplaneta americana* (Koch, 1936b; Glaser, 1946; Brooks and Richards, 1955); chlortetracyline and chloramphenicol on isolated symbionts of *Triatoma infestans* (Geigy et al., 1954); sulfanilimide and symbiont cultures from *Ernobius mollis, Xestobium plumbeum,* and *Lasioderma serricorne* (Jurzitza, 1963)], whereas others appear to show varying susceptibility (Koch, 1967).

Some protozoa of *Reticulitermes flavipes* and *Coptotermes formosanus* are not eliminated, whereas other species are, when fed in no-choice trials on heartwoods of different tree species (Carter et al., 1981; Mauldin et al., 1981). This suggests that some species are more resistant to elimination than others.

6. Microorganisms should rapidly acquire resistance to plant allelochemicals. This has been shown to occur with chlortetracycline in *Blatella germanica* (Brooks and Richards, 1955), streptomycin and chloramphenicol in *Periplaneta orientalis* (Frank, 1956), and oxytetracycline and chlortetracycline in *Rhizopertha dominica* (Huger, 1956). In the latter case resistant symbionts appeared in the first, second, and third insect generations. The studies of Carter et al. (1981) and Mauldin et al. (1981) showed that on some tree species protozoan populations initially declined and then increased. This could involve acquired resistance, selection against susceptible species or individuals, or overall adjustment. Populations of mutualist microorganisms are considerable in some cases (Buchner, 1965), but in others microbial reproduction is closely synchronized to the host and populations may be small (Buchner, 1965; Trager, 1970), perhaps an insufficient pool for selection of resistant individuals; so other mechanisms should exist.

7. Insects should evolve morphological adaptations to shield microorganisms from allelochemicals. Alimentary mutualists may be physically protected in crypt-guts, sacs, and chambers that are found in many insects and serve as fermentation chambers and for retention of the organism (Koch, 1967; Steinhaus, 1967). If these structures are located in the hind gut they may permit insect midgut oxidase or pH-induced activity to modify or detoxify compounds before they reach the microorganism. Intracellular mycetomes located in gut walls, fat bodies, gonads, and Malpighian tubules (see Table 1; Koch, 1967) serve to prevent washing out and facilitate transfer to progeny (Steinhaus, 1967). However, they may also buffer the microorganism from ingested allelochemicals. Studies on symbiont elimination from mycetomes (Brooks and Richards, 1955; Koch, 1967) show that elimination is not simultaneous from all parts of the mycetome.

8. Insects should evolve mechanisms to minimize the adverse consequences of mutualist loss. This may involve the following. (a) Reduced reliance on a single species of microorganism. Plurisymbiosis is common (Table 1; Cleveland et al., 1934; Kirby, 1937; Koch, 1967). Defaunation studies show that sequential elimination of some species can occur, although growth and survivorship of the insect is not necessarily impaired (Cleveland, 1925a,b, 1926; Yamasaki, 1931; Cleveland et al., 1934; Koch, 1967; Steinhaus, 1967; Carter et al., 1981; Mauldin et al., 1981). This indicates differential susceptibility and/or resistance, and the value of functional redundancy as an insurance. (b) A shift in degree of dependence to a less obligate relationship could occur, or evolution may progress to a facultative stage and no further. No evidence for or against allelochemical involvement is available. (c) Behavioral and morphological mechanisms ensuring continuity and refaunation should occur. Such phenomena are common because of the

loss of symbionts at molting, and the necessity of transfer between generations (Koch, 1967; Steinhaus, 1967). They include retention of part of the peritrophic membrane (Cleveland et al., 1934), cyst and spore production into the habitat (Koch, 1967; Steinhaus, 1967; Honigberg, 1970), social stomatodoel and proctodoel feeding (Krishna, 1969; Honigberg, 1970), and transport and transfer of mutualist inoculum (Francke-Grosmann, 1967; Koch, 1967; Steinhaus, 1967; Sands, 1969). Although these mechanisms may not have evolved in relation to allelochemicals, it is obvious that they could play a role.

9.   Insects should avoid plants containing allelochemicals that are toxic to their mutualists or avoid provisioning toxic substrates, even if these compounds are not toxic to the insect. Attines are generalists but do exhibit distinct preferences (Cherrett, 1972; Fowler and Robinson, 1977; Littledyke and Cherrett, 1978; Fowler and Stiles, 1980). *Acromyrmex octospinosus* removes leaf surface waxes of coffee by licking the surface during food preparation, before adding to the fungal garden. These waxes are fugitoxic to the attine fungal garden (Lampard and Carter, 1973; Quinlan and Cherrett, 1977). Cherrett (1972) suggests that repellency equates to fungitoxicity in attines, and others have concurred with this view (Schade, 1973; Lewis et al., 1974). Studies by Carter et al. (1981) and Mauldin et al. (1981) on *R. flavipes* and *C. formosanus* strongly suggest that some woods repellent or deterrent to these lower termites are not directly termiticidal, but cause elimination of protozoa and subsequent starvation. It may well be that the microbial basis for insect repellency, deterrency, or other selective behaviors in these cases is a more general phenomenon in other insects with mutualistic microorganisms.

3 INDIRECT MICROBIAL MEDIATION

A variety of plant–microbial and insect–microbial pathogen interactions may also affect insect resource exploitation, and these will be briefly considered.

3.1 Plant Mutualists

Microbial mutualists of plants may alter the nutritional status of the plant to the insect. Nitrogen content in N_2-fixing plants (Stewart, 1966) tends to be higher than in nonfixers (Mattson, 1980). Xylem of N_2 fixers contains amino acids absent in nonfixers, which have NO_3^- (Baker, 1978). Enhanced concentrations, improvement in nitrogen quality, and reduction in N_2 variability may improve nutritional status to herbivores. Alternatively, reduced nitrogen stress in N_2 fixers may not bring about the associated stress-induced increases in nitrogen found in other plants (these changes appear to enhance herbivore susceptibility in nonfixers; for refs., see Mattson, 1980) and may decrease herbivore suitability. Similarly ecto- and endosymbiotic mycorrhizal associations of roots and seeds involving mineral balance (Harley, 1959) may also change nutritional quality and quantity. Plant mutualists may also extend the distribution and abundance of plants into marginal habitats. These changes are likely to affect herbivore distribution and abundance. Little or nothing is known about the impact of plant–microbial mutualism on insect herbivores, although some studies on N_2 fixation and microbial disease resistance indicate N_2 fixation and plant allelochemical interaction with microorganisms (Pankhurst and Briggs, 1980; Viands et al., 1980).

3.2 Plant Pathogens

Plant pathogens may also bring about marked changes in proximal nutrient quality and quantity. These include increases in plant nitrogen, protein from phytoalexin enzyme production, phenolics and other allelochemicals (for refs., see Mattson, 1980; see also Section 2.5.1), and a variety of plant and/or microbial induced changes (Hancock and Huisman, 1981). Although the role of plant pathogenic or saprophytic fungal mutualists of bark- and wood-feeding insects is apparent (Francke-Grosmann, 1967; Batra, 1979; Martin, 1979), overall significance of other interactions, particularly of folivorous insects (e.g., Lewis, 1979), is not known. Changes in the distribution and abundance of plants due to pathogens is well known [e.g., Dutch elm disease (Barnes, 1976; Karnosky, 1982)], and these changes must also affect the distribution and abundance of insect herbivores over short and long evolutionary time scales. Again, these possibilities have not been examined.

3.3 The Phylloplane

Plants have a large surface area that may act as a pool for microbial mutualists and pathogens of insects and plants. The microbiology of plant surfaces is complex and poorly understood, but involves a diversity of bacterial, fungal, and plant interactions (Preece and Dickinson, 1971; Billing, 1976; Blakeman and Brodie, 1976; Davenport, 1976; Dickinson, 1976; Fokkema, 1976). Many of these interactions could affect insect herbivores. For example, nitrogen fixing by leaf surface microorganisms may be as high as 5.2 mg N/g leaf (Jones, 1976; Sucoff, 1979), and may enhance N_2 levels for insects. A large number of fungi and bacteria produce toxins (Turner, 1971; Buchanan and Gibbons, 1974; Lampe, 1979; Steyn, 1980), some of which are contact insecticidal (Dobias et al., 1979), and these could affect folivorous insects. Chemical changes on the leaf surface may change host-recognition cues important in food selection (Chapman, 1977).

3.4 Insect Pathogens

Obviously these organisms may have a marked impact on insects. I shall briefly consider only one aspect. There is an increasing body of evidence that one advantage of plant allelochemicals to insects that may balance their normally adverse effects is inhibition of insect pathogens. This is perhaps one further consequence of broad spectrum allelochemical activity, and an example of the complexity of inter- and intratrophic level interactions among insects, microorganisms, and plants. Hedin et al. (1978) demonstrated that a number of cotton constituents, notably phenolics, suppressed gut bacteria in the boll weevil. Smirnoff (1972) showed that *Abies balsamae* volatile terpenoids were inhibitory to a number of entomopathogenic bacteria used in biological control. Antibacterial activity of host plants has been shown to decrease natural pathogen susceptibility in a number of Lepidoptera (Kuschner and Harvey, 1960, 1962; Smirnoff and Hutchison, 1965; Afify and Merdan, 1969; Maksymiuk, 1970; Merdan et al., 1975; Kunimi and Aruga, 1974), and simple phenolic acids from the host plant inhibit pathogens in *Bombyx mori* (Matsubara and Hayashiya, 1969; Iizuka et al., 1974; Kinoshita and Inoue, 1977; Hayashiya, 1978; Koike et al., 1979). These discoveries should result in a reevaluation of control strategies as well as basic assumptions about the effects of plant allelochemicals on insects. The coupling of plant allelochemicals, insect mutualists, and insect pathogens is particularly intriguing and an area that is unexplored.

4 CONCLUSION: CONSEQUENCES OF MICROBIAL MEDIATION TO INSECT–PLANT RELATIONSHIPS

I have examined a number of aspects of nutritional and allelochemical interactions among insects, microorganisms, and plants. What specific and general consequences, testable hypotheses, and practical approaches emerge that are significant to insect–plant interactions?

1. If a heterogenous resource is an acceptable description of the host plant, then a primary consequence of insect–microbial mutualism may be to even out resource heterogeneity. Thus, nutritional indices (e.g., ECI) should show less variation between plants for insects with mutualists than for insects without, even if absolute values are different. This should be eminently testable in closely related insects with and without microbial mutualists (see Table 1).

2. Insect–microbial symbiosis may permit marginal resource exploitation, but will not necessarily make that resource optimal in comparison to others. Thus, nutritional indices for insect mutualists on poor resources should be lower than insects without microorganisms on better resources. If so, the evolutionary stimuli for microbial associations (i.e., which comes first: poor resource or microbial association and competitive displacement?) may be approached from the comparative nutritional ecology of these two types of insect.

3. If microbial mediation evens out resource heterogeneity and permits marginal resource exploitation, then insects with mutualists may be expected

to show considerable flexibility in host plant range and subsequent adaptive radiation. One may speculate that the evolution of host specificity in these insects is primarily due to the status of the tissues of the first plant resources responsible for selection for mutualism. It may be that host plant specificity may decrease but tissue specificity remains constant because of microbial associations (e.g., some termites are generalist wood feeders).

4. Conversely, insects losing symbionts over evolutionary time (e.g., Egyptian *Sitophilus granaria, Hippeococcus montana, Formica* sp.; see Table 1) should show a corresponding shift in host plant and/or tissue, or evolve replacement metabolic capacity. Given the likely adverse consequences of mutualist loss, these shifts may precede loss rather than be caused by them. Obviously, these insects would make excellent investigative tools.

5. The apparent unimportance of insect and microbial taxonomy indicates that microbial mediation is an underlying ecological phenomenon and not just a series of exceptions. The limits to associations are therefore probably not determined at these levels and this should facilitate cross-taxonomic comparisons.

6. Acceptance of resource nutritional heterogeneity as a primary determinant of association requires the extensions of Buchner's (1953) hypothesis to be tested over a broader range of resource suitability (e.g., leaf chewers) and down to the within-plant tissue level, for example, the niche of the herbivore. The patterns of microbial associations clearly demonstrate that detailed analysis of the nutritional characteristics of the resource utilized is paramount. If this can be accomplished for insects with and without microorganisms on the same or similar resources, the relative importance of plant vs. insect characteristics can be established. Furthermore, it should give strong clues as to the likely functions fulfilled by the microorganisms, rather than screening for every conceivable microbial activity.

7. The apparent unimportance of insect specialism and generalism is intriguing and should be thoroughly tested through resource comparison. If true, tissue-level operation of microbial mediation will change our perception of the level at which interaction occurs, the relevance of community and competitive phenomena, and niche overlap. Whether generalism is a different way of specializing to resource heterogeneity (i.e., increasing availability, selection of the most optimal) and whether the inferences concerning unimportance of specialism and generalism to insect–microbial systems can be extended to all insect herbivores remains to be seen.

8. Microbial characteristics present options to insect herbivores that may not be otherwise available, apart from obvious nutritional benefits. Allelochemical processing is one such area that is unexplored, and the proposed adaptations are all eminently testable. Discovery of extensive microbial detoxification, for example, could change the way we view plant defense and insect herbivore adaptation.

9. Similarly, the broad spectrum activity of plant allelochemicals may present a dual target to the plant (insect and/or its microorganisms). What

is the trade-off between advantage and disadvantage? Do the adaptations suggested occur? Is the net result for insect mutualists the same as for insects without mutualists? What is the result of combinations of insect, microorganism, and plant when allelochemical interactions make take place in pathogens and mutualists and commensals of the insect and the plant? Does this substantiate a claim for diffuse selection? Such questions can only be addressed by considering many of the microbial, nutritional, and allelochemical interactions involved in a given system, or by studying those in which one or more components are missing.

10. The functional diversity of microorganisms extends beyond nutritional and allelochemical considerations of the plant to encompass effects on distribution and abundance, microbial production of defenses, attractants, and stimulants. It would seem reasonable to take these factors into account.

I consider microbial mediation of plant resource exploitation by insect herbivores to have had a significant effect on the evolution and ecology of the insect–plant relationship, since the added dimensions present solutions and dilemmas that do not exist in one-on-one interactions. I have presented a broad overview of these interactions in the hope that it will serve as a stimulus for discussion and research in an underexplored area that will require interdisciplinary collaboration.

5 SUMMARY

Plants are heterogenous resources varying in suitability to insect herbivores and constraining exploitation. Microorganisms have characteristics that give them a greater capacity for adjusting to resource heterogeneity. The extensive utilization of microorganisms by insects changes the constraints on the insect and the insect–plant relationship. Insects show functional and taxonomic diversity in microbial associations. The primary determinant of the degree, specificity, and function of associations is the nutritional status of the host plant, which probably operates at the within-plant tissue level. Insect specialism and generalism probably influence the location and species specificity of these associations, but not whether or not they occur. The ability of the insect herbivore to select the most optimal resources from a mixture, and the availability of these resources in space and time is probably more important than degree of specialization. Constraints of insect taxonomy and microbial taxonomy and function do not appear to limit associations. Plant allelochemicals, as a subset of overall resource nutritional status, may adversely affect insect–microbial mutualism because of their broad spectrum activity. In this, insect–microbial systems are different from other insects. However, microorganisms possess characteristics that may facilitate allelochemical processing. Adaptations for this are proposed to involve insect and microbial detoxification; compatible metabolites; a lack of inhibitory microbial toxins; inherent and acquired microbial resistance; morphological protection of microorganisms; morphological, physiological, social, and behavioral minimization of microbial loss; and behavioral selection of nontoxic substrates. Indirect microbial mediation, such as changes in nutrient and allelochemical quantity and quality, and plant distribution and abundance are considered as further modifications of plant–insect relationships. These include plant–microbial mutualism and pathogenicity, interactions on the phylloplane, and allelochemical effects on insect pathogens. Overall consequences to insect–plant relationships are suggested to include evening out of resource heterogeneity, marginal resource exploitation, flexibility in host plant range and adaptive radiation, shifts in host plants, restriction at the tissue level, and changes in evolution and ecology of allelochemical interactions between insects and plants.

ACKNOWLEDGMENTS

I thank the following for information, suggestions, and comments on an earlier manuscript: P. Barbosa, B. Campbell, F. Carter, H. Fowler, W. Freeland, D. Janzen, H. Larew, M. Martin, J. Mauldin, P. Price, D. Rhoades, R. Scheffrahn, and A. Weis.

LITERATURE CITED

Afify, M. A., and A. I. Merdan. 1969. Reaktionsunterschiede von der Noctuidenarten bei bestimmten *Bacillus* praparaten in abhangigkeit von der nahrung und art der behandlung. *Anz. Schadlingskde. V. Planzenchutz.* **42**:102–104.

Alexander, M. 1964. Biochemical ecology of soil microorganisms. *Annu. Rev. Microbiol.* **18**:217–252.

Alstad, D. N., and G. F. Edmunds, Jr. 1983. Adaptation, host specificity, and gene flow in the black pineleaf scale. In R. F. Denno and M. S. McClure (eds.), *Variable plants and herbivores in natural and managed systems,* pp. 413–426. Academic Press, New York.

Anderson, E. S. 1968. The ecology of transferable drug resistance in the enterobacteria. *Annu. Rev. Microbiol.* **22**:131–180.

Annison, D. F., and D. Lewis. 1959. *Metabolism in the rumen.* Methuen, London. 281 pp.

Applebaum, S. W., and Y. Birk. 1979. Saponins. In G. A. Rosenthal and D. H. Janzen (eds.), *Herbivores—their interaction with secondary plant metabolites,* pp. 539–566. Academic Press, New York.

Archer, L. J. (ed.). 1973. *Bacterial transformation.* Academic Press, New York. 413 pp.

Arnold, G. W., and J. L. Hill. 1972. Chemical factors affecting selection of food plants by ruminants. In J. B. Harborne (ed.), *Phytochemical ecology* (Annu. Proc. Phytochem. Soc. 8), pp. 72–101. Academic Press, London.

Atsatt, P. R., and D. J. O'Dowd. 1976. Plant defense guilds. *Science* **193**:24–29.

Baker, D. A. 1978. *Transport phenomena in plants.* Chapman and Hall, London. 80 pp.

Baker, I., and H. G. Baker. 1979. Chemical constituents of the nectars of two *Erythrina* species and their hybrid. *Ann. Missouri Bot. Gard.* **66**:446–450.

Baker, K. F. 1980. Microbial antagonism—the potential for biological control. In D. C. Ellwood, J. N. Hedger, M. J. Latham, J. M. Lynch and J. H. Slater (eds.), *Contemporary microbial ecology*, pp. 327–347. Academic Press, London.

Barker, H. A. 1956. *Bacterial fermentations*. Wiley, New York. 95 pp.

Barnes, B. U. 1976. Succession in deciduous swamp communities of southeastern Michigan formerly dominated by American elm. *Can. J. Bot.* **54**:20–24.

Bate-Smith, E. C. 1972. Attractants and repellents in higher animals. In J. B. Harborne (ed.), *Phytochemical ecology* (Annu. Proc. Phytochem. Soc. 8), pp. 45–46. Academic Press, London.

Batra, L. R. (ed.). 1979. *Insect–fungus symbiosis. Nutrition, mutualism, and commensalism*. Wiley, New York. 276 pp.

Batra, L. R., and W. S. Batra. 1979. Termite-fungus mutualism. In L. R. Batra (ed.), *Insect–fungus symbiosis. Nutrition, mutualism, and commensalism*, pp. 117–163. Wiley, New York.

Baumberger, J. P. 1919. A nutritional study of insects, with special reference to microorganisms and their substrata. *J. Exp. Zool.* **28**:1–81.

Beal, R. H., F. L. Carter, and C. R. Southwell. 1974. Survival and feeding of subterranean termites on tropical woods. *For. Prod. J.* **24**:44–48.

Beck, S. D., and J. C. Reese. 1976. Insect plant interactions: Nutrition and metabolism. *Recent Adv. Phytochem.* **10**:41–92.

Behrenz, W., and G. Technau. 1959. Versuche zur bekampfung von *Anobium punctatum* mit symbionticiden. *Z. Angew. Entomol.* **44**:22–28.

Bell, A. A. 1974. Biochemical bases of resistance of plants to pathogens. In F. G. Maxwell and F. A. Harris (eds.), *Proceedings of the Summer Institute on Biological Control of Plant Insects and Diseases*, pp. 403–462. University of Mississippi Press, Jackson.

Belovsky, G. E. 1981. Food plant selection by a generalist herbivore: The moose. *Ecology* **62**:1020–1030.

Berenbaum, M. 1979. Adaptive significance of midgut pH in larval Lepidoptera. *Am. Nat.* **115**:138–146.

Bernays, E. A. 1981. A specialized region of the gastric caeca in the locust *Schistocerca gregaria*. *Physiol. Entomol.* **6**:1–6.

Bernays, E. A., and S. Woodhead. 1982. Plant phenols utilized as nutrients by a phytophagous insect. *Science* **216**:201–203.

Bernays, E. A., J. A. Edgar, and M. Rothschild. 1977. Pyrrolizidine alkaloids sequestered and stored by the aposematic grasshopper, *Zonocerus variegatus*. *J. Zool. London* **182**:85–87.

Bernays, E. A., D. J. Chamberlain, and E. M. Leather. 1981. Tolerance of acridids to ingested condensed tannin. *J. Chem. Ecol.* **7**:247–256.

Bewley, J. D., and M. Black. 1978. *Physiology and biochemistry of seeds in relation to germination*, Vol. 1. Springer-Verlag, Berlin. 306 pp.

Billing, E. 1976. The taxonomy of bacteria on the aerial parts of plants. In C. H. Dickinson and T. F. Preece (eds.), *Microbiology of aerial plant surfaces*, pp. 223–273. Academic Press, London.

Blakeman, J. P., and I. D. S. Brodie. 1976. Inhibition of pathogens by epiphytic bacteria on aerial plant surfaces. In C. H. Dickinson and T. F. Preece (eds.), *Microbiology of aerial plant surfaces*, pp. 529–557. Academic Press, London.

Blewett, N., and G. S. Fraenkel. 1944. Intracellular symbiosis and vitamin requirements of two insects, *Lasioderma serricorne* and *Sitodrepa panicea*. *Proc. R. Soc. London, Ser. B* **132**:212–221.

Bolten, A. B., P. Feisinger, H. G. Baker, and I. Baker. 1979. On the calculation of sugar concentration in flower nectar. *Oecologia* **41**:301–304.

Borror, D. J., D. M. DeLong, and C. A. Triplehorn. 1976. *An introduction to the study of insects.* Holt, New York. 852 pp.

Bottger, G. T., and R. Patana. 1966. Growth development and survival of certain Lepidoptera fed gossypol in the diet. *J. Econ. Entomol.* **59**:1166–1169.

Boucher, D. H., S. James, and K. H. Keeler. 1982. The ecology of mutualism. *Annu. Rev. Ecol. Syst.* **13**:315–347.

Bournier, A. 1961. Sur l'existence et l'évolution d'un mycétome au cours de l'embryologenèse de *Caudothrips buffai karny.* Verhandl. *XI Int. Congr. Entomol.* **1**:352–354.

Bourquin, A. W., and Pritchard, P. H. (eds.). 1979. *Microbial degradation of pollutants in marine environments.* U.S. Environmental Protection Agency, Gulf Breeze, Florida. 551 pp.

Boyd, N. D., and M. M. Martin. 1975. Faecal proteinases of the fungus-growing ant, *Atta texana*: Their fungal origin and ecological significance. *J. Insect Physiol.* **21**:1815–1820.

Bracke, J. W., and A. J. Markovetz. 1980. Transport of bacterial end products from the colon of *Periplaneta americana.* *J. Insect Physiol.* **26**:85–89.

Brand, J. M., J. C. Young, and R. M. Silverstein. 1979. Insect pheromones: a critical review of recent advances in their chemistry, biology and application. *Prog. Chem. Org. Nat. Prod.* **37**:1–190.

Brattsten, L. B. 1979a. Ecological significance of mixed-function oxidations. *Drug. Metab. Rev.* **10**:35–58.

Brattsten, L. B. 1979b. Biochemical defense mechanisms in herbivores against plant allelochemicals. In G. A. Rosenthal and D. A. Janzen (eds.), *Herbivores—Their interaction with secondary plant metabolites,* pp. 199–270. Academic Press, New York.

Breznak, J. A. 1975. Symbiotic relationships between termites and their intestinal microbiota. *Symp. Soc. Exp. Biol.* **29**:559–580.

Breznak, J. A., W. J. Brill, J. W. Mertins, and H. C. Coppel. 1973. Nitrogen fixation in termites. *Nature (London)* **244**:577–580.

Breznak, J. A., J. W. Mertins, and H. C. Coppel. 1974. Nitrogen fixation and methane production in a wood-eating cockroach, *Cryptocercus punctulatus* Scudder (Orthoptera: Blattidae), *Univ. Wisconsin Forestry Research Notes,* no. 184. 2 pp.

Brock, T. D. 1966. *Principles of microbial ecology.* Prentice-Hall, Englewood Cliffs, New Jersey. 306 pp.

Brooks, M. A. 1963. The microorganisms of healthy insects. In E. A. Steinhaus (ed.), *Insect pathology,* Vol. 1, pp. 215–250. Academic Press, New York.

Brooks, M. A., and A. G. Richards. 1955. Intracellular symbiosis in cockroaches. 1. Production of aposymbiotic cockroaches. *Biol. Bull.* **109**:22–39.

Brooks, M. A., and W. B. Kringen. 1972. Polypeptides and proteins as growth factors for aposymbiotic *Blatella germanica* (L.). In J. G. Rodriguez (ed.), *Insect and mite nutrition,* pp. 353–364. North-Holland, Amsterdam.

Buchanan, R. E., and N. E. Gibbons (eds.). 1974. *Bergey's manual of determinative bacteriology,* 8th ed. Williams and Wilkins, Baltimore. 1272 pp.

Buchner, P. 1921. *Tier und pflanze in intracellularer symbiose.* Gerbrüder Borntraeger, Berlin. 462 pp.

Buchner, P. 1953. *Endosymbiose der tiere mit pflanzlichen mikroorganismen.* Birkhauser, Basel. 771 pp.

Buchner, P. 1957. Endosymbiosestudien on schildlausen IV. *Hippeococcus,* eine myrmekophile pseudococcine. *Z. Morphol. Oekol. Tiere* **45**:379–410.

Buchner, P. 1958. Eine neue form der endosymbiose bei aphiden. *Zool. Anz.* **160**:222–230.

Buchner, P. 1961. Endosymbiosestudien an ipiden. 1. Die gattung *Coccotrypes.* *Z. Morphol. Oekol. Tiere* **50**:1–80.

Buchner, P. 1965. *Endosymbiosis of animals with plant microorganisms.* Interscience, New York. 909 pp.

Cameron, G. R. 1934. Inflammation in the caterpillars of Lepidoptera. *J. Pathol. Bacteriol.* **38**:441–466.

Campbell, B. C., and S. S. Duffey. 1979. Tomatine and parasitic wasps: potential incompatibility of plant antibiosis with biological control. *Science* **205**:700–702.

Campbell, B. C., and W. D. Nes. 1983. A reappraisal of sterol biosynthesis and metabolism in aphids. *J. Insect Physiol.* **29**:149–156.

Carlile, M. J., and J. J. Skehel (eds.). 1974. *Evolution in the microbial world.* Symp. Soc. Gen. Microbial. 24. Cambridge University Press, London. 450 pp.

Carter, F. L., and R. V. Smythe. 1973. Effect of sound and *Lenzites*-decayed wood on the amino acid composition of *Reticulotermes flavipes. J. Insect Physiol.* **19**:1623–1629.

Carter, F. L., L. A. Dinus, and R. V. Smythe. 1972. Effect of wood decayed by *Lenzites trabea* on the fatty acid composition of the eastern subterranean termite, *Reticulitermes flavipes. J. Insect Physiol.* **18**:1387–1393.

Carter, F. L., R. H. Beal, and J. D. Bultman. 1975. Extraction of antitermitic substances from 23 tropical hardwoods. *Wood Sci.* **8**:406–410.

Carter, F. L., J. K. Mauldin, and N. M. Rich. 1981. Protozoan populations of *Coptotermes formosanus* Shiraki exposed to heartwood samples of 21 American wood species. *Mater. Org.* **16**:29–38.

Cates, R. G. 1980. Feeding patterns of monophagous. oligophagous, and polyphagous herbivores: The effect of resource abundance and plant chemistry. *Oecologia* **46**:22–31.

Chapman, R. F. 1977. The role of the leaf surface in food selection by acridids and other insects. In V. Labeyrie (ed.), *Insect behavior and trophic environments,* pp. 133–149. Inst. Colloq. CNRS No. 265, Paris.

Cherret, J. M. 1972. Chemical aspects of plant attack by leaf-cutting ants. In J. B. Harborne (ed.), *Phytochemical ecology* (Annu. Proc. Phytochem. Soc. 8), pp. 13–24. Academic Press, London.

Chew, F. S., and J. E. Rodman. 1979. Plant resources for chemical defense. In G. A. Rosenthal and D. H. Janzen (eds.), *Herbivores—Their interaction with secondary plant metabolites,* pp. 271–307. Academic Press, New York.

Clarke, P. H. 1974. The evolution of enzymes for the utilization of novel substrates. In M. J. Carlile and J. J. Skehel (eds.), *Evolution in the microbial world* (Symp. Soc. Gen. Microbial 24), pp. 183–217. Cambridge University Press, London.

Cleveland, L. R. 1925a. The feeding habit of termite castes and its relation to their intestinal flagellates. *Biol. Bull.* **48**:295–308.

Cleveland, L. R. 1925b. The effect of oxygenation and starvation on the symbiosis between the termite, *Termopsis,* and its intestinal flagellates. *Biol. Bull.* **48**:309–326.

Cleveland, L. R. 1926. Symbiosis among animals with special reference to termites and their intestinal flagellates. *Q. Rev. Biol.* **1**:51–60.

Cleveland, L. R., S. R. Hall, E. P. Sanders, and J. Collier. 1934. The woodfeeding roach *Cryptocercus,* its protozoa, and the symbiosis between protozoa and roach. *Mem. Am. Acad. Sci.* **17**:185–342.

Conn, E. E. 1979. Cyanide and cyanogenic glycosides. In G. A. Rosenthal and D. H. Janzen (eds.), *Herbivores—Their interaction with secondary plant metabolites,* pp. 387–412. Academic Press, New York.

Cook, M. T., and J. J. Taubenhaus. 1911. The relation of parasitic fungi to the contents of the cells of the host-plants. 1. The toxicity of tannin. *Del. Coll. Agric. Exp. Stn. Bull.* no. 91. 77 pp.

Cowdry, E. V. 1923. The distribution of *Rickettsia* in the tissues of insects and arachnids. *J. Exp. Med.* **37**:431–456.

Cowdry, E. V. 1925. The occurence of *Rickettsia*-like microorganisms in adult "locusts" (*Tibicen septemdecim* Linn). *Biol. Bull.* **48**:15–18.

Cruickshank, I. A. M. 1963. Phytoalexins. *Annu. Rev. Phytopathol.* **1**:351–374.

Dadd, R. H. 1973. Insect nutrition: current developments and metabolic implications. *Annu. Rev. Entomol.* **18**:382–419.

Davenport, R. R. 1976. Ecological concepts in studies of micro-organisms on aerial plant surfaces. In C. H. Dickinson and T. F. Preece (eds.), *Microbiology of aerial plant surfaces*, pp. 199–215. Academic Press, London.

de Bary, A. 1879. *Die erscheinungen der symbiose.* Tegebl. 51. Vers. Deut. Naturforsch. und Aertze zu Cassel. Trubner, Strassburg. 30 pp.

Denno, R. F. 1983. Tracking variable plants in space and time. In R. F. Denno and M. S. McClure (eds.), *Variable plants and herbivores in natural and managed systems*, pp. 291–341. Academic Press, New York.

Denno, R. F., and H. Dingle (eds.). 1981. *Insect life history patterns—Habitat and geographic variation.* Springer-Verlag, New York. 225 pp.

Denno, R. F. and M. S. McClure (eds.). 1983. *Variable plants and herbivores in natural and managed systems.* Academic Press, New York. 717 pp.

Dethier, V. G. 1980. Evolution of receptor sensitivity to secondary plant substances with special reference to deterrents. *Am. Nat.* **115**:45–66.

Deverall, B. J. 1972. Phytoalexins. In J. B. Harborne (ed.), *Phytochemical ecology* (Annu. Proc. Phytochem. Soc. 8), pp. 217–233. Academic Press, London.

Dickinson, C. H. 1976. Fungi on the aerial surfaces of higher plants. In C. H. Dickinson and T. F. Preece (eds.), *Microbiology of aerial plant surfaces*, pp. 293–324. Academic Press, London.

Dickman, A. 1933. Studies on the wax moth, *Galleria mellonella*, with particular reference to the digestion of wax by the larvae. *J. Cell. Comp. Physiol.* **3**:223–246.

Dixon, A. F. G. 1975. Aphids and translocation. In M. H. Zimmerman and J. A. Milburn (eds.), *Transport in plants I. Phloem transport. Encyclopedia of plant physiology.* New Ser., Vol. 1. pp. 154–170. Springer-Verlag, Berlin.

Dobias, J., P. Nemec, and M. Podova. 1979. Contact insecticidal effect of fungi of the class Fungi Imperfecti. *Biologia (Bratislava)* **34**:971–974.

Duffey, S. S. 1976. Arthropod allomones: Chemical effronteries and antagonists. *Proc. 15 Int. Congr. Entomol.*, 323–394.

Duffey, S. S., M. S. Blum, M. B. Isman, and G. G. E. Scudder. 1978. Cardiac glycosides: A physical system for their sequestration by the milkweed bug. *J. Insect Physiol.* **24**:639–645.

Ehrhardt, P. 1968a. Der vitaminbedarf einer siebröhrensaugenden aphide, *Neomyzus circumflexus* Buckt. *Z. Vgl. Physiol.* **60**:416–426.

Ehrhardt, P. 1968b. Einfluss von ernährungsfaktoren auf die entwicklung von säfte saugenden insekten unter besonderer berücksichtigung von symbioten. *Z. Parasitenkd.* **31**:38–66.

Eisner, T., L. B. Hendry, D. B. Peakall, and J. Meinwald. 1971. 2,5-Dichlorophenol (from ingested herbicide ?) in defensive secretion of grasshopper. *Science* **172**:277–278.

Feeny, P. P. 1968. Effect of oak leaf tannins on larval growth of the wintermoth, *Operophtera brumata*. *J. Insect. Physiol.* **14**:805–817.

Feeny, P. P. 1970. Seasonal changes in oak leaf tannins and nutrients as a cause of spring feeding by winter-moth caterpillars. *Ecology* **51**:656–681.

Feeny, P. P. 1975. Biochemical coevolution between plants and their insect herbivores. In L. E. Gilbert and P. H. Raven (eds.), *Coevolution of animals and plants*, pp. 3–19. University of Texas Press, Austin.

Feeny, P. P. 1976. Plant apparency and chemical defense. *Recent Adv. Phytochem.* **10**:1–40.

Feeny, P. P. 1977. Defensive ecology of the Cruciferae. *Ann. Missouri Bot. Gard.* **64**:221–234.

Fokkema, N. J. 1976. Antagonism between fungal saprophytes and pathogens on aerial plant surfaces. In C. H. Dickinson and T. F. Preece (eds.), *Microbiology of aerial plant surfaces,* pp. 487–506. Academic Press, London.

Forbes, S. A. 1892. Bacteria normal to digestive organs of Hemiptera. *Bull. Ill. State Lab. Nat. Hist.* **4**:1–7.

Fowler, H. G., and S. W. Robinson. 1977. Foraging and grass selection by the grass-cutting ant *Acromyrmex landoti fracticornis* (Forel) (Hymenoptera: Formicidae) in habitats of introduced forage grasses in Paraguay. *Bull. Entomol. Res.* **67**:659–666.

Fowler, H. G., and E. W. Stiles, 1980. Conservative resource management by leaf-cutting ants? The role of foraging territories and trails, and environmental patchiness. *Sociobiology* **5**:25–41.

Fox, L. R. 1981. Defense and dynamics in plant-herbivore systems. *Am. Zool.* **21**:853–864.

Fox, L. R., and P. A. Morrow. 1981. Specialization: species property or local phenomenon? *Science* **211**:887–893.

Fraenkel, G. S. 1959. The raison d'être of secondary plant substances. *Science* **129**:1466–1470.

Fraenkel, G. S. 1969. Evaluation of our thoughts on secondary plant substances. *Entomol. Exp. Appl.* **12**:473–486.

Francke-Grosmann, H. 1967. Ectosymbiosis in wood-inhabiting insects. In S. M. Henry (ed.), *Symbiosis,* Vol. 2, pp. 142–205. Academic Press, New York.

Frank, W. 1956. Entfernung der intrazellularen symbionten der kuchenschabe (*Periplaneta orientalis* L.) durch einwirkung verschiedener antibiotica, unter besonderer beruchksicktingung der veranderungen am wirtstier und and den bakterien. *Z. Morphol. Oekol. Tiere* **44**:329–366.

Freeland, W. J., and D. H. Janzen. 1974. Strategies in herbivory by mammals: the role of plant secondary compounds. *Am. Nat.* **108**:269–289.

Friend, W. G., E. H. Salkeld, and I. L. Stevenson. 1959. Nutrition of onion maggots, larvae of *Hylemya antigua* (Meig.), with reference to other members of the genus *Hylemya. Ann. N.Y. Acad. Sci.* **77**:384–393.

Futuyma, D. J. 1976. Food plant specialization and environmental predictability in Lepidoptera. *Am. Nat.* **110**:285–292.

Gambetta, L. 1926. Richerche sulla simbiosi ereditaria di alcuni coleotteri silofagi. *Richerche Morfol. Biol. Anim. Napoli* **1**:105–119.

Garling, L. 1979. Origin of ant-fungus mutualism: A new hypothesis. *Biotropica* **11**:284–291.

Geigy, R., L. A. Halff, and V. Kocher. 1954. L'acide folique comme élément important dans la symbiose intestinale de *Triatoma infestans. Acta Tropica* **11**:163–166.

Geiman, Q. M. 1932. The intestinal protozoa of the larvae of the crane fly. *Tipula abdominalis. J. Parasitol.* **19**:173.

Gilbert, L. E. 1979. Development of theory in insect-plant interactions. In D. J. Horn, J. R. Stairs, and R. D. Mitchell (eds.), *Analysis of ecological systems,* pp. 117–154. Ohio State University Press, Columbus.

Gilbert, L. E., and P. H. Raven (eds.). 1975. *Coevolution of animals and plants.* University of Texas Press, Austin. 246 pp.

Glaser, R. W. 1925. Acquired immunity in silkworms. *J. Immunol.* **10**:651–662.

Glaser, R. W. 1930. Cultivation and classification of "bacteroids," "symbionts," or "rickettsiae" of *Blattella germanica, J. Exp. Med.* **51**:903–907.

Glaser, R. W. 1946. The intracellular bacteria of the cockroach in relation to symbiosis. *J. Parasitol.* **32**:483–489.

Glasgow, H. 1914. The gastric caeca and the caecal bacteria of the Heteroptera. *Biol. Bull.* **26**:101–170.

Gordon, H. T. 1961. Nutritional factors in insect resistance to chemicals. *Annu. Rev. Entomol.* **6**:27–54.

Gordon, H. T. 1972. Interpretations of insect quantitative nutrition. In J. G. Rodriguez (ed.), *Insect and mite nutrition,* pp. 73–105. North-Holland, Amsterdam.

Haddock, B. A. and W. A. Hamilton (eds.). 1977. *Microbial energetics* (Soc. Gen. Microbial. Symp. 27). Cambridge University Press, Cambridge. 423 pp.

Hagen, K. S., and R. L. Tassan. 1972. Exploring nutritional roles of extracellular symbiotes on the reproduction of honeydew feeding adult chrysopids and tephritids. In J. G. Rodriguez (ed.), *Insect and mite nutrition,* pp. 323–351. North-Holland, Amsterdam.

Hancock, J. G., and O. C. Huisman. 1981. Nutrient movement in host-pathogen systems. *Annu. Rev. Phytopathol.* **19**:309–331.

Harborne, J. B. 1979. Variation in and functional significance of phenolic conjugation in plants. *Recent Adv. Phytochem.* **12**:457–474.

Harley, J. L. 1959. *The biology of mycorrhiza.* Leonard Hill, London. 233 pp.

Hartzell, A. 1967. Insect ectosymbiosis. In S. M. Henry (ed.), *Symbiosis,* Vol. 1, pp. 107–140. Academic Press, New York.

Hayashiya, K. 1978. Red fluorescent protein in the digestive juice of the silkworm larva fed on host plant mulberry leaves. *Entomol. Exp. Appl.* **24**:228–236.

Haydak, M. H., and A. Dietz. 1972. Cholesterol, panthothenic acid, pyridoxine and thiamine requirements of honey bees for brood rearing. *J. Apicult. Res.* **11**:105–109.

Hedin, P. A. (ed.). 1977. Host plant resistance to pests. *Am. Chem. Soc. Symp. Ser.* **62**. 286 pp.

Hedin, P. A., F. G. Maxwell, and J. N. Jenkins. 1974. Insect plant attractants, feeding stimulants, repellents, deterrents, and other related factors affecting insect behavior. In F. G. Maxwell and F. A. Harris (eds.), *Proceedings of the Summer Institute on Biological Control of Plant Insects and Diseases,* pp. 494–527. University of Mississippi Press, Jackson.

Hedin, P. A., O. H. Lindig, P. P. Sikorowski, and M. Wyatt. 1978. Suppressants of gut bacteria in the boll weevil from the cotton plant. *J. Econ. Entomol.* **71**:294–296.

Henry, S. M., and T. W. Cook. 1964. Amino acid supplementation by symbiotic bacteria in the cockroach. *Contrib. Boyce Thompson Inst.* **22**:507–508.

Hering, M. 1926. *Die oekologie der blattminierenden insectlarven.* Gebruder Borntrager, Berlin. 253 pp.

Hertig, M., and S. B. Wolbach. 1924. Studies on rickettsia-like microorganisms in insects. *J. Med. Res.* **44**:329–374.

Hervey, A., C. T. Rogerson, and I. Leong. 1977. Studies on fungi cultivated by ants. *Brittonia* **29**:226–236.

Hill, I. R. 1978. Microbial transformation of pesticides. In I. R. Hill and S. J. L. Wright (eds.), *Pesticide microbiology,* pp. 137–202. Academic Press, London.

Hodge, C. 1933. Growth and nutrition of *Melanoplus differentialis* Thomas (Orthoptera, Acrididae), 1. Growth on a satisfactory mixed diet and on diets of single food plants. *Physiol. Zool.* **6**:306–328.

Honigberg, B. M. 1970. Protozoa associated with termites and their role in digestion. In K. Krishna and F. M. Weesner (eds.), *Biology of termites,* Vol. 2, pp. 1–36. Academic Press, New York.

Horsfield, D. 1977. Relationship between feeding of *Philaenus spumarius* (L.) and the amino acid concentration in the xylem sap. *Ecol. Entomol.* **2**:259–266.

Hough, J. A., G. E. Harman, and C. J. Eckenrode. 1981. Microbial stimulation of onion maggot oviposition. *Environ. Entomol.* **10**:206–210.

Houk, E. J., and G. W. Griffiths. 1980. Intracellular symbiotes of the Homoptera. *Annu. Rev. Entomol.* **25**:161–187.

Houk, E. J., G. W. Griffiths, and S. D. Beck. 1976. Lipid metabolism in the symbiotes of the pea aphid, *Acyrthosiphon pisum*. *Comp. Biochem. Physiol. B* **54**:427–431.

Hungate, R. E. 1975. The rumen microbial ecosystem. *Annu. Rev. Ecol. Syst.* **6**:39–66.

Huger, A. 1956. Experimentelle untersuchungen uber die kunstliche symbiontene limination bei vorratsschadlingen: *Rhizopertha dominica* F. (Bostrychidae) und *Oryzaephilus surinamensis* L. (Cucujidae). *Z. Morphol. Oekol. Tiere* **44**:626–701.

Iizuka, T., S. Koike, and J. Mizutani. 1974. Antibacterial substances in feces of silkworm larvae reared on mulberry leaves. *Agric. Biol. Chem. Jpn.* **38**:1549–1550.

Ikeda, T., F. Matsumara, and D. M. Benjamin. 1977. Chemical basis for feeding adaptation of pine sawflies *Neodiprion rugifrons* and *Neodiprion swainei*. *Science* **197**:497–498.

Ito, T. 1969. Rearing of the silkworm under aseptic conditions. *Jpn. Agric. Res. Qt.* **4**:33–35.

Ito, T., and M. Tanaka. 1962. Rearing of the silkworm by means of aseptic technique. *J. Sericult. Sci. Jpn.* **31**:7–10.

Iverson, T. M. 1974. Ingestion and growth in *Sericostoma personatum* (Trichoptera) in relation to the nitrogen content of ingested leaves. *Oikos* **25**:278–282.

Janzen, D. H. 1971. Escape of juvenile *Dioclea megacarpa* (Leguminosae) vines from predators in a deciduous tropical forest. *Am. Nat.* **105**:97–112.

Janzen, D. H. 1977. Why fruits rot, seeds mold, and meat spoils. *Am. Nat.* **111**:691–713.

Jones, C. G. 1983. Phytochemical variation, colonization, and insect communities: the case of bracken fern (*Pteridium aquilinum*). In R. F. Denno and M. S. McClure (eds.), *Variable plants and herbivores in natural and managed systems*, pp. 513–559. Academic Press, New York.

Jones, C. G., and R. D. Firn. 1978. The role of phytoecdysteroids in bracken fern, *Pteridium aquilinum* L. Kuhn, as a defense against phytophagous insect attack. *J. Chem. Ecol.* **4**:117–138.

Jones, C. G., J. R. Aldrich, and M. S. Blum. 1981a. 2-Furaldehyde from bald cypress. A chemical rationale for the demise of the Georgia silkworm industry. *J. Chem. Ecol.* **7**:89–101.

Jones, C. G., J. R. Aldrich, and M. S. Blum. 1981b. Baldcypress allelochemics and the inhibition of silkworm enteric microorganisms. Some ecological considerations. *J. Chem. Ecol.* **7**:103–114.

Jones, C. W. 1980. Unity and diversity in bacterial energy conservation. In D. C. Ellwood, J. N. Hedger, M. J. Latham, J. M. Lynch, and J. H. Slater (eds.), *Contemporary microbial ecology*, pp. 193–213. Academic Press, London.

Jones, D. A. 1972. Cyanogenic glycosides and their function. In J. B. Harborne (ed.), *Phytochemical ecology* (Annu. Proc. Phytochem. Soc. 8), pp. 103–124. Academic Press, London.

Jones, K. 1976. Nitrogen fixing bacteria in the canopy of conifers in a temperate forest. In C. H. Dickinson and T. F. Preece (eds.), *Microbiology of aerial plant surfaces*, pp. 451–463. Academic Press, London.

Jurzitza, G. 1963. Die wirkung des sulfanilids auf die symbioten einiger anobiiden. *Z. Angew. Entomol.* **52**:302–306.

Jurzitza, G. 1979. The fungi symbiotic with anobiid beetles. In L. R. Batra (ed.), *Insect-fungus symbiosis. Nutrition, mutualism, and commensalism*, pp. 65–76. Wiley, New York.

Kaplan, D. L., and R. Hartenstein. 1980. Decomposition of lignins by microorganisms. *Soil Biol. Biochem.* **12**:65–75.

Kareiva, P. 1983. Influence of vegetation texture on herbivore populations: resource concentration and herbivore movement. In R. F. Denno and M. S. McClure (eds.), *Variable plants and herbivores in natural and managed systems*, pp. 259–289. Academic Press, New York.

Karnosky, D. F. 1982. Double jeopardy for elms: Dutch elm disease and phloem necrosis. *Arnoldia* **42**:70–77.

Keeler, K. H. 1981. A model of selection for facultative nonsymbiotic mutualism. *Am. Nat.* **118**:488–498.

Khokhlacheva, V. E., and I. M. Azimdzhanov. 1977. Mycoflora of Chinese silkworm caterpillars and mulberry leaves. *Mikol. Fitopatol.* **11**:248–249.

Kimura, M. T. 1980. Evolution of food preferences in fungus-feeding *Drosophila:* an ecological study. *Evolution* **34**:1009–1018.

Kinoshita, T., and K. Inoue. 1977. Bactericidal activity of the normal cell-free haemolymph of silkworms (*Bombyx mori*). *Infect. Immunol.* **16**:32–36.

Kirby, H. 1937. Host parasite relationships in the distribution of protozoa in termites. *Univ. Calif. (Berkeley) Publ. Zool.* **41**:189–212.

Kirby, H. 1942. Devescovinid flagellates of termites. II. The genera *Coduceia* and *Macrotrichomonas*. *Univ. Calif. (Berkeley) Publ. Zool.* **45**:96–166.

Kirby, H. 1944. The structural characteristics and nuclear parasites of some species of *Trichonympha* in termites. *Univ. Calif. (Berkeley) Publ. Zool.* **49**:185–282.

Kirby, H. 1946. Protozoa in termites. In C. A. Kofod (ed.), *Termites and termite control,* pp. 89–98. University of California Press, Berkeley.

Kirst, G. O., and H. Rapp. 1974. On the physiology of the gall of *Mikiola fagi* Htg. on leaves of *Fagus silvatica* L. 2. Translocation of ^{14}C labelled assimilates from the host leaf and adjacent leaves into the gall. *Biochem. Physiol. Pflanzen* **165**:445–455.

Klun, J. A., C. L. Tipton, and T. A. Brindley. 1967. 2,4-dihydroxy-7-methoxy-1, 4-benzoxazin-3-one (DIMBOA), an active agent in the resistance of maize to the European corn borer. *J. Econ. Entomol.* **60**:1529–1533.

Koch, A. 1933. Uber das verhalten symbiontenfreier *Sitodrepa*-larven. *Biol. Zentralbl.* **53**:199–203.

Koch, A. 1936a. Symbiosestudien. I. Die Symbioses des splintkäfers, *Lyctus linearis* Goez. *Z. Morphol. Oekol. Tiere* **32**:92–136.

Koch, A. 1936b. Symbiosestudien. II. Experimentelle Untersuchungen an *Oryzaephilus surinamensis* L. (Cucujidae, Coleopt.). *Z. Morphol. Oekol. Tierre* **32**:137–180.

Koch, A. 1967. Insects and their endosymbionts. In S. M. Henry (ed.), *Symbiosis,* Vol. 2, pp. 1–106. Academic Press, New York.

Koch, A. L. 1979. Microbial growth in low concentrations of nutrients. In M. Shilo (ed.), *Strategies of microbial life in extreme environments,* pp. 261–279. Verlag Chemie, Weinheim, Germany.

Koike, S., T. Iizuka, and J. Mizutani. 1979. Determination of caffeic acid in the digestive juice of silkworm larvae and its antibacterial activity against the pathogenic *Streptococcus faecalis* AD-4. *Agric. Biol. Chem. Jpn.* **43**:1727–1731.

Kok, L. T. 1979. Lipids of ambrosia fungi and the life of mutualistic beetles. In L. R. Batra (ed.), *Insect–fungus symbiosis. Nutrition, mutualism, and commensalism,* pp. 33–52. Wiley, New York.

Konigs, W. N., and H. Veldkamp. 1980. Phenotypic responses to environmental change. In D. C. Ellwood, J. N. Hedger, M. J. Latham, J. M. Lynch, and J. H. Slater (eds.), *Contemporary microbial ecology,* pp. 161–191. Academic Press, London.

Kreiger, R. I., P. P. Feeny, and C. F. Wilkinson. 1971. Detoxification enzymes in the guts of caterpillars: An evolutionary answer to plant defenses. *Science* **172**:579–580.

Krischik, V. A., and R. F. Denno. 1983. Individual, population, and geographic patterns in plant defense. In R. F. Denno and M. S. McClure (eds.), *Variable plants and herbivores in natural and managed systems,* pp. 463–512. Academic Press, New York.

Krishna, K. 1969. Taxonomy, phylogeny, and distribution of termites. In K. Krishna and F. M. Weesner (eds.), *Biology of termites,* Vol. 2, pp. 127–152. Academic Press, New York.

Krishna, K., and F. M. Weesner (eds.). 1969. *Biology of termites*, Vol. 1. Academic Press, New York. 598 pp.

Krishna, K., and F. M. Weesner (eds.). 1970. *Biology of termites*, Vol. 2. Academic Press, New York. 643 pp.

Kuc, J., and L. Shain. 1977. Antifungal compounds associated with disease resistance in plants. In M. R. Siegel and H. D. Sisler (eds.), *Antifungal compounds*, Vol. 2, *Interactions in biological and ecological systems*, pp. 497–535. Dekker, New York.

Kunimi, Y., and H. Aruga. 1974. Susceptibility to infection with nuclear and cytoplasmic polyhedrosis virus of the fall webworm, *Hyphantria cunea* Drury, reared on several artificial diets. *Jpn. J. Appl. Ent. Zool.* **18**:1–4.

Kuschner, D. S., and G. T. Harvey. 1960. Antibacterial substances in foliage and gut contents of phytophagous insects. *Can. Dept. Agric. Bi-monthly Prog. Rep.* **16**:2–3.

Kuschner, D. S., and G. T. Harvey. 1962. Antibacterial substances in leaves: Their possible role in insect resistance to diseases. *J. Insect Pathol.* **4**:155–184.

Lampard, J. F., and G. A. Carter. 1973. Chemical investigations on resistance of coffee berry disease in *Coffea arabica*. An antifungal compound in coffee cuticular wax. *Ann. Appl. Biol.* **73**:31–37.

Lampe, K. F. 1979. Toxic fungi. *Annu. Rev. Pharmacol. Toxicol.* **19**:85–104.

Langenheim, J. H., W. H. Stubblebine, D. E. Lincoln, and C. E. Foster. 1978. Implications of variation in resin composition among organs, tissues and populations in the tropical legume *Hymenaea*. *Biochem. Syst. Ecol.* **6**:299–213.

Levin, D. A. 1976. The chemical defenses of plants to pathogens and herbivores. *Annu. Rev. Ecol. Syst.* **7**:121–159.

Lewis, A. C. 1979. Feeding preference for diseased and wilted sunflower in the grasshopper, *Melanoplus differentialis*. *Entomol. Exp. Appl.* **26**:202–207.

Lewis, T., G. V. Pollard, and G. C. Dibley. 1974. Micro-environmental factors affecting diel patterns of foraging in the leaf-cutting ant *Atta cephalotes* (L.) (Formicidae: Attini). *J. Anim. Ecol.* **43**:143–153.

Littledyke, M., and J. M. Cherrett. 1978. Defense mechanisms in young and old leaves against cutting by the leaf-cutting ants *Atta cephalotes* (L.) and *Acromyrmex octospinosus* (Reich.) (Hymenoptera: Formicidae). *Bull. Entomol. Res.* **68**:263–271.

Lynch, J. M., and N. J. Poole (eds.). 1979. *Microbial ecology—A conceptual approach*. Blackwell, Oxford. 266 pp.

McKey, D. 1979. Distribution of secondary compounds within plants. In G. A. Rosenthal and D. H. Janzen (eds.), *Herbivores-their interaction with secondary plant metabolites*, pp. 55–134. Academic Press, New York.

McLean, E. K. 1970. The toxic actions of pyrrolizidine (*senecio*) alkaloids. *Pharmacol. Rev.* **22**:429–483.

Madden, J. L., and M. P. Couts. 1979. The role of fungi in the biology and ecology of woodwasps (Hymenoptera: Siricidae). In L. R. Batra (ed.), *Insect–fungus symbiosis. Nutrition, mutualism, and commensalism*, pp. 165–174. Wiley, New York.

Maksymiuk, B. 1970. Occurrence and nature of antibacterial substances in plants affecting *Bacillus thuringiensis* and other entomogenous bacteria. *J. Invert. Pathol.* **15**:356–371.

Mani, M. S. 1964. *Ecology of plantgalls*. Junk, The Hague, Netherlands. 434 pp.

Mansour, K. 1934a. On the intracellular micro-organisms of some bostrychid beetles. *Q. J. Microscop. Sci.* **77**:243–254.

Mansour, K. 1934b. On the so-called symbiotic relationship between coleopterous insects and intracellular micro-organisms. *Q. J. Microscop. Sci.* **77**:255–272.

Martin, M. M. 1979. Biochemical implications of insect mycophagy. *Biol. Rev.* **54**:1–21.

Martin, M. M., and J. S. Martin. 1978. Cellulose digestion in the midgut of the fungus growing termite, *Macrotermes natalensis*: The role of acquired digestive enzymes. *Science* **199**:1453–1455.

Martin, M. M., and J. S. Martin. 1979. The distribution and origins of the cellulolytic enzymes of the higher termite, *Macrotermes natalensis*. *Physiol. Zool.* **52**:11–21.

Martin, M. M., M. J. Gieselmann, and J. S. Martin. 1973. Rectal enzymes of attine ants. α-amylase and chitinase. *J. Insect Physiol.* **19**:1409–1416.

Martin, M. M., J. S. Martin, J. J. Kukor, and R. W. Merritt. 1980. The digestion of protein and carbohydrate by the stream detritivore, *Tipula abdominalis* (Diptera, Tipulidae). *Oecologia* **46**:360–364.

Martin, M. M., J. J. Kukor, J. S. Martin, T. E. O'Toole, and M. W. Johnson. 1981. Digestive enzymes of fungus-feeding beetles. *Physiol. Zool.* **54**:137–145.

Matsubara, F., and K. Hayashiya. 1969. The susceptibility to the infection with nuclear polyhedrosis virus in the silkworm reared on artificial diet. *J. Sericult. Sci. Jpn.* **38**:43–48.

Matsubara, F., M. Kato, K. Hayashiya, R. Kodama, and Y. Hanamura. 1967. Aseptic rearing of silkworm with prepared food. *J. Sericult. Sci. Jpn.* **36**:39–45.

Matsuda, M., and Y. Matsuura. 1967. The germ free rearing of the silkworm (*Bombyx mori* L.) in the flexible isolator. *J. Sericult. Sci. Jpn.* **36**:403–408.

Mattocks, A. R. 1972. Toxicity and metabolism of *Senecio* alkaloids. In J. B. Harborne (ed.), *Phytochemical ecology* (Annu. Proc. Phytochem. Soc. 8), pp. 179–200. Academic Press, London.

Mattson, W. J., Jr. 1980. Herbivory in relation to plant nitrogen content. *Annu. Rev. Ecol. Syst.* **11**:119–161.

Mattson, W. J., N. Lorimer, and R. A. Leary. 1982. Role of plant variability (trait vector dynamics and diversity) in plant/herbivore interactions. In H. M. Heybroek, B. R. Stephan and K. von Weissenberg (eds.), *Resistance to diseases and pests in forest trees*, pp. 295–303. Pudoc, Wageningen, Netherlands.

Mauldin, J. K. 1977. Cellulose catabolism and lipid synthesis by normally and abnormally faunated termites, *Reticulitermes flavipes*. *Insect Biochem.* **7**:27–31.

Mauldin, J. K., R. V. Smythe, and C. C. Baxter. 1972. Cellulose catabolism and lipid synthesis by the subterranean termite, *Coptotermes formosanus*. *Insect Biochem.* **2**:209–217.

Mauldin, J. K., and R. V. Smythe. 1973. Protein-bound amino acid content of normally and abnormally faunated Formosan termites, *Coptotermes formosanus*. *J. Insect Physiol.* **19**:1955–1960.

Mauldin, J. K., F. L. Carter, and N. M. Rich. 1981. Protozoan populations of *Reticulitermes flavipes* (Kollar) exposed to heartwood blocks of 21 American wood species. *Mater. Org.* **16**:15–28.

Mauldin, J. K., N. M. Rich, and D. W. Cook. 1978. Amino acid synthesis from ^{14}C-acetate by normally and abnormally faunated termites, *Coptotermes formosanus*. *Insect Biochem.* **8**:105–109.

Meitz, A. K. 1975. *Alimentary tract microbiota of aquatic invertebrates*. M.S. thesis. Michigan State University, East Lansing. 64 pp.

Merdan, A. I., H. Abdel-Rahman, and A. Soliman. 1975. On the influence of host plants on insect resistance to bacterial diseases. *Z. Angew. Entomol.* **78**:280–285.

Milne, D. L. 1961. The mechanism of growth retardation by nicotine in the cigarette beetle, *Lasioderma serricorne*. *S. Afr. J. Agric. Sci.* **4**:277–278.

Mirsalis, J. C., T. E. Hamm, Jr., J. M. Sherrill, and B. E. Butterworth. 1982. Role of gut flora in the genotoxicity of dinitrotoluene. *Nature* (*London*) **295**:322–323.

Mittler, T. E. 1953. Amino-acids in phloem sap and their excretion by aphids. *Nature* (*London*) **172**:207.

Mittler, T. E. 1958. Studies on the feeding and nutrition of *Tuberolachnus salignus*. III. The nitrogen economy. *J. Exp. Biol.* **35**:626–638.

Mortlock, R. P. 1976. Catabolism of unnatural carbohydrates by microorganisms. *Adv. Microbiol. Physiol.* **13**:1–53.

Mulkern, G. B. 1967. Host selection by grasshoppers. *Annu. Rev. Entomol.* **12**:59–78.

Muller, C. H., and C-H. Chou. 1972. Phytotoxins: An ecological phase of phytochemistry. In J. B. Harborne (ed.), *Phytochemical ecology* (Annu. Proc. Phytochem. Soc., 8), pp. 201–216. Academic Press, London.

Nagy, J. G., H. W. Steinhoff, and G. M. Ward. 1964. Effects of essential oils of sagebrush on deer rumen microbial function. *J. Wildl. Manage.* **28**:785–790.

Nakatsugawa, T., and M. A. Morelli. 1976. Microsomal oxidation and insecticide metabolism. In C. F. Wilkinson (ed.), *Insecticide biochemistry and physiology*, pp. 61–114. Plenum Press, New York.

Niemierko, S. 1959. Some aspects of lipid metabolism in insects. *Proc. IV Int. Congr. Biochem.* **12**:185–197.

Noirot, C. H. 1970. The nests of termites. In K. Krishna and F. M. Weesner (eds.), *Biology of termites*, Vol. 2, pp. 73–125. Academic Press, New York.

Norris, D. M. 1972. Dependence of fertility and progeny development of *Xyleborus ferrugineus* upon chemicals from its symbiotes. In J. G. Rodriguez (ed.), *Insect and mite nutrition*, pp. 299–310. North-Holland, Amsterdam.

Oh, H. K., T. Sakai, M. B. Jones, and W. W. Longhurst. 1967. Effects of various essential oils isolated from Douglas fir needles upon sheep and deer rumen microbial activity. *Appl. Microbiol* **15**:777–784.

Otte, D. 1975. Plant preference and plant succession. A consideration of evolution of plant preference in *Schistocerca*. *Oecologia* **18**:129–144.

Owen-Smith, N., and P. Novellie. 1982. What should a clever ungulate eat? *Am. Nat.* **119**:151–178.

Paillot, A. 1933. *L'infection chez les insectes*. G. Patissier, Trevoux, France. 535 pp.

Pankhurst, C. E., and D. R. Biggs. 1980. Sensitivity of *Rhizobium* to selected isoflavanoids. *Can. J. Microbiol.* **26**:542–545.

Pant, N. C., and K. Dang. 1972. Physiology and elimination of intracellular symbionts in some stored product beetles. In J. G. Rodriguez (ed.), *Insect and mite nutrition*, pp. 311–322. North-Holland, Amsterdam.

Parsons, J., and M. Rothschild. 1964. Rhodanase in the larva and pupa of the common blue butterfly (*Polyommatus icarus* [Rott.]) (Lepidoptera). *Entomol. Gaz.* **15**:589.

Petri, L. 1910. Utersuchungen über die darm-bakterien der olivefliege. *Zentr. Bakt. Parasitenk. Infekt. II* **26**:357–367.

Portier, P. 1911. Passage de l'asepsie à l'envahissement symbiotique humoral et tissulaire par les microorganismes dans la série des larves insectes. *Compt. Rend. Soc. Biol.* **70**:914–917.

Potrikus, C. J., and J. A. Breznak. 1977. Nitrogen-fixing *Enterobacter agglomerans* isolated from guts of wood-eating termites. *Appl. Environ. Microbiol.* **33**:392–399.

Potrikus, C. T., and J. A. Breznak. 1980a. Uric acid-degrading bacteria in guts of termites [*Reticulotermes flavipes* (Kollar)]. *Appl. Environ. Microbiol.* **40**:117–124.

Potrikus, C. J., and J. A. Breznak. 1980b. Anaerobic degradation of uric acid by gut bacteria of termites. *Appl. Environ. Microbiol.* **40**:125–132.

Potrikus, C. J., and J. A. Breznak. 1980c. Uric acid in wood-eating termites. *Insect Biochem.* **10**:19–27.

Preece, T. F., and C. H. Dickinson (eds.). 1971. *Ecology of leaf surface microorganisms*. Academic Press, London. 640 pp.

Prestwich, G. D., and B. L. Bentley. 1981. Nitrogen fixation by intact colonies of the termite *Nasutitermes corniger. Oecologia* **49**:249–251.

Prestwich, G. D., B. L. Bentley, and E. J. Carpenter. 1980. Nitrogen sources for neotropical nasute termites: fixation and selective foraging. *Oecologia* **46**:397–401.

Price, P. W., C. E. Bouton, P. Gross, B. A. McPheron, J. N. Thompson, and A. E. Weis. 1980. Interactions among three trophic levels: Influence of plants on interactions between insect herbivores and natural enemies. *Annu. Rev. Ecol. Syst.* **11**:41–65.

Quinlan, R. J., and J. M. Cherrett. 1977. The role of substrate preparation in the symbiosis between the leaf-cutting ant *Acryomyrmex octospinosus* (Reich) and its food fungus. *Ecol. Entomol.* **2**:161–170.

Ragsdale, D. W., A. D. Larson, and L. D. Newsom. 1979. Microorganisms associated with feeding and from various organs of *Nezara viridula. J. Econ. Entomol.* **72**:725–727.

Raupp, M. J., and R. F. Denno. 1983. Leaf age as a predictor of herbivore distribution and abundance. In R. F. Denno and M. S. McClure (eds.), *Variable plants and herbivores in natural and managed systems,* pp. 91–124. Academic Press, New York.

Rauscher, M. D. 1983. Ecology of host-selection behavior in phytophagous insects. In R. F. Denno and M. S. McClure (eds.), *Variable plants and herbivores in natural and managed systems,* pp. 223–258. Academic Press, New York.

Reese, J. C. 1979. Interactions of allelochemicals with nutrients in herbivore food. In G. A. Rosenthal and D. H. Janzen (eds.), *Herbivores—Their interaction with secondary plant metabolites,* pp. 309–330. Academic Press, New York.

Rhoades, D. F. 1979. Evolution of plant chemical defense against herbivores. In G. A. Rosenthal and D. H. Janzen (eds.), *Herbivores—Their interaction with secondary plant metabolites,* pp. 3–54. Academic Press, New York.

Rhoades, D. F., and R. G. Cates. 1976. Toward a general theory of plant antiherbivore chemistry. *Recent Adv. Phytochem.* **10**:168–213.

Rodriguez, E., G. H. N. Towers, and J. G. Mitchell. 1976. Review. Biological activities of sesquiterpene lactones. *Phytochemistry* **15**:1573–1580.

Roeske, C. N., J. N. Seiber, L. P. Brower, and C. M. Moffitt. 1976. Milkweed cardenolides and their comparative processing by monarch butterflies (*Danaus plexippus* L.) *Recent Adv. Phytochem.* **10**:93–167.

Rosazza, J. P. 1978. Microbial transformations of natural antitumor agents. *Lloydia* **41**:297–311.

Rosenthal, G. A., and E. A. Bell. 1979. Naturally occurring, toxic nonprotein amino acids. In G. A. Rosenthal and D. H. Janzen (eds.), *Herbivores—Their interaction with secondary plant metabolites,* pp. 353–386. Academic Press, New York.

Rosenthal, G. A., and D. H. Janzen (eds.). 1979. *Herbivores—Their interaction with secondary plant metabolites.* Academic Press, New York. 718 pp.

Rosenthal, G. A., D. L. Dahlman, and D. H. Janzen. 1978. L-Canaline detoxification: A seed predator's biochemical mechanism. *Science* **202**:528–529.

Rosenthal, G. A., C. G. Hughes, and D. H. Janzen. 1982. L-Canavanine, a dietary nitrogen source for the seed predator *Caryedes brasiliensis* (Bruchidae). *Science* **217**:353–355.

Rumbold, C. T. 1941. A blue stain fungus, *Ceratostomella montium* n. sp., and some yeasts associated with two species of *Dendroctonus. J. Agric. Res.* **62**:589–601.

Russell, G. B., O. R. W. Sutherland, R. F. N. Hutchins, and P. E. Christmas. 1978. Vestitol: A phytoalexin with insect feeding-deterrent activity. *J. Chem. Ecol.* **4**:571–579.

Sands, W. A. 1969. Association of termites and dungi. In K. Krishna and F. M. Weesner (eds.), *Biology of termites,* Vol. 1, pp. 495–542. Academic Press, New York.

Schade, F. H. 1973. The ecology and control of the leaf-cutting ants of Paraguay. In J. R.

Gorham (ed.), *Paraguay Ecological Essays*, pp. 77–95. Academy of Arts Sciences. Americas, Florida.

Schmeltz, I. 1971. Nicotine and other tobacco alkaloids. In M. Jacobson and D. G. Crosby (eds.), *Naturally occurring insecticides*, pp. 99–136. Dekker, New York.

Schoonhoven, L. M. 1972. Secondary plant substances and insects. *Recent Adv. Phytochem.* **5**:197–224.

Schultz, J. C. 1983. Habitat selection and foraging tactics of caterpillars in heterogeneous trees. In R. F. Denno and M. S. McClure (eds.), *Variable plants and herbivores in natural and managed systems*, pp. 61–90. Academic Press, New York.

Schultz, J. E., and J. A. Breznak. 1978. Heterotrophic bacteria present in hindguts of wood-eating termites [*Reticulotermes flavipes* (Kollar)]. *Appl. Environ. Microbiol.* **35**:930–936.

Schwartz, C. C., J. G. Nagy, and W. L. Regelin. 1980a. Juniper oil yield, terpenoid concentration and antimicrobial effects on deer. *J. Wildl. Manage.* **44**:107–113.

Schwartz, C. C., W. L. Regelin, and J. C. Nagy. 1980b. Deer preference for juniper forage and volatile oil treated food. *J. Wildl. Manage.* **44**:114–120.

Scriber, J. M. 1983. Evolution of feeding specialization, physiological efficiency, and host races in selected Papilionidae and Saturniidae. In R. F. Denno and M. S. McClure (eds.), *Variable plants and herbivores in natural and managed systems*, pp. 373–412. Academic Press, New York.

Scriber, J. M., and P. P. Feeny. 1979. Growth of herbivorous caterpillars in relation to feeding specialization and to the growth form of their food plants. *Ecology* **60**:829–850.

Seigler, D. S. 1979. Toxic seed lipids. Pages 449–470 in G. A. Rosenthal and D. H. Janzen (eds.), *Herbivores—Their interaction with secondary plant metabolites*, pp. 449–470. Academic Press, New York.

Sijpesteijn, K., H. M. Dekhuijzen and J. W. Vonk. 1977. Biological conversion of fungicides in plants and microorganisms. In M. R. Siegel and H. D. Sisler (eds.), *Antifungal compounds*, Vol. 2, *interactions in biological and ecological systems*, pp. 91–147. Dekker, New York.

Slama, K., M. Romanuk, and F. Sorm. 1974. *Insect hormones and bioanalogues*. Springer-Verlag, Vienna. 477 pp.

Slansky, F., Jr., and P. P. Feeny. 1977. Stabilization of the rate of nitrogen accumulation by larvae of the cabbage butterfly on wild and cultivated plants. *Ecol. Monogr.* **47**:209–228.

Slater, J. H., and D. Godwin. 1980. Microbial adaptation and selection. In D. C. Ellwood, J. N. Hedger, M. J. Latham, J. M. Lynch, and J. H. Slater (eds.), *Contemporary microbial ecology*, pp. 137–160. Academic Press, London.

Smirnoff, W. A. 1972. Effects of volatile substances released by foliage of *Abies balsamea*. *J. Invert. Pathol.* **19**:32–35.

Smirnoff, W. A., and P. M. Hutchison. 1965. Bacteriostatic and bacteriocidal effects of extracts of foliage from various plant species on *Bacillus thuringensis* var. *thuringiensis* Berliner. *J. Invert. Pathol.* **7**:273–280.

Sondheimer, E., and J. B. Simeone (eds.). 1970. *Chemical ecology*. Academic Press, New York. 336 pp.

Southwood, T. R. E. 1973. The insect/plant relationship—an evolutionary perspective. In H. F. van Emden (ed.), *Insect/plant relationships* (Symp. Roy. Entomol. Soc. 6), pp. 3–30. Wiley, New York.

Stammer, H.-J. 1929. Die symbiose der largiiden (Coleoptera). *Z. Morphol. Oekol. Tiere* **15**:1–34.

Stammer, H.-J. 1935. Studien an symbiosen zwischen käfern und mikroorganismen. I. Die symbiose der donaciinen (Coleopter. Chrysomel.). *Z. Morphol. Oekol. Tiere* **29**:585–608.

Stammer, H.-J. 1936. Studien an symbiosen zwischen käfern und mikroorganismen II. Die symbiose des *Bromius obscurus* L. und der cassida-arten (Coleopt. Chrysomel.). *Z. Morphol. Oekol. Tiere* **31**:682–697.

Steffan, A. W. 1967. Ectosymbiosis in aquatic insects. In S. M. Henry (ed.), *Symbiosis,* Vol. 2, pp. 207–289. Academic Press, New York.

Steinhaus, E. A. 1941. A study of the bacteria associated with thirty species of insects. *J. Bact.* **42**:757–790.

Steinhaus, E. A. 1967. *Insect microbiology.* Hafner, New York. 763 pp.

Stewart, W. D. P. 1966. *Nitrogen fixation in plants.* Athlone Press, London. 168 pp.

Steyn, P. S. (ed.). 1980. *The biosynthesis of mycotoxins. A study in secondary metabolism.* Academic Press, New York. 432 pp.

Sucoff, E. 1979. Estimates of nitrogen fixation on leaf surfaces of forest species in Minnesota and Oregon. *Can. J. For. Res.* **9**:474–477.

Suflita, J. M., A. Horowitz, D. R. Shelton, and J. M. Tiedje. 1982. Dehalogenation: A novel pathway for the anaerobic biodegradation of haloaromatic compounds. *Science* **218**:1115–1117.

Sussman, A. S. 1965. Dormancy of soil microorganisms in relation to survival. In K. F. Baker and W. C. Snyder (eds.), *Ecology of soil-borne plant pathogens. Prelude to biological control,* pp. 99–110. University of California Press, Berkeley.

Sutherland, O. R. W., G. B. Russell, D. R. Biggs, and G. A. Lane. 1980. Insect feeding deterrent activity of phytoalexin isoflavonoids. *Biochem. Syst. Ecol.* **8**:73–75.

Swain, T. 1976. Angiosperm reptile coevolution. *Linn. Soc. Sym.* **3**:107–122.

Swain, T. 1977. Secondary compounds as protective agents. *Annu. Rev. Plant Physiol.* **28**:479–501.

Swezy, O., and H. P. Severin. 1930. A risckettsia-like microorganism in *Eutettix tenellus* (Baker), the carrier of curly top of sugar beets. *Phytopathology* **20**:169–178.

Thakur, S. S., and B. K. Rao. 1976. Action of certain antibiotics on the tree hopper, *Oxyrhachis tarandus* Fabr., its mycetomes and symbionts. *Indian J. Entomol.* **38**:333–336.

To, L. P., L. Margulis, D. Chase, and W. L. Nutting. 1980. The symbiotic microbial community of the Sonoran desert termite: *Pterotermes occidentis. Biosystems* **13**:109–137.

Trager, W. 1970. *Symbiosis.* Van Nostrand–Reinhold, New York. 100 pp.

Troyer, K. 1982. Transfer of fermentative microbes between generations in a herbivorous lizard. *Science* **216**:540–542.

Turner, W. B. 1971. *Fungal metabolites.* Academic Press, New York. 446 pp.

van Beneden, P. J. 1875. *Les commensaux et les parasites.* Bibliothèque Scientifique Internationale, Paris. 238 pp.

van Emden, H. F. 1972. Aphids as phytochemists. In J. B. Harborne (ed.), *Phytochemical ecology* (Annu. Proc. Phytochem. 8), pp. 25–43. Academic Press, London.

Vaughan, G. L., and A. M. Jungreis. 1977. Insensitivity of lepidopteran tissues to ouabain. Physiological mechanisms for protection from cardiac glycosides. *J. Insect Physiol.* **23**:585–589.

Vaughan, T. A. 1982. Stephen's woodrat, a dietary specialist. *J. Mammol.* **63**:53–62.

Veivers, P. C., R. W. O'Brien, and M. Slaytor. 1980. The redox state of the gut of termites. *J. Insect Physiol.* **26**:75–77.

Viands, D. R., D. K. Barnes, and F. I. Frosheiser. 1980. An association between resistance to bacterial wilt and nitrogen fixation in alfalfa. *Crop Sci.* **20**:699–703.

Virtanen, A. I. 1965. Studies on organic sulphur compounds and other labile substances in plants. *Phytochemistry* **4**:207–228.

Waldbauer, G. 1968. The consumption and utilization of food by insects. *Adv. Insect Physiol.* **5**:229–288.

Wallace, J. W., and R. L. Mansell (eds.). 1976. *Biochemical interaction between plants and insects. Recent Adv. Phytochem.* **10**. Plenum Press, New York. 425 pp.

Waterman, P. G., C. N. Mbi, D. B. McKey, and J. S. Gartlan. 1980. African rainforest vegetation and rumen microbes: Phenolic compounds and nutrients as correlates of digestibility. *Oecologia* **47**:22–33.

Weber, N. A. 1972. *Gardening ants the attines.* Am. Philosophical Society. Philadelphia. 146 pp.

Weber, N. A. 1979. Fungus culturing by ants. In L. R. Batra (ed.), *Insect-fungus symbiosis. Nutrition, mutualism, and commensalism,* pp. 77–116. Wiley, New York.

Weis, A. E. 1982. Use of a symbiotic fungus by the gall maker *Asteromyia carbonifera* to inhibit attack by the parasitoid *Torymus capite. Ecology* **63**:1602–1605.

Werner, E. 1926a. Die ernahrung der larve von *Potosia cuprea,* Frb. (*Cetonia floricola* Hbst). Ein beitrag zum problem der cellulose-verdauung bei insecten-larven. *Z. Morphol. Oekol. Tiere* **6**:150–206.

Werner, E. 1926b. *Bacillus cellulosam fermentans. Zentr. Bakt. Parasit Enk. Infekt. II.* **67**:297.

Whitham, T. G. 1981. Individual trees as heterogenous environments: adaptation to herbivory or epigenetic noise? In R. F. Denno and H. Dingle (eds.), *Insect life history patterns— habitat and geographic variation,* pp. 9–27. Springer-Verlag, New York.

Whitham, T. G. 1983. Host manipulation of parasites: Within-plant variation as a defense against rapidly evolving pests. In R. F. Denno and M. S. McClure (eds.), *Variable plants and herbivores in natural and managed systems,* pp. 15–41. Academic Press, New York.

Whitham, T. G., and Slobodchikoff, C. N. 1981. Evolution by individuals, plant-herbivore interactions, and mosaics of genetic variability: The adaptive significance of somatic mutations in plants. *Oecologia* **49**:287–292.

Whittaker, R. H., and P. P. Feeny. 1971. Allelochemics: Chemical interactions between species. *Science* **171**:757–770.

Wigglesworth, V. B. 1936. Symbiotic bacteria in a blood-sucking insect, *Rhodnius prolixus* Stal. (Hemiptera, Triatomidae). *Parasitology* **28**:284–289.

Wolin, M. J. 1981. Fermentation in the rumen and human large intestine. *Science* **213**:1463–1468.

Wollman, E. 1921. La méthode des élevages aseptiques en physiologie. *Arch. Intern. Physiol.* **18**:194–199.

Yamasaki, M. 1931. Studies on the intestinal protozoa of termites. II. Oxygenation experiments under the influence of temperature. *Mem. Coll. Sci. Kyoto Imperial Univ. Ser. B* **7**:179–188.

CHAPTER 4

Role of the Host Plant and Parasites in Regulating Insect Herbivore Abundance, with an Emphasis on Gall-Inducing Insects

GORDON W. FRANKIE
Department of Entomological Sciences
University of California
Berkeley, California

DAVID L. MORGAN
Research and Extension Center
Texas A&M University
Dallas, Texas

CONTENTS

1 INTRODUCTION

It is well documented that parasitic and predatory insects are important in determining levels of abundance of insect herbivores. In fact, the applied field of biological control is based on this reality (Clausen, 1940, 1978; Debach, 1964, 1974; Huffaker and Kennett, 1966; Huffaker, 1971; Varley et al., 1973; Price, 1975, 1980; Huffaker and Messenger, 1976; Ridgway and Vinson, 1977; Lawton and McNeill, 1979; Hall et al., 1980; Metcalf and Luckman, 1982; and many others).

It is becoming increasingly evident that the host plant also influences the regulation of insect herbivore populations. Host plant effects are often evaluated in terms of colonization, growth, development, and/or fecundity of the herbivore (van Emden and Way, 1973). Interactions between the herbivore and host plant range from simple (Osborne, 1973; Southwood, 1973) to complex (see case histories and reviews in Watson, 1970; Schoonhoven, 1972; McNeill, 1973; Ryan, 1973; van Emden, 1973; Ryan and Green, 1974; Gilbert and Raven, 1975; Beck and Reese, 1976; Rhoades and Cates, 1976; Webb and Moran, 1978; Whitham, 1978; Whitham et al., Chapter 2; Johnson and Slobodchikoff, 1979; Rosenthal and Janzen, 1979; Price et al., 1980; Whitham and Slobodchikoff, 1981; Harborne, 1982; Schultz et al., 1982; Schultz, 1983). Interactions between natural enemies of the herbivore and the host plant are also important (van Emden and Way, 1973; Vinson, 1975; Price, 1980; Price et al., 1980; Schultz, 1983; and many others).

Although there is ample documentation in the literature on the individual roles of natural enemies and the host plant in determining insect herbivore abundance, very few studies have examined these two factors when they operate simultaneously. Perhaps the dearth of such studies is attributable, in part, to the difficulty of recognizing or readily measuring often subtle host plant effects (Lawton and McNeill, 1979). A few examples do exist and these are discussed.

Wyatt (1970) offered convincing evidence that chrysanthemums could be grown under greenhouse conditions with low aphid populations [*Myzus persicae* (Sulz.)] provided that resistant varieties and parasitic wasps were employed together (see also Huffaker, 1971, p. 206). Starks et al. (1972) also demonstrated under glass that resistant varieties of barley and sorghum complemented the activity of a braconid parasite in reducing populations of greenbugs, *Schizaphis graminum* (Rondani), and diminishing plant damage. This relationship was not dependent on plant age. Washburn and Cornell (1981) found that the cynipid gall wasp, *Xanthoteras politum* (Bassett), on oak may actually become locally extinct in the New Jersey Pine Barrens due to the combined action of host plant changes and parasites. Finally, Lawton and McNeill (1979) offer a theoretical perspective on this subject in which they state that natural enemies and plant chemistry are more important than many other factors in determining insect herbivore abundance. To expand on this point they provide a generalized model on the relationships among herbivores, their natural enemies, and the host plant. In a series of equations they stress that plant chemistry may have a profound effect on the natural rate of increase, r, of the herbivores. They also provide a discussion of the variables that must be taken into consideration in assessing host plant quality.

In this chapter we assess the importance of host plant quality and parasites in determining abundance levels of the cynipid gall wasp, *Disholcaspis cinerosa* Bassett, on live oak (*Quercus* spp.) in natural and urban Texas environments. The current study, which is part of a larger investigation that extended from 1971–1982, overlaps unavoidably with two others. The first of these (Morgan et al., 1983) deals with differences in host plant quality as they relate to the capacity of live oak to produce *Disholcaspis* galls. The second (Frankie et al., 1984) concerns the role of parasitic wasps in regulating cynipid abundance in different environments. In the current treatment, we briefly review past results and offer new experimental findings on selected characteristics of host plant quality and its relationship to *Disholcaspis* abundance. The following three hypotheses are critically examined:

1. Live oaks susceptible to *Disholcaspis* lose their susceptibility to the wasp with increasing age.
2. Host plant factors fix the gall-carrying capacity on *Disholcaspis*-susceptible trees.
3. Changes in host plant susceptibility and parasites determine long-term population trends of *Disholcaspis* on live oak.

Our main goal is to use this case history to call attention to the interplay between the host plant and natural enemies in determining abundance levels of insect herbivores. To describe and assess our survey and experimental findings, we largely depart from the density-dependent/density-independent population framework since our system cannot be readily categorized as either. Rather, it falls into the "density-vague" category of Strong (Chapter 11).

2   STUDY SITES AND SEASONAL HISTORY OF *DISHOLCASPIS*

Selected trees at two urban sites in Dallas were used for population monitoring and experimental caging studies. One site, NorthPark Mall, was a 35 ha shopping plaza located in the northern section of the city (Fig. 1). Ap-

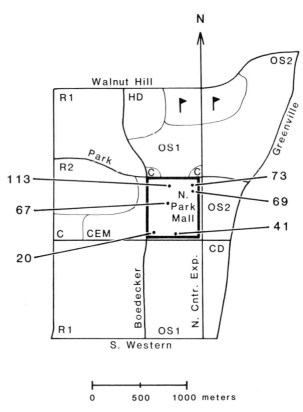

Figure 1. NorthPark Shopping Mall and surrounding urban environment in 1977. Symbols: ▶, golf courses; no live oaks; C, church and grounds; few live oaks of mixed ages; CEM, cemetary grounds; numerous older (40+ years) live oaks; CD, commercial development with limited number of small live oaks; HD, housing development just beginning; OS1, open space; no live oak; OS2, mostly open space; commercial development just beginning; very few small live oaks; R1, Older residential; numerous older (30+ years) live oaks; R2, new and old residential; new and old live oak plantings.

Figure 2. Tree 113 at NorthPark Mall in 1982.

proximately 200 live oaks, all *Quercus fusiformis* Small., were planted throughout the periphery of the centrally located shopping complex; most were 12–15 years old in 1972 (Figs. 2 and 3). Live oak and all other ornamental trees at the plaza were watered and fertilized regularly, and occasionally lightly pruned. In 1979 only, they were substantially pruned after a severe winter ice storm caused damage to many trees. During the period of

Figure 3. Tree 41R at NorthPark Mall in 1979.

our study, 1971–1982, pesticides were used on the live oaks in 1981–1982 for aphid control. The application did not correspond with the emergence of the adult stage of either generation of gall wasps.

Live oaks at NorthPark Mall and in surrounding areas were surveyed for *Disholcaspis*-infested trees. From 1972 to 1982, the NorthPark trees were examined annually for new or continuing infestations. Neighborhood blocks and open space surrounding NorthPark were surveyed casually for *Disholcaspis* in 1972. In subsequent years only specific sections that contained young trees were surveyed. In 1982, all blocks surrounding NorthPark were again surveyed. The 1977 date in Fig. 1 corresponds to the last time that high gall densities were recorded on any study tree at NorthPark.

The second site, Samuell Grand Park, was a city-owned garden and recreational facility (32 ha) located in southeast Dallas. Large numbers of 7- to 10-year-old live oaks, all *Q. virginiana* Mill., were planted during the 1950s in the park. We estimate that three of our study trees at this site (nos. 3, 4, and 16) were approximately 30 years old in 1972 (Fig. 4); one tree (no. 10) was about 20–25 years old in 1972. All ornamental plants at this park were watered and fertilized regularly. None of the study trees was pruned or treated with pesticides during the 11-year study.

Disholcaspis cinerosa annually passes through two alternating generations (heterogony) on live oak, which in Texas consists principally of *Q. fusiformis* and *Q. virginiana*. No differences in distribution and abundance of *Disholcaspis* between these two closely related species have been observed (Frankie et al., 1984). The asexual generation develops within beige

Figure 4.   Tree 16 at Samuell Grand Park in 1982.

or pink spherical galls (5–25 mm diameter) on branches. These galls first appear in late July and August, and developing larvae can be found from this period through most of October; pupation occurs in November. Parthenogenetic females, which emerge from galls in December and early January, live 3–6 weeks and oviposit in swollen leaf and flower buds. Abandoned asexual galls persist on trees for 3 or more years, turning gray and black as they age.

The sexual generation remains in the egg stage until mid to late March. Larval and pupal stages pass very rapidly from late March to early April, and resulting sexual galls, which resemble kernels of wheat (3–5 mm), develop concurrently at the bases of new leaf shoots and inflorescent stalks. Usually, one gall or rarely two form per leaf or flower bud. Sexual adults emerge from these galls in early April. Mated females live about a week and oviposit at the bases of leaf shoots giving rise to the asexual generation to start the cycle again (Frankie et al., 1977, 1984).

3   METHODS

3.1   Population Monitoring

Asexual and sexual galls were sampled annually from 10 heavily infested host trees: six at NorthPark Mall and four at Samuell Grand Park. These individuals were chosen since they supported representative high gall densities in relatively new (NorthPark) and established (Samuell Grand) plantings. A representative sample of twelve 0.5-m limbs with asexual galls and eight 0.5-m limbs with sexual galls (respectively, three and two randomly selected per cardinal direction) was removed from each marked tree. Sample branches were generally taken from the outer 50% of the crown. This was considered an adequate sampling scheme for monitoring gall densities (Frankie et al., 1984). Galls pooled from each tree were used to compute a mean per branch for each generation. Variations around the means were computed from representative trees sampled in 1977/1982 (asexual generation) and 1978 (sexual generation).

Up to 150 asexual and 50 sexual galls were dissected annually from each marked tree to determine percent mortality by parasitic wasps and to a lesser extent by other factors. When < 10 galls (pooled) were sampled per generation, a question mark follows the respective mortality value (see Figs. 5–14). Numerous collections of representative parasites were reared in the laboratory and sent to the U.S. National Museum for identification.

3.2   Experimental Cagings

Live oaks were experimentally exposed to ovipositing asexual females in 1978 and 1979. Using the criteria of presence or absence of asexual galls,

Table 1. Periods of Caging, Branch Removal, and Type of Examination for A-SUS and A-RES Trees

Location and Year of Caging	Number per Tree[a] Branches Caged	Number per Tree[a] Cohorts Removed	Cohort: Type Exam:	Date[b] of Cohort Examination[c] I E	II E/IG	III E/IG	IV MG
Northpark[d]							
1978	18	3		Mid F 1979	N/A	Mid M 1979	Early A 1979
1979	16	4		Late J 1980	Late F 1980	Mid M 1980	Early A 1980
Samuell Grand[e]							
1979	8	2		Late J 1980	N/A	N/A	Early A 1980

[a] No branches per cohort = no. branches caged/no. cohorts removed.

[b] J, January; F, February; M, March; A, April.

[c] E, eggs; IG, incipient galls (sexual); MG, mature galls (sexual).

[d] Four A-SUS and four A-RES trees caged each year.

[e] Three A-SUS trees in 1979.

we selected seven apparently susceptible (A-SUS) and four apparently resistant (A-RES) live oaks. Four A-SUS and four A-RES trees were located at NorthPark; three A-SUS trees were located at Samuell Grand Park. All A-SUS trees, which were also being monitored annually for their gall populations, were selected because of their past histories of high gall densities. Trees directly adjacent to A-SUS trees and with a past record of few or no galls were selected as A-RES trees. The same procedure for selecting the two tree types was used by Morgan et al. (1983).

We exposed trees to asexual wasps by confining them on branches within cotton organdy bags (Fig. 3). Branches (0.5 m in length) were rendered gall-free and then covered with cloth before wild adults emerged (to prevent oviposition). Current year asexual galls had stopped growing about a month prior to their removal. To obtain adult wasps for caging we collected mature galls in late November from about 12 infested trees scattered widely throughout north Dallas. Asexual adults were reared from galls kept outside in cardboard boxes (beneath a roofed shelter). Eight wasps were placed in each organdy bag in December to permit oviposition. After emergence of the wild population had ceased and caged wasps had died (4–6 weeks), we removed the bags to allow branches and galls to develop naturally.

We determined frequency of oviposited sexual eggs, incipient galls, and mature galls by periodically removing cohorts of caged branches from the study trees during January through March in 1979 and 1980. The schedule of branch removal and type of examination is presented in Table 1.

During the first winter of caging, branches of some trees were lost due to a severe ice storm; no branches were lost to inclement weather or to any other factor during the second winter. In 1979, 75 buds per branch were sampled from cohort I, with a few having slightly less than 75. Fifty buds per branch were sampled from cohort III in 1979 and from all cohorts in 1980. Buds were dissected under magnification.

Acorns produced by each tree were recorded from 1975 to 1982 to determine if a correlation existed between acorn production and gall density. Weather records were examined for possible correlations between gall and acorn production and selected climatic factors, in particular temperature, rainfall, and periods of freeze.

4   SUMMARY OF PREVIOUS WORK

An extensive and systematic survey of live oaks in seven urban and six natural Texas habitats revealed that asexual galls of *Disholcaspis* were unequally distributed among host trees (Frankie et al., 1984). That is, a few trees produced high gall densities, most produced low to moderate gall numbers, and a few had no evidence of galls. Trees with the highest densities were exclusively associated with urban environments, and all of these were young. Live oaks in natural environments were never observed supporting high gall densities.

Numerous parasitic wasp species were found associated with the larval, pupal, and occasionally adult stages of each generation. The guild of principal parasites associated with the asexual generation consisted of about 11–15 species. With possibly two rare exceptions, all of these were different from those (8–10 spp.) associated with the sexual generation (Table 2). Overall, parasites constituted the principal mortality factor in *Disholcaspis* in the nonegg stages (Frankie et al., 1984). Mortality in the egg stage of either generation was not investigated in the early studies.

Natural Habitats. In a natural live oak habitat (Austin), asexual galls were sampled over a consecutive 4-year period (1976–1979) from trees having "relatively high" gall numbers. In each year, gall densities were low (~1 gall per 0.5 m branch) compared to that recorded for urban habitats (see below). These low densities were considered characteristic of susceptible trees in natural areas of Texas. Percent mortality in the asexual galls at Austin ranged from 77 to 97%, depending on year. Almost all of this mortality was due to the combined action of several parasitic wasp species. However, a regression of percent mortality vs. gall density for the 4-year period produced no significant correlation, suggesting a lack of density dependent mortality due to parasites in this population (Frankie et al., 1984).

Mortality in the sexual galls was not investigated owing to the great difficulty of finding adequate numbers of these small galls in natural areas. However, as indicated from the urban samples of sexual galls (see below),

Table 2. Principal Parasites Reared from Asexual and Sexual Generations[a]

Taxa from Asexual Generation	Taxa from Sexual Generation
Cynipidae	Eulophidae
Synergus spp. (1 or 2)	*Tetrastichus* sp. 1
Eulophidae	*Tetrastichus* sp. 2
Tetrastichus racemariae Ashmead	*T. racemariae* Ashmead
Eurytomidae	Eupelmidae
Eurytoma spp. (3–5)	*Eupelmus* sp. 1
Eudecatoma disholcaspidis (Balduf)	*Eupelmus* sp. 2
Ormyridae	Eurytomidae
Ormyrus spp. (1 or 2)	*Eurytoma* spp. (1 or 2)
Torymidae	Ormyridae
Torymus warreni (Cockerell)	*Ormyrus* spp. (1 or 2)
T. racemariae (Ashmead)	Pteromalidae
T. memnonius Grissell	*Ormocerus* sp.
T. tubicola (Osten Sacken)	*Zatropis* sp.

[a] Modified from Frankie et al. (1984).

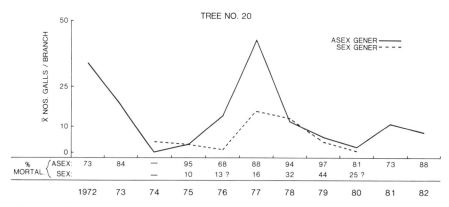

Figure 5. Density and mortality of wild asexual and sexual galls on tree 20 at NorthPark Mall.

the percent mortality of larval through adult stages can be as high as that observed in the asexual generation. It is assumed, therefore, that mortality in this generation in natural habitats is also high.

Urban Habitats.   At NorthPark and Samuell Grand Park 10 live oaks were monitored annually for asexual and sexual galls for extended periods, mostly from 1972/1973 to 1982. The trees selected at these sites showed evidence of a high tendency to produce *Disholcaspis* galls. Monitoring coupled with gall dissections produced information on long-term population trends, percentage, mortality, and percent parasitization.

Asexual gall densities on urban trees were generally much higher than that recorded for trees at the Austin natural area (Figs. 5–14). Most of the younger urban trees had periods (1–3 years) of high gall densities followed by low densities, with asexual gall densities on given trees during particular years almost always in excess of sexual densities (Figs. 5–14). Represent-ative variations around the means for both generations are provided in Tables 3 and 4. With a few exceptions, percent mortality in the asexual generation (larval to adult stages) was consistently high at both sites, and most of this was accounted for by a complex of parasitic wasp species. In contrast, mortality in the sexual generation was usually low at both sites; however, in some years it reached high levels among older trees at Samuell Grand Park. Other studies (Frankie et al., 1984) suggested that percent mortality in the sexual generation may be generally high in dense stands of urban live oaks. All mortality observed in the sexual galls was accounted for by parasitic wasps.

Regressions of percent mortality vs. gall density failed to establish sig-nificant correlations for either generation on any marked tree (Figs. 5–14), suggesting a lack of density dependent mortality. However, a significant correlation between asexual and sexual gall densities was recorded for 8 of the 10 trees (Frankie et al., 1984). This suggested that density in one gen-

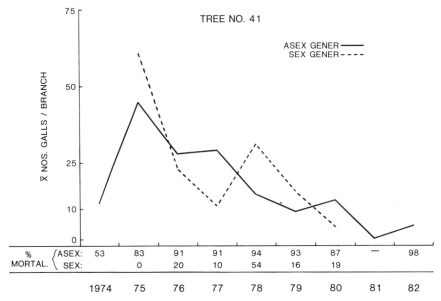

Figure 6. Density and mortality of wild asexual and sexual galls on tree 41 at NorthPark Mall.

eration is responsive to changes in the other, which in turn may be mediated by alternating high (asexual generation) and low (sexual generation) periods of parasite activity.

The following theoretical scenario was developed in Frankie et al. (1984) to account for occasional high gall densities on susceptible urban live oaks

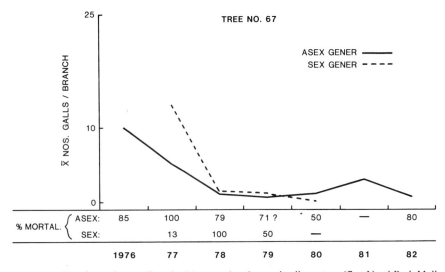

Figure 7. Density and mortality of wild asexual and sexual galls on tree 67 at NorthPark Mall.

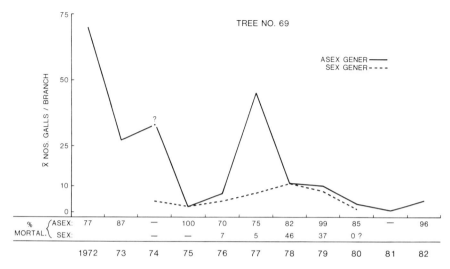

Figure 8. Density and mortality of wild asexual and sexual galls on tree 69 at NorthPark Mall. [Note: in 1974, asexual galls from tree 69 and 73 (Fig. 9) were accidentally pooled which resulted in an estimation of asexual gall density for both trees during that year.]

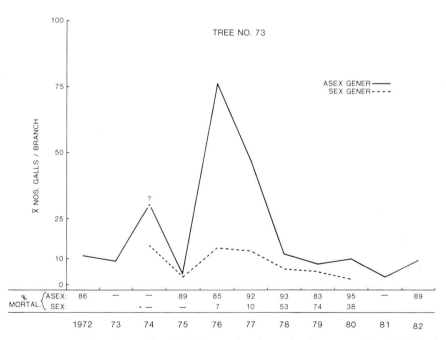

Figure 9. Density and mortality of wild asexual and sexual galls on trees 73 at NorthPark Mall (see Fig. 8 for explanation of asexual gall density in 1974).

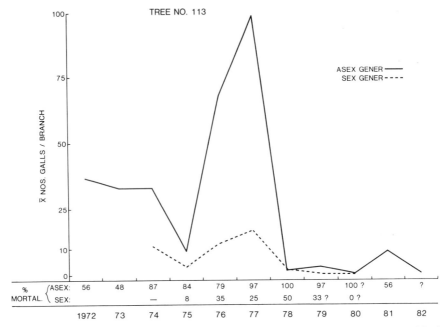

Figure 10. Density and mortality of wild asexual and sexual galls on tree 113 at NorthPark Mall.

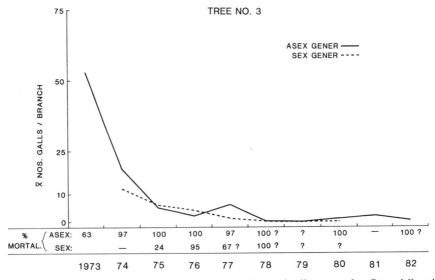

Figure 11. Density and mortality of wild asexual and sexual galls on tree 3 at Samuel Grand Park.

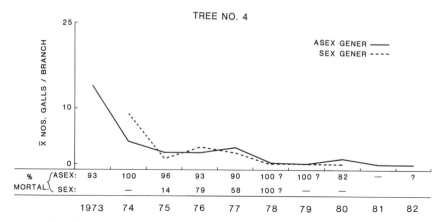

Figure 12. Density and mortality of wild asexual and sexual galls on tree 4 at Samuell Grand Park.

and consistently low densities on susceptible oaks in natural environments. Upon arrival at a newly planted, young oak, *Disholcaspis* populations may increase to relatively high levels before parasites are able to regulate their numbers. Escape in time from parasites is undoubtedly enhanced by planting trees in isolated urban situations (see Figs. 2 and 3 and figures in Frankie et al., 1984), and this seems particularly true for individuals of the sexual generation. Thus, release of sexual *Disholcaspis* from heavy parasitization leads to production of high numbers of asexual galls. Parasite guilds of both generations eventually catch up with *Disholcaspis* and regulate these populations; however, this may take several years. By way of contrast, in natural environments parasite populations may be omnipresent in live oak stands (including associated vegetation) because of their assumed greater and more

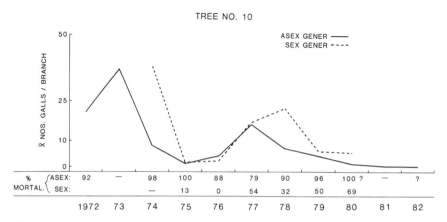

Figure 13. Density and mortality of wild asexual and sexual galls on tree 10 at Samuel Grand Park.

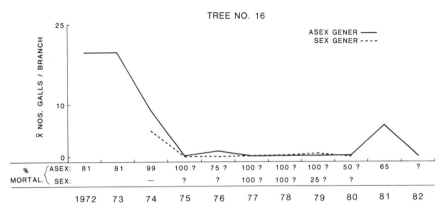

Figure 14. Density and mortality of wild asexual and sexual galls on tree 16 at Samuell Grand Park.

diverse herbivore loads. The parasites are thus available to regulate *Dish-olcaspis* before their numbers ever reach high levels.

Finally, a 5-year caging study provided important insights into how changes in host plant quality may affect the degree of susceptibility on individual trees from year to year (Morgan et al., 1983). Experimental cagings of asexual and sexual wasps on A-SUS and A-RES trees were conducted from 1974 to 1978 in north Dallas; trees were 7 years old in 1974. Significantly more galls of both generations were produced by A-SUS as compared to A-RES trees; most trees in the latter group produced few or no galls. Many A-SUS trees changed significantly in their susceptibility from year to year.

Table 3.   Mean Numbers of Asexual Galls Sampled[a] from 20 Live Oaks Scattered Widely in North Dallas

Tree No.	Mean No. Galls (±SE) per Branch	Tree No.	Mean No. Galls (±SE) per Branch
1	1.3 ± 0.5	11	10.6 ± 2.5
2	3.3 ± 1.0	12	11.8 ± 3.4
3	3.6 ± 0.9	13	14.8 ± 3.4
4	4.4 ± 1.3	14	15.6 ± 2.1
5	4.8 ± 1.4	15	16.0 ± 2.7
6	6.5 ± 2.1	16	16.9 ± 2.2
7	6.9 ± 1.9	17	23.3 ± 3.3
8	9.3 ± 4.0	18	26.3 ± 2.7
9	9.5 ± 2.0	19[b,c]	37.6 ± 15.0
10	9.5 ± 2.0	20[b]	59.6 ± 21.0

[a] Twelve 0.5-m branches sampled from each tree in 1982, unless otherwise indicated.

[b] Tree sampled in 1977.

[c] Same tree as No. 4.

Table 4. Mean Numbers of Sexual
Galls Sampled[a] from Six Live Oaks
in 1978 at NorthPark Mall[b]

Tree No.	Mean No. Galls ($\pm$ SE) per Branch
20	12.6 $\pm$ 3.0
41	31.3 $\pm$ 5.9
67	1.4 $\pm$ 0.4
69	10.6 $\pm$ 2.5
73	5.5 $\pm$ 1.0
113	3.3 $\pm$ 1.5

[a] Eight 0.5-m branches sampled per tree.
[b] See also Figures 5–10.

5 RESULTS

In this section, we present results of 2 years of experimental cagings of asexual *Disholcaspis* on A-SUS and A-RES trees at NorthPark and Samuell Grand Park. The *Disholcaspis* histories on these trees had been monitored for several years prior to the first cagings in late 1978 (Figs. 5–14).

5.1 Gall Bioassay on Susceptible Trees

The standardized caging procedure can also be used to bioassay a tree's potential to produce sexual galls during a particular period. Using past results as a guide (see Table 2 in Morgan et al., 1983), an arbitrary scheme was used in the current study to evaluate this potential (Table 5). When asexual wasps were caged on live oaks in 1978 and 1979, it was predicted that most, if not all, A-SUS trees would produce low numbers of sexual galls since

Table 5. Categories of Potential
Gall-Forming Capacity Based
on Production of Sexual Galls
from Standard Caging[a]

Potential Capacity	Mean No. Sexual Galls per Branch
Low	<21
Moderate	21–50
High	>50

[a] Eight asexual females caged per 0.5-m branch. Early study used same number of wasps on 0.75-m branches (Morgan et al., 1983).

their recent historical records (past 2–3 years) suggested that they may be declining in capacity to produce galls (Figs. 5, 6, 9–12, 14). Further, their advanced age (~20 years for NorthPark trees and 35+ years for Samuell Grand trees) suggested possibly diminished potential to produce galls (Frankie et al., 1984).

The caging study demonstrated that all A-SUS trees at both sites, except no. 41 (Fig. 6), produced relatively low gall numbers despite earlier records that indicated a much higher potential. The relatively moderate gall production in no. 41, within the context of its past history, may suggest that this tree is declining gradually in its capacity to produce galls. Differences between years on A-SUS trees were not significant. If the same trees had been caged earlier or later (e.g., in 1972/1973, 1976/1977, or 1981) we believe that the NorthPark trees would have produced considerably higher numbers of galls.

The caging of A-RES trees produced few or no galls per branch. Of particular interest was the fact that 1980 A-SUS trees 20 and 16 had gall densities that were nearly identical to the A-RES trees; no. 73 in both years also produced relatively few galls (Table 6).

Similar differences in gall production on A-SUS and A-RES trees were observed from 1974 to 1978 when asexual females were caged on the same two tree types at another Dallas location (Morgan et al., 1983). However, this early study did not examine possible differences in oviposition that might explain differences in gall densities.

Table 6.   Mean Numbers of Sexual Galls Produced per Branch[a] at NorthPark Mall and Samuell Grand Park

Tree Type and No.	Mean No. Galls/Branch ($\pm$ SE)	
	1979	1980
NorthPark Mall		
A-SUS 20	19 $\pm$ 15.0	0.3 $\pm$ 0.3
41	24 $\pm$ 9.0	37 $\pm$ 6.2
73	3 $\pm$ 2.0	2 $\pm$ 0.5
113	7 $\pm$ 3.1	6 $\pm$ 2.3
A-RES 20R	—	0.3 $\pm$ 0.3
41R	0	0
73R	0.3 $\pm$ 0.2	0.5 $\pm$ 0.3
113R	—	0
Samuell Grand Park		
A-SUS 3	—	20.3 $\pm$ 5.2
4	—	4.8 $\pm$ 1.5
16	—	0.5 $\pm$ 0.5

[a] Based on caging of eight asexual females per 0.5-m branch.

Table 7.   Mean Numbers of Sexual Eggs Sampled from Cohort I Branches of A-SUS and A-RES Trees at NorthPark Mall and Samuell Grand Park

	Mean No. Eggs/Bud ($\pm$ SE)	
Tree Type	1979[a]	1980[b]
NorthPark Mall		
A-SUS	0.99 $\pm$ 0.04	1.46 $\pm$ 0.05
A-RES	0.81 $\pm$ 0.04	1.51 $\pm$ 0.05
Samuell Grand Park		
A-SUS	N/A	1.09 $\pm$ 0.05

[a] A-SUS: $N = 1489$ dissected buds; A-RES: $N = 900$ dissected buds.

[b] NorthPark: $N = 800$ dissected buds in each tree group. Samuell Grand: $N = 600$ dissected buds.

5.2   Oviposition Activity

The numbers of eggs per bud and condition of the eggs was assessed for the two tree types; egg counts (Table 7) preceded tabulations of sexual galls (Table 6). Female wasps inserted their ovipositors beneath the bases of buds such that their eggs were deposited in the meristematic tissue without breaking surrounding budscale tissue. The bulbous eggs, which were attached to the meristem by thin stalks, were easily counted by teasing apart scales from the tops of buds. Large numbers of eggs were oviposited in all branches of A-SUS and A-RES trees, and almost identical numbers of eggs were laid in branches of both tree types at NorthPark. Somewhat higher numbers were laid in A-SUS trees at Samuell Grand Park (Table 7). This finding indicates that, although substantial numbers of eggs were laid in both tree types, oviposition did not lead to comparable numbers of galls (Table 6).

We examined the distribution of eggs from buds on A-SUS vs. A-RES trees to see if any behavioral differences among females could be inferred. We tabulated the numbers of buds with single versus multiple eggs from the NorthPark data. In 1979 we observed that about 50% of the sampled buds in both tree types had eggs; in 1980, 70% of the buds had eggs (Table 8). When the numbers of eggs per bud were counted on A-SUS and A-RES branches, no obvious differences could be detected in the distributions (Fig. 15), suggesting that asexual wasps do not oviposit differentially when confined on particular trees.

However, when numbers of expected eggs per branch (extrapolated from sampled buds of each branch) were compared with numbers of available buds per branch for the two tree types, a difference in oviposition behavior emerged. For each branch the number of potentially available buds was

Table 8. Percentages[a] of Sampled Buds with Eggs, Incipient Galls, and Mature Galls in A-SUS and A-RES Trees at NorthPark Mall and Samuell Grand Park

Year and Tree Type	Eggs (Cohort I)	Incipient Galls Cohort II	Incipient Galls Cohort III	Mature Galls (Cohort IV)
NorthPark Mall				
1979: A-SUS	50.5[b]	N/A	5.1[c]	22.9
A-RES	46.2	N/A	2.2	0.4
1980: A-SUS	68	20.3	11.5	6.9[d]
A-RES	68	12.6	4.8	0.2[e]
Samuell Grand Park				
1980: A-SUS	58.6	N/A	N/A	12.8

[a] All percentages based on 600–962 dissected and/or examined buds unless otherwise indicated.

[b] $N = 1489$ dissected buds.

[c] $N = 1044$ dissected buds.

[d] $N = 2566$ examined buds.

[e] $N = 1369$ examined buds.

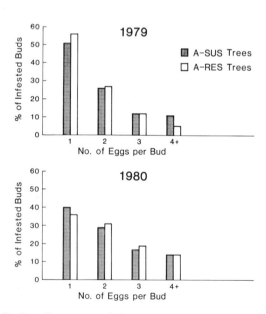

Figure 15. Distribution of eggs among infested buds from cohort I branches at NorthPark Mall (a similar distribution was also observed for cohort III branches of both tree types at NorthPark).

recorded, that is, those that were relatively large and/or swollen. Small or stubby-appearing buds were ignored since preliminary dissections indicated that these were not used by the wasps. Separating the two bud types was usually easy; only a small percentage fell between the two types. Using regression analysis, a positive and highly significant linear relationship between numbers of expected eggs and numbers of available buds was shown for A-SUS trees at NorthPark (Figs. 16–18). Although a significant corre-

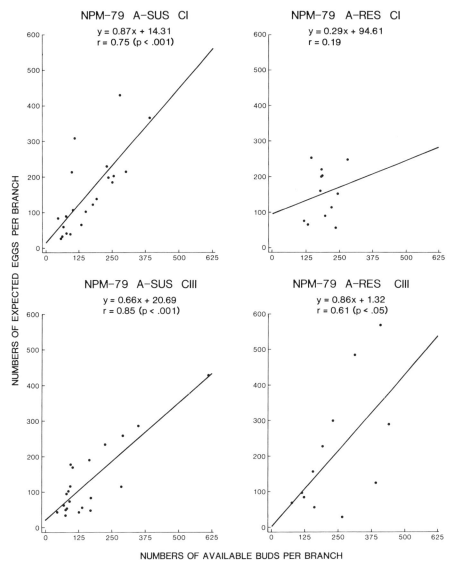

Figure 16. NorthPark Mall (NPM), 1979: regressions of expected eggs versus available buds on cohorts (C) of branches from A-SUS and A-RES trees.

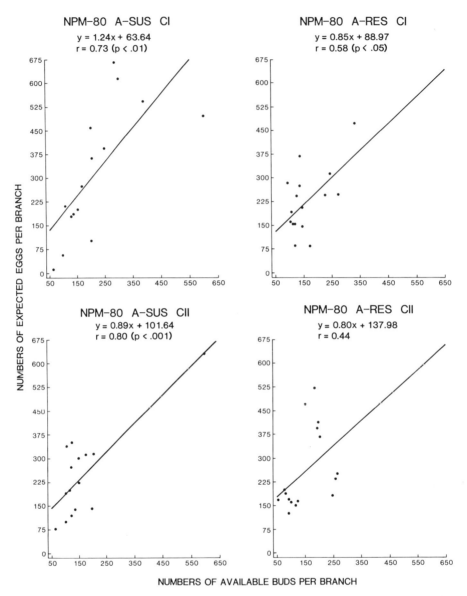

Figure 17. NorthPark Mall (NPM), 1980: regressions of expected eggs versus available buds on cohorts (C) of branches from A-SUS and A-RES trees.

lation was observed in the older A-SUS trees at Samuell Grand Park and in two of five cohorts from A-RES trees at NorthPark, it was not as pronounced as that observed in A-SUS trees at NorthPark. In the remaining three of five branch cohorts from A-RES trees, this relationship was not observed.

With each set of bud dissections we also recorded the numbers of unhealthy or morbid eggs on each tree type (Table 9). Shriveled and/or unusually small eggs were placed in this category. None of these eggs showed signs of being parasitized. The percentage of unhealthy eggs increased through time; however, the increase was far more dramatic in 1980 than 1979. There was no obvious reason for this higher level of morbidity.

During bud dissections we also observed the first evidence of gall formation. The incipient galls, which consisted of loosely arranged pink to beige colored tissue, were positioned on top of the meristem in leaf and flower

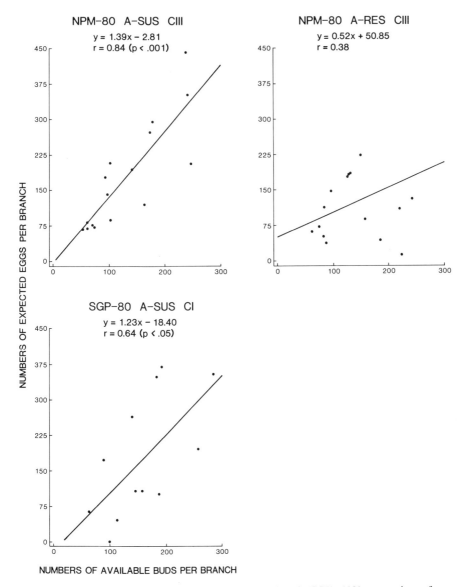

Figure 18. NorthPark Mall (NPM) and Samuell Grand Park (SGP), 1980: regressions of expected eggs versus available buds on cohorts (C) of branches from A-SUS and A-RES trees.

Table 9. Percentages[a] of Unhealthy Eggs[b] in
Sampled Buds of A-SUS and A-RES Trees
in NorthPark Mall

Year and Tree Type	Cohort		
	I	II	III
1979: A-SUS	10.6[c]	N/A	18[d]
A-RES	5.8	N/A	20
1980: A-SUS	8.8	21.8	50.6
A-RES	16.4	26.6	50.4

[a] All percentages based on 600–900 dissected buds unless otherwise
indicated.

[b] In Cohort III only, egg tabulations consisted of eggs plus a small
number (<4%) of first instar larvae that had just eclosed.

[c] 1485 dissected buds.

[d] 1044 dissected buds.

buds. They were first observed and recorded about a month after oviposition
and just prior to the production of mature sexual galls. Incipient gall frequency and the frequencies of eggs and mature galls are presented in Table
8. During both years and in both cohorts more galls (incipient and mature)
were observed on A-SUS trees. χ^2 analysis of NorthPark data revealed significant differences between all comparisons (A-SUS vs. A-RES) at the 0.001
level. The magnitude of the differences between tree types, however, was
far more pronounced in the mature gall category.

Percentages of buds that contained incipient galls and 0, 1, or multiple
eggs are recorded in Table 10. During both years, substantial numbers of

Table 10. Percentages[a] of Sampled Buds with Incipient Galls and 0, 1, or
Multiple Eggs[b] in A-SUS and A-RES Trees at NorthPark Mall

Year and Tree Type	Percentage of Buds with Incipient Galls and 0, 1, or >1 Eggs							
		Cohort II				Cohort III		
	No. Eggs:	0	1	>1	N^c	0	1	>1	N^c
1979: A-SUS				N/A		30	60	9	53[d]
A-RES				N/A		15	69	15	13
1980: A-SUS		19	32	49	152	26	52	22	92
A-RES		10	34	56	101	32	42	26	38

[a] Percentages based on 600–800 dissected buds unless otherwise indicated.

[b] In Cohort III only, egg tabulations consisted of eggs plus a small number (<9%) of first instar
larvae that had just eclosed.

[c] N = numbers of sampled buds containing incipient galls.

[d] N = 1044 dissected buds.

buds contained only incipient galls. Further, in 1980 the percent of buds with incipient galls and 0 eggs was greater in cohort III compared to cohort II branches, which suggests that formation of incipient galls may occur over an extended period during winter. A χ^2 test demonstrated a significant difference ($p < 0.05$) between these two cohorts for A-RES trees only.

Position of eggs in relation to incipient gall tissue varied. Usually the egg(s) were found in the middle of the tissue. However, occasionally they were observed to the outside of the tissue, and this usually occurred where more than one egg was laid per bud.

The overall findings of the bud dissections suggest that gall formation in the sexual generation occurs in two major stages. First, the asexual female inserts the ovipositor into the bud to test for an appropriate oviposition substrate. In the process she apparently deposits a chemical that initiates gall formation, regardless of whether an egg is laid. The second stage occurs about 3 months later when the egg hatches and developing larva secretes a chemical substance that continues the gall-forming process. This occurs only at the onset of leaf flushing in early spring (Frankie et al., 1984). Formation of asexual galls is due, at least in part, to the feeding of developing larvae (Frankie et al., 1984). The scenario for gall induction is believed to be similar in other gall wasp systems (see tenthredinid galls in McCalla et al., 1962; Mani, 1964, pp. 2, 182–183, 268). However, also see Caltagirone (1964) and Rohfritsch and Shorthouse (1982).

5.3   Other Possible Factors Influencing Gall Density

Unmarked live oaks in and around NorthPark (Fig. 1) were examined for *Disholcaspis* infestations in an effort to assess the potential reservoir of wasps that occurred in the vicinity of the six marked NorthPark trees. Surveys conducted annually among all planted live oaks at NorthPark yielded only a rare individual having moderate gall densities. None of the oaks in areas surrounding the NorthPark complex (Fig. 1) produced high gall densities, and only rarely were one or two individuals observed with moderate numbers during a given year. Overall, these few and limited reservoir foci of wasps probably were not important in contributing to high gall densities on our study trees.

Acorn numbers on the study trees from 1975 to 1982 showed no obvious correlation with gall activity. Although all trees produced relatively high acorn numbers in 1975, acorn production remained generally low in other years. Tree 10 (Fig. 13) was exceptional in that it regularly produced moderate to high numbers of acorns.

Weather data from 1971 to 1981 were examined for possible correlations with gall densities. Obvious patterns of rainfall, temperature, and freezing were examined, but none showed obvious correlations with periods of high and low gall numbers. Regarding freezing episodes, it is noteworthy that sexual egg mortality (Table 9) during the winter of 1979/1980 was significantly higher than in 1978/1979, despite a severe ice storm that caused considerable oak tree damage during the first winter. Finally, some trees at NorthPark (e.g., 41, Fig. 6) produced high gall numbers during certain years (e.g., 1975), whereas others (Figs. 5, 8–10) experienced low densities for the same years. In summary, there was no evidence that weather factors were correlated with high and low levels of gall production.

6 DISCUSSION

In this section we examine our data within the context of the three original hypotheses and discuss our conclusions in light of other studies.

6.1 Hypothesis I

Live oaks susceptible to *Disholcaspis* lose their susceptibility to this wasp with increasing age.

Evidence to support this hypothesis comes from three sources. First, extensive surveys in urban and natural live oak stands indicated that only young trees in the former environment experience high gall densities; older trees (40+ years), regardless of environment, have never been observed with high densities (Frankie et al., 1984). Second, and most importantly, knowing that all study trees (Figs. 5–14) once had a high potential to produce galls, the experimentally produced low densities in later years of the investigation (Table 6) indicate that the trees now have a diminished capacity to produce galls. Further, Morgan et al. (1983) found that one A-SUS tree caged with *Disholcaspis* for 9 consecutive years eventually declined substantially in its capacity to produce galls. Finally, continuous low production of galls on older Samuell Grand trees may be viewed as a projection in time of the eventual gall-producing capacity of the NorthPark trees. Note that relatively lower densities on most NorthPark trees from 1978 to 1982 may represent the beginning of a long-term decline in gall production.

It thus appears that young susceptible trees have the potential to produce galls at relatively high densities around age 5 to age 20, and that this potential is only realized in urban environments where *Disholcaspis* populations apparently escape regulation by parasites during the sexual generation (see Section 6.3, Hypothesis III). Most susceptible urban trees will probably produce alternating high and low gall densities during this early period (see Figs. 5, 8–10, 13). Others will produce one major burst of gall activity for a few years (e.g., tree 67, Fig. 7), and a small percentage may start high in their gall-forming capacity and then gradually decline in production over several years (e.g., tree 41, Fig. 6). Beyond age 20 or so, susceptible trees behave more like resistant individuals, with the exception of occasional small increases in gall numbers (see Section 6.2, Hypothesis II).

Several other gall inducers are also known to build up in high numbers on young trees (Cynipidae: Askew, 1962; Ejlersen, 1978). However, certain species do prefer older host trees (Cynipidae: Askew, 1962; Brown and Eads, 1965, pp. 83–84; Bailey and Stange, 1966). In the case of the cynipid *Plagiotrochus suberi* Weld on *Quercus suber* L., Bailey and Stange (1966) report that "young trees planted near older wasp-infested trees are not attacked the first 5 or 6 years, or until about the time they produce a crop of acorns." The cynipid *Xanthoteras politum* is known to be common only on sucker growth of older host trees (*Quercus stellata* Wangenh.) (Washburn and Cornell, 1981). As sucker growth ages, suitability of substrate for gall formation appears to decline. In the case of gall-inducing herbivores, especially cynipids, it seems clear that many different relationships with host plants are possible, so that generalizing about host age and susceptibility is difficult.

These observations point to a possible relationship between the phenomenon of plant juvenility and gall formation. Though its chemistry is not clearly understood, juvenility in woody plants is generally regarded as a phase change associated with physiological aging. The most clearly expressed phases are flowering and subsequent fruiting (Zimmerman, 1972; Borchert, 1976; Fortanier and Jonkers, 1976). In all woody plants there is a juvenile phase, lasting in some species up to 40 years, that precedes the first flowering. In live oak, the phase change from "juvenile" to "adult" may occur as early as the second year of growth to as late as the fifth, as evidenced by the appearance of flowers and acorns. Thus, the Bailey and Stange report of a successful cynipid attack on *Q. suber* would involve adult wood, whereas the sucker growth originating in the crown area of *Q. stellata* studied by Washburn and Cornell presumably was juvenile growth.

Although the onset of flowering is considered the end of the juvenile stage, other changes in physiological and morphological characteristics may simultaneously occur; specifically leaf shape, seasonal leaf retention, phyllotaxy, thorniness, pigmentation, and the ability to form adventitious roots (Hackett, 1976). Many changes are progressive, such that in live oak, flowers may appear on branches that bear characteristically juvenile leaves with serrated margins. *Quercus virginiana* may not produce the taxonomically identifiable rounded leaves with entire margins until 6–10 years of age, several seasons after onset of flowering (see also Morgan and McWilliams, 1976).

These growth and development phenomena suggest the possibility of eventually predicting susceptibility to gall formation in selected cynipid-plant systems, and also of using gall-inducing insects to bioassay for juvenility.

6.2 Hypothesis II

Host plant factors fix the gall-carrying capacity on *Disholcaspis*-susceptible trees.

This hypothesis represents an expansion of Hypothesis I. It is an important question since it bears on predictability of host plant quality and its short- and long-term effects on resident and immigrant *Disholcaspis*.

Considerable support for the hypothesis comes from experimental cagings in this and a previous study (Morgan et al., 1983). Overall, the findings indicated that the capacity of particular trees to produce galls may change significantly from year to year, and that over several years the capacity may drop to a level comparable to that of A-RES trees. Stated differently, the number or proportion of specific sites available for gall formation is set by the host tree (Felt, 1965, p. 32; van Emden and Way, 1973 p. 191; Webb and Moran, 1978), which in turn has important implications for establishing the herbivore density and for the associated natural enemy complex (see Section 6.3, Hypothesis III). In the case of *Disholcaspis* it was relatively easy to quantify host-carrying capacity since, in contrast to other forms of herbivory, the interaction between the cynipid and the host is an all-or-nothing phenomenon.

The population trends displayed in Figs. 5–10 provide additional support for the hypothesis. The periods of relatively high and low gall densities may represent fluctuating states of host plant quality that last for one or more years. In addition, lack of a density-dependent relationship between gall density and percent mortality ($\cong$ % parasitization) suggests that factor(s) other than parasites may, in part, regulate gall numbers (Frankie et al., 1984). The difference in numbers of unhealthy eggs between 1979 and 1980 (Table 9) is related to differences in annual gall production. Regardless of tree type and the extreme weather conditions observed during the winter of 1978/1979, the percentage of unhealthy eggs was substantially higher in 1980, suggesting a host plant response. Clearly, there is much to be learned about the relationship between host plant quality and egg mortality. For example, does the host exert a toxic reaction during some years? If so, what is the chemical nature of this reaction? Are nucleic acids (Went, 1970) and/or plant growth regulators involved in gall formation (Miles, 1968; Carter, 1973, pp. 234–246; Osborne, 1973; Southwood, 1973)?

Changes or differences in host plant quality that are known or suspected to affect growth, development, and fecundity of herbivores and their susceptibility to natural enemies have been documented for numerous insect/plant systems. Case history accounts and reviews of this topic are provided in Feeny (1970, 1976), Watson (1970), Schoonhoven (1972), Ryan (1973), van Emden (1973), Gilbert and Raven (1975), Beck and Reese (1976), Rhoades and Cates (1976), Onuf et al. (1977), Whitham (1978), and Whitham et al. (Chapter 2), Rhoades (1979), Rosenthal and Janzen (1979), Carroll and Hoffman (1980), Mattson (1980), Price et al. (1980), Barbosa et al. (1981), Greenblatt and Barbosa (1981), Schultz and Baldwin (1982), Schultz et al. (1982), Schultz (1983). Rhoades (1979, Table 1) reviews several studies that deal with host plant changes, mostly chemical, induced by actual or simulated herbivore feeding. In about half the studies, induction occurred after a short period (hours or days) (see also Carroll and Hoffman, 1980; Schultz and Baldwin, 1982). However, in several cases induction was reported 1 year later (e.g., Haukioja and Niemela, 1976); in one case 2 years later. Relaxation time of the response varied from 1 to 5 years, depending on the system.

Herbivore-induced changes in host plants may also be involved in insect gall systems (Washburn and Cornell, 1981; Whitham, 1983). However, there is little evidence to suggest that induction is involved in the resistance of live oaks to *Disholcaspis*. If it existed, one might expect predictable periods of relatively low gall densities to follow periods of high gall densities. Using the population trends in Figs. 5–14 as approximations of host carrying capacity, it is obvious that periods of high and low gall densities can vary considerably from year to year. The extreme example of this variation is observed in tree 41 (Fig. 6), which apparently declined gradually in its capacity (see also Table 6). Further, resistance to gall formation increases dramatically once a certain age is reached. Related to the question of induction is the possibility that stress (Johnson, 1968; White, 1969; Mattson and Addy, 1975; Frankie and Ehler, 1978; Rhoades, 1979) in the urban environment conditions trees to *Disholcaspis* infestations. Annual differences in population trends among trees at one location (e.g., NorthPark) suggests otherwise since all host plants received similar cultural treatments. Further, trees in more rural settings (e.g., College Station) also experience similar extreme fluctuations in gall densities. Finally, the stressful action of transplanting a host tree from one site to another also does not apparently alter host plant quality (Morgan et al., 1983).

Changes in host plant quality must be viewed from the standpoint of resident and immigrant wasps on particular trees. Limited evidence suggests that some behavioral differences exist among asexual wasps for A-SUS and A-RES trees (Figs. 16–18). In the case of A-SUS trees at NorthPark, wasps may have received enough appropriate cues from the buds to cause them to search widely within the cages to evaluate all potential oviposition sites. In the process they appear to have adjusted egg laying to the number of potentially available buds. However, if they are able to sense a previous oviposition or an attempt at ovipostion in a bud (Vinson, 1981, and personal communication), this may also result in adjusted egg numbers. Assuming that wasps do search widely for suitable buds, this would explain why galls are generally uniformly distributed over a tree's crown (Frankie et al., 1984). Uniform gall distribution should not be confused with the rare occurrence of very high gall numbers on selected branches of a tree (see Table 6 in Morgan et al., 1983; Whitham, 1978, 1981; Whitham et al., Chapter 2). In the older Samuell Grand trees and in most branch groups of A-RES trees, the cues may not have been appropriate or frequent enough, and this may have resulted in a more random pattern of oviposition. Regardless of tree type, the lack of eggs in many buds containing incipient galls (Table 10) suggests that wasps do a considerable amount of testing before ovipositing.

It is unlikely that asexual females readily disperse from parent trees since they are large-bodied, long-lived, and fly only infrequently (for short distances). However, they commonly walk quickly from branchlet to branchlet with considerable agility, apparently in search of appropriate oviposition sites.

We hypothesize that mated females of the sexual generation are the principal dispersers of this species. They are light-bodied, agile, quick-moving, and fly with the slightest provocation. Flight distances have not been recorded for either generation, but they are undoubtedly longer for sexual females. In the future, choice studies on A-SUS and A-RES plants will provide increased understanding of searching and ovipositional behavior in both generations of wasps.

Although sexual females are the most likely dispersers, the tendency for them to leave parent trees may also be relatively low (Cornell and Washburn, 1979). Survey and population monitoring data suggest that fidelity to a particular tree probably exists over several years. However, fidelity does not appear to be extreme as in the case of some herbivores such as the black pineleaf scale, *Nuculaspis californica* (Coleman), on ponderosa pine (Edmunds and Alstad, 1978, 1981; see also Wood and Guttman, 1981). Occasional and sudden increases in *Disholcaspis* galls on older once susceptible trees (e.g., see 1981 gall densities in Figs. 5, 7, 10, and 14) suggest that some wasps remain on the trees to induce these pulses. Examples of sudden gall increases have also been reported for a high proportion of susceptible trees in localized natural areas (Frankie et al., 1984).

It is possible that immigration by *Disholcaspis* into a tree may account for some increased gall numbers during pulse periods. However, if this happens it would probably involve numerous colonizing wasps since increases in gall densities during these periods may be considerable on some trees (see Fig. 14). In the case of isolated urban trees, this would mean that wasps may be attracted to these trees from some distance. Although conceivable, this explanation does not account for high gall densities that are suddenly produced on numerous trees in localized areas, natural and urban (Frankie et al., 1984). The cumulative evidence suggests that persistent low densities of resident wasps may be more important than immigrants in promoting sudden increases in gall numbers on older, susceptible trees.

6.3   Hypothesis III

Changes in host plant susceptibility and parasites determine long-term trends of *Disholcaspis* on live oak.

To test this hypothesis directly would require bioassays of the gall-producing capacity of several trees coupled with monitoring of the wild *Disholcaspis* population and its parasites on the same trees over a period of at least 4 years. Ideally, this should be done on young susceptible trees in urban and natural environments. Other ideal requisites might include using the same genetic stock in both locations, which is possible since techniques are now available for vegetatively propagating live oak (Morgan et al., 1980).

Since the above kind of data does not exist, we offer the following synthesis in support of the hypothesis. The focal point of our argument is that gall densities never reach high levels in natural habitats, and only under rare and isolated circumstances do they reach even moderate levels (see Kingsville example in Frankie et al., 1984). Based on previously reported parasite data (Frankie et al., 1984) it seems that complexes of parasitic wasps regulate *Disholcaspis* at consistently low densities in natural environments. Studies in the urban environment offered us the opportunity to view *Disholcaspis* populations (mostly sexual generation) that had been partially released from their parasites over several years. Possibly, the parasites associated with this generation simply do not fare well in urban areas because of a lack of alternate hosts, harsher more open conditions and often widely spaced host trees (Frankie et al., 1984). Further, significant correlations between gall densities of the two generations suggest strong interdependence between them that is mediated by the parasites.

Other studies on gall-inducing insects indicate that parasitic wasps play an important role in regulating gall populations (Askew, 1961, 1975; Caltagirone, 1964; Shorthouse, 1973; Varley et al., 1973; Hammel, 1973; Force, 1974; Cane and Kurczewski, 1976; Washburn and Cornell, 1979, 1981). The evolutionary importance of parasites in shaping gall wasp/plant interactions is discussed by Askew (1961) and Price (1980, pp. 160–165). Two additional studies are particularly relevant to the *Disholcaspis* parasite story. Ehler (1982) investigated the ecology of a cecidomyiid gall midge, *Rhopalomyia californica* Felt, on coyote brush, *Baccharis pilularis* DC, in an urban environment in California. The gall midge was generally more abundant in urban environments than in natural ones (Ehler, personal communication). In fact, when Ehler (1982) compared his data from the urban environment to comparable data obtained by Force (1974, unpublished) in a natural environment, the midge was from 4 to 10 times more abundant in the former setting. Although parasite guilds were structurally similar between the urban and natural habitats, comparable levels of biological control did not necessarily result. However, under certain conditions in the urban environment, parasitization was shown to be a potential regulatory factor. More recently, Ehler and associates (personal communication) have detected a major population increase of the gall midge in those areas of the natural environment that received sequential applications of malathion-bait sprays for eradication of medfly. The most likely hypothesis to account for this increase is that the sprays destroyed or greatly suppressed parasites of the midge and it increased accordingly (see also Washburn et al., 1983). Brewer (1971) in Colorado observed that the cecidomyiid *Janetiella* sp. near *coloradensis* Felt on pinyon pine caused severe needle stunting in landscape plantings; some affected trees had 60% of their current-year needles stunted. He stated that "such high levels of infestations were never found in native pinyon stands." He recorded only a 40% parasitization rate in the city by one parasitic wasp *Platygaster pini* Fouts. Unfortunately, only casual observations were made in the natural environment, and this was due to the difficulty of finding adequate biological material. It seems possible that differences in gall densities between the sites might be explained by differences in levels of parasitization between the areas (Brewer, personal communication).

As previously discussed, changes in host plant quality may dramatically affect *Disholcaspis* populations over a period of several years. These changes may in turn affect *Disholcaspis* parasites, especially in urban areas. As *Disholcaspis* populations decline on individuals or groups of urban trees, the absolute numbers of parasites in the immediate area should also decline unless some use alternate hosts. This may eventually be reflected in lowered parasitization rates. Low or reduced mortality in the asexual generation of NorthPark trees after gall densities declined over a few years offers some evidence for this possibility (see 1976 and 1981 in Fig. 5; 1980 in Fig. 7; 1976 in Fig. 8; 1981 in Fig. 10). Although this relationship does not generally hold for the older Samuell Grand trees, where the park environment may be more favorable to parasites (Frankie et al., 1984), it may apply when a lengthy period of low gall densities passes (see 1980 and 1981 in Fig. 14). The relationship may also apply when a young tree first experiences high gall densities (Figs. 5, 6, and 10). Periods of extremely low gall densities may also be viewed as periods of temporary escape for *Disholcaspis* in urban areas. A somewhat similar relationship between the cynipid *Xanthoteras politum*, its parasites and the host *Quercus stellata* was studied by Washburn and Cornell (1981). In their system, parasites progressively increased over a 3-year period as cynipid host density and host oak quality (on sucker growth) changed. These workers suggest that the action of parasites together with host plant changes eventually resulted in the local extinction of this "fugitive" cynipid. Although plant phenological characteristics (proximate) were correlated with changing gall densities, suspected chemical factors (ultimate) were not determined.

7 CONCLUDING REMARKS

Our lengthy experience with *Disholcaspis* strongly suggests that its density on susceptible trees is regulated by a combination of host plant quality and parasites. Because we bioassayed our study trees for only 2 years, it is difficult to know the relative contribution of each factor in determining annual densities. However, it seems safe to say that at one extreme *Disholcaspis* densities on young susceptible urban trees (especially in isolated locations) are largely under the control of the host plant. At the opposite extreme are young susceptible trees in natural environments whose *Dish-*

olcaspis densities are always low, and this is probably due to the predominant action of parasites.

Because of the many unique characteristics of gall insects and cynipids in particular, it is difficult to generalize to other insect herbivore/plant systems. Rather than attempt this task, we reflect on four methodological aspects of our study that were particularly useful in developing information on the *Disholcaspis*/oak system.

The development of long-term records is perhaps the most significant methodological feature of our study. The systematic collection of density and mortality data provided a strong base for making many of our observations and for evaluating our survey and experimental findings. With a few exceptions (Huffaker and Kennett, 1966; Skuhravý, 1972; Varley et al., 1973; Southwood and Reader, 1976; Strong, Chapter 11) such long-term field studies are rare; however, as stressed by many ecologists (Strong, Chapter 11; Dayton and Tegner, Chapter 17) they are sorely needed. Related to the long-term feature is the fact that we followed populations on individual plants. Even though the plants were relatively close together (Fig. 1), different developments, which affected density, occurred on each. Southwood and Reader (1976) recorded significant differences among populations of the viburnum whitefly on three closely spaced ornamental viburnums over 13 years at the Imperial College Field Station, Silwood Park.

We found it particularly useful to examine the *Disholcaspis*/live oak relationship in highly contrasting environments where the natural environment was considered as a standard. The urban environment, particularly where live oaks had been recently planted, proved to be an excellent extreme where one of the regulatory factors, parasites, was partially suppressed. Many similar interactions are known to exist in the urban environment where humans play a major role in altering biological and ecological relationships (Frankie and Ehler, 1978).

Finally, we were able to develop a means for bioassaying host plant quality. The relative ease in dealing with this aspect of our study can be traced to the basic gall-forming process, an all-or-nothing phenomenon. We expect that host plant quality can also be standardized in other systems where cultivars and varieties can be identified, as in the case of crop plants (Wyatt, 1970; Starks et al., 1972). Some of the planned studies on *Disholcaspis* will be conducted on cloned material that has already been propagated.

8 SUMMARY

The importance of host plant quality and parasitic wasps in regulating abundance of the gall wasp *Disholcaspis cinerosa* (Family: Cynipidae) on live oak (*Quercus* spp.) is assessed in natural and urban Texas environments. The investigation on this heterogonous species (alternating asexual and sexual generations) extended from 1971 to 1982.

Past studies (Morgan et al., 1983; Frankie et al., 1984) of the *Disholcaspis*/ live oak relationship are reviewed. The main part of the paper, however, deals with results of 2 years of experimental cagings of asexual wasps on *Disholcaspis*-susceptible oak trees in Dallas, which had been individually monitored for gall density and mortality of wild *Disholcaspis* for several years. Nearby non-susceptible live oaks were also caged.

8.1   Review of Past Results

1. Surveys in several natural and urban Texas environments revealed that asexual *Disholcaspis* galls were unequally distributed among host trees. A low percentage of trees had high gall densities, and these were always young individuals in urban environments; trees in natural environments never had high gall numbers. Regardless of environment, the vast majority of the live oaks had few or no galls.
2. Parasites were found to be the principal mortality factor in larval, pupal and adult stages of *Disholcaspis*. Mortality in the egg stage was not investigated.
3. Eleven to 15 species of parasitic wasps were associated with the asexual galls, and 8–10 mostly different species were associated with the sexual galls.
4. Parasitization levels appeared to be consistently high for both cynipid generations in natural environments. These levels were consistently high in asexual galls and regularly low in sexual galls on trees monitored for several years in Dallas.
5. Experimental cagings of asexual and sexual *Disholcaspis* wasps on young, apparently susceptible (A-SUS) and apparently resistant (A-RES) trees were conducted from 1974 to 1978 in north Dallas. Significantly more galls of both generations were produced by A-SUS trees. Many A-SUS trees changed significantly in their susceptibility from year to year.

8.2   Results of Current Study

Results of 2 years (1978–1980) of experimental cagings of asexual *Disholcaspis* on trees that had been monitored for wild *Disholcaspis* for ~10 years (1972/1973 to 1982) at two Dallas locations are presented:

1. More sexual galls were generally produced on A-SUS vs. A-RES trees, but levels on the former were relatively low compared to earlier recorded gall densities (wild) on these trees. This suggested that as live oaks age, they lose their capacity to produce galls.
2. Almost identical numbers of eggs were laid in the buds of A-SUS and A-RES trees.

3. Regardless of tree type, 50–70% of the sampled buds had eggs, indicating that ample eggs were laid in each tree.

4. A correlation between numbers of eggs laid and numbers of available buds per branch was observed. It was considerably more pronounced in A-SUS vs. A-RES trees. This suggested that asexual ♀♀ could discriminate between buds of the two tree types.

5. Sexual gall formation probably occurs in two steps. First, an excitant chemical is laid down at the time of oviposition, which results in incipient gall tissue. Second, after ~3 months, the eggs hatch and developing larvae secrete another chemical that continues gall growth.

6. Other factors such as acorn production, climatic factors, proximity to other *Disholcaspis*-infested trees, and stress were examined for possible correlations with gall density on our Dallas study trees. No obvious correlations were observed.

Overall survey and experimental findings suggest the following scenario of factors that determine population development of *Disholcaspis* in urban and natural environments. Unknown host plant factors set the gall-carrying capacity in susceptible live oaks, regardless of site. In natural areas, the full potential of a tree to produce galls is never realized since omnipresent parasites probably regulate asexual and sexual *Disholcaspis* at low densities. In the urban environment, a high potential to produce galls is realized when host plant quality is optimal and parasite pressure is periodically relaxed (in the sexual generation). This combination is almost always observed on young trees that have some degree of isolation. After several years (40+), susceptible trees in both environments lose their capacity to produce galls through factors that are thought to be associated with host tree aging.

ACKNOWLEDGMENTS

The research was supported by the California and Texas Agricultural Experiment Stations. We thank J. Crisp, M. Gaylor, D. Hammel, J. Frankie, H. Reed, and F. Rohner for assistance in the field. Special thanks are due to J. and M. Fraser, M. Higgins, S. Mandel, P. Smith, and J. Washburn who spent many hours dissecting live oak buds. E. Grissell at the U.S. National Museum, and J. Tucker, University of California, Davis provided parasite and tree identifications, respectively. We benefited greatly from discussions with H. Burke, R. Colwell, L. Ehler, E. Grissell, C. Huffaker, W. Getz, and S. Vinson. We are indebted to S. Mandel for assistance in data analysis and C. Tibbits for graphics. R. Askew, L. Ehler, W. Getz, W. Haber, M. Higgins, R. Robichaux, S. Vinson, J. Washburn, and T. Whitham kindly read an early draft of the manuscript. We extend our appreciation to

the directors of NorthPark Shopping Mall and the Dallas Park and Recreation
Department for permission to use their landscape trees the duration of the
study. The H. Heep family and the University of Texas at Austin kindly
allowed us to use the live oaks in their natural stands (Heep Ranch and
Breckenridge Field Laboratory, respectively).

LITERATURE CITED

Askew, R. R. 1961. On the biology of the inhabitants of oak galls of Cynipidae (Hymenoptera)
   in Britain. *Trans. Soc. Br. Entomol.* **14**:237–268.

Askew, R. R. 1962. The distribution of galls of *Neuroterus* (Hym:Cynipidae) on oak. *J. Anim.
   Ecol.* **31**:439–455.

Askew, R. R. 1975. The organization of chalcid-dominated parasitoid communities centered
   upon endophytic hosts. In P. W. Price (ed.), *Evolutionary strategies of parasitic insects
   and mites*, pp. 130–153. Plenum Press, New York.

Bailey, S. F., and L. A. Stange. 1966. The twig wasp of cork—Its biology and control. *J. Econ.
   Entomol.* **59**:663–668.

Barbosa, P., W. Cranshaw, and J. A. Greenblatt. 1981. Influence of food quantity and quality
   on polymorphic dispersal behaviors in the gypsy moth, *Lymantria dispar. Can. J. Zool.*
   59:293–296.

Beck, S. D., and J. C. Reese. 1976. Insect-plant interactions: Nutrition and metabolism. *Recent
   Adv. Phytochem.* **10**:41–92.

Bochert, R. 1976. The concept of juvenility in woody plants. *Acta Hortic.* **56**:143–154.

Brewer, J. W. 1971. Biology of the pinyon stunt needle midge. *Ann. Entomol. Soc. Am.* **64**:1099–
   1102.

Brown, L. R., and C. O. Eads. 1965. A technical study of insects affecting the oak tree in
   southern California. *Calif. Agric. Exp. Stn.* Bull. 180.

Caltagirone, L. E. 1964. Notes on the biology, parasites, and inquilines of *Pontania pacifica*
   (Hymenoptera:Tenthredinidae), a leaf-gall incitant on *Salix lasiolepis. Ann. Entomol. Soc.
   Am.* **57**:279–291.

Cane, J. H., and F. E. Kurczewski. 1976. Mortality factors affecting *Eurosta solidaginis* (Dip-
   tera:Tephritidae). *J. N.Y. Entomol. Soc.* **84**:275–282.

Carroll, C. R., and C. A. Hoffman. 1980. Chemical feeding deterrent mobilized in response to
   insect herbivory and counteradaptation by *Epilachna tredecimnotato. Science* **209**:414–
   416.

Carter, W. 1973. *Insects in relation to plant disease.* Wiley, New York.

Clausen, C. P. 1940. *Entomophagous insects.* Hafner, New York. (1974 reprint)

Clausen, C. P. (ed.) 1978. *Introduced parasites and predators of arthropod pests and weeds:
   A world reveiw.* USDA Agric. Hdbk. 480.

Cornell, H. V., and J. O. Washburn. 1979. Evolution of the richness-area correlation for cynipid
   gall wasps on oak trees: A comparison of two geographic areas. *Evolution* **33**:257–274.

Debach, P. (ed.). 1964. *Biological control of insect pests and weeds.* Reinhold, New York.

Debach, P. 1974. *Biological control by natural enemies.* Cambridge University Press, Cambridge.

Edmunds, G. F., Jr., and D. N. Alstad. 1978. Coevolution in insect herbivores and conifers.
   Science **199**:941–945.

Edmunds, G. F., and D. N. Alstad. 1981. Responses of black pineleaf scales to host plant variability. In R. F. Denno and H. Dingle (eds.), *Insect life history patterns: Habitat and geographic variation*, pp. 29–38. Springer-Verlag, New York.

Ehler, L. E. 1982. Ecology of *Rhopalomyia californica* Felt (Diptera:Cecidomyiidae) and its parasites in an urban environment. *Hilgardia* **50**:1–32.

Ejlersen, A. 1978. The spatial distribution of spangle galls (*Neuroterus* spp.) on oak (Hymenoptera:Cynipidae). *Ent. Meddr.* **46**:19–25.

Feeny, P. 1970. Seasonal changes in oak leaf tannins and nutrients as a cause of spring feeding by winter moth caterpillars. *Ecology* **51**:565–588.

Feeny, P. 1976. Plant apparency and chemical defense. *Recent Adv. Phytochem.* **10**:1–40.

Felt, E. P. 1965. *Plant galls and gall makers*. Hafner, New York and London. (reprint of 1940 ed.)

Force, D. C. 1974. Ecology of insect host-parasitoid communities. *Science* **184**:624–632.

Fortanier, E. J., and H. Jonkers. 1976. Juvenility and maturity of plants as influenced by their ontogenetical and physiological ageing. *Acta Hortic.* **56**:37–43.

Frankie, G. W., and L. E. Ehler. 1978. Ecology of insects in urban environments. *Annu. Rev. Entomol.* **23**:367–387.

Frankie, G. W., D. L. Morgan, M. J. Gaylor, J. G. Benskin, W. E. Clark, H. C. Reed, and P. J. Hamman. 1977. The mealy-oak gall on ornamental live oak in Texas. Texas Agric. Exp. Stn. MP-1315.

Frankie, G. W., D. L. Morgan, and E. E. Grissell. 1984. Effects of urbanization on the distribution and abundance of the gall wasp, *Disholcaspis cinerosa*, on ornamental live oak in Texas (Hymenoptera:Cynipidae). In J. D. Shorthouse and O. Rohfritsch (eds.), *Biology of insect and acarina induced galls*. Praeger Press, New York (in press).

Gilbert, L. E., and P. H. Raven. 1975. *Coevolution of animals and plants*. University of Texas Press, Austin and London.

Greenblatt, J. A., and P. Barbosa. 1981. Effects of host's diet on two pupal parasitoids of gypsy moth: *Brachymeria intermedia* (NEES) and *Coccygomimus turionellae* (L.). *J. Appl. Ecol.* **18**:1–10.

Hackett, W. P. 1976. Control of phase change in woody plants. *Acta Hortic.* **56**:143–154.

Hall, R. W., L. E. Ehler, and B. Bisabri-Ershadi. 1980. Rate of success in classical biological control of arthropods. *Bull. Entomol. Soc. Am.* **26**:111–114.

Hammel, D. R. 1973. *The biology and ecology of the live oak woolly leaf gall Andricus langier Ashmead (Hymenoptera:Cynipidae)*. Unpubl. M.S. thesis. Texas A&M University, College Station, Texas.

Harborne, J. B. 1982. *Introduction to ecological biochemistry*, 2nd ed. Academic Press, New York.

Haukioja, E., and P. Niemela. 1976. Does birch defend itself actively against herbivores? *Rep. Kevo Subarctic Res. Stat.* **13**:44–47.

Huffaker, C. B. (ed.). 1971. *Biological control*. Plenum Press, New York and London.

Huffaker, C. B., and C. E. Kennett. 1966. The biological control of *Parlartoria olease* (Colvee) through the compensatory action of two introduced parasites. *Hilgardia* **37**:283–334.

Huffaker, C. B., and P. S. Messenger (eds.). 1976. *Theory and practice of biological control*. Academic Press, New York.

Johnson, C. D., and C. N. Slobodchikoff. 1979. Coevolution of Cassia (Leguminosae) and its seed beetle predators (Bruchidae). *Environ. Entomol.* **8**:1059–1064.

Johnson, N. E. 1968. Insect attack in relationship to the physiological condition of the host tree. *Cornell Univ. Entomol. Limnol. Mimeo. Rev.* No. 1. 46 pp.

Lawton, J. H., and S. McNeill. 1979. Between the devil and the deep blue sea: On the problem of being a herbivore. *Symp. Br. Ecol. Soc.* **20**:223–244. Blackwell, Oxford.

McCalla, D. R., M. K. Genthe, and W. Hovanitz. 1962. Chemical nature of an insect gall factor. *Plant Physiol.* **37**:98–103.

McNeill, S. 1973. The dynamics of a population of *Leptopterna dolabrata* (Heteroptera: Miridae) in relation to its food resources. *J. Anim. Ecol.* **42**:495–507.

Mani, M. S. 1964. *Ecology of plant galls.* Junk, The Hague, Netherlands.

Mattson, W. J., Jr. 1980. Herbivory in relation to plant nitrogen content. *Annu. Rev. Ecol. Syst.* **11**:119–161.

Mattson, W. J., and N. D. Addy. 1975. Phytophagous insects as regulators of forest primary production. *Science* **190**:515–522.

Metcalf, R. L., and W. H. Luckmann. 1982. *Introduction to insect pest management*, 2nd ed. Wiley, New York.

Miles, P. W.. 1968. Insect secretions in plants. *Annu. Rev. Phytopath.* **6**:137–164.

Morgan, D. L., and E. L. McWilliams. 1976. Juvenility as a factor in propogating *Quercus virginiana* Mill. *Acta Hortic.* **56**:263–268.

Morgan, D. L., E. L. McWilliams, and W. C. Parr. 1980. Maintaining juvenility in live oak. *HortScience* **15**:493–494.

Morgan, D. L., G. W. Frankie, and M. J. Gaylor. 1983. An evaluation of the gall-forming capacity of live oak in response to the cynipid wasp, *Disholcalspis cinerosa*, in Texas. *J. Kans. Entomol. Soc.* **56**:100–108

Onuf, C. P., J. M. Teal, and I. Valiela. 1977. Interactions of nutrients, plant growth and herbivory in a mangrove ecosystem. *Ecology* **58**:514–526.

Osborne, D. J. 1973. Mutual regulation of growth and development in plants and insects. In H. F. van Emden (ed.), *Symposia of the Royal Entomological Society of London: 6. Insect/ Plant Relationships*, pp.. 33–42. Blackwell, Oxford.

Price, P. W. (ed.). 1975. *Evolutionary strategies of parasitic insects and mites.* Plenum Press, New York.

Price, P. W. 1980. *Evolutionary biology of parasites.* Monogr. Pop. Biol., Vol. 15. Princeton University Press, Princeton, New Jersey.

Price, P. W., C. E. Bouton, P. Gross, B. A. McPheron, J. N. Thompson, and A. E. Weis. 1980. Interactions among three trophic levels: Influence of plants on interactions between insect herbivores and natural enemies. *Annu. Rev. Ecol. Syst.* **11**:41–65.

Rhoades, D. F. 1979. Evolution of plant chemical defence against herbivores. In G. A. Rosenthal and D. H. Janzen (eds.), *Herbivores: Their interactions with secondary plant metabolities*, pp. 4–55. Academic Press, New York.

Rhoades, D. F., and R. G. Cates. 1976. A general theory of plant herbivore chemistry. *Recent Adv. Phytochem.* **10**:168–213.

Ridgway, R. L., and S. B. Vinson (eds.). 1977. *Biological control by augmentation of natural enemies.* Plenum Press, New York and London.

Rohfritsch, O., and J. D. Shorthouse. 1982. Insect galls. In G. Kahl and J. S. Schell (eds.), *Molecular biology of plant tumors*, pp. 131–152. Academic Press, New York.

Rosenthal, G. A., and D. H. Janzen (eds.). 1979. *Herbivores: Their interaction with secondary plant metabolites.* Academic Press, New York.

Ryan, C. A. 1973. Proteolytic enzymes and their inhibitors in plants. *Annu. Rev. Plant Physiol.* **24**:173–196.

Ryan, C. A., and T. R. Green. 1974. Proteinase inhibitors in natural plant protection. *Recent Adv. Phytochem.* **8**:123–140.

Schoonhoven, L. M. 1972. Secondary plant substances and insects. *Recent Adv. Phytochem.* **5**:197–224.

Schultz, J. C. 1983. Habitat selection and foraging tactics of caterpillars in heterogeneous trees.

In R. F. Denno and M. S. McClure (eds.), *Variable plants and herbivores in natural and managed systems*, pp. 61–90. Academic Press, New York.

Schultz, J. C., and I. T. Baldwin. 1982. Oak leaf quality declines in response to defoliation by gypsy moth larvae. *Science* **217**:149–151.

Schultz, J. C., P. J. Nothnagle, and I. T. Baldwin. 1982. Seasonal and individual variation in leaf quality of two northern hardwood tree species. *Am. J. Bot.* **69**:753–759.

Shorthouse, J. D. 1973. The insect community associated with rose galls of *Diplolepis polita* (Cynipidae, Hymenoptera). *Quaestiones Entomol.* **9**:55–98.

Skuhravý, V. 1972. Distribution and outbreaks of the gall midge *Thecodiplosis brachyntera* (Schwägr.) in Europe (Diptera, Cecidomyiidae). *Acta Entomol. Bohemoslov.* **69**:217–228.

Southwood, T. R. E. 1973. The insect/plant relationship—An evolutionary perspective. In H. F. van Emden (ed.), *Symposia of the Royal Entomological Society of London*: 6. *Insect/Plant Relationships*, pp. 3–30. Blackwell, Oxford.

Southwood, T. R. E., and P. M. Reader. 1976. Population census data and key factor analysis for the viburnum whitefly, *Aleurotrachelus jelinekii* (Frauenf.), on three bushes. *J. Anim. Ecol.* **45**:313–325.

Starks, K. J., R. Muniappan, and R. D. Eikenbary. 1972. Interaction between plant resistance and parasitism against the greenbug on barley and sorghum. *Ann. Entomol. Soc. Am.* **65**:650–655.

van Emden, H. F. (ed.). 1973. *Insect/plant relationships. Symp. Royal Entomol. Soc. London*: 6. Blackwell, Oxford.

van Emden, H. F., and M. J. Way. 1973. Host plants in the population dynamics of insects. In H. F. van Emden (ed.), *Symposia of the Royal Entomological Society of London*: 6. *Insect/Plant Relationships*, pp. 181–199. Blackwell, Oxford.

Varley, G. C., G. R. Gradwell, and M. P. Hassell. 1973. *Insect population ecology, an analytical approach*. University of California Press, Berkeley and Los Angeles.

Vinson, S. B. 1975. Biochemical coevolution between parasitoids and their hosts. In P. W. Price (ed.), *Evolutionary strategies of parsitic insects and mites*, pp. 14–48. Plenum Press, New York.

Vinson, S. B. 1981. Habitat location. In D. A. Nordlund (ed.), *Semiochemicals: Their role in pest control*, pp. 51–77. Wiley, New York.

Washburn, J. O., and H. V. Cornell. 1979. Chalcid parasitoid attack on a gall wasp population on oak (*Acraspis hirta* Bassett, on *Quercus primus*, Fagaceae). *Can. Entomol.* **111**:391–400.

Washburn, J. O., and H. V. Cornell. 1981. Parasitoids, patches, and phenology: Their possible role in the local extinction of a cynipid gall wasp population. *Ecology* **62**:1597–1607.

Washburn, J. O., R. L. Tassan, K. Grace, E. Bellis, K. S. Hagen, and G. W. Frankie. 1983. Effects of malathion sprays on the ice plant insect system. *Calif. Agric.* **37**:30–32.

Watson, A. (ed.). 1970. Animal populations in relation to their food resources. *Symp. Br. Ecol. Soc.* 10. Blackwell, Oxford.

Webb, J. W., and V. C. Moran. 1978. The influence of the host on the population dynamics of *Acizzia russellae* (Homoptera: Psyllidae). *Ecol. Entomol.* **3**:313–321.

Went, F. W. 1970. Plants and the chemical environment. In E. Sondheimer and J. B. Simeone (eds.), *Chemical ecology*, pp. 72–81. Academic Press, New York and London.

White, T. C. R. 1969. An index to measure weather-induced stress of trees associated with outbreaks of psyllids in Australia. *Ecology* **50**:905–909.

Whitham, T. G. 1978. Habitat selection by *Pemphigus* aphids in response to resource limitation and competition. *Ecology* **59**:1164–1176.

Whitham, T. G. 1981. Individual trees as heterogenous environments: adaptation to herbivory or epigenetic noise? In R. F. Denno and H. Dingle (eds.). *Insect life history patterns*, pp. 9–27. Springer-Verlag, New York.

Whitham, T. G. 1983. Host manipulation of parasites: within-plant variation as a defense against rapidly evolving pests. In R. F. Denno and M. S. McClure (eds.), *Variable plants and herbivores in natural and managed systems*, pp. 15–41. Academic Press, New York.

Whitham, T. G., and C. N. Slobodchikoff. 1981. Evolution by individuals, plant-herbivore interactions, and mosaics of genetic variability: the adaptive significance of somatic mutations in plants. *Oecologia* **49**:287–292.

Wood, T. K., and S. I. Guttman. 1981. The role of host plants in the speciation of treehoppers: An example from the *Enchenopa binotata* complex. In R. F. Denno and H. Dingle (eds.), *Insect life history patterns*, pp. 39–54. Springer-Verlag, New York.

Wyatt, I. J. 1970. The distribution of *Myzus persicae* (Sulz.) on year-round chrysanthemums. II. Winter season: The effect of parasitism by *Aphidius matricariae* Hal. *Ann. Appl. Biol.* **65**:31–41.

Zimmerman, R. H. 1972. Juvenility and flowering in woody plants: A review. *HortScience* **7**:447–455.

PART II

Life History Strategies

CHAPTER 5

Boundaries to Life History Variation and Evolution

CONRAD A. ISTOCK
Department of Biology
University of Rochester
Rochester, New York

CONTENTS

1 PROLEGOMENON

The variety of life history patterns displayed by plant and animal species offers a profound challenge to ecological and evolutionary science. Not only do we wish to categorize or systematize life history phenomena, but we want also to elucidate the evolutionary processes that shape life history patterns, and thereby better understand their origins, course of evolution, and adaptive properties. Though most biologists would not doubt that Darwinian evolution is the explanation for life history differentiation in general, few would be so bold as to assert that we can say with any precision or coherence what the common rules are by which the genetic variation of populations interacts with natural selection to fashion the exquisite and widely diverse complexes of life-span, life expectancy, mating conventions, reproduction, and seasonality so readily apparent in nature.

Quite some time ago, L. C. Birch (1960) commented that "it would be a fascinating field of study to investigate the genetics of such evolutionary changes in life history patterns". That "field of study" has now progressed considerably (Charlesworth, 1980; Dingle et al., 1982; Hegmann and Dingle, 1982, Lande, 1982a,b; Etges, 1982; Istock, 1982a,b) in both its ecological and genetic aspects, but the empirical and theoretical conundrum that the field now faces is, as a consequence, all too clear. The primary purpose of this chapter is to explore the kinds and intricacies of the problems which must now be solved, and thereby dispel the notion that simple heuristic devices, such as r and K selection, or recourse to reasoning about "strategies," will be sufficient. This is not to say that these heuristics have not been worthwhile. Indeed they are part and parcel of the recently developed framework of evolutionary ecology which brought us to the present stage. I also want to argue that a better understanding of life history evolution is central to a deeper understanding of evolution as a process.

I think the process of evolution by natural selection as currently understood can be captured in the following duality.

Character variation subject to ecolological selection

Genetic variation and control of character expression

One side of the duality is the ecological aspect where differences in character states create differences in "fitness" among individuals. The other side is the hereditary encoding of character variations modified as the consequences of selection and leading, for example, to changes of allelic frequencies, altered regulation of gene expression, revision of polygenic architecture, or rearrangement of chromosomal geometry. The arrow in the middle is meant to convey the recurrent, perhaps relentless, interaction or reciprocal mapping between the two parts of the whole conception (Istock, 1970; Lewontin,

1974, 1982). Though this duality is deeply imbedded in thinking about evolution, and all of biology for that matter, it has its sharpest rendering when we recognize how integral the genetic variation, phenotypic expression, and selection of life history patterns are to the mechanism of evolution. The spread of birth and death phenomena in time and the variable participation of individuals in the birth and death events occurring as generations pass encompass all the transforming operations of selection on the genetics of a population. Every illation we care to make about evolution as a process has some equivalent rendering within this demographic framework. The more direct and penetrating our studies of the ecology and genetics of life history patterns, the better our understanding of evolution will be.

It is instructive to examine the place of life history characters in evolutionary change more closely. When phenotypic traits other than those directly representing demographic performance evolve, they do so because of their conditioning of birth and death probabilities. Some linking of life history traits with all other evolving traits must then exist; otherwise, phenotypic variations remain neutral. This coupling of life history with all other evolutionary change means that genetic variation for the life history characters themselves is fundamental to evolution. Hence, the direct measurement of amounts and kinds of heritable variation for life history characters present in natural populations is one of the most meaningful assessments of the raw material of evolution which we can make. Typically, such variation will be polygenic. Cumulative knowledge of the last decade suggests that such variation is reasonably abundant, widespread, and in some cases maintained by stabilizing forms of selection in fluctuating environments (Istock, 1982b, for a review). To the extent that heritable life history variation exists, a population is poised to respond to certain, but not all, kinds of directional or fluctuating selection. Theory we will shortly examine argues that the direction of response, as well as the rate of response, will be determined in part by the amounts of additive variance and covariance for the different life history characters. When genetic variance is unequal across different life history characters, ages, and life cycle stages, or when negative genetic covariances among traits exist, there will be a "boundedness," or set of constraints to the evolutionary process. Not only will all possible modifications of life history be out of the question (Stearns, 1976, 1977), but the genetic architecture will condition responses to specific directional or stabilizing selective forces.

2 A LEGACY FROM POPULATION GENETICS

Provine (1971) has thoroughly reviewed the development of theoretical population and evolutionary genetics from Mendel on to the period of the formal foundations established by Haldane, Fisher, Wright, and others. His chronicle and analysis move the story to about the late 1930s just prior to the development of the "modern synthesis" of evolutionary thought. Mayr and Provine (1980) and Mayr (1982) have treated subsequent developments in evolutionary biology. These books provide a comprehensive backdrop for some derivative remarks that I think are particularly relevant to the material of this chapter.

The resolution of the conflict between the "Mendelists" and the "biometricians" by East (1910), Nilsson-Ehle (1909), and Fisher (1918, 1930) actually had little effect on the subsequent course of development in theoretical population genetics. Despite the intuitive appeal of the demonstration that relatively few variable loci with additive influences on a trait lead to near continuity in the distribution of phenotypes, the particulate approach of the "Mendelists" prevailed and became the dominant approach to population genetic theory applied to evolutionary problems, and this population genetics was blended into the "modern synthesis" of evolutionary thought of the 1930s and 1940s. With a Kafkaesque twist, one side did indeed win afterall. Mather (e.g., Mather and Harrison, 1949) was primarily responsible for carrying on the less popular "biometrical" or quantitative genetics and empirical exploration of the "manifold effects of selection," a more complicated and cumbersome view of genetics and evolution. Studies of pleiotropy (Caspari, 1952) formed a persistent bridge between the two otherwise still rather separate traditions. As the aftermath to all of this, population genetics emerged principally as a consummately particulate science. Not only was there the focus on the beadlike Mendelian gene with all of of its power as an analytical paradigm, but a rather shakier focus on "units of selection" and "units of evolution" which emerged with the "modern synthesis." Even the polygene, as an imagined subset of all loci associated with a given metric trait, has a somewhat particulate formulation. Thinking about particles and units probably, more than anything, engendered some of the vague theoretical concepts that came to rest uneasily in the modern synthesis: prominent examples include the "unity of the genotype" or "cohesion" of gene pools (Mayr, 1963) and the postulate of coadaptation within genotypes or gene pools (Dobzhansky, 1970).

Holding close to the theoretical security of the particulate gene gave clearer meaning to notions about "units" and "amounts" of variation in natural populations, genotypes and phenotypes, "rates" of evolution as allelic frequency change, and "fitness." Larger scale phenotypic evolution, beyond the one-Mendelian-character sort, and attendant problems posed in units of selection and units of evolution remained ambiguous despite much ingenuity given to resolving them. Fitness joined the ranks of the ambiguous whenever discussion moved away from the security of the "p and q models". Augmenting the confusion, most evolutionary biologists recited regularly that "selection really operates at the level of the individual" although this conclusion in due course required modification to give theoretical embodiment to the evolution of sociality through kin selection (Hamilton, 1964).

In a natural and fascinating cascade beginning in the mid-1960s, population genetics exploited the descriptive techniques of molecular genetics and biochemistry to achieve a different kind of measurement of genetic variation, one often thought to be intrinsically more fundamental and closer to the particulate gene of classical models. Too much of such variation was found to be consistent with existing selection theory, and a beautiful theory of genetic neutrality due to Kimura, Nei, and others was fashioned to explain the anomalous genetic wealth and its source in mutation (Crow and Kimura, 1970; Lewontin, 1974). Evolutionary biology now had a compelling explanation of how the "rawest" of genetic material enters with regularity and considerable though constrained variety (the differential opportunity for neutral mutations at the three different codon positions), often lingers or leaves by chance in natural populations, and in so drifting also leaves behind reliable traces of relative amounts of past phyletic radiation on a fairly grand scale. An immensely important part of evolutionary biology, with surprising results, was thus opened (Lewontin, 1982). Unfortunately, understanding of the way natural selection molds the morphology, ecology, behavior, and development of life forms on earth was not furthered. Even the potential importance of stochastic processes, so prominent in the theory of neutrality for macromolecular variations, remains largely unexplored for the phenotypic evolution of important ecologial attributes.

Studies in quantitative genetics have brought a somewhat clearer theoretical and empirical grasp of the genetics and selection of organisms and populations since the revolution in molecular evolutionary genetics broke on the scene in the mid-1960s (Istock, 1982b), but our ability to integrate levels and kinds of analysis within evolutionary biology has not increased much. Old and huge problems still remain, such as where and what within genomic structures is this genetic variation reached by selection, how much of such accessible genetic variation is there in natural populations, how is it organized through all the layers from chromosome to operon or intron, and how freely are various levels and kinds of genetic variation modified in the remolding of phenotypic distributions by selection?

Throughout the development of formal theory in population genetics, only minimal attempts to characterize "the environment" were made. Usually it was assumed constant, occasionally it came in a few distinct forms alternating in time (Haldane and Jayakar, 1963) or space (Levene, 1953). Certainly mathematical necessity played its correct role in many instances, but convenience and necessity are hard to separate in such work, and convenience may be a substitute for deeper insight and greater effort. Without more specific renderings of phenotype distributions and their interrelations with surrounding environmental settings it will be impossible to extend modeling of the central duality of evolutionary theory in one piece.

3   A LEGACY FROM POPULATION ECOLOGY

A partial history of ecological thought about the structure and dynamics of populations is represented in the consecutive theoretical writings of Euler

(1760), Verhulst (1838), Lotka (1925), Volterra (1926), Leslie (1945, 1948), Cole (1954), and Hutchinson (1957). Age structure, resources for the growth and maintenance of populations, species interaction, and abiotic environmental limits emerge as fundamental themes. Age structure is preeminent in the Euler, Lotka, Leslie, and Cole contributions. Homogenous populations growing on finite, homogenous, and time-independent resource supplies characterize the Verhulst and Volterra formulations, as well as some of Lotka's. Species interaction is a strong focus for parts of the work of Lotka and Volterra, and a rich theoretical tradition has followed from their use of simultaneous differential equations to represent interacting species, with MacArthur (1972) recently carrying this tradition somewhat further. Emphasis on either a simple resource base or interactions of only a few species at once marks another example of the difficulties population theorists have recognized in attempts to incorporate "environments" within general theoretical formulations. Only with the Lotka and Hutchinson contributions does the elusive question of meaningful environmental representation see subtantial discussion, and both include relatively simple abiotic, or physical, and biotic, or species interaction, aspects. There is some discussion of changing environments within the traditions of population ecology, but it was Levins (1968), Murphy (1968), and Schaeffer (1974a,b) who recently gave this aspect of environment a sharpened focus. Still, the worst ghost in the closet, an uncertainly fluctuating environment, was only beginning to creep out in the work of Murphy and Schaeffer.

Age structure is the *sine qua non* of a life history whether encapsulated in the $l(x)$, $m(x)$ formulation, the elements of a Leslie matrix, or a simple two-stage life cycle. The life line and progeny sequence of an individual, or the average vital statistics of a whole population, provide a compact summary of the temporal and age-specific occurrence of birth and death events. Recently, Tuljapurkar and Orzack (1980) and Tuljapurkar (1982a,b) have given dynamic form to such formulations, one which also allows environmental change and uncertainty and offers new definitions of fitness, the logaritmic growth rate of a population and related measures. Irrespective of specific mathematical form, a combining of a demographically extensive representation with a means to rigorously portray environmental variation is a bold and essential step toward sufficient ecological realism on the character variation side of the evolution duality. Is such complexity necessary? Apparently so. The simpler formulations to such problems involving: the r–K dichotomy (MacArthur and Wilson, 1967), cost–benefit models (Gadgil and Bossert, 1970), or populations with minimal age structure in fluctuating environments (Schaeffer, 1974a,b) along with many other related papers have clarified important ideas but failed as yet to engender a steady progression of theoretical advance. Though we should do it grudgingly, we are likely to have to increase the complexity of our models. In fact this has been happening even in the purely ecological models of life history evolution (Istock, 1967; Emlen, 1970; Rickleffs, 1977, 1981; Caswell, 1982a,b; though see Crandall and Stearns, 1982, for a different view). Perhaps this tendency to increasing ecological complexity recognizes that models that are too general may defy testing (Deniston, 1978; Emlen, 1980) and surely it is time to take the testing of life history theory seriously (Barclay and Gregory, 1982), though the heuristic value of highly general models also should not be underestimated.

Environments, and the impress they make on the vital rates of a population, set and reset the framework for selection. Just as the underlying genetic structure may create boundaries and directions to evolution, so may the specific coupling of life history with environment establish a second kind of "boundedness" on the opposite side of the duality, one that proscribes, at least for a time, the domain within which selection may alter the life history and correlated features of population and individual (Hairston et al., 1970), with the whole process subject to greater or lesser stochastic influence.

4 THE JOINING OF POPULATION GENETICS AND ECOLOGY

A few years ago Lewontin (1979, p. 3) opened a paper with these sentiments. "Despite the pious hopes and intellectual convictions of evolutionary geneticists and ecologists, evolutionary genetics and ecology remain essentially separate disciplines, traveling separate paths while politely nodding to each other as they pass. . . . The marriage of population genetics and evolution was declared to be the social event of the season some years ago, but it so far has been without issue and, in fact, has yet to be consummated."

Clearly, the traditions of ecology most essential to the "marriage" are those of population ecology and most specifically life history analyses. Although Lewontin's statements do reflect the impediments inherent in the legacy of early population genetics and the recent and all but consuming excursion of population genetics into molecular realms, as well as the quite separately preoccupied legacy of population ecology, they are not altogether accurate. There has, "in fact," been an accumulation of theoretical and empirical papers specifically incorporating and interrelating genetic and ecological components which extend beyond "pious hopes" (e.g., Levene, 1953; Clarke et al., 1961; Campbell, 1962; Haldane and Jayaker, 1963; Dawson, 1965a,b, 1966, 1975, 1977; Levins, 1968; Murphy, 1968; Ayala, 1967; Morris and Fulton, 1970; Prout, 1968, 1971a,b; Slatkin, 1970; Anderson and King, 1970; Falconer, 1971; Mukai and Yamazaki, 1971; Bulmer, 1971; Morris, 1971; King and Anderson, 1971; Anderson, 1971; Roughgarden, 1971; Charlesworth and Giesel, 1972; Nash and Kidwell, 1972; Perrins and Jones, 1974; Taylor et al., 1974; Schaeffer, 1974a,b; Chapco, 1977, 1979; Dingle, Chapter 6; Dingle et al., 1977, 1982; Deniston, 1978; Tauber and Tauber, 1978; Istock, 1970, 1978, 1981, 1982a,b; Lande, 1980, 1982a,b; Livdahl, 1979; Charlesworth, 1980; Emlen, 1980; Doyle and Hunte, 1981a,b; Rose and Charlesworth, 1981a,b; Doyle and Myers, 1982; King, 1982; Rose, 1982; Giesel et al., 1982; Hegmann and Dingle, 1982; Tuljapurkar, 1982b, Caswell, 1982c).

This research literature, of which I have cited only a part, is actually quite large, but presently lacks the coherence, or rigorous and consistent crossing of paths, which must eventually take place if population genetics and ecology are to be combined into a richer and deeper evolutionary theory. It does appear, however, that such a development is under way.

The magnitude, and one of the central issues, of the problem posed for a synthesis of ecology and population genetics can be made clearer by considering a case from Lewontin (1974, p. 50, Fig. 3). The original data are from studies of *Drosophila pseudoobscura* stocks from Bogota, Colombia, made by Dobzhanksy (1963). Dobzhansky extracted large numbers of individual second chromosome examples from wild material, and tested them for viability (egg to adult survival) in homozygous and heterozygous combinations. Lewontin summarizes these relative viability estimates for 208 second chromosome homozygotes and 209 second chromosome heterozygotes. The viabilities range from lethality to ones so unusually high as to be labeled supervital. Both homozygotes and heterozygotes scored at the highest levels, but overall, heterozygotes scored higher than homozygotes, and there were no lethal heterozygotes. Lewontin presents the analysis as one typical of other such findings, namely, that there is wide genetic variation for viability and that heterozygotes have higher average survival. Here is a case where genetic constitution and life history have been studied in the laboratory, albeit in a limited way. Can we readily take the resulting conclusions as an inference about nature and the relative fates of specific homozygotes or heterozygotes, or about all heterozygotes vs. all homozygotes? It would be risky because, in fact, variability arising from only one of the chromosomes has been tested, and with interaction against only one genetic background, the results pertain to only one environment with no calibration to a natural environment, and only one life history character for a part of the life cycle has been studied. An important feature of Dobyzhansky's study is that it shows that a wide field of genetic variability for survival rates might exist. There is, however, no way to gauge, from data of the form provided in Dodzhansky's study, whether the observed variability is potentially subject to selection. To assess adequately the relative life history performance of any one genotype in several characters and in many different and possibly fluctuating environments, and to make such comparisons against variable genetic backgrounds seems exceedingly difficult. Yet, something like this is required. Another possible approach lies through quantitative genetic techniques.

Many of the papers cited in the long list near the beginning of this section have used the statistical techniques of quantitative or biometrical genetics to assess the presence and response to selection of genetic variation for life history traits. As a preliminary venture these efforts have been remarkably instructive (Istock, 1982b). There does seem to be widespread heritable variation for fertility, survival, and especially developmental rates. These traits appear to be moderately heritable relative to morphological traits, and frequently show quite different heritability estimates between traits or between ages for expression of the same life history trait. Both positive and negative genetic covariances have been found. In a number of cases the responses of such variation has been studied in artificial selection experiments. A few cases can support the inference that genetic variation and variable life history responses in nature are related (Morris, 1971; Perrins and Jones, 1974; Dingle et al., 1977, 1982; Istock, 1978; Palmer, 1983).

Despite its abstract formulations, there is a naturalness to a quantitative genetic approach to life history variation. As natural selection goes about the modifying of phenotypic distributions it performs alterations of underlying genetic structure "as if" it were a husbandman with one or another set of specific phenotypic modifications as a goal. Natural selection, as husbandman, may vacillate a good bit in its goals. With metric traits, the progress of selection, like the success of the husbandman, depends not on the details of specific gene expressions, but on the aggregate consequence of all the gene expressions. The measurements of quantitative genetics, additive variance, covariance, heritability, and so on, do nothing more than assess these aggregate effects. However, the imaginary aspects of quantitative genetic theory, such as polygenes and "average effects," have already led to new theoretical suggestions about important departures from classical population genetic theory which may occur when inheritance is polygenic. Lande (1976, 1977) has postulated a potentially more powerful role for mutation and explored interesting ways in which polygenic variation may be recurrently concealed and released with the making and breaking of linkage relationships. Similarly, Turelli (1982) has produced a model suggesting that under linkage disequilibrium *cis–trans* changes in polygenic subsets can cause heterozygotes that are otherwise genetically identical to have differences in fitness or associated life history components. Rose (1982) has offered one- and two-locus models suggesting that "antagonistic pleiotropy" (specific gene expressions that physiologically create opposing effects on different life history components) may preserve additive genetic variance in such traits.

Thinking back to the results of Dobzhansky's study, a quite different way of interpreting the observed deviations in survival for different wild chromosomes is possible. We can view the viability deviations as the consequence of "many genes of small effect" in the language of quantitative genetics. Of course, allelic representatives of many such genes must be packaged together temporarily within the same chromosome. However, in the fullness of nature, each allele enters into many different suites of alleles presumably affecting viability in different ways, and as individual "genes of small effect" many alleles may alternately contribute positive and negative deviations of phenotype depending on the frequencies of specific environments in concert with particular genetic backgrounds. The biochemical individuality of the expression of each allele may not change (though pre- or posttranslational modification might cause such), but its developmental consequences can still undergo a situational fluctuation. As I have suggested elsewhere, the notations "lethal," "quasi-normal," "normal," and "supervital" are then also not soley properties of an allele or chromosome, but properties of the tripartite interaction of gene, background, and environment (Istock, 1982b), an interaction the complexity of which we are only beginning to appreciate and measure. The idea that heritabilities of fitness characters are typically lower than those for morphological traits (Falconer, 1981; O'Donald, 1971; Mather, 1973) because selection reduces the additive variance of fitness traits most rapidly is appealing, but needs some kind of experimental testing. It is possible that fitness traits are merely subject to much random deviation, which inflates phenotypic variance during immature development and the conduct of adult life. And still differently, the additive variance of fitness traits may be more tightly controlled by fluctuating or stabilizing selection in nature.

A cautionary note concludes this section. Slavish measurement of heritabilities or related statistics for their own sake is a waste of time and effort. The pressing need for a more explicit and and more coherent body of fact and theory requires that studies in life history genetics be framed in some larger and meaningful context which makes the achievment of quantitative genetic measurements worth the anguish. Many of the papers cited in this section vividly illustrate the value of conceptually well set explorations and experiments. Theoretical considerations in the next section bring out the details of some contexts of enlarged scope.

5 QUANTITATIVE GENETIC THEORIES FOR LIFE HISTORY EVOLUTION

Emlen (1980) has suggested that three types of evolutionary models be distinguished: (1) those incorporating genetic descriptions cast in terms of single gene loci and including interaction of loci through linkage and epistasis, (2) quantitative genetic models using standard statistical partitioning of variances and with natural selection patterned after the truncation process of artificial selection, and (3) phenotypic selection models with emphasis on the evolution of population growth characteristics and optimization of components of reproduction and fitness. These distinctions are, however, not easily made in practice. Models with pleiotropy join types (1) and (2) (Rose, 1982), models created by Emlen (1980), Lande (1982a,b) and Istock (1981, 1982a) contain elements of (2) and (3), whereas models produced by Charlesworth (1980), Livdahl (1979), Rose (1982), Tuljapurkar (1982b), and Orzack (unpublished) combine (1) and (3). Details of two models employing demographic representation of life history and quantitative genetics will be discussed here because they help us explore some central considerations in the study of life history evolution.

5.1 Lande's Selection Gradient Model

Lande (1982a,b) offers a theoretical exploration of the dynamics of life history evolution which in outline is as follows. The life history of any individual is represented by a column vector Z composed of a suite of life history characters many of which may be age-specific. The population is represented by a comparable mean column vector $\bar{Z}$ with elements $\bar{z}_i$. Taking the intrinsic rate of increase r as fitness, natural selection is depicted as the gradient vector of r with respect to all changes in the mean values of the life history characters or

$$\nabla r = \left(\frac{\partial r}{\partial \bar{z}_i}, \ldots, \frac{\partial r}{\partial \bar{z}_n} \right)^{\mathrm{T}}$$

where T indicates vector transposition. Then the rate of change in $\bar{z}$ will be

$$\frac{d\bar{Z}}{dt} = G\nabla r$$

with G the genetic variance–covariance matrix for $\bar{Z}$. Thus, the change of the whole life history suite at once is represented. The approximate rate of change in the ith life history character is given as

$$\frac{d\bar{z}_i}{dt} = \sum_{j=1}^{h} G_{ij} \frac{\partial r}{\partial \bar{z}_j}$$

where G_{ij} is the additive genetic covariance between life history traits i and j. On the right-hand side, the partial derivative is the magnitude of change in fitness with change in $\bar{z}_j$, or the potential for fitness change due to the jth character, including i. The G_{ij} control the extent to which these potential changes are allowable within the system of genetic variances and covariances. Thus, the elements of the selection gradient and the variance–covariance matrix respectively depict (1) the potential for change due to each life history character in relation to fitness, and (2) the pattern of the boundedness, or constraint, developed from within the system of polygenic inheritance. Together these features provide an internally consistent way of expressing the evolution duality in the context of life history evolution.

    The appealing simplicity of the above treatment is somewhat misleading. Lande's treatment requires several special assumptions prior to obtaining the above model. He assumes multivariate normality, after some scale transformation if necessary, for all the life history traits, a stable age distribution to preserve the integrity of r, and weak selection so that the demographic processes are stronger than the selection imposed, say by environmental change, from outside the described ecological–genetic system as initially constituted. The model's rather obtuse rendering of environmental influence via the gradient vector makes it less explicit from an ecological standpoint than one might wish. Put another way, there is no indication of how the effect of any particular environmental configuration might actually be measured.

In summary, Lande (1982b) concludes that "as the primary phenotypic characters evolve toward a joint optimum the additive genetic correlations between major components are expected to decrease. . . , depleting the additive genetic variance in fitness and increasing the proportion of nonadditive variance in fitness."

5.2   The Regulative Value Model

Though starting from quite different initial assumptions and a different definition of fitness, I developed a discrete time model of selection on any life history trait (Istock, 1970, 1982a). The model has similarities and differences when compared with Lande's model. The first part of the model derives the portion of total population growth restriction, or "regulative value" of each age as

$$c_i = l_i m_i + \sum_{x=i}^{\omega} l'_{ix} M_x$$

where l_i is actual average survival to age i, l'_{ix} is an "unrealized" and higher survival possible for some genotype(s), m_i is "unrealized" fertility feasible for some genotype(s) that would otherwise express maximum fertility or progeny production as M_x, x is the index for summation over ages, and ω is the age of last reproduction. The logic of the derivation ensures that c_i measures the loss of potential population addition or replacement, and the loss of fitness, in units of net reproduction specifically ascribable to the composite birth and death statistics at age i. The construction portrays the determination of the average vital rates of the population, "as presently constituted," by the surrounding environment. Thus, the complete structure of selective forces is established largely, though not entirely, from the ecological side of the evolutionary duality. Genotypic potentials that complete the characterization of the age-specific selection pattern arise from additive genetic variance. Either equilibrium or nonequilibrium populations are allowed.

Now incorporating h_{ik}^2, equal to the ratio of additive and phenotypic variances, as the usual narrow sense heritability of the kth life history trait at the ith age, and ρ the phenotypic correlation of the kth trait at age i with all other traits m at all ages x, we construct the following expression:

$$s_{ik} = h_{ik}^2 \sum_m \sum_x c_x \rho_{ix,m}, \qquad \rho_{ii,m} = 1, \qquad h_{ik}^2 = \frac{V_{a,ik}}{V_{\rho,ik}}$$

for the "selective value" operating on the kth life history character at the ith age. As with c_i, s_{ik} carries units of net reproduction. Thus, s_{ik} measures the joint effect of both ecological and genetic constraints. Genetic constraints enter either through the h_{ik}^2 values representing limits imposed by available additive genetic variance or through negative ρ as restrictive phenotypic covariances among traits. But since ρ is the phenotypic correlation, it may also include underlying genetic covariance. The mean value of any life history trait changes over one generation in proportion to $s_{ik} V_{\rho,ik}$. The important difference between this model and Lande's on this point is that it is only the heritability of the trait k at age i which conditions the evolutionary response in a defined genetic sense, whereas additional constraints may arise from the full phenotypic correlation gathered over all other ages and traits. As an example, potential genetic change tending to raise the mean character value and reduce c_i at one age may engender sufficient loss of fitness in other ways as to be blocked purely because of its ecological consequences. As in Lande's model a narrowed interpretation of the evolutionary duality expressed in (1) and (2) of the previous section is appropriate for the present model. The c_i represent the potential for evolutionary modification, whereas the s_{ik} incorporate genetic boundedness or constraint, with the added possibility that purely ecological contraints may arise concurrently. It should also be noted that positive phenotypic correlations, perhaps supported in part by underlying genetic correlations, can accelerate and direct the evolution of given traits.

From an early version of the regulative value model (Istock, 1970), I concluded that evolutionary change of life history performance could occur at any and all ages simultaneously in populations with either increasing or decreasing mean fitness, ignoring possible negative genetic covariances, and conversely that in equilibrium populations an increase in life history performance at one age must be accompanied by a decrease at another. Employing purely ecological models of life history evolution, Rickleffs (1977, 1981) has portrayed such a process. Hence, it is possible that, barring genetic limitations, life histories may continue to evolve even in populations at numerical equilibrium.

The upper graph in Figure 1 depicts schematically the way in which age-specific profiles in the s_{ik} for life history traits, both central ones such as l_x and m_x, and more peripheral ones almost certainly correlated with l_x and m_x, might be envisioned for the regulative value model. With $s_{ik} > 0$ there is some free potential for life history adaptation. When $s_{ik} = 0$ no life history evolution is allowed, that is, it is blocked by the genetic and ecological constraints jointly. When $s_{ik} < 0$ for certain age–trait combinations, evolutionary change is maladaptive, and fitness contributions may be forced lower than originally at these ages. The lower graph in Figure 1 depicts the way in which either of the two models of this section accord with the intuitive notion of an adaptive landscape and evolution toward an optimum in two life history characters.

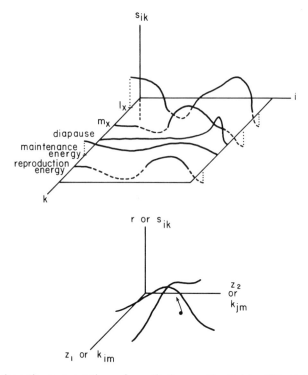

Figure 1. Schematic representations of quantitative genetic models of life history evolution. The upper panel depicts a set of possible relations between a suite of life history characters, age, and distributions of s_{ik} values. Solid portions of the distribution for each character indicate zero or positive s_{ik}. When the s_{ik} are positive the combined effect of ecological and genetic conditions promotes directional selection of that character. Negative s_{ik} indicate age–character combinations where negative ecological and genetic correlations within the whole system of ages and characters cause selection of the character in the direction counter to elevation of fitness. The lower panel sketches the adaptive topography and optimization process for two life history characters under either of the models discussed in Section 5.

Both of the above models allow the conclusion that negative genetic co-variance may influence the rate and possibly the course of life history evolution. Lande (1982b) asserts at the start of his discussion that negative genetic covariances might sometimes "prevent a net increase in adaptation," which is in conflict with the earlier quotation, given above from the same paper, in which he says that the additive genetic covariances can be expected to disappear as selection proceeds. It is of course certain that when the additive genetic variance of a trait disappears so must all additive genetic correlations with other traits. At present, scrutiny of either of the above quantitative genetic models has not gone far enough yet to resolve this point rigorously. In any event, study of the dynamics of genetic variances and covariances remains an important empirical and theoretical issue for our understanding of the genetic architecture underlying life history traits. Are there cases when the material basis of two or more polygenically inherited traits lies in such a rigidly organized physical architecture that recombination cannot take it apart and along with selection erase or reverse convariances? We know that both positive and negative genetic covariances arise (Tantawy and El-Helw, 1966; Istock, 1978, 1982b; Rose and Charlesworth, 1981a,b; Falconer, 1981), but experiments examining responses to contrary selection have not been done.

In a general sense the two models of this section fall within the framework provided by Deniston (1978) for the rate of change in the mean value of any character where

$$\frac{d\bar{x}}{dt} = CV_{rx} + \frac{\bar{d}\bar{x}}{dt}$$

and $\bar{x}$ is the mean of the character, CV_{rx} is the genetic covariance of character and fitness, and $\bar{d}\bar{x}/\bar{d}t$ is the average time derivative of the character imposed by environmental circumstance. The response in mean character value $\bar{x}$ may be positive or negative depending on CV_{rx}, as in the foregoing models.

These recent models may allow us to frame better questions in life history genetics and evolution, or at least to approach questions in evolutionary ecology with a clearer awareness of the limitations of present concepts and methods. I want to consider briefly some open questions by way of suggesting concrete entrées to future research.

6 LIFE-SPANS EVOLVING IN FLUCTUATING ENVIRONMENTS

Figure 2 shows an upper set of time-dependent, putatively important environmental variables. All but the uppermost one fluctuate with different periods, amplitudes, and serial autocorrelations. Below are life-spans or mean generation times of short to long durations. If one projects upward from each of the life-spans through the concurrent histories of the variously fluctuating environmental variables one observes that all life-spans must lead to different composite experiences with the environment as we chain succesive, constant life-spans of different duration one after another through time. What is the population to do by way of adapting to the "worst cases," or to the average impact, which these multiple buffetings by environmental fluctuation create? Where lies the optimum? The problem has not been solved, but it is probably now surmountable because of the theoretical work of Tuljapurkar and Orzack (1980), Tuljapurkar (1982a,b), and Orzack (unpublished) mentioned earlier. In principle, it should be possible to work out rather general rules for the interaction of quantitative genetic architecture and multiply fluctuating environments along the lines of the scenario in Figure 2. King (1982) has tackled the problem experimentally.

Analogous explorations having to do with dormancy or migration, re-

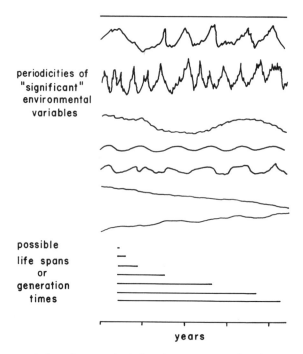

Figure 2. Representation of temporal relations between a set of environmental variables exerting significant influence on the vital statistics of populations and possible durations of lifespan or generation time. the main idea here is that life-spans of any duration coincide with a multiplicity of environmental variables differing in period, amplitude, and serial autocorrelation. The question can be raised whether optimization will occur to minimize environmentally induced demographic variation, to maximize fitness, or do something else.

source allocation along the course of a life cycle, the level and patterning of offspring production, semel vs. iteroparity, and study of the seasonality of life cycle staging may be approached anew with such a theoretical viewpoint.

7   THE EVOLUTION OF COMPLEX LIFE CYCLES

The semelparity–iteroparity divergence within life history patterns is undoubtedly one of the most enduring and intriguing of conceptual frameworks in ecology. The juxtaposition of simple, one-resource pool, life cycles with the ecologically complex life cycles with two or more distinct, temporally sequential resource pools provides another sharp dichotomy in biological life-style (Istock, 1967, 1970; Wilbur, 1980; and Chapter 7).

When I first explored theoretical aspects of complex life cycles, I implicitly made the evolutionary ecologist's usual assumption of infinite genetic variability for all required life history features (Istock, 1967). The first model suggested resource and numerical expansion as the adaptive inducement for initial development of a two-phase life cycle, and ensuing long-term instability of complex life cycles leading to subsequent reduction of the resource extraction of one phase and even complete secondary reversion to a simple form. Inevitable selection for simplification of complex life cycles, due to long-term genetic constraints, is still a reasonable expectation. We do, after all, observe many life cycle conditions in the natural world which are hard to explain except as secondary simplification (Istock, 1967). The instability conclusion was based on my inability to construct restoring feedback mechanisms within the partitioned demography of such organisms when one phase breaks away into density-independent, prolonged, directional evolution. The demographic imbalance arises because the phase breaking away becomes so much more abundant relative to the bottlenecking phase. It also seems likely that the phase undergoing the destabilizing evolution will often be the one that was added when the life cycle went initially from one to two phases. It simply goes on with too much of a good thing. However, various groups, particularly a variety of insects, cnidarians, and amphibians, display patterns suggesting reasonably long retention of the complex life cycle, certainly longer than the temporal duration of the average species, though perhaps not typically longer than phyletic development much past the family level. Admittedly, these are very rough guesses. The odonates remain my favorite exception to the instability postulate. Recently, studies of life history genetics have shed new light on the problem.

During the past decade empirical studies of the ecology and life history genetics of the pitcher-plant mosquito have uncovered persistent additive variance in development time and tendency to diapause in the latter half of the growing season when completion of one additional generation for the year becomes a tenuous enterprise depending for its success on the vagaries of year to year weather fluctuations. The variable diapause response seems to be a continually adjusted adaptation to environmental uncertainty, and coming in the latter part of larval life it consequently regulates the timing of the switch to the next phase of the complex life cycle (Istock et al., 1976; Istock, 1978, 1981). The pitcher-plant mosquito throughout its range, shows reduction of the complex life cycle in the form of a partial or complete loss of blood feeding in the adult phase, but carbohydrates from nectar are still important to adult longevity and completion of offspring production. The aquatic larvae share part of the smorgasbord trapped by the pitcher-plant. Genetic variability for characters associated with timing of the crucial transition in a complex life cycle could function to ameliorate selection imbalances created by unequal distribution of population control on opposite sides of this transition, the imbalance I postulated as the undoing of a complex life cycle (Istock, 1967).

By changing the relative durations of life proportioned over the two environments of the life cycle the proportional loading of selective forces that would otherwise change the life cycle are redressed. One has to imagine that the genetically conditioned distribution of transition, or metamorphosis, ages swings back and forth over the run of the generations in reaction to demographic and selective changes imposed by a postively autocorrelated environment. The genetic shifts in concert with fluctuating demographic changes provide a possible feedback of the kind I could not construct on purely ecological grounds. In the case of the pitcher-plant mosquito, the selective differences involved may be quite large (Istock, 1981). The suggestion, then, is that genetic variability might provide stabilization at least in the short term. There is also a suggestion here that a complex life cycle may not have a fixed optimum under some circumstances. Demonstration of the existence of additive genetic variance for the pupal duration of the fall webworm by Morris and Fulton (1970), and their interpretation of it as an adaptation to temporal variation in the time propitious for the switching of resources seems very close to the mosquito example.

On both theoretical and empirical grounds we are left with the prediction that populations with presently functional complex life cycles will also possess genetic variability with substantial, probably often dramatic and usually polygenic, influence on the way life-span is apportioned between life cycle phases. The interaction of individal and population in the continuous adjustment of diapause and life cycle are explored theoretically with a so-called lineage model in Istock (1981).

Models of the kind sketched in Section 5, or in the recent papers of Caswell (1982a,b) on the demography of complex life cycles should allow more careful analysis of this problem than has been achieved heretofore. It should be realized that Caswell does not define a complex life cycle that requires separate age-related resource pools and selective regimes (Istock, 1967; Wilbur, 1980), but his mathematical treatment is still of interest for the evolutionary problem. In the context of the regulative value model, the balance of selection on the phases of a complex life cycle is closely related to the respective proportions of the total sum of the age-specific regulative values and selective values falling to each phase.

8 DISCUSSION

The fascination to be found in life history genetics which Birch (1960) anticipated surely exists and an increasing number of population biologists find it irresistible. Preliminary models of life history evolution such as those examined here argue quite strongly that analyses in life history genetics are essential to evolutionary ecology and will perforce bring a substantial unification of population genetics and population ecology. Quantitative genetic models with demographic foundations incorporated to bring a semblence of environmental realism seem promising. Some will regret the loss of security, even a fall from grace, when models of genetic change in populations fail to include explicitly Mendelizing genes. But it is likely that important new genetic processes, not predictable from classical p and q genetics may be explored using quantitative genetics. Furthermore, there is a naturalness about the quantitative genetic formulations when one reflects on our general ideas about how natural selection and the phenotypic distributions of populations create the front line of genetic transformation. However, even if it turns out that p and q models are too simple to accommodate life history change and adaptation generally, they will probably provide important heuristic insight and approximation for some time to come just as they have in the past (Michod, 1981). Two additional observations are probably also important. Recalling the theory of Fisher (1918, 1930), which brought de jure, if not de facto, resolution of the Mendelist–biometrician spat, it is really possible to build up concrete theoretical representations of polygenic architectures of considerable complexity from discrete gene loci, alleles, and even chromosomes, at least if one is willing to use a computer. All manner of interesting processes such as recombination within and between polygenes, mutation and drift influencing polygenic content, polygenic geometry on chromosomes, tightness of linkage within polygenes, and so on through the gamut of classical considerations might thereby be studied. The second observation is that even today, and despite molecular genetic analogues, the Mendelian gene defined throughout history and population genetics by its one to one relation with some phenotypic character at the whole organism level remains an abstraction in its own right.

Enticing and important as the theoretical approaches to life history genetics and evolution are, experimental laboratory and field analyses are even more important at present. When the context for making genetic analyses of particular species and populations is clearly set in the existing conceptual framework of life history evolution we are likely to obtain valuable insight into the distribution of variable and nonvariable components of the life history and testable predictions about why we find things as they are. One of the most promising tests yet to be done would be a transplant experiment in nature with populations seeming to show meaningful divergent adaptation of life histories which are still genetically variable. Reciprocal or one-way latitudinal or altitudinal transplants seem feasible based on existing quantitative genetic analyses such as ours with the pitcher-plant mosquito (Istock et al., 1976). Many kinds of arduous laboratory investigations are possible with direct quantitative genetic analysis and selection experiments (e.g., Rose and Charlesworth, 1981a,b). Use of species amenable to field study in such studies is desirable because it will allow field testing of hypotheses.

The words "strategy" and "tactic" have become commonplace in ecological and evolutionary discourse, but their meanings and connotations as theoretical constructs are not clear. To some the word strategy is synonomous with phenotype (Eshel, 1982). Others mean tactic or strategy to imply a "set of coadapted traits, designed by natural selection, to solve particular ecological problems" (Stearns, 1976, 1980). Perhaps the most impressive use of the strategy idea in evolutionary biology is the "evolutionarily stable strategy'' or ESS of Maynard Smith and Price (1973). Reasoning by metaphor and analogy with concious human endeavors is always dangerous. If it is shown through future research that a definite genetic architecture has been molded by selection to produce coordinated sets of life history characters there might be justification for the metaphor. However, we do not yet have adequate demonstrations of such coordination, or coadaptation, in life history genetics, though I suspect that they exist. At the moment it is probably more prudent to admit that the concept of ''strategy'' usually has the same vagueness as the ''unity of the genotype'' or the coadaptation of the genotype. In fact, it is quite nearly the same idea.

The ''evolutionarily stable strategy'' based on reasoning and formal treatment from game theory (Maynard Smith and Price, 1973) is a strong form of strategy reasoning. Eshel (1982) has argued that it is sometimes supported by other forms of mathematical reasoning. Though, he has also suggested that the idea, its game-theoretic representation, and the implicit notion of optimization may break down in other instances. One puzzling aspect of ESS reasoning arises in applications to evolution because any arrangement of environment, selection, and population genetic structure closed to entry of new genetic types seems the opposite of evolution. If local populations frequently reach optimal and stable ''strategies'' in a joint genetic and ecological sense, it means that evolution is not continuous but episodic. If evolution is, to the contrary, more or less continuous, a quite different way of viewing life history phenomena is needed. Direct study of life history genetics and evolution seems the most appropriate avenue for an attack on this question.

Major features of life history such as life-span or generation time, semelparity vs. iteroparity, seasonality, dormancy, complex life cycles, and adaptation to fluctuating environments seem destined to receive far more searching theoretical and empirical analyses than ever before. The changes of genomic architecture underlying the evolution of even small suites of life history traits may be so extensive as to border on the processes of meso- and macroevolution. These too may be analyzed directly in due course. The yawning hiatus in our knowledge of what genomic structure consists of, and of the parts of it reached in various ways and times by selection, may with a good bit of luck become less of a hiatus.

The century-long legacy of population and quantitative genetics brings to life history studies a rich body of thought, experiment, and technique. It offers prospects for analyzing boundaries to life history evolution created by the genetic variation and genetic processes which condition adaptation.

As it was with Cole's (1954) brilliant early paper on the population consequences of life history phenomena, the ecologist's approach to the joining of ecology and genetics will carry the insistence that populations, life histories, and environments have many clearly delineated, conceptually interesting, and too often heterogenous properties—properties that must be encompassed by evolutionary models and empirical analyses with a disconcerting richness of variety and complexity. That is the ecologist's 200-year legacy.

9 SUMMARY

The historically separate development of population genetics and population ecology has tended to retard the development of a common science of evolution. Recently, the dramatic expansion of population genetics into areas of molecular evolution has also reduced interaction between geneticists and ecologists on a large scale. Despite these antecedent conditions, there has been a steadily increasing interest in the joining of ecological and genetic studies of evolution through the analysis of life history phenomena.

Models involving both genetic and ecological considerations suggest that constraints on life history evolution may arise from both environmental conditions and insufficient additive genetic variance or negative genetic covariances among life history traits. In any case, the evolution of life histories seems central to an increased understanding of the duality implicit in current evolutionary thought: environmentally conditioned selection on expressed phenotypes interacting reciprocally with genetic change and genetic control of these phenotypes. It appears possible to encompass this duality within moderately general models of life history evolution and to subject life history models and phenomena to increasingly good empirical analysis. Accurate and comprehensive empirical analyses will probably be most crucial to progress in the near future.

Conventional reasoning about ecological "strategies" has contributed much to the development of our current framework for studies of life histories, but the vagueness and possible theoretical weaknesses of such ideas bear careful and searching scrutiny from now on. Explicit problems such as the evolution of semelparity vs. iteroparity, the evolution of life-span, and the evolution of complex life cycles offer examples of the many, ecologically rich, and specific evolutionary problems that life history studies provide.

ACKNOWLEDGMENTS

I am grateful to William Etges, Joseph Warnick, Ernst Caspari, Tom Caraco, and Uzi Nur for many helpful discussions of the material in this chapter, and to several of them for critical reading of the manuscript. Parts of this work were supported by grants from the National Science Foundation.

LITERATURE CITED

Anderson, W. W. 1971. Genetic equilibrium and population growth under density-regulated selection. *Am. Nat.* **105**:489–498.

Anderson, W. W., and C. E. King. 1970. Age-specific selection. *Proc. Natl. Acad. Sci. USA* **66**:780–786.

Ayala, F. 1967. Evolution of fitness III. Improvement of fitness in irradiated populations of *Drosophila serrata*. *Genetics* **58**:1919–1923.

Barclay, H. J., and P. T. Gregory. 1982. An experimental test of life history evolution using *Drosophila melanogaster* and *Hyla regilla*. *Am. Nat.* **120**:26–40.

Birch, L. C. 1960. The genetic factor in population ecology. *Am. Nat.* **94**:5–24.

Bulmer, M. G. 1971. The effect of selection on genetic variability. *Am. Nat.* **105**:201–211.

Campbell, I. M. 1962. Reproductive capacity in the genus *Choristoneura* Led. (Lepidoptera: Tortricidae). I. Quantitative inheritance and genes as controllers of rates. *Can. J. Genet. Cytol.* **4**:272–288.

Caspari, E. 1952. Pleiotropic gene action. *Evolution* **6**:1–18.

Caswell, H. 1982a. Optimal life histories and the maximization of reproductive value: A general theorem for complex life cycles. *Ecology* **63**:1218–1222.

Caswell, H. 1982b. Stable population structure and reproductive value for populations with complex life cycles. *Ecology* **63**:1223–1231.

Caswell, H. 1982c. Life history theory and the equilibrium status of populations. *Am. Nat.* **120**:317–339.

Chapco, W. 1977. Correlations between chromosome segments and fitness in *Drosophila melanogaster*. I. The X-chromosome and egg production. *Genetics* **85**:721–732.

Chapco, W. 1979. Correlations between chromosome segments and fitness in *Drosophila melanogaster*. II. The X chromosome and egg viability. *Genetics* **92**:595–601.

Charlesworth, B. 1980. *Evolution in age-structured populations*. Cambridge University Press, New York.

Charlesworth, B., and J. T. Giesel. 1972. Selection in populations with overlapping generations: IV. Fluctuations in gene frequency with density-dependent selection. *Am. Nat.* **106**:402–411.

Clarke, J. M., J. Maynard Smith, and K. C. Sondhi. 1961. Asymmetrical response to selection for rate of development in *Drosophila subobscura*. *Genet. Res. Camb.* **2**:70–81.

Cole, L. C. 1954. The population consequences of life history phenomena. *Q. Rev. Biol.* **29**:103–137.

Crandall, R. E., and S. C. Stearns. 1982. Variational models of life-histories: When do solutions exist. *Theor. Pop. Biol.* **21**:11–23.

Crow, J. F., and M. Kimura. 1970. *Introduction to population genetics theory*. Harper and Row, New York.

Dawson, P. S. 1965a. Genetic homeostasis and developmental rate in *Tribolium*. *Genetics* **51**:873–885.

Dawson, P. S. 1965b. Estimates of components of phenotypic variance for development rate in *Tribolium*. *Heredity* **20**:403–417.

Dawson, P. S. 1966. Developmental rate and competitive ability of *Tribolium*. *Evolution* **20**:104–116.

Dawson, P. S. 1975. Directional versus stabilizing selection for development time in natural and laboratory populations of flour beetles. *Genetics* **80**:773–783.

Dawson, P. S. 1977. Evolutionary adjustment of development time in mixed populations of flour beetles. *Nature (London)* **270**:340–341.

Deniston, C. 1978. An incorrect definition of fitness revisited. *Ann. Hum. Genet.* **42**:77–85.

Dingle, H., C. K. Brown, and J. P. Hegmann. 1977. The nature of the genetic variance influencing photoperiodic diapause ina migrant insect. *Oncopeltus fasciatus*. *Am. Nat.* **111**:1047–1059.

Dingle, H., W. S. Blau, C. K. Brown, and J. P. Hegmann. 1982. Population crosses and the genetic structure of milkweed bug life histories. In H. Dingle and J. P. Hegmann (eds.), *Evolution and genetics of life histories*. Springer-Verlag, New York.

Dobzhansky, Th. 1963. Genetics of natural populations. XXXIII. A progress report on genetic changes in populations of *Drosophila pseudoobscura* and *Drosophila persimilis* in a locality in California. *Evolution* **17**:333–339.

Dobzhansky, Th. 1970. *Genetics of the Evolutionary Process*. Columbia Univ. Press, New York.

Doyle, R. W., and W. Hunte. 1981a. Genetic changes in the components of "fitness" and yield of a crustacean population in a controlled environment. *J. Exp. Mar. Biol. Ecol.* **52**:147–156.

Doyle, R. W., and W. Hunte. 1981b. Demography of an estuarine amphipod (*Gammarus lawrencianus*) experimentally selected for high "r": A model of the genetic effects of environmental change. *Can. J. Fish. Aquat. Sci.* **38**:1120–1127.

Doyle, R. W., and R. A. Myers. 1982. The measurement of the direct and indirect intensities of natural selection. In H. Dingle and J. P. Hegmann (eds.), *Evolution and genetics of life histories*. Springer-Verlag, New York.

East, E. M. 1910. A Mendelian interpretation of variation that is apparently continuous. *Am. Nat.* **44**:65–82.

Emlen, J. M. 1970. Age specificity and ecological theory. *Ecology* **51**:588–601.

Emlen, J. M. 1980. A phenotypic model for the evolution of ecological characters. *Theor. Pop. Biol.* **17**:190–200.

Eshel, I. 1982. Evolutionary stable strategies and viability selection in Mendelian populations. *Theor. Pop. Biol.* **22**:204–217.

Etges, W. E. 1982. "A new view of life history evolution"?—A response. *Oikos* **38**:118–122.

Euler, L. 1760. Recherches generales sur la mortalite et la multiplication du genre humain. *Mem. Acad. Sci., Berlin* **16**:144–164. In *Opera Omnia Ser. 1, 7*:79–100.

Falconer, D. S. 1971. Improvement of litter size in a strain of mice at a selection limit. *Genet. Res. Camb.* **17**:215–235.

Falconer, D. S. 1981. *Introduction to quantitative genetics*. Longman, New York. 340 pp.

Fisher, R. A. 1918. The correlation between relatives on the supposition of Mendelian inheritance. *Trans. R. Soc. Edinburgh* **52**:399–433.

Fisher, R. A. 1930. *The genetical theory of natural selection*. Oxford Univ. Press, New York. 291 pp.

Gadgil, M., and W. H. Bossert. 1970. Life historical consequences of natural selection. *Am. Nat.* **104**:104–124.

Giesel, J. T., P. A. Murphy, and M. N. Manlove. 1982. The influence of temperature on genetic interrelationships of life history traits in a population of *Drosophila melanogaster*: What tangled data sets we weave. *Am. Nat.* **119**:464–479.

Hairston, N. G., D. W. Tinkle, and H. M. Wilbur. 1970. Natural selection and the parameters of population growth. *J. Wildl. Manage.* **34**:681–689.

Haldane, J. B. S., and S. D. Jayakar. 1963. Polymorphism due to selection of varying direction. *J. Genet.* **58**:237–242.

Hamilton, W. D. 1964. The genetical theory of social behavior I, II. *J. Theor. Biol.* **7**:1–52.

Hegmann, J. P., and H. Dingle. 1982. Phenotypic and genetic covariance structure in milkweed bug life histories. In H. Dingle and J. P. Hegmann (eds.), *Evolution and genetics of life histories*, Springer-Verlag, New York.

Hutchinson, G. E. 1957. Concluding Remarks. In *Population Studies: Animal Ecology and Demography. Cold. Spring Harb. Symp. Quant. Biol., 22*. The Biological Laboratory, Cold Spring Harbor. 437 pp.

Istock, C. A. 1967. The evolution of complex life cycle phenomena: An ecological perspective. *Evolution* **21**:592–605.

Istock, C. A. 1970. Natural selection in ecologically and genetically defined populations. *Behav. Sci.* **15**:101–115.

Istock, C. A. 1978. Fitness variation in a natural population. In H. Dingle (ed.), *Evolution of Insect Migration and Diapause*. Springer-Verlag, New York.

Istock, C. A. 1981. Natural selection and life history variation: theory plus lessons from a mosquito. In R. Denno and H. Dingle (eds.), *Species and life history patterns*. Springer-Verlag, New York.

Istock, C. A. 1982a. Some theoretical considerations concerning life history evolution. In H. Dingle and J. Hegmann (eds.), *Evolution and genetics of life histories*. Springer-Verlag, New York.

Istock, C. A. 1982b. The extent and consequences of heritable variation for fitness characters. In C. R. King and P. S. Dawson (eds.), *Population biology: Retrospect and prospect*. Columbia University Press, New York.

Istock, C. A., J. Zisfein, and K. Vavra. 1976. Ecology and evolution of the pitcher-plant mosquito. 2. The substructure of fitness. *Evolution* **30**:535–547.

King, C. E. 1982. The evolution of life span. In H. Dingle and J. P. Hegmann (eds.), *Evolution and genetics of life histories*. Springer-Verlag, New York.

King, C. E., and W. W. Anderson. 1971. Age-specific selection. II. The interaction between r and K during population growth. *Am. Nat.* **105**:137–156.

Lande, R. 1976. The maintenance of genetic variability by mutation in a polygenic character with linked loci. *Genet. Res. Camb.* **26**:221–235.

Lande, R. 1977. The influence of the mating system in the maintenance of genetic variability in polygenic characters. *Genetics* **86**:485–498.

Lande, R. 1980. The genetic covariance between characters maintained by pleiotropic mutations. *Genetics* **94**:203–215.

Lande, R. 1982a. Elements of a quantitative genetic model of life history evolution. In H. Dingle and J. P. Hegmann (eds.), *Evolution and genetics of life histories*. Springer-Verlag, New York. 249 pp.

Lande, R. 1982b. A quantitative genetic theory of life history evolution. *Ecology* **63**:607–615.

Leslie, P. H. 1945. On the use of matrices in certain population mathematics. *Biometrika* **33**:183–212.

Leslie, P. H. 1948. Some further notes on the use of matrices in population mathematics. *Biometrika* **35**:213–245.

Levene, H. 1953. Genetic equilibrium when more than one niche is available. *Am. Nat.* **87**:331–333.

Levins, R. 1968. *Evolution in changing environments*. Princeton Univ. Press, Princeton, New Jersey. 120 pp.

Lewontin, R. C. 1974. *The genetic basis of evolutionary change*. Columbia University Press, New York. 346 pp.

Lewontin, R. C. 1979. Fitness, survival and optimality. In D. J. Horn, G. R. Stairs, and R. D. Mitchell (eds.), *Analysis of Ecological Systems*. Ohio State University Press, Columbus. 312 pp.

Lewontin, R. C. 1982. Prospectives, perspectives, and retrospectives. *Paleobiology* **8**:309–313.

Livdahl, T. P. 1979. Environmental uncertainty and selection for life cycle delays in opportunistic species. *Am. Nat.* **113**:835–842.

Lotka, A. J. 1925. *Elements of physical biology*. Dover, New York. 465 pp.

MacArthur, R. H. 1972. *Geographical ecology*. Harper and Row, New York. 269 pp.

MacArthur, R. H., and E. O. Wilson. 1967. *The theory of island biogeogrphy.* Princeton University Press, Princeton, New Jersey. 203 pp.

Mather, K. 1973. *Genetical structure of populations.* Chapman and Hall, London.

Mather, K., and B. J. Harrison. 1949. The manifold effect of selection. *Heredity* **3**:1–52 and 131–162.

Maynard Smith, J., and G. R. Price. 1973. The logic of animal conflict. *Nature (London)* **246**:15–18.

Mayr, E. 1963. *Animal species and evolution.* Harvard University Press, Cambridge, Massachusetts. 797 pp.

Mayr, E. 1982. *The growth of biological thought.* Harvard University Press, Cambridge, Massachusetts.

Mayr, E., and W. Provine. 1980. *The evolutionary synthesis.* Harvard University Press, Cambridge, Massachusetts.

Michod, R. E. 1981. Positive heuristics in evolutionary biology. *Br. J. Phil. Sci.* **32**:1–36.

Morris, R. F. 1971. Observed and simulated changes in genetic quality in natural populations of *Hyphantria cunea. Can. Entomol.* **103**:893–906.

Morris, R. F., and W. C. Fulton. 1970. Heritabilty of diapause intensity in Hyphantria cunea and related fitness responses. *Can. Entomol.* **102**:927–938.

Mukai, T., and T. Yamazaki. 1971. The genetic structure of natural populations of *Drosophila melanogaster.* X. Developmental time and viability. *Genetics* **69**:385–398.

Murphy, G. I. 1968. Patterns in life history and environment. *Am. Nat.* **102**:391–403.

Nash, D. J., and J. F. Kidwell. 1972. A genetic analysis of lifespan, fecundity, and weight in the mouse. *J. Hered.* **64**:87–90.

Nilsson-Ehle, H. 1909. Kreuzungsuntersuchungen an Hafer und Weizen. *Lunds Universitets Arsskrift N. S.,* Ser. 2, Vol. 5, no. 2.

O'Donald, P. 1971. Natural selection for quantitative characters. *Heredity* **27**:137–153.

Orzack, S. H. (unpublished ms.). Population dynamics in variable environments. V. The genetics of homeostasis revisited.

Palmer, J. O. 1983. Photoperiodic control of reproduction in the milkweed leaf beetle, *Labidomera clavicollis. Physiol. Ent.* (in press).

Perrins, C. M., and P. J. Jones. 1974. The inheritance of clutch size in the great tit (*Parus major* L.). *Condor* **76**:225–229.

Prout, T. 1968. Sufficient conditions for a multiple niche polymorphism. *Am. Nat.* **102**:493–496.

Prout, T. 1971a. The relation between fitness components and population prediction in *Drosophila.* I. The estimation of fitness components. *Genetics* **68**:127–149.

Prout, T. 1971b. The relation between fitness components and population prediction in *Drosphila.* II. Population prediction. *Genetics* **68**:151–167.

Provine, W. B. 1971. *The origins of theoretical population genetics.* University of Chicago Press, Chicago.

Rickleffs, R. E. 1977. On the evolution of reproductive strategies in birds: Reproductive effort. *Am. Nat.* **111**:453–478.

Rickleffs, R. E. 1981. The optimization of life-history patterns under density dependence. *Am. Nat.* **117**:403–408.

Rose, M. R. 1982. Antagonistic pleitrophy, dominance and genetic variation. *Heredity* **48**:63–78.

Rose, M. R., and B. Charlesworth. 1981a. Genetics of life history in *Drosophila melanogaster.* I. Sib analysis of adult females. *Genetics* **97**:173–186.

Rose, M. R., and B. Charlesworth. 1981b. Genetics of life history in *Drosophila melanogaster*. II. Exploratory selection experiments. *Genetics* **97**:187–196.

Roughgarden, J. 1971. Density-dependent natural selection. *Ecology* **52**:453–468.

Schaeffer, W. M. 1974a. Optimal reproductive effort in fluctuating environments. *Am. Nat.* **108**:783–790.

Schaeffer, W. M. 1974b. Selection for optimal life histories: The effects of age structure. *Ecology* **55**:291–303.

Slatkin, M. 1970. Selection and polygenic characters. *Proc. Natl. Acad. Sci. U.S.A.* **66**:87–93.

Stearns, S. C. 1976. Life-history tactics: a review of the ideas. *Q. Rev. Biol.* **51**:3–47.

Stearns, S. C. 1977. The evolution of life history traits: a critique of the theory and a review of the data. *Annu. Rev. Ecol. Syst.* **8**:145–171.

Stearns, S. C. 1980. A new view of life history evolution. *Oikos* **35**:266–281.

Tantawy, A. O., and M. R. El-Helw. 1966. Studies on natural populations of *Drosophila*. V. Correlated responses to selection in *Drosophila melanogaster*. *Genetics* **53**:97–110.

Tauber, M. J., and C. A. Tauber. 1978. Evolution of phenological strategies in insects: A comparative approach with Eco-physiological and genetic considerations. In H. Dingle (ed.), *Evolution of insect migration and diapause*. Springer-Verlag, New York.

Taylor, H. M., R. S. Gourley, and C. E. Lawrence. 1974. Natural selection of life history attributes. *Theor. Pop. Biol.* **5**:104–122.

Tuljapurkar, S. D. 1982a. Population dynamics in variable environments. II. Correlated environments, sensitivity analysis, and dynamics. *Theor. Pop. Biol.* **21**:114–140.

Tuljapurkar, S. D. 1982b. Population dynamics in variable environments. III. Evolutionary dynamics of *r*-selelction. *Theor. Pop. Biol.* **21**:141–165.

Tuljapurkar, S. D., and S. H. Orzack. 1980. Population dynamics in variable environments. I. Long-run growth rates and extinction. *Theor. Pop. Biol.* **18**:314–342.

Turelli, M. 1982. Cis-trans effects induced by linkage disequilibrium. *Genetics* **102**:807–815.

Verhulst, P. F. 1838. Notice sur la loi que la population suit dans son accroissement. *Corresp. Math. Phys.* **10**:113–121.

Volterra, V. 1926. Variazioni e fluttuazioni del numero d'individui in specie animali conviventi. *Mem. R. Acad. Naz. dei Lincei* (*Ser.* 6) **2**:31–113. (Translation in an appendix to Chapman R. 1931. *Animal Ecology*, McGraw-Hill, New York.)

Wilbur, H. M. 1980. Complex life cycles. *Annu. Rev. Ecol. Syst.* **11**:67–93.

CHAPTER 6

Behavior, Genes, and Life Histories: Complex Adaptations in Uncertain Environments

HUGH DINGLE
Program in Evolutionary Ecology and Behavior
Department of Zoology
University of Iowa
Iowa City, Iowa

CONTENTS

Current Address: Department of Entomology, University of California at Davis, Davis, California 95616.

1 INTRODUCTION

The life history of an organism is an assemblage of traits specifically contributing to survival, reproduction, and hence fitness, and for this reason is
of central importance to general Darwinism (Bell, 1980). Such assemblages
of traits have been called "complex adaptations" by Frazzetta (1975) and
"strategies" or "tactics" in much of the literature of evolutionary ecology.
They are characterized by the fact that they coevolve and function together
to make the adaption "work." Because they bring into focus so much of
evolutionary biology, a great deal of theoretical and empirical effort has been
directed at analyzing the genetic and phenotypic structure of life history
strategies (reviewed by Stearns, 1976, 1977; Bell, 1980; and papers in Denno
and Dingle, 1981a, and Dingle and Hegmann, 1982).

In this chapter I shall concentrate on two themes in the analysis of life
histories. The first concerns genetic structure. The issue is not simply
whether life history traits are genetically based. Rather the important evolutionary question is how the genome is *organized* with respect to life histories, in particular, how genes may influence how traits evolve to function
together (Dingle et al., 1982; Hegmann and Dingle, 1982). Central to this
issue is not only genetic variance for any one trait, but also the genetic
correlations tying together combinations of traits in complex adaptations.
Any ultimate explanation of a life history strategy depends on understanding
both the genetic variation for the constituent traits, since it is this variation
on which natural selection acts, and on the genetic covariance structure,
since it also will impose limits on the magnitude and direction of evolutionary
change. Traits tied together because they are influenced by the same set of
genes will each be constrained by the other; this is especially true if covariances are negative (Lande, 1982a,b). To fully understand a life history strategy or any other adaptation one must at least know something about the
interplay of genetic flexibility and genetic constraint. I briefly discuss estimates of genetic variance and genetic correlation in Section 2.

Flexibility also enters into my second theme, which is the role of behavior
in life histories. Recent attention has focused more and more on this issue
because clearly it is as important for offspring to be placed appropriately
with respect to favorable resources as it is to adjust schedules of reproduction and longevity (cf. Denno and Dingle, 1981b). Flexibility of response to
environmental uncertainties can be an important element in life history evolution, and behavior provides much of that flexibility in allowing choices
of where and when to breed (Nichols et al., 1976; Solbreck, 1978; Taylor
and Taylor, 1978; Livdahl, 1979; Dingle, 1981).

Two behaviors of critical importance to insect life histories are migration
and diapause, and I shall focus most of my attention on these since insects
are the best-studied organisms for assessing the contribution of behavior to
life history strategies. I also include, however, some examples from other
organisms for comparison where appropriate and available. Migration and

diapause are often intimately related physiologically (Kennedy, 1961; Dingle, 1978a) and integrate reproduction and resources by allowing adjustment of the life history to environmental uncertainty (Solbreck, 1978; Vepsäläinen, 1978; Denno and Grissell, 1979; Denno et al., 1980, 1981; Dingle, 1981). Thus, the timing of migration may be as important as whether migration occurs (Ziegler, 1976), and diapause may determine such timing (Dingle et al., 1977). Important here is an understanding of environmental uncertainties from the perspective of the organism concerned. I address environmental uncertainty in Section 3 before proceeding to examine the influence of these uncertainties in the shaping of life histories.

2 GENETIC VARIANCE AND GENETIC CORRELATION

Metric or continuously varying traits such as those involved in behavior and life histories are usually polygenic, that is, they are the result of the action of genes at more than one locus. The appropriate methods for analyzing such polygenic characters are available from quantitative genetics. These are covered clearly and in detail by Falconer (1981) so a brief summary will suffice here. The basic model assumes that individual differences in metric traits are the result of both genetic and environmental differences with the total (phenotypic) variance expressed by

$$V_P = V_A + V_N + V_E$$

where V_A is the additive genetic variance, V_N is the nonadditive genetic variance (due to dominance and epistasis), and V_E is the environmental variance. The additive genetic variance arises from the summation of influences across all genes contributing to a character and is the variance contributing specifically to parent–offspring resemblance. Because of this, it determines sensitivity to selection and the maximum rate at which evolutionary change in a trait can occur. The relation between additive genetic variance and phenotypic variance is expressed as the heritability (h^2) of a trait where $h^2 = V_A/V_P$. Heritability can be estimated in several ways, the most powerful being by offspring on parent regression (because V_A contributes to resemblance) or by selection (the so-called "realized heritability" or ratio of response to selection over selection differential). Strong directional selection, as presumably for a character associated with fitness, should reduce V_A and hence h^2 (Fisher's so-called "Fundamental Theorem of Natural Selection").

Metric characters can also be correlated (e.g., height and weight) with the causes of correlation being both genetic and environmental. This is expressed in the relation

$$r_P = h_x h_y r_A + e_x e_y r_E$$

where r_P is the phenotypic correlation between characters X and Y, r_A and r_E are the additive genetic and environmental correlations respectively, h_x and h_y are the respective square roots of the heritabilities, and $e^2 = 1 - h^2$ (and so includes variance due to dominance and epistasis as well as environmental variance *sensu stricto*). The genetic correlation indicates the extent to which genes influence the traits in common (pleiotropy) with the result that when selection influences one trait, it must influence the other as well. The parent–offspring relationship and selection can also be used to estimate r_A with methods analagous to those discussed above for estimating h^2. In this case, however, the appropriate characters are X in the parents and Y in the offspring. As I hope will be clear in the ensuing discussion, both genetic variance and genetic correlation are important to understanding the nature of a complex adaptation, in this case a life history.

3   UNCERTAIN ENVIRONMENTS

By environmental uncertainty I mean simply the inability of an organism to predict (in the evolutionary sense) what its current response should be to take advantage of some future state. Usually "taking advantage" means producing offspring. Natural selection will favor those individuals that can find and respond to some reliable cue that forecasts with accuracy future conditions. For many organisms such a cue is photoperiod, and indeed photoperiodism is one of the best known and most widely studied physiological responses from flowering times in plants to pelage changes in mammals. It is well known in insect and bird migration and in insect diapause (e.g., Dingle, 1978a; Murton and Westwood, 1977). In many situations, however, photoperiod is an unreliable or only partially reliable cue.

Recently, increasing attention has focused on the consequences of such uncertainty for the organisms concerned (Hoffmann, 1978; Istock, 1978, 1981, 1982 and Chapter 5). Hoffmann (1978), for example, described a range of habitats for *Colias* butterflies which differed in the degree to which photoperiod predicted thermal regime. The latter influences the amount of melanin (a pigment required for behavioral thermoregulation) found in the wing. Hoffmann hypothesized that phenotypic variance in wing coloration ought to be positively associated with variance (i.e., degree of unpredictability) in the thermal regime and assumed, in view of the nondirectional selection involved, a major additive genetic influence. The presence of high additive genetic variance was demonstrated conclusively for life history traits (diapause and development rate) in the pitcher-plant mosquito, *Wyeomyia smithii*, by Istock (1978, 1981). In this case natural selection displayed a fluctuating–stabilizing pattern arising from seasonal uncertainty and maintained such variance. From the point of view of a late summer female mosquito it was usually not obvious whether one should breed, thus producing an extra generation in a lineage, or enter an overwintering diapause. Both responses were present in the sampled population.

A nice example of the problem presented by environmental uncertainty is the situation faced by late summer adults of the milkweed leaf beetle, *Labidomera clivicollis*, in central Texas (Palmer, 1982) as illustrated in Figure 1. The generation which oversummers as adults (G_1) produces G_2 eggs from late August to late October. The late August eggs (A in Fig. 1) grow to adults before the autumnal equinox in September and in turn produce A′ eggs of G_3 since the adults are not exposed to the critical photoperiod for diapause (LD 12.5 : 11.5). The resultant A′ adults of the G_3 generation mature in plenty of time to enter an overwintering diapause barring an exceptionally unfavorable year. Conversely the postequinox C eggs of G_2 mature to adulthood in the very short days of November (at approximately the same time as A′ adults) and are also photoperiodically induced to enter overwintering diapause. If instead this group did produce C′ eggs, these would fail to reach maturity because of the lateness of the season for the third generation.

The problem comes for the B adults of G_2. On the basis of rainfall and temperature (which determine sustained food abundance and maturation rates of the beetles), Palmer estimated that 17% of the years would be "ex-

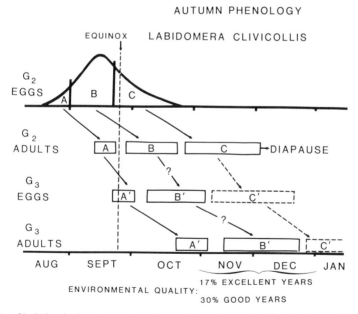

Figure 1. Variation in the autumn phenology of the milkweed leaf beetle. Eggs of the second annual generation (G_2) are distributed so that the resulting adults may either produce a third generation (A group), enter diapause (C group), or face an uncertain choice depending on environmental quality, indicated by the bracket below November and December, in the particular year (B group). See text for details. After Palmer (1982).

cellent'' for maturation of prospective offspring and 30% would be ''good,'' that is, that in about half the years a large portion of B adults would succeed in producing a G_3 generation to enter overwintering diapause. In the remaining years few would succeed and those that tried would also lose the opportunity to overwinter since reproduction precludes diapause in this generation. So how should a G_2 adult of the B group be programmed to respond, with diapause or with reproduction?

One solution would be a decision by the G_1 adults to reproduce earlier thus shifting the G_2 egg distribution to the left. Under these circumstances, if the shift were great enough, all G_2 adults could mature before the equinox and produce G_3 eggs. This is not an available option, however, because it would push egg production into the summer drought when food is usually unavailable for either adults or their offspring. A second solution would be for G_1 adults to delay and shift the G_2 egg curve to the right. In this case, all would diapause, but any female which ''cheated'' and produced earlier eggs would produce an extra generation in her lineage in the favorable years (i.e., about half the time) in the B range and in virtually every year if there were any eggs in the A range. So there seems to be no solution along these avenues of directional selection. We are thus left with the dilemma of environmental uncertainty for B adults; in some years it will pay to produce B′ eggs, in others it will be better to enter diapause. We expect the result to be high variance in photoperiodic response under these circumstances, maintained by shifting selection pressures, and in fact this is what Palmer observed, although it is as yet not known how much of this variation is genetic. Under similar circumstances and with the addition of population mixing from migration, we have demonstrated high additive genetic variance for critical photoperiod in the milkweed bug, *Oncopeltus fasciatus* (Dingle et al., 1977).

In the above instances we have some intuitive notion of the relevant environmental uncertainty, at least qualitatively. For more accurate predictions of variation in response to environmental inputs, however, we really need quantitative measures of uncertainty. Some recently proposed measures are promising (Colwell, 1974; Stearns, 1981), and a statistical test of the ''correlation'' between environmental and population variances has been developed (Derr and Ord, 1983). Discussion of these is beyond the scope of this chapter since they have yet to be applied in the situations I shall present, but it should be stressed that a critical issue is *which* environmental variables need to be assessed in each case. This whole area of evolutionary ecology is still very much in its infancy.

4   MIGRATION

One response to fluctuations in environmental conditions is migration. When such fluctuations are predictable, as with seasonal changes, migration is usually programmed by reliable cues of which the most prominent is photoperiod as indicated above. In insects, photoperiod may influence both migration and diapause which can share the same underlying physiological mechanisms (see below). Other inputs may also have important influences. For example, warm temperatures inhibit long duration flight in milkweed bugs so that when these bugs arrive in a thermally favorable habitat, they tend to stay there (Dingle, 1968). In cotton stainer bugs (*Dysdercus* spp.) the relevant inputs are food and moisture; in their absence these insects enter diapause and migrate in search of new host plants (Dingle and Arora, 1973; Derr, 1980a). Food similarly influences the lygaeid bug *Neacoryphus bicrucis* (Solbreck, 1978). In all these situations, however, uncertainty of choice (e.g., food resources may be intermediate in quality), and hence fluctuating selection pressures result in high phenotypic variance for flight. Variable flight performance is also reported for a number of other insects (Johnson, 1976). Laboratory selection experiments have demonstrated high levels of genetic variance for flight in *Oncopeltus fasciatus* (Dingle, 1968). In another milkweed bug, *Lygaeus kalmii*, which is a local migrant from diapause sites to host plants, the additive genetic variance for flight duration is on the order of 30% (Caldwell and Hegmann, 1969).

Geographic variation also exists for insect migratory flight (Dingle et al., 1980a). Table 1 demonstrates this for *O. fasciatus*. Temperate zone populations, which migrate up from the south each spring and summer, display a relatively high proportion of long duration flights (but note that flight is highly variable). Tropical populations, on the other hand, show very few long flights, and in the case of island bugs, some individuals never flew.

In vertebrates, attempts have been made to assess gene influences on geographic variation in migration for two species (Fig. 2). Rasmuson et al. (1977) examined the dispersal tendency in two Swedish populations of the vole, *Microtus agrestis*. The northern population undergoes classic microtine cycles with high dispersal rates, especially among young males at high densities; in the south there is little variation in population size, and dispersal rates are low. Dispersal was assessed in the laboratory with a device that allowed voles to pass between compartments in a large test chamber, and the behavior observed was consistent with field observations of more movement in northern animals. The populations were then crossed and the hybrids were intermediate but with a strong bias toward the northern parents, at least in the males. These results imply a significant additive component, but also a dominance effect possibly the result of directional selection for dispersal in the north.

The second vertebrate species studied is the European blackcap warbler, *Sylvia atricapilla*, examined by Berthold and Querner (1981), where populations along a north–south gradient differ in their migratory performance and in the amount of nocturnal migratory restlessness (*Zugenruhe*) they display when caged. As shown in Figure 2, offspring crosses between a northern and a southern sample were intermediate in performance between their parents indicating between population additive genetic variance. The German birds are wholly migratory (although probably varying in distance flown), whereas the Canaries warblers are only partially so, suggesting greater climatic uncertainty in the latter site. For insectivorous birds such as these, there is certainty of failure to survive if one overwinters in Germany. I would predict a high degree of genetic dominance for migration within the German groups (reflecting environmental certainty and directional selection although note that there may be uncertainty as to how *far* to migrate) but high additive genetic variance within the Canaries birds (reflecting uncertainty and fluctuating selection). The prediction remains to be tested.

Timing is also important in migration, and *when* one chooses to depart can be as important as the distance traveled (Ziegler, 1976). Uncertain environments should maintain additive genetic variance for timing. That this is indeed the case was demonstrated in an elegant selection experiment by M. A. Rankin (1978) on *Oncopeltus fasciatus* (Fig. 3). Rankin had discovered that intermediate titers of juvenile hormone promoted migratory flight in this insect. She also noted that some bugs delayed flight for several days. With an appropriate selection regime she rapidly produced a population of late flying individuals which also showed two correlated responses. First, oviposition was also delayed (suggesting reproductive diapause), and second, juvenile hormone titers remained at intermediate levels instead of rising to levels promoting reproduction. Thus, the genes in this case apparently influence migration and reproduction through their effects on the rate of increase of juvenile hormone in the blood. Rankin thus has demonstrated the genetic and physiological basis for coordinating a set of life history responses and for the relation between diapause and migration in these insects. Juvenile hormone also plays a role in such coordination in other insects (Rankin and Rankin, 1980; reviewed in Dingle, 1983).

Table 1. Distributions of Longest Daily Tethered Flight Durations for Temperate and Tropical Populations of the Milkweed Bug, *Oncopeltus fasciatus*[a]

Population	No Flight	Number Flying for			Proportion >30 min
		<5 min	5–30 min	>30 min	
Temperate					
Iowa	0	114	27	44	0.238
Maryland	0	33	7	13	0.245
Michigan	0	193	44	74	0.238
Tropical					
Guadeloupe	13	28	1	2	0.045
Mexico	0	96	2	0	0
Puerto Rico	6	48	5	3	0.048

[a] Data from Dingle et al. (1980a).

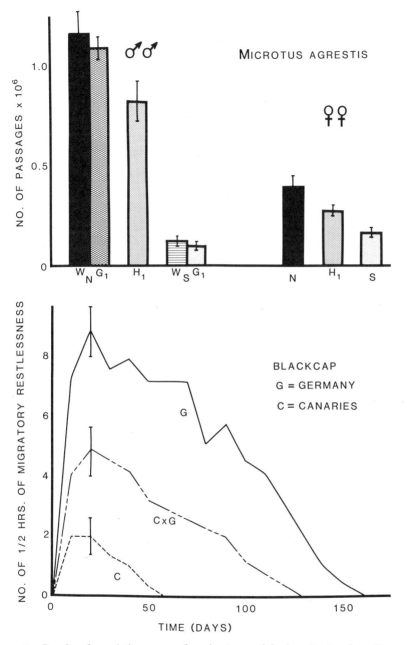

Figure 2. Results of population crosses for migratory activity in voles (top from Rasmuson et al., 1977) and blackcap warblers (bottom after Berthold and Querner, 1981). For the voles activity was assessed by determining the number of passages between compartments in a laboratory chamber. N and S refer to northern and southern Sweden, H is the laboratory reared hybrid, and W and G_1 refer to wild caught and first laboratory generation lines. For the blackcap, measurements were of restlessness in caged birds. Standard errors are indicated by vertical lines.

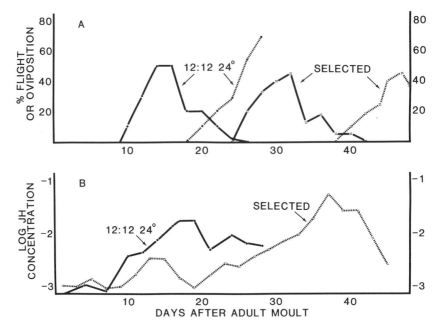

Figure 3. Relation between laboratory flight and juvenile hormone titer in the milkweed bug *Oncopeltus fasciatus*. (*a*) The percent of individuals flying over 30 min and of females ovipositing at LD 12:12 at 24°C and in a line selected for delayed flight at LD 12:12 at 23°C. Solid lines, flight; dashed lines, oviposition (note that oviposition follows flight), (*b*) Juvenile hormone titers (log JH concentration in JH III equivalents in ng/ml). Flight occurs at intermediate titers; oviposition when titers reach maximum. Data from Rankin (1978).

5 DIAPAUSE

As insect migration allows escape in space so diapause allows escape in time. As we have seen, the two behavioral responses may be physiologically related in those species for which diapause means delayed reproduction in adults. Diapause further provides additional time for migration to occur (Dingle, 1978b). It is apparent that there is much variation in diapause depth and duration (Tauber and Tauber, 1976, 1981) especially in response to uncertain environments. Some examples are illustrated in Figure 4. The distribution of ages at first reproduction under nondiapause conditions is shown for *Oncopeltus fasciatus*; similar distributions occur for the other two species when not diapausing. In two populations of *O. fasciatus* diapause occurs in short days, but with high variances presumably reflecting environmental uncertainty. In Iowa the question is whether a third summer generation can mature during the growing season, analogous to the situation in *Labidomera clivicollis* (Fig. 1), whereas in Florida some winters are favorable for reproduction and others are not (Dingle et al., 1977, 1980b; Miller and Dingle, 1982). Adult females of *Dysdercus bimaculatus* need water to feed and reproduce.

Reduced water supplies ("stress") mean that the choice between staying and reproducing or diapausing and flying to a new host is not an obvious one (Derr, 1980b). Finally, *Neacoryphus bicrucis* under short days faces a situation similar to Iowa *O. fasciatus*, that is, whether to try to fit in a third summer generation (Solbreck, 1978).

Several lines of evidence suggest that much diapause variation is polygenically based (Dingle et al., 1977; Hoy, 1978; Istock, 1978). These are chiefly of two sorts. First, numerous workers have crossed geographically separated populations differing in critical photoperiod and reported the responses of the hybrid offspring to be intermediate between those of their parents (reviews in Danilevskii, 1965; Dingle et al., 1977; Hoy, 1978). Such results imply major between population components of additive genetic variance as seen for migration in the voles and warblers cited above (Fig. 2). Second, selection experiments have demonstrated rapid responses to artificial selection either for or against diapause and estimates of additive genetic variances of around 70% of the phenotypic variances seem not unusual (Morris and Fulton, 1970; Dingle et al., 1977; Istock, 1978). In some species such

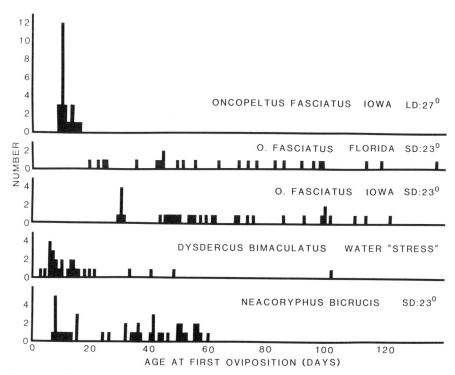

Figure 4.   Distributions of ages of first reproduction (indicating adult diapause) in three hemipteran species reared in the laboratory. LD indicates a 14:10 photoperiod and SD an 11:13 photoperiod. See text for discussion. Data for *Oncopeltus* from Dingle et al. (1980b); for *Dysdercus* from Derr (1980); and for *Neacoryphus* from Solbreck (1978).

as northern *Drosophila*, however, dominance has apparently evolved in the genetic system controlling diapause (Lumme, 1978). In *D. lummei* and the European corn borer, *Ostrinia nubilalis*, regulation is by X-chromosomal factors (Lumme and Keränen, 1978; Reed et al., 1981). The selective regimes responsible for these latter systems are presently undefined and could well repay study. The implication of the genetic data is that they are directional or highly stabilizing.

Another way to assess the impact of varying environments on life histories is through analysis of genotype–environment interactions. One way to do this is with the split litter, two environment design shown in Figure 5. In these experiments parents are reared in each of two environments and their offspring are then divided and reared in the same two conditions. Offspring thus mature either in the same environment as their parents or in one different from their parents (and half their siblings). The results of one such set of experiments on a Maryland population of *Oncopeltus fasciatus* are illustrated in Figure 6. These graphs plot the relation between the age at first reproduction (α) of offspring reared in long day against that of their siblings reared in short day. For offspring of long day parents the correlation is statistically significant indicating that a portion of the genes controlling α in short day also do so in long day. There is no significant correlation, however, for offspring whose parents were in short day. There thus seems to be a "maternal effect" of parental photoperiod on the way offspring respond to daylength so that environmental effects can be one generation removed. A possible contributing factor here was the tendency of short day offspring of short day parents to reproduce earlier than short day offspring of long day parents, whereas the same effects were not observed for long day offspring (although this interaction was not significant). What this perhaps suggests is that offspring of diapausing parents are more likely to "gamble" on earlier reproduction even in short days. If this result is real, and in view of the failure to be statistically significant it should be interpreted with extreme caution to say the least, it opens up some interesting lines of investigation for ways in which natural selection arising from environmental uncertainty could act.

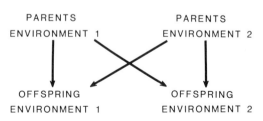

SPLIT LITTER, TWO ENVIRONMENT DESIGN

Figure 5. Design of experiment in which parents are reared in two environments and offspring of each parental pair are divided between the same two environments. Offspring are thus reared in both the same and different environments from their parents.

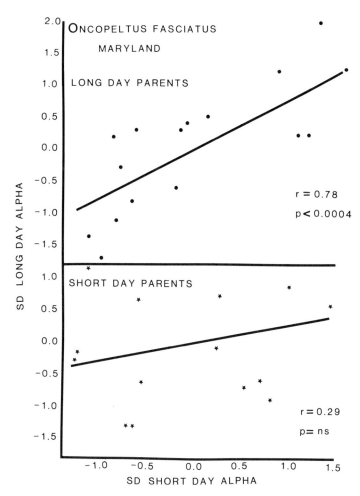

Figure 6. Results from a split litter, two environment experiment with a Maryland population of *Oncopeltus*. Offspring of long day parents show correlated ages of first reproduction (alpha) between long and short day as measured in standard deviation (SD) units (top). Offspring of short day parents fail to show a significant correlation (bottom).

A similar experimental design has been used to assess the influence of water stress on the cotton stainer bug *Dysdercus bimaculatus* (Derr, 1980a; Table 2). Heritability estimates in "nonstressful" (high H_2O) environments for characters having to do with timing (age at first reproduction or τ, interclutch interval, age at death) were not significantly different from zero, whereas characters relating to egg production had estimates of around 30% (i.e., about 30% of the variance in the trait is attributable to additive genetic variance). In the "stressful" (low H_2O) and more uncertain (cf. Derr, 1980b; Fig. 4) environment, heritability estimates were on the order of 30–40% for τ depending on whether they were based only on offspring females that

Table 2. Heritability Estimates ($\pm$ SE) by Offspring on Parent Regression for Various Life History Characters in *Dysdercus bimaculatus*[a]

Character	High H_2O	Low H_2O
Days eclosion to first clutch (τ)	0.18 ± 0.11	0.40 ± 0.035^{b} reproductives only[c]
		0.30 ± 0.067^{b} all females[d]
Interclutch interval	-0.12 ± 0.14	
Size first clutch	0.34 ± 0.089^{b}	
Clutch number	0.21 ± 0.088^{b}	
Total fecundity	0.27 ± 0.079^{b}	
Age at death	0.11 ± 0.057	
Wing length	0.51 ± 0.21^{b}	
τ between environments[e]		-0.19 ± 0.073^{b} reproductives only[c]
		0.15 ± 0.024^{b} all females[d]

[a] Data from Derr (1980b).

[b] Heritability significantly different from zero.

[c] Only females reproducing before death included in calculation of heritability.

[d] Age of death used as an estimate of τ and all females included.

[e] Parent τ in low H_2O environment, offspring τ from high H_2O environment (so regression statistic is not heritability).

reproduced or on all offspring females; in either case they were significantly greater than zero. (Heritability estimates were not made for the other traits.) If one looked at τ between environments, there was a negative correlation between stressed mothers and nonstressed offspring, using reproductives only and a positive association using all females. The former result makes a point that although obvious, needs to be emphasized, namely, that a good tactic in one environment may not be in another. It is also apparent from all the above results that genetic variance for timing of reproduction is influenced by environmental uncertainty.

6   LIFE HISTORIES

Migration and diapause are two behavioral responses that contribute flexibility to life histories in the face of environmental uncertainty. They allow escape from unfavorable conditions and colonization of favorable habitats in either space or time. Selection for colonizing abilities should produce a suite of life history traits that coevolve with behavior in successful colonizers. In other words, there should be complex adaptations of particular kinds contributing to persistence in the new habitat. A colonizing syndrome should include both the ability to disperse and the ability to persist (Simberloff, 1981). Since ability to persist is likely to involve early and rapid reproduction by colonizers (MacArthur and Wilson, 1967; Safriel and Ritte, 1980), these characters should be phenotypically correlated with dispersal and/or diapause. The important issue for evolutionary biologists is the extent to which the phenotypic correlations contributing to a colonizing syndrome are genetically based (see Section 2).

In her study of *Dysdercus bimaculatus* life histories under water stress, Derr (1980a) also examined the question of the phenotypic association of stress tolerance (which is presumably related to dispersal) with other life history attributes. On the basis of median age at first reproduction in low H_2O, parent females were divided into an early and a late group. These groups were then compared with each other and with offspring reared in a nonstress environment with respect to four life history traits (Table 3). There was no evidence that division on the basis of τ resulted in division on the basis of other characters or on division of the offspring on the basis of other characters. In other words, there was no evidence in these bugs of a characterizable life history syndrome associated with stress tolerance. There does, however, seem to be a suite of traits associated with body size, including flight (Derr et al., 1981). Regrettably, there is no direct test involving genetic correlations between flight and life history characters, a test that would be most interesting in view of the above results.

Two independent sets of selection experiments have been used to assess the genetics of dispersal syndromes in *Tribolium* (Ritte and Lavie, 1977; Lavie and Ritte, 1978; Wu, 1981). In both cases beetles were selected for high and low dispersal rates, and then after several generations the selected lines were compared with respect to an array of life history variables. In Wu's experiments beetles could move between flour filled shell vials by means of pipe cleaners; dispersal was determined as a function of the number

Table 3.   Phenotypic Comparisons in *Dysdercus bimaculatus* between Parents of High and Low τ in Low H_2O Environment and Other Life History Characters with the Same Population and with Their High H_2O Offspring[a]

Character	Value of U[b] for Pairwise Comparison of Parent τ with:	
	Other Parents	High H_2O Offspring
Wing length	56.5	58.0
Size first clutch	64.5	51.0
Total fecundity	36.0	46.0
Age at death	40.0	56.0

[a] Comparisons evaluated using Mann–Whitney U test; there was no significant association with low H_2O parent. Data from Derr (1980b).

[b] Critical U score is 30 for significance.

Table 4. Comparison between *Tribolium castaneum* and *T. confusum* with Respect to Relation between Dispersal and Life History Characters[a]

	T. castaneum		*T. confusum*	
Character	High Dispersal	Low Dispersal	High Dispersal	Low Dispersal
Dispersal rate[b]	6.83	1.59[c]	6.46	1.61[c]
Development time[d]	29.03	30.51[c]	31.02	30.83
Egg production in 9-day period	124.8	159.7	112.0	119.6[c]
Egg size	0.601	0.586[c]	0.621	0.610
Body weight	22.6	19.7[c]	20.6	20.9

[a] Data from Wu (1981).

[b] Mean number of vials traversed in generation 17.

[c] Difference between high and low lines statistically significant.

[d] Median number of days from egg to adult.

of vials traversed by individuals. The results for *T. castaneum* (a dispersing species) and *. confusum* (generally considered low or nondispersing) are summarized in Table 4. In *T. castaneum* beetles from the high selected line were larger, developed more rapidly, and produced larger eggs than those from the low line. Egg production was not significantly different, largely because of high variances in a few of the replicates. In *T. confusum*, on the other hand, the only significant difference between the lines was in egg production. There thus seems to be a dispersal life history syndrome in *T. castaneum*, characterized by larger and more rapidly maturing beetles, but not in *T. confusum*.

These results for *T. castaneum* were essentially confirmed by Lavie and Ritte (1978) in their selection experiment (Table 5). In this case dispersal was measured as movement along a string between vials of flour, and "dispersants" were those beetles that moved to the outer rim of vials in the test apparatus at least twice. Dispersing individuals developed more rapidly than nondispersers and produced more fertile eggs. In contrast to Wu's results (Table 4) they also produced more eggs, at least over the 4-day interval sampled. The important point is that there was a distinct set of life history characteristics in the line selected for high dispersal, and these included rapid development and high fecundity, traits predicted a priori to be characteristic of colonizers (MacArthur and Wilson, 1967; Safriel and Ritte, 1980).

In an attempt to understand the genetic architecture underlying the migration and life history strategies of *Oncopeltus fasciatus*, we have recently completed a sibling analysis of Iowa populations of this insect (Hegmann and Dingle, 1982). The data in Table 6 are extracted from that analysis. The leading diagonal of the matrix gives the heritabilities of the traits. The bugs in this experiment were reared under long day conditions which should preclude diapause. Note that in these circumstances the heritability of age at first reproduction is only about one-third that produced under the greater environmental uncertainty of short days (see above). Values for other traits are similar to heritabilities estimated for comparable traits by Derr for *Dysdercus bimaculatus* in stable environments (Table 2). What is particularly interesting is the high heritability of development time. It is hard to know exactly what to make of this although it is worth pointing out that development time displays high negative correlations with most other traits and additive genetic variance could be maintained for this reason (cf. Lande, 1982a,b; Istock, 1982).

These negative correlations plus the positive genetic correlations of fecundity related traits with body size appear to define the *O. fasciatus* life history strategy. Large bugs develop more rapidly and produce larger clutches of higher fertility. Note also the rather high heritability estimate for wing length. To the extent that this variable correlates with flight, there is likely to be considerable genetic variation for the latter, as indeed is implied by the direct experimental data (Dingle, 1968, and above). There is a strong positive genetic correlation of wing length with body size and in fact larger bugs seem to be migrants, at least when comparisons are made among populations (Dingle et al., 1980a). This possible association between body size and flight is now being tested directly via selection experiments (J. O. Palmer, in preparation). The implication from these data is that there is a complex life history adaptation in *O. fasciatus* geared for migration and colonization, characterized by gene influences producing large, rapidly developing, long flying, fecund bugs.

A major exception to this array of genetic correlations is age at first reproduction (the interval between eclosion and oviposition and thus the same as Derr's τ) which appears to be uncorrelated with any of the other traits. This trait is also a function of diapause which in turn permits a higher portion

Table 5. **Relation between Dispersal and Some Life History Characters in *Tribolium castaneum*[a]**

Character	High Dispersal Line	Low Dispersal Line
Dispersal rate[b]		
Generation 1	0.34 ± 0.12	0.08 ± 0.02
Generation 5	0.50 ± 0.10	0.06 ± 0.04
Generation 10	0.56 ± 0.06	0.02 ± 0.01
Development time[c]	0.084 ± 0.0042	0.14 ± 0.0057
Egg production in 4-day period	47.8 ± 1.8	42.2 ± 1.8
Fertility	0.70 ± 0.0081	0.62 ± 0.0078

[a] All differences between lines are statistically significant. Data from Ritte and Lavie (1977) and Lavie and Ritte (1978).

[b] Frequency of beetles that dispersed at least twice.

[c] Proportion of larvae remaining on day 21.

Table 6. Heritabilities (on Leading Diagonal) and Genetic Correlations (± SE) for Some Life History Traits of *Oncopeltus fasciatus* Based on Analysis of Half-Sibs and Full-Sibs[a]

	WL	BL	α	CS	PH	DT
Wing length (WL)	0.55 ± 0.22					
Body length (BL)	0.67 ± 0.21	0.20 ± 0.14				
Age at first reproduction (α)[b]	−0.29 ± 0.28	−0.41 ± 0.34	0.25 ± 0.12			
Size of first clutch (CS)	0.73 ± 0.13	0.91 ± 0.06	−0.22 ± 0.29	0.25 ± 0.10		
Percent hatch first clutch (PH)	0.60 ± 0.21	0.77 ± 0.17	0.07 ± 0.34	0.71 ± 0.16	0.25 ± 0.10	
Development time (DT)[c]	−0.55 ± 0.19	−0.56 ± 0.24	−0.13 ± 0.28	−0.57 ± 0.18	−0.52 ± 0.22	0.89 ± 0.32

[a] Data from Hegmann and Dingle (1982).

[b] Interval between adult eclosion and first clutch. Same as τ of Derr (see Table 2).

[c] Interval between oviposition and adult eclosion.

of the life cycle to be devoted to migratory flight (Dingle, 1978b). It is thus a character directly involved with life history flexibility in the face of environmental uncertainty. The relative independence of the genetic variance influencing age at first reproduction is consistent with our hypothesis that photoperiodically induced diapause will "provide optimal association of diapause, migration, and photoperiod in the face of varying association of photoperiod with temperature and food" (Dingle et al., 1977). If there were strong genetic correlations between age at first reproduction and other life history characters, they would tend to "tie up" the genetic variance contributing to life history flexibility (Hegmann and Dingle, 1982).

Migration can also influence life history strategies via direct feedback (reviewed in Dingle, 1983). For example, in both the frit fly, *Oscinella frit*, a pest of grains, and the milkweed bug, tethered flight shortened the preoviposition period relative to those individuals that were not flown (Rygg, 1966; Slansky, 1980), and in the latter the interval over which the first five clutches were oviposited was also abbreviated. In the black bean aphid, *Aphis fabae*, flight promotes the array of settling responses on a new host leaf culminating eventually in insertion of the stylets and larviposition (Kennedy and Booth, 1963). In a number of insects arrival at a suitable habitat following migration results in loss of wings, wing muscle histolysis, or both followed by rapid development of eggs and oviposition. In an experimental test of the effect of de-alation, Tanaka (1976) removed the wings of the macropterous morph of the cricket *Pteronemobius taprobanensis*. In this species micropters begin laying eggs sooner and produce more eggs than macropters. The effect of de-alation of macropters immediately upon emergence was a significant increase in egg production which was a consequence of more rapid oocyte growth and not simply the stimulation of reproductive behavior in the presence of an already mature reproductive system. In all the above cases the events surrounding migration contribute directly to a colonization syndrome and add additional flexibility to the life history strategy. An interesting and largely unexplored area for research would be the relation between gene influences on wing polymorphism and the associated phenotypic and genotypic structure of life histories.

7 BEHAVIOR AND LIFE HISTORIES IN UNCERTAIN ENVIRONMENTS

In this chapter I have tried to outline two themes: the role of behavior in life histories and how genes may influence how life history traits evolve to function together. It should be apparent at this point that the two themes are related. Behavior provides flexibility to life histories, and it is important to our understanding of strategies to gain insight into how behavior, specifically migration and diapause in the examples cited here, is genotypically and phenotypically associated with the array of life table variables usually considered the major constituents of fitness.

In the case of *Tribolium castaneum* we have a clear case of a migration–life history syndrome based on underlying genetic structure demonstrated by artificial selection experiments (Tables 4 and 5). This syndrome seems to involve high dispersal rates combined with large size, rapid development, early reproduction, and high fecundity, an association predicted on theoretical grounds for colonizing species (MacArthur and Wilson, 1967; Safriel and Ritte, 1980). Although not so compelling, there is also suggestive evidence for such a syndrome in two hemipterans, *Oncopeltus fasciatus* and *Dysdercus bimaculatus*. Particular phenotypic arrays of traits associated with dispersal are also frequently reported in the small mammal literature and in some fish (reviewed in Dingle, 1980), but the genetic correlation structure remains to be studied in detail. It would seem a worthwhile area of research in these (and other) organisms.

What turns out to be particularly interesting is the genetic correlation structure in the face of environmental uncertainty, especially with respect to age at first reproduction (α). This character has been shown repeatedly to display high proportions of additive genetic variance (V_A) when environments fluctuate in unpredictable ways (e.g., Dingle et al., 1977; Istock, 1978; Table 2), while displaying low V_A under regimes that are predictable and strongly stabilizing, or impose directional selection (e.g., Tables 2 and 6, and papers in Dingle and Hegmann, 1982). The situation is made even more complicated in cases where an environmentally related maternal effect can influence the degree of resemblance among sibs reared under different regimens (Fig. 6). Finally, superimposed on this can be the further variation produced by interbreeding among migratory individuals derived from differing habitats and selection pressures (Dingle et al., 1977; Derr, 1980b). In all these instances high V_A for α allows a range of responses to span the limits of uncertain environmental contingencies and functions as a "genetic rheostat" (Dingle et al., 1977).

A further notable aspect of α is its lack of genetic correlation with any of the other life history variables sampled, at least in milkweed bugs (Table 6). The life history strategy of this insect includes a set of genetically correlated traits, but the set does not include α. It is nevertheless variation in α which allows response to environmental uncertainty, and this variation is legitimately part of such a strategy. To understand the reason for this, it is important to consider the consequences of strong genetic correlations between α and other traits. If these were present, variation in α might result in detrimental variation in traits with which it was correlated (positive or negative depending on the sign of the correlation). Conversely, α might be prevented from varying if, for example, a sufficiently large negative correlation with another character made it impossible for natural selection to alter both in the same direction. This and similar consequences of genetic correlations have been termed genetic slippage (Dickerson, 1955) and are discussed further by Lande (1982a,b) and Hegmann and Dingle (1982). With respect to α and other life history traits, slippage and detrimental correlated responses can be considered a "cost of correlation," and we would predict that natural selection would act to reduce the genetic correlations and hence slippage between α and other fitness characters. Such seems to be the case in *Oncopeltus fasciatus* and in view of the failure of other variables to divide on the basis of τ (equivalent to α) in *Dysdercus bimaculatus* (Table 3), it is likely true of this species as well.

Other predictions follow from the notion of a "cost of correlation" and would seem to be worth testing with further research. Following on the above discussion, there should in general be little genetic correlation between life history characters and diapause or factors that specifically influence the timing of migration where there are unpredictable environments and genetic flexibility is a consequence of selection. On the other hand, there should be stronger correlations where migration or diapause occur in predictable environments. A test of this might be those situations where dominant Mendelian units for diapause or nondiapause have evolved (e.g., Lumme, 1978, on nothern *Drosophila*). Correlations with wing polymorphism would also be interesting to investigate in view of the phenotypic relationships between wing morphs and life histories and the apparent complexity of the genetic mechanism (Harrison, 1980; Dingle, 1980, 1983). In general there should be stronger correlations of α with life history traits in micropterous than in macropterous forms because the latter morph has the behavioral capability to respond flexibly to environmental uncertainty. In any event, the relation between the genetic and phenotypic structure of life histories in wing polymorphic species would seem to be an area ripe for investigation.

Additive genetic variance can also be maintained by other means than environmental uncertainty. Examples are polygenic mutations (Lande, 1975) and "hidden" variability masked because combinations of closely linked alleles are negatively correlated with respect to their phenotypic effects (Lande, 1975; 1982a,b; Istock, 1982). These are not unrelated to V_A resulting from environmental uncertainty, but the precise nature of such relationships have yet to be worked out. It should be apparent that both genetic flexibility and genetic constraints (Istock, Chapter 5) are important to the evolution of life histories. We have only the vaguest adumbration of what the complex adaptations arising from these two factors might be, but I hope I have succeeded in outlining some elements of a strategy to go about finding out.

8 SUMMARY

This chapter explores two related themes; the role of behavior in life histories and the genetic structure of complex life history adaptations. The behaviors discussed are migration and diapause, which provide flexibility in the face of uncertain environments, especially in choices of where and when to breed. There is a good deal of evidence from insects and some from vertebrates (with respect to migration) that under conditions of environmental unpredictability, fluctuating and nondirectional selection maintains in populations high additive genetic variance (V_A) for these traits. These results largely confirm theoretical predictions. There is also some evidence that where selection is strongly directional, dominance may evolve in the genetic systems.

In some insects, with *Tribolium castaneum* being the best example, there seems to be a characteristic genetically based migration–life history syndrome. Selection experiments in *T. castaneum* indicate that beetles with high dispersal tendencies are larger, develop faster, reproduce earlier, and are more fecund, an association of traits predicted on theoretical grounds for colonizing species. Although not as strong, there is also evidence for a similar colonizing syndrome in two hemipterans, *Oncopeltus fasciatus* and *Dysdercus bimaculatus*.

A particularly interesting aspect of life histories is that age at first reproduction (α) varies with respect to V_A depending on the environment and that, at least in *O. fasciatus*, it fails to show genetic correlations with other life history traits which do display such correlations among themselves. Age at first reproduction is a trait that tends also to vary as a function of migration and diapause, thus allowing responses to environmental variation. Because of this there would be a "cost of correlation" if α were genetically correlated with other life history characters, and this would "tie up" the variation in α. The genetic uncoupling of α from other fitness traits by natural selection would maintain the flexibility necessary for appropriate responses to environmental uncertainty.

ACKNOWLEDGMENTS

The manuscript was read by Francis R. Groeters, Jane L. Hayes, Edward Klausner, James F. Leslie, and James O. Palmer. To all I am most grateful. J. O. Palmer kindly lent me an advance copy of the manuscript from which Fig. 1 is drawn. E. Klausner, Deena Staub, and Carol Wolvington aided in preparing the manuscript. My own research on insect behavior and life histories is supported by the National Science Foundation.

LITERATURE CITED

Bell, G. 1980. The costs of reproduction and their consequences. *Am. Nat.* **116**:45–76.

Berthold, P., and U. Querner. 1981. Genetic basis of migratory behavior in European warblers. *Science* **212**:77–79.

Caldwell, R. L., and Hegmann, J. P. 1969. Heritability of flight duration in the milkweed bug. *Lygaeus kalmii*. *Nature (London)* **223**:91–92.

Colwell, R. K. 1974. Predictability, constancy, and contingency of periodic phenomena. *Ecology* **55**:1148–1153.

Danilevskii, A. S. 1965. *Photoperiodism and seasonal development of insects.* Oliver & Boyd, Edinburgh.

Denno, R. F., and H. Dingle (eds). 1981a. *Insect life history patterns: Habitat and geographic variation.* Springer-Verlag, New York.

Denno, R. F., and H. Dingle. 1981b. Considerations for the development of a more general life history theory. In R. F. Denno and H. Dingle (eds.), *Insect life history patterns: Habitat and geographic variation*, pp. 1–6. Springer-Verlag, New York.

Denno, R. F., and E. E. Grissell. 1979. The adaptiveness of wing-dimorphism in the salt marsh-inhabiting planthopper, *Prokelesia marginata* (Homoptera: Delphacidae). *Ecology* **60**:221–236.

Denno, R. F., M. J. Raupp, D. W. Tallamy, and C. F. Reichelderfer. 1980. Migration in heterogeneous environments. Differences in habitat selection between the wing forms of the dimorphic planthopper. *Prokelesia marginata* (Homoptera: Delphacidae). *Ecology* **61**:859–867.

Denno, R. F., M. J. Raupp, and D. W. Tallamy. 1981. Organization of a guild of sap-feeding insects: Equilibrium vs. nonequilibrium coexistence. In R. F. Denno and H. Dingle (eds.), *Insect life history patterns: Habitat and geographic variation*, pp. 151–182. Springer-Verlag, New York.

Derr, J. A. 1980a. Coevolution of the life history of a tropical seed-feeding insect and its food plants. *Ecology* **61**:881–892.

Derr, J. A. 1980b. The nature of variation in life history characters of *Dysdercus bimaculatus* (Heteroptera: Pyrrhocoridae), a colonizing species. *Evolution* **34**:548–557.

Derr, J. A., and J. K. Ord. 1983. Trends in variance: A test of an evolutionary theory. *Biometrics* (in press).

Derr, J. A., B. Alden, and H. Dingle. 1981. Insect life histories in relation to migration, body size, and host plant array: A comparative study of *Dysdercus*. *J. Anim. Ecol.* **50**:181–193.

Dickerson, G. E. 1955. Genetic slippage in response to selection for multiple objectives. *Cold Spring Harbor Symp. Quant. Biol.* **20**:213–224.

Dingle, H. 1968. The influence of environment and heredity on flight activity in the milkweed bug *Oncopeltus. J. Exp. Biol.* **48**:175–184.

Dingle, H. (ed.). 1978a. *Evolution of insect migration and diapause.* Springer-Verlag, New York.

Dingle, H. 1978b. Migration and diapause in tropical, temperate, and island milkweed bugs. In H. Dingle (ed.), *Evolution of insect migration and diapause*, pp. 254–276. Springer-Verlag, New York.

Dingle, H. 1980. Ecology and evolution of migration. In S. A. Gauthreaux, Jr. (ed.), *Animal migration, orientation, and navigation*, pp. 1–101. Academic Press, New York.

Dingle, H. 1981. Geographical variation and behavioral flexibility in milkweed bug life histories. In R. F. Denno and H. Dingle (eds.), *Insect life history patterns: Habitat and geographical variation*, pp. 55–73. Springer-Verlag, New York.

Dingle, H. 1983. Migration. In G. A. Kerkut and L. I. Gilbert (eds.), *Comprehensive insect physiology, biochemistry and pharmacology*, Vol. 9, Ch. 14. Pergamon Press, Oxford (in press).

Dingle, H., and Arora, G. 1973. Experimental studies of migration in bugs of the genus *Dysdercus*. *Oecologia* **12**:119–140.

Dingle, H., and J. P. Hegmann (eds.). 1982. *Evolution and genetics of life histories.* Springer-Verlag, New York.

Dingle, H., Brown, C. K., and Hegmann, J. P. 1977. The nature of genetic variance influencing photoperiodic diapause in a migrant insect, *Oncopeltus fasciatus. Am. Nat.* **111**:1047–1059.

Dingle, H., N. R. Blakley, and E. R. Miller. 1980a. Variation in body size and flight performance in milkweed bugs (*Oncopeltus*). *Evolution* **34**:371–385.

Dingle, H., B. M. Alden, N. R. Blakley, D. Kopec, and E. R. Miller. 1980b. Variation in photoperiodic response within and among species of milkweed bugs (*Oncopeltus*). *Evolution* **34**:356–370.

Dingle, H., W. S. Blau, C. K. Brown, and J. P. Hegmann. 1982. Population crosses and the genetic structure of milkweed bug life histories. In H. Dingle and J. P. Hegmann (eds.), *Evolution and Genetics of Life Histories*, pp. 209–229. Springer-Verlag, New York.

Falconer, D. S. 1981. *Introduction to quantitative genetics*, 2nd ed. Ronald, New York.

Frazetta, T. H. 1975. *Complex adaptations in evolving populations*. Sinauer, Sunderland, Massachusetts.

Harrison, R. G. 1980. Dispersal polymorphisms in insects. *Annu. Rev. Ecol. Syst.* **11**:95–118.

Hegmann, J. P., and H. Dingle. 1982. Phenotypic and genetic covariance structure in milkweed bug life history traits. In H. Dingle and J. P. Hegmann (eds.), *Evolution and genetics of life histories*, pp. 177–184. Springer-Verlag, New York.

Hoffmann, R. J. 1978. Environmental uncertainty and evolution of physiological adaptation in *Colias* butterflies. *Am. Nat.* **112**:999–1015.

Hoy, M. A. 1978. Variability in diapause attributes of insects and mites: Some evolutionary and practical implications. In H. Dingle (ed.), *Evolution of insect migration and diapause*, pp. 101–126. Springer-Verlag, New York.

Istock, C. A. 1978. Fitness variation in a natural population. In H. Dingle (ed.), *Evolution of insect migration and diapause*, pp. 171–190. Springer-Verlag, New York.

Istock, C. A. 1981. Natural selection and life history variation: Theory plus lessons from a mosquito. In R. F. Denno and H. Dingle (eds.), *Insect life history patterns: Habitat and geographic variation*, pp. 113–128. Springer-Verlag, New York.

Istock, C. A. 1982. The extent and consequences of heritable variation for fitness characters. In C. R. King and P. S. Dawson (eds.), *Population biology: Retrospect and prospect* (Oregon State Univ. Colloq. Pop. Biol.), pp. 61–96 Columbia University Press, New York.

Johnson, C. G. 1976. Lability of the flight system: A context for functional adaptation. *R. Entomol. Soc. Symp.* **7**:217–234.

Kennedy, J. S. 1961. A turning point in the study of insect migration. *Nature (London)* **189**:785–791.

Kennedy, J. S., and C. O. Booth. 1963. Co-ordination of successive activities in an aphid. The effect of flight on the settling responses. *J. Exp. Biol.* **40**:351–369.

Lande, R. 1975. The maintenance of genetic variability by mutation in a polygenic character with linked loci. *Genet. Res.* **26**:221–234.

Lande, R. 1982a. A quantitative genetic theory of life history evolution. *Ecology* **63**:607–615.

Lande, R. 1982b. Elements of a quantitative genetic model of life history evolution. In H. Dingle and J. P. Hegmann (eds.), *Evolution and genetics of life histories*, pp. 21–29. Springer-Verlag, New York.

Lavie, B., and U. Ritte. 1978. The relation between dispersal behavior and reproductive fitness in the flour beetle *Tribolium castaneum*. *Can. J. Genet. Cytol.* **20**:589–595.

Livdahl, T. P. 1979. Environmental uncertainty and selection for life-cycle delays in opportunistic species. *Am. Nat.* **113**:835–842.

Lumme, J. 1978. Phenology and photoperiodic diapause in northern populations of *Drosophila*. In H. Dingle (ed.), *Evolution of insect migration and diapause*, pp. 145–170. Springer-Verlag, New York.

Lumme, J., and L. Keränen. 1978. Photoperiodic diapause in *Drosophila lummei* Hackman is controlled by an X-chromosomal factor. *Hereditas* **89**:261–262.

MacArthur, R. H., and E. O. Wilson. 1967. *The theory of island biogeography.* Princeton University Press, Princeton, New Jersey.

Miller, E. R., and H. Dingle. 1982. The effect of host plant phenology on reproduction of the milkweed bug, *Oncopeltus fasciatus* in tropical Florida. *Oecologia* **52**:97–103.

Morris, R. F., and W. C. Fulton. 1970. Heritability of diapause intensity in *Hyphantria cunea* and correlated fitness responses. *Can. Entomol.* **102**:927–938.

Murton, R. K., and N. H. Westwood. 1977. *Avian breeding cycles.* Clarendon Press, Oxford.

Nichols, J. D., W. Conley, B. Batt, and A. R. Tipton. 1976. Temporally dynamic reproductive strategies and the concept of *r-* and *k*-selection. *Am. Nat.* **110**:995–1005.

Palmer, J. O. 1982. Photoperiodic control of reproduction in the milkweed leaf beetle *Labidomera clivicollis. Physiol. Entomol.***8**:187–194.

Rankin, M. A. 1978. Hormonal control of insect migratory behavior. In H. Dingle (ed.), *Evolution of insect migration and diapause,* pp. 5–32. Springer-Verlag, New York.

Rankin, S. M., and M. A. Rankin. 1980. The hormonal control of migratory flight behaviour in the convergent ladybird beetle *Hippodamia convergens. Physiol. Entomol.* **5**:175–182.

Rasmuson, B., M. Rasmuson, and J. Nygren. 1977. Genetically controlled differences in behaviour between cycling and non-cycling populations of field vole (*Microtis agrestis*). *Hereditas* **87**:33–42.

Reed, G. L., W. D. Guthrie, W. B. Showers, B. D. Barry, and D. F. Cox. 1981. Sex-linked inheritence of diapause in the European corn borer: Its significance to diapause physiology and environmental response of the insect. *Ann. Entomol. Soc. Am.* **74**:1–8.

Ritte, U., and B. Lavie. 1977. The genetic basis of dispersal behavior in the flour beetle *Tribolium castaneum. Can. J. Genet. Cytol.* **19**:717–722.

Rygg, T. D. 1966. Flight of *Oscinella frit* L. (Diptera, Chloropidae) females in relation to age and ovary development. *Entomol. Exp. Appl.* **9**:74–84.

Safriel, U. N., and U. Ritte. 1980. Criteria for the identification of potential colonizers. *Biol. J. Linn. Soc.* **13**:287–297.

Simberloff, D. 1981. What makes a good island colonist? In R. F. Denno and H. Dingle (eds.), *Insect life history patterns: Habitat and geographic variation,* pp. 195–206. Springer-Verlag, New York.

Slansky, F. 1980. Food consumption and reproduction as affected by tethered flight in female milkweed bugs (*Oncopeltus fasciatus*). *Entomol. Exp. Appl.* **28**:277–286.

Solbreck, C. 1978. Migration, diapause, and direct development as alternative life histories in a seed bug, *Neacoryphus bicrucis.* In H. Dingle (ed.), *Evolution of insect migration and diapause,* pp. 195–217. Springer-Verlag, New York.

Stearns, S. C. 1976. Life history tactics: A review of the ideas. *Q. Rev. Biol.* **51**:3–47.

Stearns, S. C. 1977. The evolution of life history traits. *Annu. Rev. Ecol. Syst.* **8**:145–172.

Stearns, S. C. 1981. On measuring fluctuating environments: Predictability, constancy, and contingency. *Ecology* **63**:185–199.

Tanaka, S. 1976. Wing polymorphism, egg production and adult longevity in *Pteronemobius taprobanensis* Walker (Orthoptera, Gryllidae). *Kontyu* **44**:327–333.

Tauber, C. A., and M. J. Tauber. 1981. Insect seasonal cycles: Genetics and evolution. *Annu. Rev. Ecol. Syst.* **12**:281–308.

Tauber, M. J., and C. A. Tauber. 1976. Insect seasonality: Diapause maintenance, termination, and postdiapause development. *Annu. Rev. Entomol.* **21**:81–107.

Taylor, L. R., and R. A. J. Taylor. 1978. The dynamics of spatial behaviour. In F. J. Ebling and D. M. Stoddart (eds.), *Population control by social behaviour,* pp. 181–212. Institute of Biology, London.

Vepsäläinen, K. 1978. Wing dimorphism and diapause in *Gerris*: Determination and adaptive significance. In H. Dingle (ed.), *Evolution of insect migration and diapause*, pp. 218–253. Springer-Verlag, New York.

Wu, A.-C. 1981. *Life History Traits Correlated with Emigration in Flour Beetle Populations.* Ph.D. Thesis, University of Illinois at Chicago Circle.

Ziegler, J. R. 1976. Evolution of the migration response: Emigration by *Tribolium* and the influence of age. *Evolution* **30**:579–592.

CHAPTER 7

Complex Life Cycles and Community Organization in Amphibians

HENRY M. WILBUR
Department of Zoology
Duke University
Durham, North Carolina

CONTENTS

1   INTRODUCTION

I define a complex life cycle (CLC) as a life history in which there is an abrupt, usually irreversible change in an individual's ecology and morphology, usually with changes in physiology and behavior, which are sufficient for an ecologist to declare that the organism sequentially occupies two niches (Wilbur, 1980). Most amphibians have CLCs with a free-living, aquatic larval stage followed by an abrupt metamorphosis to a terrestrial adult. The larvae of pond-breeding amphibians often strongly interact in communities in which the structure of the food web frequently changes.

The ecological implications of CLSs can be studied effectively in amphibians using both comparative and experimental methods. Amphibian larvae range from omnivorous filter feeders to selective predators. The comparative method is useful because the life histories of many species vary geographically, there is a wide diversity of life histories within some genera, and the class as a whole has extant taxa displaying a range of reproductive modes including reproduction in individuals with larval-like morphology and direct development. In the latter case the "larval" stage is passed in the egg capsule, often with parental care (Dent, 1968; Frazer, 1977; Goin and Goin, 1962; Heyer, 1969; Kirtisinghe, 1946; Lutz, 1947, 1948; McDiarmid, 1978; Martin, 1967; Menzies and Taylor, 1977; Poynton, 1964; Salthe and Mecham, 1974; Van Dijk, 1971). A few taxa are even viviparous (Frazer, 1977; Hogarth, 1976; Parker, 1956; Wake, 1977).

Amphibians are especially suited for experimental study of the ecology and maintenance of CLCs. Many species are easily cultured in the laboratory where food, density, and temperature can be rigorously controlled throughout the larval period (Wilbur, 1977a, 1982). Enclosures in natural ponds can be used to control initial densities and species composition of larvae in replicated experiments (Wilbur, 1972, 1976) set in an arena where chemical, physical, and most biological variables, except the species being manipulated, are free to vary naturally. Artificial ponds can be constructed to provide replicated, nearly natural ecosystems for the study of colonization and species interactions (Morin, 1981; Wilbur and Travis, 1982). In artificial ponds all aspects of the initial ecosystem are under experimental control but replicates may diverge due to their location or chance historical effects, such as the sequence of colonization. Finally, enclosed ponds can be used to control nearly all components of a pond's trophic structure. This chapter reviews much of this new experimental methodology and suggests how its results can be used to address general issues in community theory and to interpret the diversity of life histories that occur in amphibians.

This chapter advocates and demonstrates the experimental approach to community ecology. The evaluation of the roles of species and of the strength of species interactions requires detailed understanding of natural history. Only field experience can point to the relevant questions and suggest appropriate species to use in testing predictions from theory. An early section of this chapter describes the sampling methods we developed to assay population densities in amphibian communities. Observational studies can contribute to realistic theories and can sometimes be used to test predictions. There are many important issues that can only be answered by comparative or long-term studies of undisturbed communities. A danger of reliance only on observation is that explanations are often post facto and limited only by the cleverness of the natural historian. Statistical analysis of observations made in nature yields correlations that may support explanations, but manipulative experiments, when possible, are usually far more powerful in rigorously testing predictions from theory. The misuses of so-called natural experiments, which are never properly controlled, is discussed by Connell (1979). A proper statistical design for an experiment forces the investigator to state the null hypothesis being tested and to plan the proper statistical analyses before data are collected (Hairston, 1980, 1981). Recent pleas for testing null hypotheses (e.g., Connor and Simberloff, 1979) are a healthy tonic that should be taken seriously by ecologists. I find it frustrating that most papers addressing the "null hypothesis problem" are post facto examinations of old data. I have not seen increased attention to the appropriate use of null hypotheses and the power of statistical designs in the planning of ecological experiments.

Most of this chapter discusses generalizations about the importance of competition and predation in determining the success of larval amphibians based on field and laboratory experiments. These experiments help to clarify the interaction among body size, competition, and predation in determining the structure of food webs and relative abundance of species. Only frequent visits to the field and detailed observations of the dynamics of undisturbed populations can maintain the realism of these studies and lead to an understanding of the evolution of life history diversity in amphibians.

2 METHODS

2.1 Field Sampling Techniques

Small temporary ponds that are too transient to support fish populations are an important breeding habitat of amphibians with free-living larval stages. Sampling these structurally complex habitats poses special problems. Even the construction of a faunal list can be biased by choice of sampling gear and by different risks of capture of the various species present. Many ecological questions require not only a list of species present but also some measure of their relative abundances or densities. I require estimates of natural densities of amphibian larvae and aquatic insects to calibrate the realism of experimental treatments in artificial ponds. Such data are difficult to obtain.

My students and I have settled on a 0.5-m by 1.0-m box sampler, made of 24 ga galvanized sheet steel (a variation of the device used by Turnipseed and Altig, 1975), as a device to estimate density and microhabitat distribution of aquatic insects and amphibians accurately. By walking slowly through the pond and plunging the sampler horizontally into the pond's bottom, a section of the water column 0.5 m² in cross section can be "captured." The sampler's metal edge forms a seal with the pond bottom and the solid metal sides prevent escape of the captured animals. The enclosed water column can then be sampled with a net the same width as the box for either a standard number of sweeps or until no animals have been captured after a preset number of sweeps (Fig. 1). The risk of capture per individual increases with

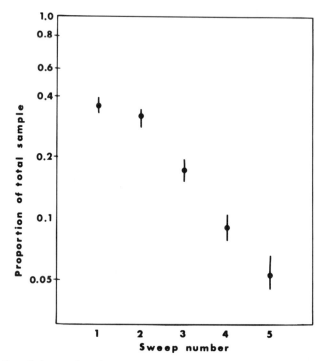

Figure 1. Cumulative number of *Notophthalmus viridescens* larvae captured in five successive sweeps with a net 0.5×0.33 m² with 2 mm mesh of the water column enclosed in a 1.0×0.5 m² box sampler. Twenty samples were made on each of three dates (9 June, 21 June, and 6 July 1982). There were no significant differences in the number of larvae caught on the three dates in each of the five sweeps per sample (MANOVA, Wilk's Criterion, $p = 0.69$).

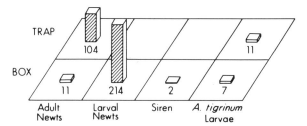

Figure 2. Number of *Ambystoma tigrinum* larvae, *Notophthalmus viridescens* adults and larvae, and *Siren intermedia* adults captured by two sampling devices. The box sample was 0.50 × 1.00 m² and was cleared by five sweeps with a 0.50 × 0.33 m² net with 2 mm mesh. Twenty box samples were made between 1345 and 1650 EST on 27 May 1982. Forty unbaited minnow traps were evenly distributed about the pond and left undisturbed between 1300 and 2100 EST on the same date.

successive sweeps, probably because of removal of emergent vegetation, filamentous algae, and litter, as well as an increase in the turbidity of the water. Haphazard sampling with a dipnet, seining, or trapping may be easier methods to obtain a species list, but short of poisoning or other destructive techniques we know of no less biased method than our box sampler. Estimates of relative abundance taken on the same day in one pond by 50 plastic minnow traps and 20 box samples yielded very different estimates of relative abundance (Fig. 2). Traps alone cannot measure density except in conjunction with a mark–recapture study. Traps may yield overestimates of density and biased sex ratios if the capture of one individual changes the likelihood of capturing additional individuals.

Box samples are used to estimate abundance of species in natural ponds to provide information for planning and interpretation of experiments. In laboratory experiments I have used about five levels of initial density ranging from slightly above to below natural densities to obtain the form of the relationship between initial density of hatchlings and survival, body size at metamorphosis, and length of larval period (Wilbur, 1976, 1977a,b, 1982). Studies of competition and predation have also used realistic initial densities of both predators and prey (Morin, 1981; Wilbur et al., 1983). One outstanding advantage of amphibians is that many species have clutch sizes of several thousand eggs and breeding is often synchronized within local populations. This allows experiments with both realistic densities and sufficient replication to examine statistically all interactions among factors controlled by the experimental design.

2.2 Artificial Ponds

We have used galvanized steel tanks designed for watering livestock and available in several sizes and shapes as experimental ponds. Painting them with epoxy prevents leaching of toxic ions from the gavanizing and allows successful establishment of a wide range of phytoplankton, periphyton, zooplankton, freshwater sponges, hydras, molluscs, aquatic insects, and amphibians. These tanks can be placed in natural habitats to observe colonization and food web dynamics under nearly natural, but replicated, conditions. Alternatively they can be used as containers for artificially constructed pond communities for use in controlled experiments with adequate replication. In North Carolina, precipitation exceeds evaporation in all months of most years so that rate of decrease in water volume in tanks can be controlled as an experimental factor by the use of overflow pipes attached to drains in each tank. Without this intervention tanks remained filled within about 5 cm of capacity even in an unusually dry year such as 1981. The straight sides and hence constant surface area is the most obvious artificiality of these ponds.

In the late summer of 1977, 16 tanks, each 2.13 m in diameter and 0.63 m deep, were sunk flush with the ground in four sets of four tanks each on a study area in the sandhills region of Scotland County, North Carolina. Two sets were near permanent streams not used by pond-breeding amphibians and two sets were near temporary ponds. Within each of the four sets, two ponds were placed in a longleaf pine (*Pinus palustris*)–turkey oak (*Quercus laevis*) savannah and two were placed at the edge of evergreen shrub bogs (pocosins). Tanks were filled by pumping water through a plankton net from the nearest stream or pond. Tanks were open to colonization by amphibians and insects and were not disturbed except by sampling. Over 130 species of insects, about 25 species of microcrustacea, and 11 species of amphibians colonized the tanks within the first 2 years of their establishment (Wilbur and Travis, 1982).

Each tank was sampled 50 times between August 1977 and October 1979. A sample consisted of one sweep around the top perimeter with a D-shaped aerial insect net and a second sweep around the bottom perimeter with a rectangular net 0.5 m wide by 0.33 m high with 2-mm mesh. The contents of both nets were examined separately with all individuals immediately identified, counted, and returned to the tank. Samples were preserved only as required to verify field identifications.

On several occasions this sampling procedure was calibrated by taking repeated cumulative samples until no further organisms were captured in 20 sweeps. These exhaustive samples were used to compute the fraction of the total population that was captured on the first "standard" sample. Risks of capture per sample per individual ranged from 0.004 to 0.522 among a wide range of insect and amphibian taxa. The average value for all individuals was 0.35 in two experiments (Wilbur, in preparation). A mean value of 0.35 indicates that we are 95% confident that any species with a population size greater than eight individuals will be recorded as present by our standard sample.

To understand the problems of estimating population size in *Notophthalmus viridescens*, the top predator in these communities, 20 or 100 adult newts were released into each of four tanks that had well established communities but had not had newts in any of the previous 50 samples. About 2 hr after release, before emigration was likely, the four tanks were repeatedly sampled without replacement. Zippen (1956) estimates of populations size and risk of capture were made after each sample (Fig. 3). The Zippen technique treats the number captured in each sample as a multinomial distribution and computes maximum likelihood estimates of the risk of capture and the size of the population being sampled. It took 10–15 samples to obtain population estimates with standard errors less than 10% of the estimate. This intensity of sampling is very destructive to the structural and biological complexity of these communities. Therefore, we use input–output experiments in which we introduce exactly known numbers of individuals into artificial ponds with covers that prevent colonization and retain the survivors until they are captured as they attempt to leave the ponds.

This sampling exercise dramatized the difficulty of estimating parameters of populations in natural ponds that do not have the structural simplicity of our cylindrical tanks with straight sides and no emergent vegetation. Our

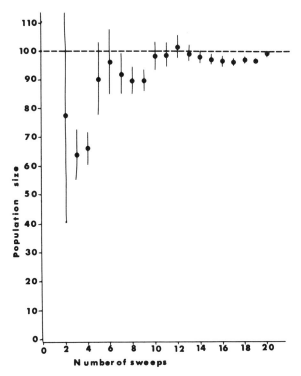

Figure 3. Successive Zippen (1956) estimates ($\pm$ SE) of the number of adult *Notophthalmus viridesens* in a 2.13 m diameter by 0.61 m deep artificial pond. Samples were made without replacement by sweeping once around the bottom perimeter of the pond with a net 0.5 m wide and 0.33 m high. One hundred newts were released in the pond about 2 hr before sampling began.

confidence in the efficiency of the box sampler described above rests on the fact that the width of our net is the same as the width of the sampler so that rapid swimmers cannot avoid capture.

Our observational study has demonstrated that pond communities have a dynamic foodweb with dramatic changes in trophic structure as colonization proceeds. Differences in the terrestrial habitat (location of the tank) and successional effects between years are nearly as important as seasonal differences within a year in contributing to a rejection of a null hypothesis for random structure of the communities of insects and amphibians (Wilbur and Travis, 1982, and in preparation). More importantly, this study is my touchstone and source of ideas for studies in which colonization is rigorously controlled to allow replicated experiments.

An array of 50 tanks, each 1.52 m in diameter and 0.63 m deep and covered with a tight-fitting lid of fiberglass window screening, is used for experiments in which immigration and emigration are controlled. Tanks are emptied, scrubbed, and refilled with water each winter. One kg of dried grass raked from the margin of a natural pond, and 50 g of commercial trout food are added to each tank as a nutrient base for decomposers. A constant number of stems of a rooted aquatic plant (*Myriophyllum*) are added to each tank to increase structural complexity. Randomized aliquots from a well-mixed pool of zooplankton and microorganisms in pond water collected from several natural ponds are added to ensure realistic levels of primary producers and consumers. These replicated pond communities are the standard unit of our experiments (Morin, 1981, 1982, 1983a,b; Morin et al. 1983; Wilbur et al. 1983; Wilbur and Alford, in preparation). Exactly known numbers of hatchlings of predator and prey species of amphibians can be added to the tanks and allowed to complete their larval stages in these seminatural ponds. Results of experiments are assayed for the survival, length of larval period, and body sizes at metamorphosis of each species.

Small newly excavated ponds, 3 m in diameter and a standard depth, divided into quarters and encircled by fences that allow complete counts of the arriving and departing adult and juvenile amphibians are being used by my student R. C. Chambers in experiments in which insects, plankton, and microorganisms have natural rates of colonization, but amphibians can be manipulated using each pond as a statistical "block" with the four compartments as experimental cells.

3 POPULATION BIOLOGY

I have argued elsewhere (Wilbur, 1980) that growth is the key to understanding the larval ecology of amphibians. The larval period is thought to be an adaptation to exploit transient opportunities for rapid growth that occur after ponds fill (Seale, 1980; Slade and Wassersug, 1975; Wassersug 1975; Wilbur and Collins, 1973; Wilbur, 1980). Advantages of exploiting these opportunities are weighed against two liabilities: predation and desiccation (Fig. 4). Temporary ponds are also exploited by a wide range of predators that, under some conditions, may preclude successful metamorphosis of some species of anurans (Channing, 1976; Grubb, 1972; Jameson, 1956; Voris and Bacon, 1966; Werschkul and Christensen, 1977). Rapid growth may provide a size escape from gape-limited predators (Morin, 1981). Fish may exclude tadpoles from all but the littoral zone of permanent ponds (Macan, 1966). Ponds that dry often enough to preclude fish but retain water long enough to allow completion of the larval period are the principal habitat of anurans (Channing, 1976; Collins and Wilbur, 1979; Crump, 1971; Heyer et al., 1975). Within this range of lengths of expected habitat duration, there may be considerable variation between years in the length of time that a particular pond has water. Death by desiccation is a common source of mass mortality in many species (Bannikov, 1948; Black, 1974; Blair, 1957; Creusere and Whitford, 1976; Martof, 1952; Peckham and Dineen, 1955; Tevis, 1966). Individuals with rapid growth may escape drying ponds in which slower growing tadpoles are doomed.

3.1 Lognormal Body Size Distributions

Growth and differentiation are mutiplicative processes that can generate lognormal distributions (see Johnson and Kotz, 1970, or Aitchison and Brown, 1969, for mathematical discussions of the lognormal). The early stages of growth of most organisms can be fit by the exponential equation

$$M(t) = M(0) \exp(gt)$$

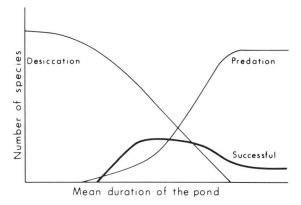

Figure 4. Diagrammatic model of how the number of species in a pond (heavy line) is determined by the mean length of time the pond holds water. The left descending curve represents the decrease in the number of species that are excluded because the pond does last long enough for them to complete their larval period. The right ascending curve represents the increase in the number of species eliminated by predators in longer-lasting ponds.

where $M(t)$ is mass at time t; $M(0)$ is mass at hatching; and g is a constant, the instantaneous growth rate. The initial stage of popular asymptotic growth models (e.g., Gompertz, Bertalanffy, and logistic) are all exponential. They differ primarily in location of the inflection point and form of their approach to the maximum size. Measurements of the early phases of growth are rarely accurate enough to distinguish among these models (Ricklefs, 1967).

If a cohort of animals has a constant initial body size, $M(0)$, but a normal distribution of growth rates, g's, then $M(t)$ will have a lognormal distribution at each time during the exponential phase of growth. Two individuals growing according to their particular values of g will maintain a constant difference in log $M(t)$ and hence a constant ratio of masses on an arithmetic scale. Differences among individual growth rates may be due to genetic endowment or maternal effects, such as yolk quality, or chance differences in the environment in which the egg develops (Travis, 1981).

The assumption of a constant size at hatching, $M(0)$, can easily be relaxed if $M(0)$, the result of embryonic growth, is assumed to have a lognormal distribution. The logarithm of size at time t is then the sum of two normally distributed variates, log $M(0)$ and gt. Log $M(t)$ would also be normally distributed, with a variance equal to the sum of the variances of log $M(0)$ and (gt) plus twice their covariance,

$$\text{Var } [\log M(t)] = \text{Var } [\log M(0)] + t^2 \text{ Var}(g) + 2\text{Cov } [gt, \log M(0)].$$

As a cohort of isolated individuals grows, the body size distributions of the cohort at successive times can be described by a family of lognormal distributions determined by the values of the individuals' initial body sizes and growth coefficients. This effect was demonstrated by raising 21 *Acris gryllus* larvae, all full sibs that hatched on the same day, in individual dishes with 250 ml of well water. The water was changed and 0.10 g of a dry mixture of powdered rabbit fodder and high protein baby cereal was added to each dish twice a week. Tadpoles were individually weighed once a week from the time they could be safely handled (44 days after hatching) until some individuals began to lose weight prior to metamorphosis (80 days after hatching). Growth curves of individuals were used to estimate the mean ($g = 0.0259$), variance (9.330×10^{-5}), and to test for normality ($P = 0.657$) of growth coefficients. These estimates were then used to simulate the body size trajectory of a population of 500 isolated individuals assuming size at hatching was a constant (Fig. 5).

This exercise demonstrated that skewed distributions developed in cohorts of individuals reared apart as a consequence of normal variation in intrinsic growth rates. These studies establish a null hypothesis for competition. To test for effects of competition among individuals raised together, two experiments were run concurrently with the experiment reported above. Pairs of tadpoles from the same sibship as the isolated tadpoles were raised in the same type of containers and with the same total food ration per con-

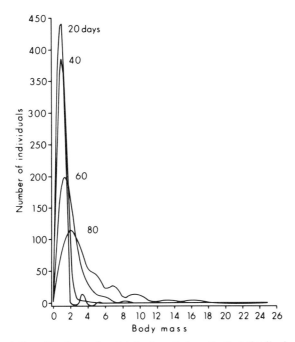

Figure 5. Simulation of the body size distribution of a hypothetical distribution of 500 tadpoles raised in individual containers. The growth rate for each tadpole was drawn from a normal distribution with parameters (mean = 0.259, variance = 0.00966) equal to those observed in a cohort of 20 isolated tadpoles fed 0.10 g of rabbit fodder and high protein baby cereal twice a week and kept in 250 ml of well water at about 22°C. The cohort passes through a family of lognormal distributions as individuals grow. The smallest size at metamorphosis observed was 0.132 g.

tainer. A group of 20 full sibs from a second clutch, laid on the same night as the first clutch, was also raised in the same size container and with the same total food ration.

All pairs of larvae developed a significant difference between the two members by the first weighing, 44 days after hatching (t test for paired observations with equal variances, $t = 4.12$, 5 df, $P < 0.01$). There was an Allee effect in which the larger member of the pair grew faster than would have been expected if it were alone, a well-known effect in tadpoles (Gromko et al., 1973; Rose, 1960; Steinwascher, 1978a,b; Wilbur, 1977b). Smaller members of the pairs began losing weight in the middle of their larval period (Fig. 6). The cohort of 20 larvae developed an asymmetrical distribution because most individuals grew slowly and the population was dominated by one individual, which grew more slowly than expected if it had grown alone.

Wilbur and Collins (1973) argued that metamorphosis occurs between a lower and upper size threshold of body size. The smallest and largest sizes at metamorphosis of *Acris gryllus* in the experiment are 0.132 and 0.432 g, respectively. Figure 6 also shows the number of individuals that will have

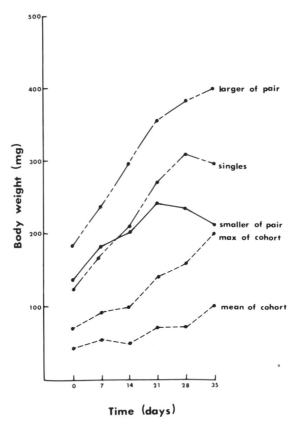

Figure 6. Body masses of *Acris gryllus* tadpoles raised at about 22°C in 250 ml of pond water and fed 0.10 g of rabbit fodder and high protein baby cereal twice a week. All were weighed individually weekly from the start of the experiment (day 0), which was 44 days after hatching. Three groups of larvae are represented: the mean size of 20 isolated individuals, the mean maximum and mean minimum of size pairs, and the maximum and mean of 20 larvae raised together. All larvae were the same age and each dish received the same total food ration.

exceeded the lower size threshold as a function of time. The retarded growth imposed on individuals at high density results in a very long larval period because the probability of exceeding the lower size threshold for metamorphosis would remain low until very late in the season. This prolonged larval period increases the cumulative risk of predation, parasitism, and desiccation. The effects of population density on growth are sufficient to explain the exponential decreases of body mass at metamorphosis and survival to metamorphosis as a function of intial density. Density effects also explain variation in mean length of larval period found in several field and laboratory experiments with larval frogs and salamanders (Crump, 1981; Wilbur, 1976, 1980, 1982; Wilbur and Collins, 1973).

4   COMMUNITY ECOLOGY

4.1   Predation

4.1.1   Growing Predators and Prey.   Trophic relationships between species of amphibians, and also between amphibians and insects, are continuously changing as both predators and prey grow (Brockleman, 1969; Walters, 1975). A frog larva's risk of predation is certainly a function of its body size. Rapid growth may provide an escape from some gape-limited predators, such as adult newts (*Notophthalmus viridescens*) and many insects. Although higher survival may accrue to large tadpoles, they may still be exposed to mutilation by gape-limited predators. High rates of tail damage and signs of regeneration are often seen in populations of tadpoles exposed to newts or fish (Worthington, 1968).

Morin's (in preparation) experiments demonstrated different rates of tail damage in populations of *Scaphiopus holbrooki* tadpoles exposed to four densities of adult newts. The conspicuous tails of some species (e.g., jet black tips in *Acris* and bright red in *Hyla femoralis*) may in fact be an adaptation to such "grazing" (Altig, 1972; Bragg, 1957; Heyer, 1974). How the cost of tail regeneration affects size at metamorphosis or length of the larval period has not been studied.

Figure 7 suggests how trophic relationships may vary with the sizes of predators and prey. Only detailed experiments can determine the boundaries of the region of likely mutilation. The boundaries may be curved due to behavioral changes correlated with body size or ontogenetic changes in prey defences.

4.1.2   Cannibalism.   Many species of salamanders and a few species of frogs are exposed to cannibalism either by larger larvae or adults. Figure 7 can apply to two members of the same population, even full sibs. The lognormal distributions described in Section 3.1 could lead to size differentials permitting cannibalism within cohorts of the same age. Such may be the

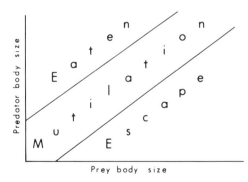

Figure 7. Diagrammatic model of the change in trophic relationship between species of different body sizes. As predators and prey grow they can move from predation to mutilation to escape (or competition). This space can also be used to represent the transition from competition to cannibalism within populations of growing predators with skewed body size distributions.

case in "cannibal" morphs that occur in southwestern populations of *Ambystoma tigrinum*. In these morphs head allometry and dentition are modified to create distinctive individuals that are trophic specialists in low frequency in populations (Collins, 1981; Rose and Armentrout, 1976). In the Rocky Mountains overwintering *A. tigrinum* can eliminate younger age classes by cannibalism (Burger, 1950; but see also Collins, 1981). A similar polymorphism occurs in the spadefoot toad genus *Scaphiopus* (Bragg, 1956, 1964).

Cannibalism occurs in natural populations of newts, *Notophthalmus viridescens*. Adults are voracious egg predators and also may eat small larvae (Bishop, 1941; Christman and Franz, 1973; Hamilton, 1932; Wood and Goodwin, 1954). Large newt larvae may eat small larvae, at least they are capable of cannibalism in the laboratory (Harris, in preparation). Cannibalism has been reported in several species of frogs. Adult bullfrogs (*Rana catesbeiana*) eat conspecific larvae (Korschben and Baskett, 1963) and eat many metamorphs as they leave the pond (personal observation). Frog larvae are generally thought to be herbivores and opportunistic carrion feeders, although tadpoles eat frog eggs (Heusser, 1970) and other larvae (Borland and Rugh, 1943; Orton, 1954). Several exceptional species are clearly evolved carnivores that are often cannibalistic [e.g., *Scaphiopus* (Bragg 1956, 1964) and *Hyla rosenbergii* (Kluge, 1981)].

In spite of recent reviews of cannibalism by Fox (1975) and Polis (1981), and the fascination of evolutionary ecologists with brood reduction (fratricide) in birds (Ricklefs, 1965) and salamanders (Kaplan and Sherman, 1980) the role of cannibalism as a mechanism of population regulation is not widely considered in theoretical models, except those for *Tribolium* (Mertz, 1972).

4.2 Shifts Between Competition and Predation

Differences in growth rates of predators and prey may lead to complex interactions involving switches from competition to mutilation to predation depending on size differences among interacting individuals. These possible modes of interaction can occur within a single population of *Ambystoma* or *Notophthalmus* or between different species resulting in a trophic web that is constantly changing in structure as interacting species colonize a pond, grow, and leave the pond.

I showed (Wilbur, 1972), using enclosures in a natural pond, that the relationship between *Ambystoma tigrinum* larvae and three smaller congeners (*A. maculatum, A. laterale,* and *A. tremblayi*) changes from competition, if the four species were raised in the absence of *Rana sylvatica* tadpoles, to predation if *A. tigrinum* obtained a size advantage by feeding on *Rana* tadpoles. This experimental result was confirmed by observations of *Ambystoma tigrinum* feeding in a single pond in 2 years with *Rana sylvatica* present in one and absent in the other (personal observation). Similar processes may occur in New Jersey (Hassinger et al., 1970) and in California (Anderson, 1968).

The ecological relationship between *Ambystoma opacum* and *Ambystoma maculatum* may shift from competition to predation depending on weather patterns and food availability. *Ambystoma opacum* migrates to pond sites in early fall and lays eggs under cover in the dry basin (Dunn, 1917). Females attend the eggs for several weeks, sometimes until they hatch as rising water floods the nest site (Kaplan and Crump, 1978). Larvae are free swimming predators of zooplankton and aquatic insects during late fall and winter. In late winter *Ambystoma maculatum* migrates to the ponds, which are now full of water and often have high densities of *A. opacum* larvae. *Ambystoma maculatum* eggs are laid in clumps surrounded by a tough outer envelope and not attended by parents. In early spring the *A. maculatum* hatch. If the size difference between the species is small, perhaps due to delayed hatching of *A. opacum* because of a dry autumn, cold temperatures in the winter, or high density of larvae relative to the food supply, then the two species may be competitors (Worthington, 1968, 1969). If *A. opacum* has reached a large size during winter, then it may be an effective predator of *A. macultum* almost precluding its successful metamorphosis (Shoop, 1974). Stein et al. (1954) report a reversal in which *A. opacum* breeds first and preys on small *A. triginum* but since *A. tigrinum* grows faster it obtains a size advantage and then preys on *A. opacum*. R. C. Chambers is manipulating dates of hatching of *A. opacum* and *A. maculatum* experimentally and raising them together and apart in experimental ponds described in Section 2.2. His work will elucidate the mechanics of shifts between competition and predation in these two species.

4.3   Competition

Food availability may determine both the body size distribution of a population and the level of competition between individuals and between species. Even isolated individuals raised in carefully controlled experiments will develop skewed distributions of body sizes due to intrinsic variation in growth rates. A simple model of this process (see Section 3.1) leads to the population passing through a family of lognormal distributions of body sizes. This model can be used to generate null hypotheses for competition. Amount of food available per individual will determine the parameters of the family of lognormals traced by the population as growth continues.

I have recently (1980) reviewed experimental studies of competition among species of amphibians. The outcome of competition can determine both risk of desiccation and of predation. Competition slows growth and, since a minimum size threshold is required for metamorphosis (Wilbur and Collins, 1973), it may also determine the length of exposure to gape-limited predators and the probability that metamorphosis will not be completed before the pond drys.

An experiment was done in 1982 (Wilbur and Alford, in preparation) in which the order of arrival of three species (and therefore their relative body sizes) is directly controlled in a replicated experiment using 30 artificial communities in cattle tanks. The design includes three replicates of all nine combinations of *Rana sphenocephala* and *Bufo americanus* being added on day 1, on day 7, or not at all. This design includes a control treatment of no species added on either date. After a month, *Hyla chrysoscelis* was added in equal densities to all 27 tanks plus three newly started communities. Preliminary analyses show dramatic effects of the order of arrival on both timing and body size at metamorphosis. A simple change in order of arrival can influence the relative body sizes of two species and hence their competitive position and their relative abundance as recruits into the terrestrial phase of the population. Such "historical" or "priority" effects are often discussed in the context of plant succession and biogeography, but they have not penetrated deeply into general ecological theory.

4.4   The Interaction Between Competition and Predation

There has been much speculation as to whether competition or predation is more important in determining survival of larval anurans (Brockelman, 1969; Calef, 1973; Cecil and Just, 1979; Dickman, 1968; Heyer et al., 1975; Licht, 1974; Wilbur, 1980). The interaction between competition and predation in assemblages of amphibians can become complex as both predators and prey may be entering ponds at different times and growing at different rates. I shall make a first approach to the problem by examining the interaction between adult newts, *Notophthalmus viridescens*, and *Rana sphenocephala* tadpoles.

4.4.1   Number of Metamorphs, a Model.   The first biological assumption of my model is that all adult newts have the same appetite for anuran eggs and hatchlings. This simplifying assumption ignores size structure in adult newts. In ponds we study in the Sandhills of North Carolina, adult newts are year-round residents that remain in the basins of drying ponds by burrowing into the litter or mud cracks and forage for terrestrial insects on moist nights. Newts therefore can prey on amphibian eggs as soon as they are laid after the ponds fill.

The second biological assumption is that newt predation happens during egg and hatchling stages because tadpoles rapidly obtain a size escape from predators before the trophic demands of the population exceeds the food supply. This is tantamont to defining a two-step process in which egg laying is followed by a period of risk of predation and then survivors of predation are exposed to competition. The sequential action of predation and then the possibility of competition are supported by our field observations and experimental studies (Morin, 1981, 1983a; Wilbur et al., 1983).

The first mathematical assumption is that an exponential function mimics the relationship between number of hatchlings escaping predation and density of predators. Let

E = number of eggs laid by all females in the population in a given bout of breeding

T = number of hatchlings that escape predation by obtaining a size refuge

M = number of tadpoles that successfully metamorphose

P = number of predators (adult newts)

v = a coefficient of predation

s = a coefficient of competition

Then assume that

$$\frac{T}{E} = \exp\left(-\frac{vP}{E}\right) \tag{1}$$

that is, the probability of surviving predation is an exponentially decreasing function of the ratio of predators to prey. This assumption would hold, for example, under the simple model that each predator has an equal and independent chance of killing each egg or hatchling.

The second mathematical assumption is that the probability of metamorphosis is exponentially related to the number of individuals that escape predation. I (Wilbur, 1976) derived the relationship

$$\frac{M}{T} = \exp(-sT) \tag{2}$$

from several empirical studies of density dependent survival to metamorphosis in the absence of predation. This model fits a wide variety of data involving frog and salamander larvae in both laboratory and field experiments and can be modified to include effects of interspecific competition (Wilbur, 1982).

Solving Equation (1) for T and substituting into Equation (2) yields an expression for the number of metamorphs to be expected from the initial number of predators and prey

$$M = E \exp\left(-\frac{vP}{E}\right) \exp\left[-sE \exp\left(-\frac{vP}{E}\right)\right] \tag{3}$$

This equation has only two parameters: v, the predation coefficient, and s, the competition coefficient. These parameters were estimated experimentally for *Rana sphenocephala* by manipulating densities of adult newts and densities of hatchling tadpoles independently in a 3×3 factorial design replicated three times using 27 cattle tank communities (Wilbur et al., 1983).

The parameter estimates were then used in Equation (3) to generate values graphed in Figure 8.

Figure 8 demonstrates that for simple realistic conditions, namely, predation acts before competition and simple exponential relationships mimic density dependence, the number of survivors is enhanced by predation if initial egg densities are high. This result is likely to be general in nature since amphibians often breed in very large aggregations and females often have large clutch sizes. Without moderate predation these eggs would have a very low probability of eventual metamorphosis. The study of natural colonization outlined in Section 2.2 and box samples from natural ponds yielded many estimates of natural densities of frog larvae and adult *Notophthalmus*. Densities observed in nature are in the region of Figure 8 where predation dominates competition (Morin, 1983b). The high fecundity and relatively small eggs of aquatic frogs are probably adaptations to a high risk of predation rather than to intense competition in the larval stage.

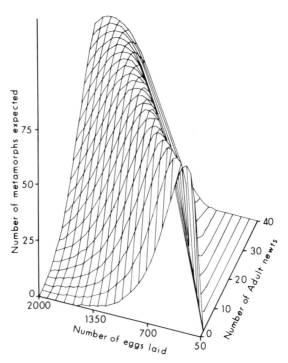

Figure 8. Simulation of the expected number of metamorphs of *Rana sphenocephala* in populations raised in 1.52 m tanks with various initial densities of tadpoles and adult newts. The model assumes that the risk of predation is an exponential function of the ratio of predators to prey, that predation occurs before competition, and that the probability that a tadpole that has escaped predation will survive competition is an exponentially decreasing function of the density of competitors. The two parameters of the model were fit from the experimental data of Wilbur et al. (1983).

4.4.2 Body Size at Metamorphosis. Equation (3) models the expected number of survivors for various initial densities of frog eggs and adult newts. Body size at metamorphosis decreases exponentially with initial density of amphibian populations in experimental systems without predators (reviewed by Wilbur, 1980). An empirically determined relationship between W, wet mass at metamorphosis, and T, initial density of tadpoles, has three parameters that can be estimated from the data: b, minimum size at which metamorphosis occurs, c, range of sizes at metamorphosis, and a, a coefficient of density dependence (Wilbur, 1976).

$$W = b + c \exp(-aT) \tag{4}$$

This model can be modified to include effects of interspecific competition (Wilbur, 1982). If, as my model assumes, predation does act before competition, then body size at metamorphosis will reflect number of survivors from predation rather than number of eggs laid. The parameter a of Equation (4) was estimated from the data of Wilbur et al. (1983) using sizes at metamorphosis of *Rana sphenocephala* in nine tanks that had 50, 100 or 200 tadpoles (three replicates each) but no adult newts. Equation (1) was used to estimate T of Equation (4) to predict mean size of metamorphs in tanks with different combinations of initial densities of adult newts and frog eggs (Fig. 9),

$$W = b + c \exp\left\{ -a\left[E \exp\left(-\frac{dP}{E} \right) \right] \right\}$$

There is an individual benefit from predation to those individuals that escape being eaten because they do not suffer the costs of competition: low survival and small size at metamorphosis. A large size at metamorphosis may result in attainment of sexual maturity earlier than in individuals that metamorphose at small sizes (Bayless, 1969; Clarke, 1974, 1977).

4.4.3 Length of Larval Period. Timing of metamorphosis may also be modified by predation (Bragg, 1966). In laboratory and field experiments, mean length of larval period is an increasing function of population density. In the field experiments of Wilbur et al. (1983), the rate of increase in larval period with initial density for *Rana sphenocephala* that metamorphosed was 18 days per 100 tadpoles in a range of experimental densities from 50 to 200 conspecific tadpoles. Predation reduces competition and increases the likelihood that survivors can escape the pond after a short larval period. This opportunity is obviously important in temporary habitats, but it may also permit escape from large predators in permanent ponds. Arnold and Wassersug (1978) argued that synchronized metamorphosis, which is often observed in toads, may be a mechanism to satiate terrestrial predators.

In permanent ponds, individuals that have grown to a size at which risk

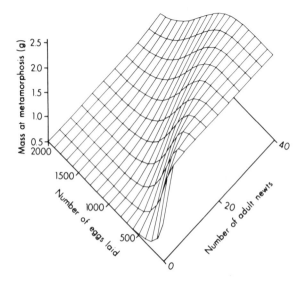

Figure 9. Simulation of the expected body size at metamorphosis of *Rana sphenocephala* tadpoles raised at various initial densities of tadpoles and adult newts. The model assumes that predation acts before competition and that the body size at metamorphosis is an exponentially decreasing function of the density of competitors.

of predation is low have the option of remaining in the pond until either an optimal size or optimal season for metamorphosis is reached. Overwintering 1 or 2 years is common in north temperate species of *Rana* (Collins, 1979; Hedeen, 1975; Martof, 1956; Ryan, 1953; Willis et al., 1956). Our 1981 experiment also gave information on the effect of population density on the probability of overwintering in *Rana sphenocephala*. Of the 27 populations, tadpoles only overwintered in the three tanks with the highest density (200 hatchling *Rana*) and without predators. Three replicate populations had 39, 53, and 58% of initial larvae overwintering. My explanation is that at high density mean growth rate was retarded such that only about half the population was able to obtain the minimum size threshold for the initiation of metamorphosis before cold temperatures arrived during October. These overwintering tadpoles would have remained in the tanks and metamorphosed early the following summer.

As initial population density is increased in experimental populations, the number of survivors increases until resources are in short supply, then the number of survivors decreases asymptotically toward zero. In species that have the capacity to overwinter, survivors are partitioned between those that metamorphose in the fall and those that remain in ponds until the following spring (Fig. 10).

Apparently not all species have the physiological mechanisms required to overwinter successfully. In Morin's (1982) experiments, some *Rana sphenocephala* overwintered in the tanks, but all *Hyla gratiosa* that had not metamorphosed by mid-September died when water temperature dropped below 10°C.

In temporary ponds the option of prolonging the larval period until a large body size is obtained is not viable. A high level of phenotypic plasticity allows metamorphosis over a wide range of size, which permits continued exploitation of a pond during wet years. Thus, in wet years, metamorphosis can occur at a large size without loss of the ability to escape drying ponds in other years by metamorphosing at a small size (Conant, 1940). The Wilbur and Collins (1973) model suggests that the cue to determine the initiation of metamorphosis could be growth rate itself. Individuals growing rapidly under favorable conditions by assaying their own growth rate would delay metamorphosis until a large size is obtained. Other individuals in the population that are growing slowly, perhaps because of competition, parasitism, or loss of pieces of tail to predators, would metamorphose as soon as a minimum size threshold is exceeded. The model was developed to explain variation in size of metamorphosis and length of larval periods among members of a single population exposed to the same temperature and chemical cues. The mechanism we proposed could also generate differences in sizes at metamorphosis between populations; in nature, however, comparisons among populations are nearly always confounded by temperature differences (Atlas, 1935; Berven et al., 1979; Moore, 1939, 1940, 1942; Ryan, 1941; Stewart, 1956; Zweifel, 1968). Smith-Gill and Berven (1979) criticized our model because we did not explicitly include the effect of temperature. Certainly temperature will be the factor explaining the greatest amount of variation in size at metamorphosis among populations; but, I maintain that within populations

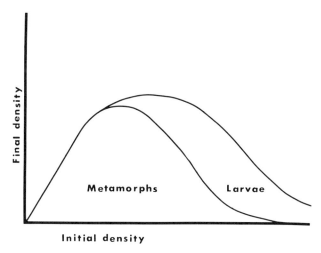

Figure 10. Diagrammatic model of the number of tadpoles that metamorphose in the first year (lower curve) and overwinter (difference between the two curves) as a function of initial density of tadpoles. The shape of the curves are deduced from experimental studies in which competition rather than predation determines larval survival.

growth rate may be an important predictor of the timing and size of meta-morphosis. Growth is the result of ecological processes that may also influ-ence differentiation, but the prediction of metamorphosis from differentia-tion rate is circular whereas the use of growth rate leads to testable hypotheses. Our model has been accepted by ecologists (Blakley, 1981; Blak-ley and Goodner, 1978; Brown, 1979; Healy, 1975; Kaplan, 1980; Fitzgerald et al., 1979; Patterson, 1978; Sweeney and Vannote, 1978; Bruce, 1976) more readily than by developmental biologists (Gilbert and Frieden, 1981).

5 CONCLUSIONS

Species with complex life cycles challenge contemporary ecological theory in several ways. A species with a CLC may sequentially occupy two, often radically different, ecological niches. Competition or predation may be in-tense in one phase of the life cycle without resulting in regulation of the population about a steady state in the other phase. If the larval stage is an adaptation to exploit transient opportunities for growth, then the notion of population regulation may be valid only at the higher level of the metapo-pulation (*sensu* Gill, 1978). At this level an assemblage of populations might exchange immigrants in a density dependent process.

Each time a pond fills, a new episode in population and community ecol-ogy begins. We used the methods outlined in Section 2 to approach the problem of how competition and predation affect the growth, survival, size at metamorphosis, and length of larval period of amphibians. Annual vari-ation in timing and intensity of breeding activity determine initial densities of natural populations. We used experimental ponds to eliminate much of the variation in hydrology and biological complexity of natural ponds. In these ponds it is possible to use replicated, experimental designs to study the roles of competition, predation, and their interaction. We plan also to use these systems to evaluate the importance of the biological interactions in relation to the disruptive influences of annual variation in climate.

The dominant conclusion of our studies is that body sizes of individuals determine their risk of predation, competition, and desiccation. Growth rate determines the velocity of an individual through stages of these risks. Even groups of individuals raised in isolation, but with replicated food rations and identical temperatures, develop lognormal distributions of body sizes due to the multiplicative effects of genetic differences, maternal factors, and chance environmental conditions. Skewness of these distributions is en-hanced by exploitative or interference mechanisms of competition. Large individuals dominate resources and have the ability to inhibit growth of smaller members of the population or of other species. Annual variation in climate can generate many different patterns of breeding dates and growth rates in an assemblage of species. Relative abundance of metamorphs of the species may reflect size dependent reversals of competitive ability triggered by differences in time of breeding, egg size, and growth rate of competing species.

Intraspecific density can effect radical changes in life history. Some species of *Rana* may switch from a larval period of 2–5 months in low density populations to one of 13–15 months in high density populations because insufficient food is available for many individuals to reach a minimum size for the initiation of metamorphosis before the onset of cold temperatures. The option of overwintering in the pond is open, of course, only to species that live in ponds that retain water for more than a year at a time. These species can often coexist with predators because they are large, fast, and distasteful.

Larval life may be prolonged for the opposite reason as well. A preliminary experiment demonstrated that newt larvae raised at low density with abundant food became neotenic and successfully reproduced when 1 year old compared to the delayed maturity that accompanies the typical terrestrial eft stage (Morin et al., 1983). Paedogenesis has arisen independently several times in salamanders that exploit permanent ponds, usually in habitats without the risk of predation by fish and often in ponds surrounded by environments hostile to terrestrial amphibians (Bruce, 1979; Sexton and Bizer, 1978; Sprules, 1974a,b; Wilbur and Collins, 1973). Reproduction with larval-like morphology is facultative in some species, such as *Notophthalmus viridescens* and *Ambystoma tigrinum*.

The result that the regime of competition and predation can determine the life cycle of a species, such as a developmental switch to paedogenesis or an order of magnitude increase in the length of the larval period, argues for the power of conventional microevolutionary theory to explain some macroevolutionary events that would appear as saltational steps, even in a very complete fossil record (Gould, 1977).

Growth rates may also determine risks from predation, but this is complicated by the fact that most predators of anuran larvae are themselves growing. Anuran eggs and hatchlings are at a very high risk of predation by adult newts in many ponds. As tadpoles grow they escape predation but may still risk mutilation. Experimental results suggest that newt predation may reduce tadpole densities to levels at which competition is unimportant. Experiments in artificial ponds demonstrate that newts may, in fact, enhance success of tadpole populations. Without predation, initial densities are so high that competition slows growth resulting in low survival and small body size at metamorphosis.

This conclusion is not new. The Hairston, Smith, and Slobodkin (1960) hypothesis articulated clearly that prey do not overexploit their resources in terrestrial systems because herbivores, as a group, are regulated by predation. The experiments reviewed in Section 4 demonstrate that in replicated aquatic ecosystems approaching natural ponds in physical and biological complexity, predation can preclude the competition observed in ponds without predators. Surveys of densities of predatory newts and the densities of breeding adults and metamorphosing frogs suggest that in undisturbed ponds that have a long hydrologic cycle, predation is more powerful than competition in determining survival and body size at metamorphosis of frogs. Competition may be the most important mechanism of population regulation in short-lived communities in which predators are not able to colonize and overexploit tadpole populations within the first several days after ponds have filled. These biological mechanisms may frequently be obscure to the naturalist because of variation among ponds and among years in density-independent selective agents. Experimentation can control these confounding variables and uncover the biological mechanisms that determine long-term trajectories of communities as they are buffetted about by environmental variation and chance historical effects.

ACKNOWLEDGMENTS

I benefited greatly from the discussions of my paper, both public and private, at the symposium. R. Chris Chambers, Jim Collins, Paul Dayton, and Conrad Istock commented on the manuscript. The experiments on which the model is based were done in collaboration with Peter Morin and Reid Harris. My recent graduate students Ross Alford, Chris Chambers, Reid Harris, Peter Morin, and Joe Travis have helped me develop many of my ideas about amphibian ecology. Nelson Hairston was most important in teaching me the value of experimentation and the dangers of wishful thinking about my observations. My research has been supported by the National Science Foundation grant DEB 7911539.

LITERATURE CITED

Aitchison, J., and J. A. C. Brown. 1969. *The lognormal distribution with special reference to its uses in economics*. Cambridge University Press, Cambridge. 176 pp.

Altig, R. 1972. Notes on the larvae and postmetamorphic tadpoles of four *Hyla* and three *Rana* with notes on tadpole color patterns. *J. Elisha Mitchell Sci. Soc.* **88**:113–119.

Anderson, J. D. 1968. A comparison of the food habits of *Ambystoma macrodactylum croceum* and *Ambystoma tigrinum californiense*. *Herpetologica* **24**:273–284.

Arnold, S. J., and R. J. Wassersug. 1978. Differential predation on metamorphic anurans by garter snakes (*Thamnophis*): Social behavior as a possible defense. *Ecology* **59**:1014–1022.

Atlas, M. 1935. The effect of temperature on the development of *Rana pipiens*. *Physiol. Zool.* **8**:290–310.

Bannikov, A. G. 1948. On the fluctuations of anuran populations. *Dokl. Akad. Nauk USSR* **61**:131–134.

Bayless, L. E. 1969. Post-metamorphic growth of *Acris crepitans*. *Am. Midl. Nat.* **81**:590–592.

Berven, K. A., D. E. Gill, and S. J. Smith-Gill. 1979. Countergradient selection in the green frog, *Rana clamitans*. *Evolution* **33**:609–623.

Bishop, S. C. 1941. The salamanders of New York. *N. Y. State Mus. Bull.* **324**:1–365.

Black, J. H. 1974. Larval spadefoot survival. *J. Herpetol.* **8**:371–373.

Blakley, N. 1981. Life history significance of size-triggered metamorphosis in milkweed bugs (*Oncopeltus*). *Ecology* **62**:57–64.

Blakley, N., and S. R. Goodner. 1978. Size-dependent timing of metamorphosis in milkweed bugs (*Oncopeltus*) and its life history implications. *Biol. Bull.* **155**:499–510.

Blair, W. F. 1957. Changes in vertebrate populations under conditions of drought. *Cold Spring Harbor Symp. Quant. Biol.* **22**:273–275.

Borland, J. R., and R. Rugh. 1943. Evidences of cannibalism in the tadpole of the frog *Rana pipiens*. *Am. Nat.* **77**:282–285.

Bragg, A. N. 1956. Dimorphism and cannibalism in tadpoles of *Scaphiopus bombifrons* (Amphibia, Salientia). *Southwestern Nat.* **1**:105–108.

Bragg, A. N. 1957. Variation in colors and color patterns in tadpoles in Oklahoma. *Copeia* **1957**:36–39.

Bragg, A. N. 1964. Further study of predation and cannibalism in spadefoot tadpoles. *Herpetologica* **20**:17–24.

Bragg, A. N. 1966. Longevity of the tadpole stage in the plains spadefoot. (Amphibia:salentia). *Wasmann J. Biol.* **24**:71–73.

Brockelman, W. Y. 1969. An analysis of density effects and predation in *Bufo americanus* tadpoles. *Ecology* **50**:632–644.

Brown, K. M. 1979. The adaptive demography of four freshwater pulmonate snails. *Evolution* **33**:417–432.

Bruce, R. C. 1976. Population structure, life history and evolution of paedogenesis in the salamander *Eurcyea neotenes*. *Copeia* **1976**:242–249.

Bruce, R. C. 1979. Evolution of paedomorphosis in salamanders of the genus *Gyrinophilus*. *Evolution* **33**:998–1000.

Burger, W. L. 1950. Novel aspects of the life history of two Ambystomas. *J. Tenn. Acad. Sci.* **25**:252–257.

Calef, G. W. 1973. Natural mortality of tadpoles in a population of *Rana aurora*. *Ecology* **54**:741–758.

Cecil, S. G., and J. J. Just. 1979. Survival rate, population density and development of a naturally occurring anuran larvae (*Rana catesbeiana*). *Copeia* **1979**:447–453.

Channing, A. 1976. Life histories of frogs in the Namib Desert. *Zool. Afr.* **11**:299–312.

Christman, S. P., and L. R. Franz. 1973. Feeding habits of the striped newt, *Notophthalmus perstriatus*. *J. Herpetol.* **7**:133–135.

Clarke, R. D. 1974. Postmetamorphic growth rates in a natural population of Fowler's toad, *Bufo woodhousei fowleri*. *Can. J. Zool.* **52**:1489–1498.

Clarke, R. D. 1977. Postmetamorphic survivorship in Fowler's toad, *Bufo woodhousei fowleri*. *Copeia* **1977**:594–597.

Collins, J. P. 1979. Intrapopulation variation in the body size at metamorphosis and timing of metamorphosis in the bullfrog, *Rana catesbeiana*. *Ecology* **60**:738–749.

Collins, J. P. 1981. Distribution, habitats, and life history variation in the tiger salamander, *Ambystoma tigrinum*, in East-central and Southeast Arizona. *Copeia* **1981**:666–675.

Collins, J. P., and H. M. Wilbur. 1979. Breeding habits and habitats of the amphibians of the Edwin S. George Reserve, Michigan, with notes on the local distribution of fishes. *Occ. Pap. Mus. Zool. Univ. Mich.* **686**:1–34.

Conant, R. 1940. *Rana virgatipes* in Delaware. *Herpetologica* **1**:176–177.

Connell, J. 1979. Some mechanisms producing structure in natural communities; a model and evidence from field experiments. In M. L. Cody and J. M. Diamond (eds.), *Ecology and evolution of communities*, pp. 460–490. Belknap, Cambridge, Massachusetts.

Connor, E. F., and D. S. Simberloff. 1979. The assembly of species communities: Chance or competition? *Ecology* **60**:1132–1140.

Creusere, F. M., and W. G. Whitford. 1976. Ecological relationships in a desert anuran community. *Herpetologica* **32**:7–18.

Crump, M. L. 1971. Quantitative analysis of the ecological distribution of tropical herpetofauna. *Occ. Pap. Mus. Nat. Hist. Univ. Kans.* **3**:1–62.

Crump, M. L. 1981. Energy accumulation and amphibian metamorphosis. *Oecologia* **49**:167–169.

Dent, J. N. 1968. Survey of amphibian metamorphosis. In W. Etkin and L. I. Gilbert (eds.), *Metamorphosis*, pp. 271–311. Appleton-Century-Crofts, New York.

Dickman, M. 1968. The effect of grazing by tadpoles on the structure of a periphyton community. *Ecology* **49**:1188–1190.

Dunn, E. R. 1917. The breeding habits of *Ambystoma opacum. Copeia* **1917**:40–43.

Fitzgerald, K. T., L. J. Guillette, Jr., and D. Duval. 1979. Notes on birth, development and care of *Gastrotheca riobambae* tadpoles in the laboratory (Amphibia, Anura, Hylidae). *J. Herpetol.* **13**:457–460.

Fox, L. R. 1975. Cannibalism in natural populations. *Annu. Rev. Ecol. Syst.* **6**:87–106.

Frazer, J. F. D. 1977. Growth of young vertebrates in the egg or uterus. *J. Zool. (London)* **183**:189–201.

Gilbert, L. I., and E. Frieden. 1981. *Metamorphosis, a problem in developmental biology*, 2nd ed. Plenum Press, New York.

Gill, D. E. 1978. The metapopulation ecology of the red-spotted newt, *Notophthalmus viridescens* (Rafinesque). *Ecol. Monogr.* **48**:145–166.

Goin, O. B., and C. J. Goin. 1962. Amphibian eggs and the montane environment. *Evolution* **16**:364–371.

Gould, S. J. 1977. *Ontogeny and Phylogeny*. Belknap, Cambridge, Massachusetts.

Gromko, M. H., F. S. Mason, and S. J. Smith-Gill. 1973. Analysis of the crowding effect in *Rana pipiens* tadpoles. *J. Exp. Zool.* **186**:63–72.

Grubb, J. C. 1972. Differential predation by *Gambusia affinis* on the eggs of seven species of anuran amphibians. *Am. Midl. Nat.* **88**:102–108.

Hairston, N. G. 1980. An experimental test of an analysis of field distributions: competition in terrestrial salamanders. *Ecology* **61**:817–826.

Hairston, N. G. 1981. An experimental test of a guild: Salamander competition. *Ecology* **65**:65–72.

Hairston, N. G., F. E. Smith, and L. B. Slobodkin. 1960. Community structure, population control, and competition. *Am. Nat.* **94**:421–425.

Hamilton, W. 1932. The food and feeding habits of some eastern salamanders. *Copeia* **1932**:83–86.

Hassinger, D. D., J. D. Anderson, and G. H. Dalrymple. 1970. The early life history and ecology of *Ambystoma tigrinum* and *Ambystoma opacum* in New Jersey. *Am. Midl. Nat.* **84**:474–495.

Healy, W. R. 1975. Breeding and postlarval migrations of the red-spotted newt, *Notophthalmus viridescens*, in Massachusetts. *Ecology* **56**:673–680.

Hedeen, S. E. 1975. Premetamorphic growth of *Rana catesbeiana* in southwestern Ohio. *Ohio J. Sci.* **75**:182–183.

Heusser, H. 1970. Spawn eating by tadpoles as a possible cause of specific biotype preferences and short breeding times in European anurans (Amphibia, Anura). *Oecologia* **4**:83–88.

Heyer, W. R. 1969. The adaptive ecology of the species groups of the frog genus *Leptodactylus* (Amphibia: Leptodactylidae). *Evolution* **23**:421–428.

Heyer, W. R. 1974. Niche measurements of frog larvae from a seasonal tropical location in Thailand. *Ecology* **55**:651–656.

Heyer, W. R., R. W. McDiarmid, and D. L. Weigmann. 1975. Tadpoles, predation, and pond habitats in the tropics. *Biotropica* **7**:100–111.

Hogarth, P. J. 1976. *Viviparity*. The Institute of Biology's Studies in Biology, Arnold, London. 68 pp.

Jameson, D. L. 1956. Growth, dispersal and survival of the Pacific treefrog. *Copeia* **1956**:25–29.

Johnson, N. I., and S. Kotz. 1970. *Continuous Univariate Distributions*, Vol. 1. Houghton Mifflin, Boston.

Kaplan, R. H. 1980. Ontogenetic energetics in *Ambystoma*. *Physiol. Zool.* **53**:43–56.

Kaplan, R. H., and M. L. Crump. 1978. The non-cost of brooding in *Ambystoma opacum*. *Copeia* **1978**:99–103.

Kaplan, R. H., and P. Sherman. 1980. Intraspecific oophagy in California newts. *J. Herpetol.* **14**:183–185.

Kirtisinghe, P. 1946. The presence in Ceylon of a frog with direct development on land. *Ceylon J. Sci.* **23**:109–112.

Kluge, A. G. 1981. The life history, social organization and parental behavior of *Hyla rosenbergi*, a nest-building gladiator frog. *Misc. Pub. Mus. Zool. Univ. Mich.* **160**:1–170.

Korschben, L. T., and T. S. Baskett. 1963. Adult bullfrogs eat tadpoles in nature. *Herpetologica* **19**:89–99.

Licht, L. E. 1974. Survival of embryos, tadpoles, and adults of the frogs *Rana aurora* and *Rana pretiosa* pretiosa sympatric in southwestern British Colombia. *Can. J. Zool.* **52**:613–627.

Lutz, B. 1947. Trends towards non-aquatic and direct development in frogs. *Copeia* **1947**:242–252.

Lutz, B. 1948. Ontogenetic evolution in frogs. *Evolution* **2**:29–39.

Macan, T. T. 1966. The influence of predation on the fauna of a moorland fishpond. *Arch. Hydrobiol.* **61**:432–452.

McDiarmid, R. W. 1978. Evolution of parental care in frogs. p. 127–147 In G. M. Burhardt and M. Bekoff (eds.), *The development of behavior: Comparative and evolutionary aspects*, pp. 127–147. Garland STPM Press, New York.

Martin, A. A. 1967. Australian anuran life histories: some evolutionary and ecological aspects. In A. H. Weatherly (ed.), *Australian inland waters and their fauna*, pp. 175–191. Australian National University Press, Canberra.

Martof, B. 1952. Early transformation of the greenfrog, *Rana clamitans* Latreille. *Copeia* **1952**:115–116.

Martof, B. 1956. Growth and development of the green frog, *Rana clamitans*, under natural conditions. *Am. Midl. Nat.* **55**:101–117.

Menzies, J. L., and M. J. Taylor. 1977. The systematics and adaptations of some Papuan microhylid frogs which life underground. *J. Zool. London* **183**:431–464.

Mertz, D. B. 1972. The *Tribolium* model and the mathematics of population growth. *Annu. Rev. Ecol. Syst.* **3**:51–78.

Moore, J. A. 1939. Temperature tolerance and rates of development in the eggs of Amphibia. *Ecology* **20**:459–478.

Moore, J. A. 1940. Adaptive differences in the egg membranes of frogs. *Am. Nat.* **74**:89–93.

Moore, J. A. 1942. Embryonic temperature tolerance and rate of development in *Rana catesbeiana*. *Biol. Bull.* **83**:375–388.

Morin, P. J. 1981. Predatory salamanders reverse the outcome of competition among three species of anuran tadpoles. *Science* **212**:1284–1286.

Morin, P. J. 1982. *Predation and the composition of temporary pond communities*. Ph.D. thesis. Department of Zoology, Duke University, Durham, North Carolina. 171 pp.

Morin, P. J. 1983a. Predation, competition, and the composition of larval anuran guilds. *Ecol. Monogr.* **53**:119–138.

Morin, P. J. 1983b. Competitive and predatory interactions in natural and experimental populations of *Notophthalmus viridescens dorsalis* and *Ambystoma tigrinum*. *Copeia* **1983** (in press).

Morin, P. J., H. M. Wilbur, and R. N. Harris. 1983. Salamander predation and the structure of experimental communities: responses of *Notophthalmus* and microcrustacea. *Ecology* **64** (in press).

Orton, G. L. 1954. Dimorphism in larval mouthparts in spadefoot toads of the *Scaphiopus hammondi* group. *Copeia* **1954**:97–100.

Parker, H. W. 1956. Viviparous caecilians and amphibian phylogeny. *Nature* (*London*) **178**:250–252.

Patterson, K. K. 1978. Life history aspects of paedogenetic populations of the mole salamander, *Ambystoma talpoideum*. *Copeia* **1978**:649–655.

Peckham, R. S., and C. F. Dineen. 1955. Spring migrations of salamanders. *Proc. Indiana Acad. Sci.* **64**:278–280.

Polis, G. A. 1981. The evolution and dynamics of intraspecific predation. *Annu. Rev. Ecol. Syst.* **12**:225–251.

Poynton, J. C. 1964. Relationships between habitat and terrestrial breeding in amphibians. *Evolution* **18**:131.

Ricklefs, R. E. 1965. Brood reduction in the curved-billed thrasher. *Condor* **67**:505–510.

Ricklefs, R. E. 1967. A graphical method of fitting equations to growth curves. *Ecology* **48**:978–983.

Rose, F. I.., and D. Armentrout. 1976. Adaptive strategies of *Ambystoma tigrinum* Green inhabiting the Llano Estacado of West Texas. *J. Anim. Ecol.* **45**:713–729.

Rose, S. M. 1960. A feedback mechanism of growth control in tadpoles. *Ecology* **41**:188–199.

Ryan, F. J. 1941. Temperature changes and the subsequent rate of development. *J. Exp. Biol.* **88**:25–54.

Ryan, R. A. 1953. Growth rates of some ranids under natural conditions. *Copeia* **1954**:73–80.

Salthe, S. N., and J. S. Mecham. 1974. Reproduction and courtship patterns. In B. Lofts (ed.), *Physiology of the amphibia*, Vol. 2, pp. 309–521. Academic Press, New York. 592 pp.

Seale, D. B. 1980. Influence of amphibian larvae on primary production, nutrient flux, and competition in a pond ecosystem. *Ecology* **61**:1531–1550.

Sexton, O. J., and J. R. Bizer, 1978. Life history patterns of *Ambystoma tigrinum* in montane Colorado. *Am. Midl. Nat.* **99**:101–118.

Shoop, C. R. 1974. Yearly variation in larval survival of *Ambystoma maculatum*. *Ecology* **55**:440–444.

Slade, N. A., and R. J. Wassersug. 1975. On the evolution of complex life cycles. *Evolution* **29**:568–571.

Smith-Gill, S. J., and K. A. Berven. 1979. Predicting amphibian metamorphosis. *Am. Nat.* **113**:563–585.

Sprules, W. G. 1974a. The adaptive significance of paedogenesis in North American species of *Ambystoma* (Amphibia: Caudata): An hypothesis. *Can. J. Zool.* **52**:393–400.

Sprules, W. G. 1974b. Environmental factors and the incidence of neoteny in *Ambystoma gracile* (Baird) (Amphibia: Caudata). *Can. J. Zool.* **52**:1545–1552.

Stein, C. J., J. A. Fowler, and R. S. Simmons. 1954. Occurrence of the eastern tiger salamander, *Ambystoma tigrinum tigrinum* (Green), in Maryland with notes on its life history. *Ann. Carnegie Mus.* **3**:145–148.

Steinwascher, K. F. 1978a. Interference and exploitation competition among tadpoles of *Rana utricularia*. *Ecology* **59**:1039–1046.

Steinwascher, K. F. 1978b. The effect of copraphagy on the growth of *Rana catesbeiana* tadpoles. *Copeia* **1978**:130–134.

Stewart, M. M. 1956. The separate effects of food and temperature differences on development of marbled salamander larvae. *J. Elisha Mitchell Sci. Soc.* **72**:47–56.

Sweeney, B. W., and R. L. Vannote. 1978. Size variation and the distribution of hemimetabolus aquatic insects: Two thermal equilibrium hypotheses. *Science* **200**:444–446.

Tevis, L. 1966. Unsuccessful breeding by desert toads (*Bufo punctatus*) at the limit of their ecological tolerance. *Ecology* **47**:766–775.

Travis, J. 1981. Control of larval growth variation in a population of *Pseudacris triseriata* (Anura: Hylidae). *Evolution* **35**:423–432.

Turnipseed, G., and R. Altig. 1975. Population density and age structure of three species of hylid tadpoles. *J. Herpetol.* **9**:287–291.

Van Dijk, D. E. 1971. Anuran ecology in relation to oviposition and development out of water. *Zool. Afr.* **6**:119–132.

Voris, H. K., and J. P. Bacon. 1966. Differential predation on tadpoles. *Copeia* **1966**:594–598.

Wake, M. H. 1977. Fetal maintenance and its evolutionary significance in the amphibia: Gymnophiona. *J. Herpetol.* **11**:379–386.

Walters, B. 1975. Studies of interspecific predation within an amphibian community. *J. Herpetol.* **9**:267–279.

Wassersug, R. J. 1975. The adaptive signficance of the tadpole stage with comments on the maintenance of complex life cycles in anurans. *Am. Zool.* **15**:405–417.

Werschkul, D. F., and M. T. Christensen. 1977. Differential predation by *Lepomis macrochirus* on the eggs and tadpoles of *Rana*. *Herpetologica* **33**:237–240.

Wilbur, H. M. 1972. Competition, predation, and the structure of the *Ambystoma-Rana sylvatica* community. *Ecology* **53**:3–21.

Wilbur, H. M. 1976. Density-dependent aspects of metamorphosis in *Ambystoma* and *Rana sylvatica*. *Ecology* **57**:1289–1296.

Wilbur, H. M. 1977a. Density-dependent aspects of growth and metamorphosis in *Bufo americanus*. *Ecology* **58**:196–200.

Wilbur, H. M. 1977b. Interactions of food level and population density in *Rana sylvatica*. *Ecology* **58**:206–209.

Wilbur, H. M. 1980. Complex life cycles. *Annu. Rev. Ecol. Syst.* **11**:67–93.

Wilbur, H. M. 1982. Competition between tadpoles of *Hyla femoralis* and *Hyla gratiosa* in laboratory experiments. *Ecology* **63**:278–282.

Wilbur, H. M., and J. P. Collins. 1973. Ecological aspects of amphibian metamorphosis. *Science* **182**:1305–1314.

Wilbur, H. M., and J. Travis. 1983. An experimental approach to understanding pattern in natural communities. In D. Simberloff, D. R. Strong, and L. G. Abele (eds.), *Community ecology: The issues and the evidence*. Princeton University Press, Princeton, New Jersey.

Wilbur, H. M., P. J. Morin, and R. N. Harris. 1983. Salamander predation and the structure of experimental communities: Anuran responses. *Ecology* (in press).

Willis, T. L., D. L. Moyle, and T. S. Baskett. 1956. Emergence, breeding, hibernation, movements, and transformation of the bullfrog, *Rana catesbeiana* in Missouri. *Copeia* **1956**:30–41.

Wood, J. T., and O. K. Goodwin. 1954. Observations on the abundance, food, and feeding behavior of the newt, *Notophthalmus viridescens viridescens* (Rafinesque) in Virginia. *J. Elisha Mitchell Sci. Soc.* **70**:27–30.

Worthington, R. D. 1968. Observations on the relative sizes of three species of salamander larvae in a Maryland pond. *Herpetologica* **24**:242–246.

Worthington, R. D. 1969. Additional observations on sympatric species of salamander larvae living in a Maryland pond. *Herpetologica* **25**:227–229.

Zippen, C. 1956. An evaluation of the removal method of estimating animal populations. *Biometrics* **12**:163–189.

Zweifel, R. G. 1968. Reproductive biology of anurans of the arid southwest with emphasis on adaptations of embryos to temperature. *Bull. Am. Mus. Nat. Hist.* **140**:1–64.

PART III

Ecology of
Social Behavior

CHAPTER **8**

Resources and the Evolution of Social Behavior

C. N. SLOBODCHIKOFF
Department of Biological Sciences
Northern Arizona Univerity
Flagstaff, Arizona

CONTENTS

1 INTRODUCTION

Although there are perhaps one million species of animals in the world, relatively few are social (Wilson, 1975a). Even among some groups of animals that are well known for their sociality, there are closely related species that are not social. The ants and honeybees in the insect order Hymenoptera are recognized as having a complex form of sociality, but other species of Hymenoptera are not social. The parasitic wasps are generally not social. Wasps in families closely related to the ants are not social. Wild bee species that are related to the social bees are solitary. Similar situations can be found among other animal groups. Some are social, others are not. We may then ask the question, why are so few animals social? In attempting to answer this question, we must also ask the question, how did sociality evolve?

Attempts to explain sociality can be classified into descriptive models and evolutionary models. Descriptive models attempt to infer a pathway for the development of social behavior from successive stages of sociality found among different animal groups. The familial and the parasocial pathways (Vehrencamp, 1979; Lin and Michener, 1972) are examples of descriptive models. The stages of the familial pathway (Vehrencamp, 1979; Wilson, 1975a) are solitary, subsocial, intermediate subsocial, and eusocial. In the *solitary* stage, the adults merely come together for the purposes of breeding. The male leaves or dies immediately after mating, and the female provides no parental care for the offspring. In the *subsocial* stage, the adults mate, and one or both parents care for the young for a certain length of time, so that the generations overlap. The young usually leave the care of the parents before the onset of sexual maturity. In the *intermediate subsocial* stage, there is also extended parental care, and the group members of at least two generations cooperate in a variety of activities such as foraging and defense. In the *eusocial* stage, there are the overlapping generations, extended parental care, and cooperation in a variety of tasks, but also there is a reproductive division of labor, with some individuals producing offspring while others do not breed. Since it is based on the premise of increasing parental care, as well as increasing levels of cooperation, the familial pathway is primarily applicable to situations in which relatives, parents and offspring, cooperate with one another.

Various stages of sociality along the familial pathway have been ascribed to a variety of animal groups. Among the mammals, the primates, carnivores, cervids, and squirrels provide examples of different familial stages of sociality (Wilson, 1975a). The naked mole rat, the one mammal that seems to have a totally eusocial stage, probably belongs within this pathway (Jarvis, 1981). In the birds, species that have helpers (Woolfenden, 1976; Zahavy, 1976) belong to the familial pathway. Among the insects, the ants, termites, and many of the bees are believed to represent stages of the familial pathway (Wilson, 1971).

The parasocial pathway is based on the premise that cooperation in groups evolves through aggregations of unrelated individuals, rather than by aggregations of offspring through increased parental care (Lin and Michener, 1972). The stages of the parasocial pathway are as follows (Vehrencamp, 1979): solitary, communal, quasisocial, semisocial, and eusocial. The *solitary* stage is the same as in the familial pathway. Adults come together to breed. In the *communal* stage adults of the same generation form noncooperative aggregations, often centering around some resource such as limited nest sites or clumped food. The *quasi-social* stage involves cooperation among the individuals in the aggregation. Such cooperation may involve communal nest construction, foraging, or defense against predators. The *semisocial* stage involves cooperation, and also a reproductive division of labor. Up until this point, having two or more generations overlapping in the group is not a necessary condition. In the *eusocial* stage, however, just as in the familial pathway, two or more generations must be present, as well as cooperation and a reproductive division of labor.

Examples of the parasocial pathway are somewhat sparser than examples of the familial pathway. Among the mammals, some antelope (Jarman, 1974), pinnipeds (Bartholomew, 1970), and bats (Bradbury, 1977) aggregate. Some bats form stable colonies of unrelated females (McCracken and Bradbury, 1977; Bradbury and Vehrencamp, 1976, 1977), and some buffalo form groups that have antipredator defenses (Sinclair, 1977). Among the birds, a number of species form aggregations. These may have the function of allowing the birds to exploit limited nesting sites (Lack, 1968), exploit patchily distributed food (Krebs, 1974), or provide a passive defense against predators (Page and Whitacre, 1975). In a few species unrelated helpers cooperate with breeding individuals (Gaston, 1973; Fry, 1972). In the anis, several females contribute eggs to a communally built nest (Vehrencamp, 1978). Among the insects, the vespid wasps are examples of stages in the parasocial pathway (Wilson, 1971).

Both the familial pathway and the parasocial pathway describe several stages in the evolution of sociality, but do not really address the "why" of social behavior. Both models stress cooperation, with the cooperation based on parental care and parent–offspring association in the familial pathway, and on aggregations of unrelated individuals in the parasocial pathway. The essential question of why the cooperation develops is not addressed. Many other animal species that are related to the ones cited above have not developed any stage of sociality, and remain solitary.

Evolutionary models of sociality attempt to answer the "why" question. Most of the evolutionary models have emphasized aspects of population genetics, following the elegant arguments of Hamilton (1964) on the evolution of sterile workers among the bees and wasps. Hamilton suggested that the haplodiploid system of the Hymenoptera plays a large role in the evolution of sociality. In the Hymenoptera, males are haploid with N chromosomes, whereas females are diploid with $2N$ chromosomes. A female can choose whether or not to fertilize an egg, from sperm stored in her spermatheca. If she does not fertilize the egg, the egg develops parthenogenetically into a male with a haploid set of chromosomes (Slobodchikoff and Daly, 1971). Hamilton suggested that in normal diploid systems parents and offspring, and also two siblings, are related to each other, on the average, by one-half of their genes. However, because of the haplodiploid system of the Hymenoptera, a mother and her daughter are still related by one-half, but two sisters are related to each other by three-quarters. Hamilton reasoned that such a high level of relatedness would promote cooperation among sisters, and the system would evolve toward a preponderance of cooperating females. The relatedness argument has been the basis for the concept of the selfish gene (Dawkins, 1976), with the assumption that genes promote behavior ensuring the production of more copies of those genes. Since relatives are more likely to have copies of the same genes than are strangers, the inclusive fitness (the fitness of the animal and its relatives) is expected to be important by these arguments.

Unfortunately, although Hamilton's inclusive fitness arguments are elegant, they do not explain sociality in the general case. All of the Hymenoptera are haplodiploid, but only a few species are social. Although suggestions have been made that inbreeding (Hamilton, 1972) or genetic linkage (Lacy, 1980) would increase the relatedness of termite siblings, none of the termites is haplodiploid, yet they are social. Even among the social honeybees, queens may mate with a number of males, lowering the genetic relatedness between sisters having the same mother but different fathers (Page and Metcalf, 1982). Also, the haplodiploid argument does not explain the behavior of slave ants, who cooperate with their masters but are not related to them (Wilson, 1975b). In both ants and honeybees, transfers of large numbers of individuals from one colony to another are known (Wilson, 1971). The individuals transferring to a new colony are unlikely to be sisters, yet they cooperate with the resident colony members.

Other evolutionary models attempt to isolate the conditions under which cooperation can develop. A model by Baylis and Halpin (1982) suggests three circumstances in which animals can develop cooperation: aggregations, gamete transfer, and parturition. All three circumstances have the potential for bringing animals into close proximity with one another, setting up the conditions necessary for the development of cooperative behavior. The assumption is that once animals are brought into close proximity, they can evolve social behaviors. However, some animals do not evolve social behaviors under those circumstances, so proximity is clearly a necessary condition, but not a sufficient condition in itself.

2   A RESOURCE-BASE MODEL OF SOCIALITY

I would like to propose an evolutionary model of sociality that also contains some elements of descriptive models. I suggest that sociality develops in response to an unequal distribution of resources, a need to defend those resources, and an inability to exploit those resources except as a group. These, I suggest, are the ultimate factors causing sociality. Proximate mechanisms in this model, corresponding to the evolutionary stages of the descriptive models, are (1) habitat variability and mating systems; (2) cooperation and aggression; (3) group size and dominance; and (4) division of labor in resource extraction.

3   HABITAT VARIABILITY AND MATING SYSTEMS

Let us suppose that in a variable habitat, resources are patchily distributed, with some patches having high-quality resources and others having low-quality resources. As a further constraint, let us suppose that a criterion of low-quality patches is a uniform distribution, within the patch, of a subset of the necessary resources, whereas high-quality patches have within them a variable distribution of a subset of resources. For example, an animal feeding on dietary items A, B, C, D, and E may find items A, B, and C uniformly distributed in a low-quality patch, and in a high-quality patch may find items A–E, but with each item clumped into a subpatch, and the subpatches randomly distributed within the high-quality patch. I predict that low-quality uniform patches will support monogamous animals, whereas high-quality variable-distribution patches will support polygynous animals (Fig. 1).

The reasons for this prediction are as follows. High-quality patches should have higher carrying capacities, and should be able to support more animals. If a high-quality patch has a uniform distribution of resources, animals occupying the patch have two alternatives. One is to shrink their use area of the patch to one that can be defended or exploited by a single, or at most two, animals. The other is to collect a polygynous group of animals. Because of the potential costs of interanimal aggression (see below), I predict that under these circumstances the animals will remain solitary. If a high-quality patch has unequally distributed resources, the animal has the same two choices discussed above. Shrinking its use area to one that can be defended or exploited by one or two animals may leave those animals without access to one or more resources. The alternative is to form a group of animals that

Figure 1. A model showing the relative fitness of monogamous (M) and polygynous (P) animals in response to habitat quality. The habitat changes from poor, uniform (P,U) to good, patchy (G,P).

could collectively function in defending or exploiting all the resources of the patch.

At least in the beginning stages of group formation, I predict that such animal assemblages would be polygynous groups, rather than groups with an equal number of males and females. This is because of the costs of aggression between males, competing for females. As Trivers (1972) has shown, females that have a large investment in the mating and postmating process (e.g., larger egg, postparturition costs) would tend to be choosy in mating with males. Such female choice leads to a large variance in male reproductive success, causing intermale competition. In the beginning stages of group formation, such intermale competition would prevent a male from tolerating another breeding male within the group. However, if the relative investment of males and females into the mating and postmating process does not follow this pattern, the above condition would not be expected to hold.

A model somewhat similar to the above one was proposed by Wittenberger (1979). In his model, Wittenberger considered the relationship between environmental quality and female fitness, in a variant of the Verner–Willson–Orians polygyny threshold model (Orians, 1969; Verner and Willson, 1966). Wittenberger and also Altmann et al. (1977) suggest that female cooperation may be an important component of the polygyny threshold model, and that monogamous females should have a higher fitness than polygynous ones when females compete for resources, but polygynous females should have a higher individual fitness when females cooperate (Altmann et al., 1977). In low-quality territories, Wittenberger suggests that competition would counteract the effects of cooperation so that monogamous females have a higher fitness, whereas in high-quality territories the negative effects of competition would be less important than the positive benefit of cooperation.

My model does not depend on competition, and it predicts that resource quality and distribution affect both male and female fitness. The conditions for maximizing male and female fitness are not necessarily the same. Downhower and Armitage (1971) showed that among the yellow-bellied marmots, the average female fitness was highest in monogamous groups, whereas the average male fitness was highest in polygynous groups. Svendsen (1974) has pointed out that three kinds of females occur in marmot societies: aggressive, social, and avoiders. Aggressive females tend to live as solitary individuals or monogamous pairs, and have the highest fitness when they are monogamous. Social females tend to live in social groups and have their highest fitness in a larger group. Avoider females tend to live by themselves at peripheral burrows. Armitage (1977) has suggested that these relationships can be complicated by the population density, the age–sex structure of the population, and the number of years that residents have lived together. Since the quality of marmot territories is highly variable (Anderson et al., 1976), the Downhower and Armitage (1971) results do not provide a specific test of the relationship among fitness, mating system, and habitat quality.

3.1 A Test of the Model: Gunnison's Prairie Dogs

Prairie dogs offer an excellent test of the habitat variability–mating system model. Blacktailed and Gunnison's prairie dogs are social, colonial animals that feed on plants and defend territories (King, 1955; Fitzgerald and Lechleitner, 1974). The territories are defended by cooperative groups involving one or more males and one or more females (King, 1955). Cooperative behavior also extends to the use of alarm calls, warning other prairie dogs of the approach of predators (Slobodchikoff and Coast, 1980). The behavior of the blacktailed prairie dog, *Cynomys ludovicianus*, has been studied extensively by King (1955) and Hoogland (1977, 1979).

Our work has concentrated on the Gunnison's prairie dog, *Cynomys gunnisoni*. This prairie dog lives in colonial groups in Arizona, New Mexico, Colorado, and Utah (Hall and Kelson, 1959). The towns or colonies tend to have fewer animals than the blacktailed prairie dog towns (Fitzgerald and Lechleitner, 1974). Plant species diversity and habitat variability are high both between towns and within towns (Shalaway, 1976).

To test the model, my associates Ken Paige, Mike Schwartz and I manipulated the resources within a single town. The town is located on the grounds of the Museum of Northern Arizona. Three other towns are within 2 km. Our manipulations have been confined to only a single town because the variability in plant species present and plant abundances between even adjacent towns is extreme, precluding the comparison of treatment effects involving manipulation of the same plant species in different towns. Experimental manipulations were carried out from May to September 1982.

Experimental procedures involved observation, removal, and supplementation of resources to create more uniform territories. The town was divided into a $120 \times 80\text{-m}^2$ grid, with stakes placed at 10-m intervals. Prairie dogs were live-trapped, marked with Lady Clairol hair dye, and released. The behavior and movements of the animals were observed with a spotting scope from a 4-m high observation tower positioned at one edge of the grid. Movements and aggressive interactions were noted daily, allowing feeding territories to be mapped. Observations were made from a blind on the tower. The animals were observed under normal conditions until 12 July, when the field was mowed with a tractor-mower, and the plants were mulched into the ground. Two weeks later, each territory was supplemented with 800 g (5600 kcal) of unshelled sunflower seeds, placed in the territory center. Sunflower seeds are part of the normal diet of these prairie dogs (Shalaway, 1976), and have a caloric value of 7 kcal/g, unlike the green forbs and grasses comprising the rest of the diet that have caloric values of 3–4 kcal/g. Plant diversity and distribution were sampled with line transects in each territory every 2 weeks.

The prairie dogs established feeding territories, which they actively defended against intruders (Table 1). Although some territories had only a single male and a single female, most territories early in the season had at least one male and more than one female (Table 2). The territories were very variable in the caloric content of food plants, that is, the actual plants utilized by the animals for food (Table 3). As the resources increased in terms of the quantity of available energy (Table 3), the feeding territories

Table 1. Response of Mean Territory Size of Prairie Dogs to Experimental Treatment of Resources

Treatment	Date	Mean Territory Size (m²)	F
None	May 14–31	444	$F = 2.41$
None	June 1–July 2	231	$F(1,10)$, NS
None	June 1–July 2	212	$F = 7.18$
Mowed	July 13–19	603	$F(1,10)^*$
Mowed	July 13–19	548	$F = 0.51$
Seeds	July 21–August 5	439	$F(1,12)$, NS
Seeds	July 21–August 5	356	$F = 3.03$
Seeds	August 6–17	245	$F(1,13)$, NS
Mowed	July 13–19	601	$F = 4.96$
Seeds	August 6–17	268	$F(1,10)^*$

* Significant at $p < 0.05$

Table 2. Mean Number of Prairie Dogs Occupying Territories in Response to Experimental Treatment of Resources

Time (Treatment)	$\bar{x}$	s	n	p
May (none)	2.5	0.84	6	NS
June (none)	2.8	0.94	6	NS
Early July (mowed)	1.9	0.99	8	<0.001
Late July–early August (seeds)	2.0	0.53	8	NS
Mid-August (seeds)	1.9	0.69	7	NS

contracted in size (Table 1), and on the average there were more animals per territory (Table 2).

The experimental manipulation of removing the plant resources caused both an increase in the size of the feeding territory and a decrease in the number of animals per territory. Most territories became monogamous, in the sense that they were occupied by a single male and a single female (Table 2). Supplementing the territories with a single food resource (sunflower seeds), caused a contraction in the size of the territory to approximately the pretreatment levels, but did not create a corresponding increase in the number of animals occupying the territory.

These results suggest that although the energetic content of the resources influences the size of the feeding territory, energy alone does not necessarily influence the formation of monogamous or polygynous groups of prairie dogs. Since the groups remained monogamous when the energetic content of the territories was boosted with a uniform resource, I suggest that an

Table 3. Caloric Content of Food Plants on Gunnison Prairie Dog Territories[a]

Territory	28 May 1982	10 July 1982	14 September 1982
80–70	431	727	1038
154	263	2434	3607
7	23	1572	2581
20	118	759	3607
70	—	2097	3607
5	347	1364	421
17	411	1520	1038
38	—	691	3227
72	—	3719	205
12	402	987	387
90	796	444	387

[a] Values are in kcal/m^2.

important component of group formation in the prairie dogs is the presence of a diversity of resources, perhaps due to contraints of nutrition and water balance. Prairie dog towns characteristically have the food resources very patchily distributed (Shalaway, 1976).

The above data suggest that Gunnison's prairie dogs respond to both the quantity of their resources and the distribution of the resources. When the resources are abundant, diverse, and patchily distributed, the prairie dogs form polygynous groups. When the resources are uniform, the animals form monogamous groups. At present, data on relative fitnesses on the territories are not yet available. If, however, the prairie dogs are responding in a way that would maximize their fitness, these results suggest that group formation in prairie dogs follows the predictions of the model.

This pattern may reflect a general trend in prairie dog sociality. The black-tailed prairie dog, *Cynomys ludovicianus*, lives in territories that are highly variable in patch quality (Halpin, personal communication; Kelso, 1938), and the predominant mating system is polygyny (King, 1955). The white-tailed prairie dog, *Cynomys leucurus*, lives in fairly uniform, low-quality patches, and its predominant mating system is monogamy (Clark et al., 1971).

3.2 Evidence from Other Animal Groups

The same relationship between resource quality and group formation also seems to hold true in the case of marmots (which are closely related to the prairie dogs). Yellow-bellied marmots, *Marmota flaviventris*, have a variable social system (Armitage, 1977), unlike the more solitary eastern wood-chucks, *Marmota monax* (Barash, 1974a). The resource patches used by the yellow-bellied marmots are extremely variable (Andersen et al., 1976) and most territorial groups of yellow-bellied marmots are polygynous. The east-ern woodchuck tends to live semicolonially in a more uniform habitat and is mostly monogamous (Merriam, 1971). The Olympic and arctic marmots live in rather uniform, poor habitats in family groups of one male, two fe-males, and the young. But, only one female bears young per year, so the system is functionally monogamous (Barash, 1974a).

Similar shifts from monogamy to polygyny as a function of resource dis-tribution can be seen among the African antelope. Jarman (1974) classified antelope into five social groups, depending on the type of habitat they lived in. Class A consists of the duikers and the dik-diks, animals with small body size. They are monogamous, feed selectively, and stay in one vegetation type. They take items such as flowers, twig tips, fruits, and seed pods, food that is relatively uniformly distributed in space and time. Class B consists of reedbucks, gerenuks, and lesser kudus, which are polygynous, have a group size of three to six, with at least two or more adult females in a group. They feed on a range of grass species or browse, remaining in one or several vegetation types and one home range throughout the year. Their food tends to be somewhat clumped spatiotemporally. Class C includes the waterbucks, kobs, gazelles, and impalas, which are polygynous and have a group size of six to sixty. They feed on a range of grasses and browse rather selectively. Their diet changes seasonally, as does their preference for vegetation type, caused by even more clumping of resources spatiotemporally. Class D involves the wildebeests and topis, which are polygynous and have a group size of six to several hundred thousand. They feed on grasses, and select specific plant parts or growth stages, leading to even greater spatiotemporal clumping. Class E consists of buffalo and elands. They are polygynous, feed on a wide range of grasses or browse, and move seasonally. Their diets are diverse and the patches vary spatiotemporally.

Other animals form larger or smaller groups in response to abundance and distribution of resources. Coyotes form packs of several animals when large prey such as elk are available. Otherwise, they hunt as solitary animals (Bekoff and Wells, 1981). When the coyotes form groups, helpers are present that assist in territorial and food defense (Bekoff and Wells, 1982). Acorn woodpeckers seem to have a similar response to resource levels (Stacey and Bock, 1978). Some woodpeckers within a population form social, cooperative groups that store acorns in caches, and collectively utilize and defend the stored acorns. Other woodpeckers in the same population live as temporary monogamous pairs that do not store acorns. These pairs migrate during the winter, whereas the social groups do not. Monogamous individuals seem to be less efficient than groups at harvesting acorns. Primate studies (Clutton-Brock and Harvey, 1977; Milton and May, 1976; Southwick and Siddiqi, 1974) relating ecology to social organization generally suggest that monogamy occurs in patches with uniformly distributed resources, whereas polygyny occurs in patchy resource distributions. Monogamous species tend to occupy large home ranges (Clutton-Brock and Harvey, 1977). Baboon groups tend to be larger when resources have high densities but very patchy distributions (Altmann, 1974).

4 COOPERATION AND AGGRESSION

If animals are to join groups in order to exploit or defend essential resources, the animals must be able to cooperate with one another. Noncooperative aggregation may be important to some animals such as colonial sea birds (Lack, 1968) for exploiting limiting nesting sites, but without cooperation the evolution of social behavior is not possible. Cooperation in turn requires limiting or controlling aggressive interactions within the group, so that the animals are relatively free to pursue other activities in reasonably close proximity to other members of their group. This means the development of mechanisms that allow a curtailment of personal space and other aggressive tendencies that may result in noncooperation and group disintegration. In this section I would like to consider how cooperation may arise, and the general relationship between cooperation and aggression in social groups.

4.1   A Game-Theoretic Approach to Cooperation

One promising approach to the question of cooperation has been a game-theoretic one. Game theory has been applied to biological problems in a number of contexts, principally in an analysis of the strategies of aggressive behavior (Maynard Smith and Price, 1973). A basic component of the approach is a payoff matrix that expresses what each player is expected to win by pursuing a given strategy. If a player pursues a single strategy, this is known as a pure strategy. If the player uses more than one strategy based on a certain probability or expectation of winning, this is called a mixed strategy. In a game, a player usually can play a pure or a mixed strategy that consistently provides the highest payoff for the conditions of the game. In other words, for the particular game, no other strategy will yield better results (Maynard Smith, 1982). Maynard Smith and Price (1973) originally considered that an animal would be genetically programmed to play a particular strategy, and defined an evolutionary stable strategy (ESS) as a strategy that cannot be beaten by a mutant playing another strategy. In behavioral terms, animals are rarely genetically programmed to play only one particular strategy, but an analogous situation occurs. The animal could play a behaviorally stable strategy (BSS) that cannot be beaten by another animal playing another strategy. Game theory analysis often involves equilibrium points, most notably the Pareto equilibrium and the Nash equilibrium. The other authors on social behavior in this volume also deal extensively with a game-theoretic approach.

Axelrod and Hamilton (1981) have formalized some of the requirements for cooperation, using a game-theoretic approach. Figure 2 shows a payoff matrix for a game situation called the Prisoner's Dilemma (Rapoport and Chammah, 1965). In this game, the Prisoner (Player 1) has several choices. One choice is to play C (cooperation), and if Player 2 also plays C, that is, if they mutually cooperate, Player 1 receives a payoff of 3, in arbitrary units. On the other hand, if Player 1 plays C and Player 2 plays D (defection), Player 1 receives a payoff of 0, the "sucker's payoff." Player 1 can get the highest payoff in this game by playing D when Player 2 plays C. As part of the Dilemma, Player 1 receives a payoff of only 1 if both players play D.

One BSS is D-D, that is, neither animal cooperates, because the costs of consistently playing C to a D are too high (Axelrod and Hamilton, 1981). This may explain from a game-theoretic point of view why so few animals are social. The costs of cooperation may be too high in many circumstances. However, the payoff for C-C may be profitable when a group is necessary for extracting or defending a resource. Then the penalty of noncooperation for an individual animal may be considerably more than the "sucker's payoff" of trying and failing to elicit cooperation. This would be particularly true when interindividual aggression within the group follows a pattern of decline with increasing sociality, for example, intragroup aggression of primates (Bernstein, 1976), lions (Bygoff et al., 1979), and bees (Lin and Mich-

PLAYER 2

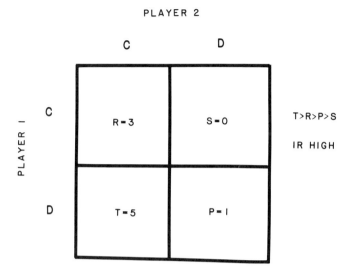

PRISONER'S DILEMMA

Figure 2. The payoff matrix of the Prisoner's Dilemma. The payoff is expressed in arbitrary units for Player 1. C designates cooperation, D designates defection. Player 1 receives a payoff of T when Player 1 plays D to Player 2's C, a payoff of R for C-C, a payoff of P for D-D, and S for C-D. Payoffs are structured so that T pays the most and S pays the least. A necessary condition for the development of cooperation is high individual recognition. (Figure is modified from Axelrod and Hamilton, 1981. Copyright 1981 by the American Association for the Advancement of Science.)

ener, 1972), and when resource extraction increases with sociality, for example, bird flock integration (Macdonald and Henderson, 1977; Krebs, 1974) (see Fig. 3).

4.2   The Influence of Individual Recognition

Axelrod and Hamilton (1981) show that another BSS is to play C first, and then follow the other player's lead, a tit-for-tat strategy. If Player 2 plays D, then Player 1 plays D, but if Player 2 plays C, then Player 1 continues to play C until Player 2 switches. The prediction of Axelrod and Hamilton is that this strategy would work best when there is individual recognition among animals, and when individuals play the game a number of times against other individuals that they know, so that they can predict the response. Frequent contact between animals may occur initially when home ranges overlap, or when territorial animals meet frequently at boundaries. For example, the home ranges of grey squirrels overlap, and individuals can recognize their neighbors (Thompson, 1978). Neighbor recognition has been described between two species of carnivores (Barash, 1974b), as well as for a variety of conspecific birds, mammals (Wilson, 1975a), some amphibians (Jaeger, 1981) and ants (Jutsum et al., 1979). When such neighbor recognition occurs, it may take the form of the "dear enemy" response, where animals show less aggression toward neighbors than toward strangers (Wilson, 1975a).

The mechanisms through which animals may recognize other individuals, particulary kin, have been reviewed by Bekoff (1983), who suggested several ways through which such recognition can develop: (1) genetic mechanisms; (2) recognition of phenotypic comparisons; and (3) recognition based on familiarity and spatial association. Genetic mechanisms of identifying related individuals from the same colony have been described for some bees (Breed, 1981; Greenberg, 1979). Recognition by phenotypic comparisons may occur through learning to distinguish the phenotypic differences of other individual animals. Recognition based on familiarity and spatial associations seems to be a common way of identifying individuals (Marler, 1976).

4.3 Philopatry and Cooperation

The above mechanisms for individual recognition obviously are not mutually exclusive, and at least two of the mechanisms may be promoted by philopatry. Philopatry is the tendency of animals of one sex or the other to stay near their parents, or to stay on or near a particular plot of land where they were born. Philopatry is a common phenomenon, with examples known from a number of animal groups (Shields, 1982). The nondispersing sex varies with the animal group. Among the mammals, males tend to disperse while females tend to settle close to their mothers (Waser and Jones, personal communication). In the birds, the females tend to disperse, whereas the males settle close to their father (Vehrencamp, 1979).

Such associations among nondispersing individuals provide the raw materials for the development of individual recognition and cooperation. Kin selection may start off the cooperative process, although it is not a necessary condition for cooperation. An example of a probably kin-selected system of philopatry is that of the ground squirrel (Michener, 1982). The typical pattern is for female juveniles to settle near their mother's territory. The mother may contract her territory or the female juvenile may inherit her mother's territory. While in the early evolutionary stages of this system each animal defends the perimeter of its own territory. The development of cooperation, as in the prairie dogs, allows a shared defense of the perimeter of a common territory. An example of a probably nonkin selected system is that of the neotropical fruit bat, *Phyllostomus hastatus* (McCracken and Bradbury, 1977). Females form cohesive stable groups that cluster in potholes of limestone caves in Trinidad. The female groups maintain relatively exclusive foraging areas, and when they find a rich food source, they vocalize to attract their roost mates. New groups of females are formed by the ousted youngsters from all the harems in the cave. Once a new group is assembled, it stays stable in composition for many years.

4.4   Costs and Benefits of Aggression and Resource Extraction

On a very simple level, the process of forming groups through cooperation can be treated as a cost–benefit analysis. Individual fitness in a group is equal to the benefit from resource extraction minus the cost of aggression (Fig. 3). If the benefit is greater than the cost, the animal should stay in the group. If the cost is greater, the animal should leave. This can be stated more formally by the following equations:

$$IF_0 = f(RE) \tag{1}$$

$$IF_g = f(RE) - f(A) \tag{2}$$

$$IF_g = B - C \tag{3}$$

$$\text{if } B < C, \text{ then leave}$$

$$\text{if } B > C, \text{ then stay}$$

where IF is individual fitness, RE is resource extraction, A is aggression, B is benefits, and C is costs. Equation (1) shows that the individual fitness of a solitary animal (IF_0) is a function of the ability of the animal to extract resources alone, whereas Equation (2) shows that the individual fitness of an animal in a group (IF_g) is a function of its ability to extract resources in a group, minus a function of the costs of aggression in the group. This trans-

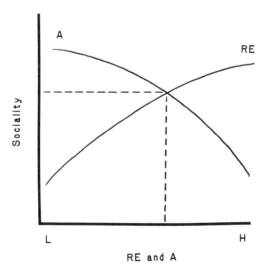

Figure 3.   A hypothetical relationship between sociality and resource extraction (RE) and aggression (A) at low (L) and high (H) levels of RE and A. Dotted line shows how a balance between resource extraction and aggression may influence the degree of sociality.

lates into Equation (3), the individual fitness in a group is related to the benefits minus the costs of being in the group, with the animal staying in the group if the benefits exceed the costs.

5 GROUP SIZE AND DOMINANCE

Group size and dominance are discussed extensively by Caraco and Pulliam in Chapter 10, so the discussion here will center on a few of the relevant details of the relationship among group size, dominance, and the development of sociality.

As animals form groups, the costs and benefits to each animal of being in a group may potentially change with the size of the group. Too few animals in the group may not be able to extract or defend a resource in such a way as to provide sufficient benefit, per individual animal, for being a group. Too many animals may increase the costs of aggression, or deplete the resource. Krebs (1974) shows that heron flocks have a resource extraction curve that peaks at a certain group size, then falls as more flock numbers are added. This leads to the question, asked by Caraco and Pulliam: Is there an optimal group size for animals?

Caraco and Pulliam show that in a group, not all the animals may have the same fitness. Dominance relationships may develop, and the dominants may acquire access to more resources. For example, dominant vespid wasps tend to become reproductive queens more frequently than subordinates (West, 1967). In multiple-queen nests, the dominant vespid wasp queen prevents reproduction of subordinates by eating their eggs, and also forces subordinates to pursue riskier and energetically more expensive tasks (Noonan, 1981). A similar process occurs in bumblebees (Heinrich, 1979). In bird flocks, dominant birds may use subordinates to ensure a better supply of resources (Rohwer, 1977).

As Caraco and Pulliam point out, group size can be considered from the standpoint of Fretwell's (1972) Habitat Selection model. Dominants do better in better-quality habitats. Subordinates do worse, *but* they can do better in a poorer-quality habitat only if they can become dominant in that habitat. If they switch to the poorer habitat and remain low-ranking subordinates, they do much worse than if they had stayed in their original group. The net result is that the group size would be shifted to maximize the mean fitness of the group, and not necessarily the fitness of the dominants (Fig. 4). An example is the analysis of lion group size by Caraco and Wolf (1975). Although the group size for hunting small prey was determined by the physiological limits of the lions, the group size of lions hunting larger prey was offset in the direction of the mean fitness of the group. Thus, when the lions hunt larger prey that can provide more food per lion than the required physiological minimum, the group size of the hunting females is not at the optimum that would maximize the caloric return. Instead, the group size is

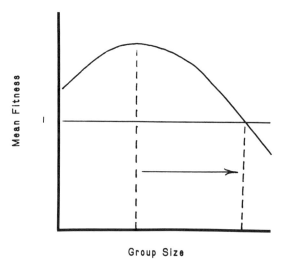

Figure 4. The mean fitness of group sizes under the conditions of dominance and Fretwell's (1972) Habitat Selection model. Joiners collect within the group, driving the group size away from the optimum and toward the point where each animal receives a distribution of resources that just satisfies the minimum physiological requirements of the animal (I).

increased, so that each lion receives, on the average, only enough food to satisfy the physiological minimum caloric intake.

I suggest that in a variable, patchy habitat, a social unit will tend to collect group members. If the costs of aggression are high for the dominants to eject newcomers, the group will collect more members. The dominants may be willing to suffer a small drop in fitness by having more group members, rather than face a large drop in fitness by incurring the costs of ejecting joiners. On the other hand, joiners in a colony may stay in the group because of the low probability of surviving as a solitary individual. This may explain the initial stages in the formation of large colonies of termites and honeybees, and does not have to be a kin-selected argument.

5.1 The Probability of Breeding and the Evolution of Castes

If an animal has no probability of breeding as a solitary individual, and a very low probability of breeding in a group, I predict that the animal will join the group. For example, if an animal had the following options:

$$p(IF_0) = 0$$

$$p(IF_g) = 10^{-6}$$

where the $p(IF_0)$ is a measure of the probability of breeding as a solitary individual, and the $p(IF_g)$ represents the probability of breeding in a group, selection should favor choosing a group over choosing to be solitary.

This is essentially the choice open to the social insects that have non-reproductive castes. A honeybee, ant, or termite that abandons the nest or hive would not be able to survive to the point of producing viable offspring. At that stage, its fitness is zero. However, a nonreproductive in the social insects has a certain probability, albeit small, during a narrow window in its life-span of becoming a reproductive and reaping enormous benefits in terms of fitness. For example, honeybees become workers or queens as a function of diet, that is, whether they are fed royal jelly throughout their larval life or only for a few days. Each egg potentially can become a queen. In fact, few do so, but the probability of becoming a breeding individual is still present. Even as adult workers, nonreproductive females are capable of occasionally laying male eggs and contributing minutely to their potential fitness (Wilson, 1971). Among the ants and the termites similar windows occur (Wilson, 1971), so that each individual can potentially become a reproductive. Even among the socially more primitive bees and wasps, where queens are often the largest females, larval diet determines to a large extent the size of the individual (Noonan, 1981), and, consequently, the probability of breeding.

The scenario that I suggest for the evolution of nonreproductive castes is as follows. Initially, individuals form groups, eventually increasing the group size beyond the optimal level. These individuals develop some measure of cooperation among themselves. As a group, they can extract and defend resources much more efficiently than they can as solitary individuals. In the group, dominance relationships arise. Cooperative behavior and a larger group size allow a modification of the immediate habitat for the purpose of food storage and shelter, and the modifications ameliorate the daily physiological stresses placed on each animal in the group. The animals become specialists in cooperative resource extraction, perhaps in response to competition from solitary, nonsocial animals. Although they may be more efficient at resource extraction and defense as a group, their efficiency at performing the entire gamut of tasks required for resource extraction declines, compared with that of solitary animals. Modification of habitat and efficiency of resource extraction reduce the probability of breeding as a solitary individual, so that at this point the animals become locked into a social existence. The dominance relationships allow manipulation of colony members [this is analogous to the manipulation arguments of Alexander (1974)], in such a way that the dominants breed while the subordinates do not. This dominant–subordinate relationship, however, can only be supported if all group members have a chance, however small, of becoming a dominant at some point in their lifetime.

Among many of the social mammals and birds, the earlier stages of this scenario are evident. In the insects, some of the later intermediate stages leading to eusociality and caste formation can be seen, particularly among the halictid bees (Breed and Gamboa, 1977; Michener and Brothers, 1974; Michener et al., 1971). Mammals and birds have apparently not developed true nonreproductive castes, with the exception of the naked mole rat, which requires extensive cooperation for modifying its habitat and extracting resources (Jarvis, 1981). Partial nonreproductives in the form of helpers, however, are known in both mammals and birds (Bekoff, 1983; Emlen, 1982a,b).

Perhaps mammals and birds have not evolved social systems with nonreproductive castes because of the interplay of body size, resource partitioning, and competition. May (1978) has shown that generally as body size among taxa decreases, the number of species increases. With more species present, we can assume that there is more subdivision of the available resources in response to competition, that is, more specialization on subsets of resources. Increased body size and fewer species imply a broader partitioning of resources in response to competition. Perhaps because of more broadly partitioned resources, the birds and mammals have not become cooperative specialists at the expense of individual efficiency to the same extent as the social insects.

6   DIVISION OF LABOR IN RESOURCE EXTRACTION

Once social animals cross the threshold and become cooperative specialists, they can extract a set of resources by having individual animals specialize on one resource in the set. This is the basis for flower constancy in honeybees and majoring in bumblebees (Heinrich, 1979). Specialization on one type of resource by each animal in the group can increase the animal's efficiency

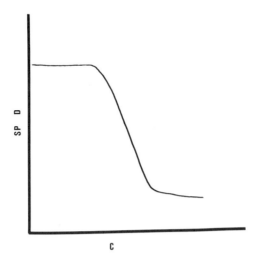

Figure 5. A hypothetical curve showing the impact of cooperative societies on species diversity of competitors in the immediate habitat. As cooperative sociality develops (from left to right), species diversity of competitors decreases. The rapid decrease in species diversity occurs at the stage in which the social group develops cooperative specialists through division of labor in resource extraction.

at collecting that resource. It can also increase the efficiency of the group in extracting the entire set of available resources. At this stage, each individual in the group may not be as good a competitor with a solitary animal over the entire range of resources, but the group, by being composed of specialists, becomes a good competitor with solitary animals at exploiting the entire resource set. This in turn can affect the diversity of species found in the vicinity of eusocial groups (Fig. 5). Such an effect can have a profound impact on the dynamics of communities. Honeybees, for example, can quickly replace a number of wild bee species as major pollinators of some agricultural communities (Wilson, 1971). Ant species can depress the abundance of their closest competitors (Hölldobler and Wilson, 1977).

7 SUMMARY

I suggest that in evolutionary terms, social behavior has developed in response to the distribution and abundance of resources. I develop a hypothesis that proposes four stages in the evolution of sociality: (1) habitat variability and mating systems; (2) cooperation and aggression; (3) group size and dominance; and (4) division of labor in resource extraction. Initial conditions for group formation involve a patchy habitat, ranging from low-quality uniform patches to high-quality patches where subsets of resources are clumped and distributed randomly. I suggest that low-quality uniform patches promote monogamy or solitary, nonsocial behavior, whereas high-quality patches with variable distributions of resource subsets promote polygynous associations that yield more efficient resource extraction and resource defense.

In forming polygynous associations, animals have the opportunity to cooperate with each other. Development of such cooperation may be promoted by individual recognition and philopatry. As cooperation develops, resource extraction or defense increases in efficiency while interanimal aggression within the group decreases.

Within cooperative groups, dominance relationships often develop which have a profound effect on the group size. Development of such dominance relationships allows the group size to increase, since subordinates or joiners often cannot improve their fitness by moving to another group. This tends to increase the group size to a point where each animal in the group receives not the optimal access to resources, but access to the required physiological minimum.

Through cooperation, social groups can modify their immediate surroundings (e.g., build nests) and ameliorate the physiological stress placed on individuals by abiotic factors. At the same time, through cooperation, a social group can become collectively very efficient at extracting resources. Habitat modification and efficiency at resource extraction may affect the probability of breeding of an animal in a social group, to the point where it becomes unable to reproduce as a solitary animal. At that point, selection may favor the animal's participation in the formation of nonreproductive castes, as long as the animal has even a small probability of being able to reproduce at some point in its life cycle.

Once social groups develop to the point of dividing up the labor of resource extraction, individuals within the group may become specialists on subsets of resources (e.g., the flower constancy of honeybees). As a group of specialists, the social group may then be able to outcompete solitary species that are generalists on a set of resources, and, as such, social species may profoundly affect the diversity of other species in their immediate habitats.

ACKNOWLEDGMENTS

I thank Ken Paige and Mike Schwartz, who patiently coped with the difficulties of prairie dog research, often beyond the call of duty. Model analyses and Tables 1, 2, and 3 were prepared by Ken Paige. I thank Meredith Clough, Charles and Dolores Biggerstaff for helpful discussions of their preliminary research on the Museum of Northern Arizona prairie dog town. I also thank Drs. R. Pulliam, S. Altmann, M. Bekoff, C. Michener, and P. Waser and the Fern Mountain Field Ecology Group, who provided many useful criticisms of the ideas presented here.

LITERATURE CITED

Alexander, R. D. 1974. The evolution of social behavior. *Annu. Rev. Ecol. Syst.* **5**:325–383.

Altmann, S. A. 1974. Baboons, space, time, and energy. *Am. Zool.* **14**:221–248.

Altmann, S. A., S. S. Wagner, and S. Lenington. 1977. Two models for the evolution of polygyny. *Behav. Ecol. Sociobiol.* **2**:397–410.

Andersen, D. C., K. B. Armitage, and R. S. Hoffman. 1976. Socioecology of marmots: Female reproductive strategies. *Ecology* **57**:552–560.

Armitage, K. B. 1977. Social variety in the yellow-bellied marmot: A population-behavioural system. *Anim. Behav.* **25**:585–593.

Axelrod, R., and W. D. Hamilton. 1981. The evolution of cooperation. *Science* **211**:1390–1396.

Barash, D. P. 1974a. The evolution of marmot societies: A general theory. *Science* **185**:415–420.

Barash, D. P. 1974b. Neighbor recognition in two "solitary" carnivores: The racoon (*Procyon lotor*) and the red fox (*Vulpes fulva*). *Science* **185**:794–796.

Bartholomew, G. A. 1970. A model for the evolution of pinniped polygyny. *Evolution* **24**:546–559.

Baylis, J. R., and Z. T. Halpin. 1982. Behavioral antecedents of sociality. In H. C. Plotkin (ed.), *Learning, development and culture*, pp. 255–272. Wiley, New York.

Bekoff, M. 1983. The development of behavior from evolutionary and ecological perspectives: Towards a generic social biology. In E. S. Gollin (ed.), *The comparative development of adaptive skills: Evolutionary implications*. Academic Press, New York (in press).

Bekoff, M., and M. C. Wells. 1981. Behavioural budgeting by wild coyotes: The influences of food resources and social organization. *Anim. Behav.* **29**:794–801.

Bekoff, M., and M. C. Wells. 1982. Behavioural ecology of coyotes: social organization, rearing patterns, space use, and resource defense. *Z. Tierpsychol.* **60**:281–305.

Bernstein, I. S. 1976. Dominance, aggression and reproduction in primate societies. *J. Theor. Biol.* **60**:459–472.

Bradbury, J. W. 1977. Social organization and communication. In W. A. Winsatt (ed.), *Biology of bats*, Vol. 3, pp. 1–73. Academic Press, New York.

Bradbury, J. W., and S. L. Vehrencamp. 1976. Social organization and foraging in emballonurid bats. II. A model for the determination of group size. *Behav. Ecol. Sociobiol.* **1**:383–404.

Bradbury, J. W., and S. L. Vehrencamp. 1977. Social organization and foraging in emballonurid bats. IV. Parental investment patterns. *Behav. Ecol. Sociobiol.* **2**:19–29.

Breed, M. D. 1981. Individual recognition and learning of queen odors by worker honeybees. *Proc. Natl. Acad. Sci. U.S.A.* **78**:1635–1637.

Breed, M. D., and G. J. Gamboa. 1977. Behavioral control of workers by queens in primitively eusocial bees. *Science* **195**:694–696.

Bygoff, J. D., B. R. C. Bertram, and J. P. Hanby. 1979. Male lions in large coalitions gain reproductive advantages. *Nature (London)* **282**:839–841.

Caraco, T., and L. L. Wolf. 1975. Ecological determinants of group sizes of foraging lions. *Am. Nat.* **109**:343–352.

Clark, T. W., R. S. Hoffman, and C. F. Nadler. 1971. *Cynomys leucurus (Mammalian Species* No. 7). American Society of Mammalogy, Lawrence, Kanses

Clutton-Brock, T. H., and P. H. Harvey. 1977. Primate ecology and social organization. *J. Zool.* **183**:1–39.

Dawkins, R. 1976. *The selfish gene.* Oxford University Press, Oxford.

Downhower, J. F., and K. B. Armitage. 1971. The yellow-bellied marmot and the evolution of polygyny. *Am. Nat.* **105**:355–370.

Emlen, S. T. 1982a. The evolution of helping. I. An ecological constraints model. *Am. Nat.* **119**:29–39.

Emlen, S. T. 1982b. The evolution of helping. II. The role of behavioral conflict. *Am. Nat.* **119**:40–53.

Fitzgerald, J. P., and R. R. Lechleitner. 1974. Observations on the biology of Gunnison's prairie dog in central Colorado. *Am. Midl. Nat.* **92**:146–163.

Fretwell, S. D. 1972. *Populations in a seasonal environment.* Princeton University Press, Princeton, New Jersey.

Fry, C. H. 1972. The social organization of Bee-eaters (Meropidae) and cooperative breeding in hot-climate birds. *Ibis* **114**:1–14.

Gaston, A. J. 1973. The ecology and behavior of the long-tailed tit. *Ibis* **115**:330.

Greenberg, L. 1979. Genetic component of bee odor in kin recognition. *Science* **206**:1095–1097.

Hall, E. R., and K. R. Kelson. 1959. *The mammals of North America*, Vol. 1. Ronald Press, New York.

Hamilton, W. D. 1964. The genetical evolution of social behavior, I and II. *J. Theor. Biol.* **7**:1–52.

Hamilton, W. D. 1972. Altruism and related phenomena, mainly in social insects. *Annu. Rev. Ecol. Syst.* **3**:193–232.

Heinrich, B. 1979. *Bumblebee economics.* Harvard University Press, Cambridge, Massachusetts.

Hölldobler, B., and E. O. Wilson. 1977. The number of queens: An important trait in ant evolution. *Naturwissenschaften* **64**:8–15.

Hoogland, J. L. 1977. The evolution of coloniality in white-tailed and black-tailed prairie dogs (Sciuridae: *Cynomys leucurus* and *C. ludovicianus*). Ph.D. dissertation, University of Michigan, Ann Arbor.

Hoogland, J. L. 1979. The effect of colony size on individual alertness of prairie dogs (Sciuridae: *Cynomys* spp.). *Anim. Behav.* **27**:394–407.

Jaeger, R. G. 1981. Dear enemy recognition and the costs of aggression between salamanders. *Am. Nat.* **117**:962–974.

Jarman, P. J. 1974. The social organization of antelope in relation to their ecology. *Behaviour* **58**:215–267.

Jarvis, J. U. M. 1981. Eusociality in a mammal: Cooperative breeding in naked-mole rat colonies. *Science* **212**:571–573.

Jutsum, A. R., T. S. Saunders, and J. M. Cherrett. 1979. Intraspecific aggression in the leaf-cutting ant *Acromyrmex octospinosus*. *Anim. Behav.* **27**:839–844.

Kelso, L. H. 1938. Food habits of prairie dogs. *USDA Circ.* No. 529.

King, J. A. 1955. Social behavior, social organization and population dynamics in a black-tailed prairie dog town in the Black Hills of South Dakota. *Contr. Lab. Vert. Biol., Univ. Mich.* **67**:1–123.

Krebs, J. R. 1974. Colonial nesting and social feeding as strategies for exploiting food resources in the great blue heron (*Ardea herodias*). *Behaviour* **51**:99–134.

Lack, D. 1968. *Ecological adaptations for breeding in birds*. Methuen, London.

Lacy, R. L. 1980. The evolution of eusociality in termites: A haplodiploid analogy? *Am. Nat.* **116**:449–451.

Lin, N., and C. D. Michener. 1972. Evolution of sociality in insects. *Q. Rev. Biol.* **47**:131–159.

McCracken, G. F., and J. W. Bradbury. 1977. Paternity and genetic heterogeneity in the polygynous bat, *Phyllostomus hastatus*. *Science* **198**:303–306.

Macdonald, D. W., and D. G. Henderson. 1977. Aspects of the behaviour and ecology of mixed-species bird flocks in Kashmir. *Ibis* **119**:481–491.

Marler, P. 1976. On animal aggression: The roles of strangeness and familiarity. *Am. Psychol.* **31**:239–246.

May, R. M. 1978. The evolution of ecological systems. *Sci. Am.* **239**:161–175.

Maynard Smith, J. 1982. *Game theory and evolution*. Cambridge University Press, Cambridge.

Maynard Smith, J., and G. R. Price. 1973. The logic of animal conflict. *Nature (London)* **246**:15–18.

Merriam, H. G. 1971. Woodchuck burrow distribution and related movement patterns. *J. Mammal.* **52**:732–746.

Michener, C. D., D. J. Brothers, and D. R. Kamm. 1971. Interactions in colonies of primitively social bees: Artifical colonies of *Lasioglossum zephyrum*. *Proc. Natl. Acad. Sci. USA* **68**:1241–1245.

Michener, C. D., and D. J. Brothers. 1974. Were workers of eusocial Hymenoptera initially altruistic or oppressed? *Proc. Natl. Acad. Sci. USA* **71**:671–674.

Michener, G. R. 1982. Kin identification, matriarchies, and the evolution of sociality in ground-dwelling sciurids. In J. F. Eisenberg and D. Kleiman (eds.), *Recent advances in the study of mammalian behavior* (Special Publ. No. 7). American Society of Mammalogy

Milton, K., and M. L. May. 1976. Body weight, diet and home range area in primates. *Nature (London)* **259**:459–462.

Noonan, K. M. 1981. Individual strategies of inclusive-fitness; maximizing in *Polistos fuscatus* foundresses. In R. D. Alexander and D. W. Tinkle (eds.), *Natural selection and social behavior*, pp. 18–44. Chiron Press, New York.

Orians, G. H. 1969. On the evolution of mating systems in birds and mammals. *Am. Nat.* **103**:589–603.

Page, E., Jr., and R. A. Metcalf. 1982. Multiple mating sperm utilization, and social evolution. *Am. Nat.* **119**:263–281.

Page, G., and Whitacre, D. F. 1975. Raptor predation on wintering shorebirds. *Condor* **77**:73–83.

Rapoport, A., and A. M. Chammah. 1965. *Prisoner's dilemma*. University of Michigan Press, Ann Arbor.

Rohwer, S. 1977. Status signalling in Harris' sparrows: Some experiments in deception. *Behaviour* **61**:107–129.

Shalaway, S. 1976. Habitat diversity and diet of the Zuni prairie dog in northern Arizona. M.S. Thesis, Northern Arizona University, Flagstaff.

Shields, W. M. 1982. *Philopatry, inbreeding, and the evolution of sex*. State Univ. New York Press, Albany.

Sinclair, A. R. E. 1977. *The African buffalo: A study of resource limitation of populations*. University of Chicago Press, Chicago.

Slobodchikoff, C. N., and R. Coast. 1980. Dialects in the alarm calls of prairie dogs. *Behav. Ecol. Sociobiol.* **7**:49–53.

Slobodchikoff, C. N., and H. V. Daly. 1971. Systematic and evolutionary implications of parthenogenesis in the Hymenoptera. *Am. Zool.* **11**:273–282.

Southwick, C. H., and M. F. Siddiqi. 1974. Contrasts in primate social behavior. *Bioscience* **24**:398–405.

Stacey, P. B., and C. E. Bock. 1978. Social placticity in the acorn woodpecker. *Science* **202**:1298–1300.

Svendsen, G. E. 1974. Behavioral and enviornmental factors in the spatial distribution and population dynamics of a yellow-bellied marmot population. *Ecology* **55**:760–771.

Thompson, D. C. 1978. The social system of the grey squirrel. *Behaviour* **64**:305–328.

Trivers, R. L. 1972. Parental investment and sexual selection. In B. Campbell (ed.), *Sexual selection and the descent of man*, pp. 136–179, Aldine, Chicago.

Vehrencamp, S. L. 1978. The adaptive significance of communal nesting in groove-billed anis (*Crotophaga sulcirostris*). *Behav. Ecol. Sociobiol.* **4**:1–33.

Vehrencamp, S. L. 1979. The roles of individual, kin, and group selection in the evolution of sociality. In P. Marler and J. G. Vandenbergh (eds.), *Handbook of behavioral neurobiology*, Vol. 3, pp. 351–394. Plenum Press, New York.

Verner, J., and M. F. Willson. 1966. The influence of habitats on the mating systems of North American birds. *Ecology* **47**:143–147.

West, M. J. 1967. Foundress associations in polistine wasps: Dominance hierarchies and the evolution of social behavior. *Science* **157**:1584–1585.

Wilson, E. O. 1971. *The insect societies*. Belknap Press, Cambridge, Massachusetts.

Wilson, E. O. 1975a. *Sociobiology*. Harvard University Press, Cambridge, Massachusetts.

Wilson, E. O. 1975b. *Leptothorax duloticus* and the beginnings of slavery in ants. *Evolution* **29**:108–119.

Wittenberger, J. F. 1979. The evolution of mating systems in birds and mammals. In P. Marler and J. G. Vandenbergh (eds.), *Handbook of behavioral neurobiologyf*, Vol. 3, pp. 271–349. Plenum Press, New York.

Woolfenden, G. E. 1976. Cooperative breeding in American birds. *Proc. Int. Ornithol. Congr.* **16:**674–684.

Zahavy, A. 1976. Cooperative breeding in Eurasian birds. *Proc. Int. Ornithol. Congr.* **16:**685–69.

CHAPTER 9

Constraints on Adaptation, with Special Reference to Social Behavior

RICHARD E. MICHOD
Department of Ecology and Evolutionary Biology
University of Arizona
Tucson, Arizona

CONTENTS

1   INTRODUCTION

1.1   Survival of the Fittest

At the encouragement of Wallace (1866), Darwin (1866) began using Spencer's (1864) now notorious phrase "survival of the fittest" to summarize the process of natural selection. In this phrase, by "fittest" Darwin meant most adapted to survive and reproduce and by "survival" Darwin meant not simply survival through a single generation but rather survival over evolutionary time. In other words, Darwin assumed that there was a direct relationship between the adaptedness of traits for individual survival and reproduction and their increase in frequency over time. Indeed, almost all analyses in contemporary evolutionary ecology embrace such a view. [In this chapter, I use the word "adaptedness" (instead of "fitness") to mean "the quality or state of being adapted" to carry out a specified activity (Oxford English Dictionary). I use the word "fitness," as it is used in discrete generation models in population genetics, to mean the expected number of gametes (or offspring) produced on the average by a genotype. All things being equal, a genotype's fitness measures its adaptedness at surviving to reproduce.]

While it is clearly true that the adaptedness of a trait for survival and reproduction is an important component of an evolutionary explanation, it is equally true that knowledge of adaptedness alone is not a complete and logically sound explanation. As Darwin realized, the process of natural selection is based on two essential postulates: (1) variation in adaptedness for survival and reproduction (some traits aid in the "struggle to survive" more than others) and (2) heritability. Thus, there are two main components to an evolutionary explanation. The first, adaptedness, is the mainstay of evolutionary ecology. The second, heritability, concerns the transmission process by which the trait is passed on through the generations and this component is the mainstay of population genetics.

In the case of neutral traits in which there are no differences in adaptedness, the transmission process alone can produce evolutionary results. However, the results of random genetic drift are described by the tautology "survival of the survivor," as Eigen (1971) has pointed out. In other words, in the case of genetic drift, nothing can be said before the fact about which genotype will be successful to increase frequency. The genotype that is successful is simply the one that is successful. The claim implicit in Spencer's phrase "survival of the fittest" is that success can be predicted before the fact from a consideration of differences in adaptedness at survival and re-production alone. This is the basis of the "adaptationist program" (Gould and Lewontin, 1979), which takes Spencer's phrase literally instead of as a catchphrase for the more complicated logic that Darwin used to deduce the process of natural selection from the postulates discussed above. The ad-aptationist program basically assumes that the process by which a trait is transmitted from parent to offspring has little effect on its evolutionary dy-namics. The fact that Darwin so approvingly embraced Spencer's phrase suggests that he too felt that the genetic transmission process, although nec-essary for organic evolution, was of secondary importance to adaptedness in shaping the results of evolution.

In this chapter I discuss evolutionary constraints by considering how adaptedness and success at increasing in frequency are actually connected. Second, in the case of social behaviors, I examine the evolutionary modi-fication of some of these genetic constraints.

1.2 On Evolutionary Constraints

It is important to have a clear notion in the present context of what is meant by the term "constraint."

Most work in modern ecology focuses on the adaptedness of a trait for survival and reproduction of individual organisms that possess the trait. Of course, a trait will have effects on survival and reproduction through more immediate effects on, for example, the organism's capacity to accrue re-sources or its capacity to obtain a mate. Because of its genetic constitution, an organism has certain *intrinsic* capacities or propensities to carry out these and other activities (see Bernstein et al., 1983; Mills and Beatty, 1979; Den-niston, 1978, for more discussion of these points). I will represent these intrinsic capacities of a particular genotype by the lower case English letters a_i, b_i, c_i, . . . , etc., where i subscripts the genotype and the letters stand for various adaptive capacities such as the capacity to obtain a mate, find a resource, and so on. In most explicit models of the evolution of a trait, these intrinsic capacities will be represented by genotype-specific constants. It is the intrinsic capacities that are transmitted from parent to offspring. The capacities delineated, in any particular case, will be directed in large part by the particular problem at hand and several examples will be discussed shortly. The units of each intrinsic capacity will vary and these capacities will of necessity be measured under certain environmental conditions. For a specified activity, these conditions must be identical for all genotypes under consideration. Under these environmental conditions, it is, of course, impossible to partition a genotype's capacity into an intrinsic and environmental component. However, differences between genotypes in their intrinsic capacities as defined above result from intrinsic differences alone. Thus, differences between genotypes in their intrinsic capacities measure differences in adaptedness at carrying out the specified activity in the specified environment.

Of course, the organism's intrinsic capacities or abilities may be expressed in an environment which may differ from that in which they were measured. These resulting capacities of genotype i will be designated by capital English letters, say A_i, B_i, C_i, ... The resulting capacities may be complex functions of physical and biotic factors in the environment as well as of the intrinsic capacities and frequencies of other genotypes in the population or community. (In this chapter, with two exceptions, I also use capital English letters, but without subscripts, to denote environmental variables and parameters. The two exceptions are in Section 3 in which X and Y refer to specific behaviors and in Section 4 in which A, S, and C refer to mutant hemoglobin molecules. Throughout the chapter Greek letters are as defined and no convention applies to their use.)

The resulting capacities most often considered are survivorship and fecundity, or simply "resulting fitness." Resulting fitness is defined in the discrete generation model considered later as the expected number of gametes produced by a genotype in its lifetime. The analogous concept in evolutionary ecology is an organism's expected reproductive success. However, this concept as usually employed is insufficient to study evolution in a sexually reproducing population, since an organism's offspring may have different genotypes and hence different intrinsic capacities. Since the transmission process is not specified, the effects of sex cannot be addressed by the concept of expected reproductive success as it is usually employed. In a genetical model, which specifies the transmission process, the effects of sexual reproduction are included because Mendel's laws with recombination partitions the total gametic output into the appropriate gametic types.

The distinction between resulting and intrinsic capacities is basic to a correct understanding of the evolutionary process, since it is only the intrinsic capacities that are passed on from parent to offspring. However, there is nothing new about this distinction. It is common in quantitative genetics to partition the observed (or resulting) trait into components due to the genotype, the environment, and the genotype–environment interaction. Such a distinction is necessary because it is only the genetic component of the trait (the so called breeding value) which figures in the mating process and is passed from parents to offspring.

In most explicit, dynamical models for the evolution of a trait it is obvious which capacities are intrinsic and which result from interaction with the environment. In the so-called "constant fitness" models of population genetics, the two capacities are defined to be the same (see Section 1.3). However, these models are of restricted interest from an ecological point of view. Most ecological processes, especially social behaviors, lead to density- or frequency-dependent selection, in which the distinction between the intrinsic capacities (which are genotype specific constants) and the resulting fitnesses (which are density or frequency dependent) arises naturally (see Section 2.3).

Finally, through reproduction the intrinsic capacities are passed on to offspring. The success of a genotype at surviving to reproduce will not depend on the transmission process, although as already discussed it will be some complex function of the genotype's various intrinsic capacities and the environment in which it exits. However, the success of a genotype at increasing in frequency over evolutionary time will depend on the process by which its intrinsic capacities are transmitted from parent to offspring. Following Fisher (1930), and Bernstein et al. (1983), I will measure the potential for evolutionary success of a genotype by its per capita rate of increase ($\dot{\chi}_i/\chi_i$, or $\Delta\chi_i/\chi_i$ in the case of discrete time). The various issues discussed so far are summarized in Figure 1.

The phrase "survival of the fittest" can now be seen as a mapping of the adaptedness of a genotype at carrying out various activities involved in survival and reproduction (a_i, b_i, c_i, etc.) onto its per capita rate of increase ($\dot{\chi}_i/\chi_i$). In particular, the phrase proposes that there is a direct, monotonic relationship between these two quantities. In other words, if, for example, we focused on one intrinsic adaptive capacity, say b_i, and ranked the various genotypes in the population according to their relative magnitude of b_i, then by "survival of the fittest" we would mean that the per capita rate of increase of the various genotypes would be similarly ranked. However, as we will see, the effects of a genotype's adaptedness at some activity on its rate of increase may be *masked* either by the environment or by the transmission process (Fig. 1). In some cases, such as a totally random environment or the example of molecular replicators to be discussed shortly, this may lead

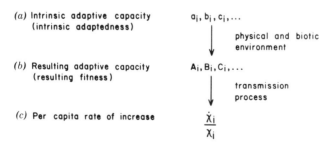

Figure 1. Diagram of the relationship between the intrinsic adaptedness of a trait and the trait's per capita rate of increase in frequency. The relationship may not be direct because of masking of intrinsic adaptedness by the environment or by the transmission process. See text for further explanation.

to a total decoupling of evolutionary success from intrinsic adaptedness. In other cases, the environment or the transmission process interacts with a genotype's intrinsic capacities so that their mapping onto the genotype's rate of increase is not direct and monotonic (see, e.g., Fig. 2). Those factors that lead to this masking of a genotype's adaptedness will be said to "constrain" the evolutionary process, since the genotypes most intrinsically adapted are not necessarily the ones that increase in frequency over evolutionary time.

1.3 Heterozygote Superiority with Constant Fitnesses

At the risk of being pedantic, I consider here the case of constant viability selection for a single, diallelic locus with heterozygote superiority. The purpose of doing so is to illustrate how the genetic transmission process may mask a genotype's intrinsic capacities. The case considered here of constant fitness blurs the distinction between intrinsic adaptedness and resulting fitness (Fig. 1), since the resulting fitnesses are simply assumed to be constants independent of any environmental effect. Nevertheless, this example illustrates the idea of an evolutionary constraint as defined in the last section.

Let w_0, w_1, and w_2 be the constant (resulting and intrinsic) fitnesses and χ_0, χ_1 and χ_2 the frequencies among offspring of the three genotypes $\alpha_1\alpha_1$, $\alpha_1\alpha_2$ and $\alpha_2\alpha_2$, respectively. Let γ be the frequency of the α_1 allele, the evolutionary dynamics of which are given by the standard equation

$$\Delta\gamma = \frac{\gamma(1 - \gamma)}{2\bar{W}} \frac{d\bar{W}}{d\gamma} \tag{1}$$

with $\bar{W} = \sum_i \chi_i w_i$. For the case of heterozygote superiority ($w_1 > w_0, w_2$; without lack of generality I will assume that $w_0 > w_2$), the graph of $\bar{W}$ is given in Figure 2a. In this case, there is a single interior equilibrium at $\hat{\gamma} = (w_1 - w_2)/(2w_1 - w_0 - w_2)$, which is globally stable as can be seen from Equation (1) and the graph in Figure 2a. From the assumption of random mating, $\chi_0 = \gamma^2$, $\chi_1 = 2\gamma(1 - \gamma)$ and $\chi_2 = (1 - \gamma)^2$. The change in frequency of the three genotypes ($\Delta\chi_i = \chi_i' - \chi_i$, where superscript prime denotes the next generation) can then be expressed in terms of the frequencies of the alleles as follows:

$$\Delta\chi_0 = \Delta\gamma(\gamma' + \gamma)$$
$$\Delta\chi_1 = 2\Delta\gamma(1 - \gamma' - \gamma) \tag{2}$$
$$\Delta\chi_2 = -\Delta\gamma(1 - \gamma' + 1 - \gamma)$$

The relationship of the per capita change in frequency ($\Delta\chi_i/\chi_i$) and fitness (w_i) is given in Figure 2b and c for the two regions of gene frequency given in Figure 2a (Region I: $\gamma < \hat{\gamma}$; Region II: $\gamma > \hat{\gamma}$). For each region, the fitnesses and per capita changes in frequency are ranked from top to bottom according

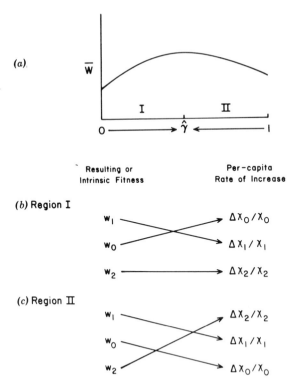

Figure 2. Heterozygote superiority in constant fitness. (*a*) Graph of average individual fitness, $\bar{W}$, as a function of the frequency of the α_1 allele, γ. As indicated, there is a stable interior equilibrium at $\hat{\gamma}$. (*b*) and (*c*) Relationship of intrinsic adaptedness or resulting fitness of a genotype (they are one and the same in a constant fitness model) to the genotype's per capita rate of increase, for the two regions of gene frequency given in (*a*). In both cases the w_i's and the $\Delta\chi_i/\chi_i$'s are ranked from top to bottom in order of decreasing magnitude. See text for further explanation and definition of terms.

to their relative magnitudes. The relationships between the $\Delta\chi_i/\chi_i$'s are found from Equation (2) using Equation (1) and Figure 2*a*. "Survival of the fittest" predicts a direct (monotonic) relationship between fitness and per capita rate of increase (i.e., all arrows would be horizontal in Fig. 2*b* and *c*). However, as can be seen in Figure 2, there is no simple relationship between these two quantities in this case. The reason for this masking of fitness is the fact that the fittest genotype, the heterozygote, cannot breed true because of sexual recombination, which segregates out the less fit homozygotes.

2   ENVIRONMENTAL MASKING OF INTRINSIC CAPACITIES

In this section, several examples are given of the masking of a genotype's intrinsic capacities due to environmental factors outside the genotype's control. The presentations are somewhat abbreviated, since the supporting details are available elsewhere. The first example considers the process of natural selection as it originated in the origin of life and is based on the work of Bernstein et al. (1983), Eigen and co-workers (1971; Eigen and Schuster, 1979), and Michod (1983a). The second example discusses the somewhat standard results of the theory of kin selection from the present perspective. The reader is referred to the work of Hamilton (1964, 1972), Wilson (1980), and Michod (1982) for the necessary background information. Finally, I discuss some results of the theory of density-regulated selection and age-specific selection from the standpoint of environmental masking.

2.1 Prebiotic Molecular Replicators

The simplest replicating system relevant to the origin of life involves non-enzymatic, template-directed replication of single-strand RNA (Bernstein et al., 1983; Eigen, 1971; Eigen and Schuster, 1979). The experimental evidence for many aspects of such a system is provided by Orgel and co-workers at the Salk Institute (Lohrman et al., 1980; Lohrman and Orgel, 1980). Replication in such a system is not highly accurate but generally follows current base pairing rules (adenine-uracil, guanine-cytosine). There are three intrinsic capacities of the replicating molecule which have been discussed in detail elsewhere and which measure the intrinsic adaptedness of the replicating molecule (Bernstein et al., 1983): b_i the maximum rate of template-mediated replication, d_i the rate of decomposition of the molecule in a hydrolytic environment (d_i is inversely related to the capacity of the molecule to avoid "death"), and r_i a specific combination of rate coefficients which is inversely related to the capacity of the molecule to obtain and process its resources, which are primarily mononucleotide building blocks. These three intrinsic capacities are ultimately determined by the nucleotide sequence (the genotype indexed by i) of the molecule through the physical–chemical properties of its three-dimensional folded structure (the phenotype). Transmission of these intrinsic capacities is essentially asexual, involving direct copying of the nucleotide sequence via an intermediary complementary strand. As shown in Bernstein et al. (1983), the dynamics of such a system are given by

$$\frac{\dot{\chi_i}}{\chi_i} = b_i \frac{M}{r_i + M} - d_i \qquad (3)$$

where M is the overall concentration of mononucleotide resources in the environment. From analysis of Equations (3), a replicator, i say, will increase in a population dominated by another replicator, j say, if (Bernstein et al., 1983)

$$\frac{b_i - d_i}{r_i d_i} > \frac{b_j - d_j}{r_j d_j} \qquad (4)$$

Thus, there is a direct relationship between a molecular replicator's intrinsic adaptive capacities [Equation (4): high b_i; low d_i, and r_i] and its evolutionary success $\dot{\chi}_i/\chi_i$, so that "survival of the fittest" accurately describes the evolution of such a system.

The next stage in the evolution of life probably involved the production by a molecular replicator of an enzyme which aided its replication. The enzyme may also have aided in the replication of other replicators and cooperative groups of such replicators may form under certain situations. Eigen (1971) has termed a replicator and its enzyme, or the multimembered cooperative groups which may form, "hypercycles" and has extensively studied their evolution (Eigen, 1971; Eigen and Schuster, 1979). The intrinsic adaptive capacities of a hypercycle are similar to a template-directed, non-enzyme-mediated replicator except that b_i is now the maximum rate of enzyme-mediated replication. Consequently b_i measures the capacity of the enzyme (which is coded for by the nucleotide sequence) to replicate the nucleotide sequence. At the earliest stage, the hypercycle is assumed *not* to be encapsulated in a cellular membrane so that the chance of a replicator encountering its protein is proportional to χ_i (the concentration of the enzyme is assumed to be proportional to the concentration of the RNA molecule which produces it). This creates problems for the initial increase of the first hypercyclic replicator with a new favorable mutation in the sequence encoding the enzyme, since the probability of encountering its improved enzyme is vanishingly small. Thus, its *realized* capacity for birth is essentially independent of b_i. As discussed by Michod (1983a), passive localization of the enzyme products of translation in rock crevices or suspended water droplets may help overcome these difficulties when the replicators are rare. However, these issues will be ignored here, since my main purpose is simply to illustrate how total decoupling of evolutionary success from intrinsic adaptedness can occur.

The following example of the simplest possible hypercycle serves to illustrate the points just discussed. Consider a replicating molecule that produces a protein that only aids replication of the nucleotide sequence that produced it. Let b_i be the intrinsic capacity of the enzyme to replicate the nucleotide sequence under high concentrations of enzyme. The death rate will be ignored in what follows, although its inclusion would not affect the conclusions drawn. The concentration of the protein is assumed to be directly proportional to the concentration of the replicator which produced it, χ_i. Consequently, the probability of a replicator encountering its protein is proportional to χ_i, if there is no structure (such as a cell) to keep the protein close by the replicator that produced it. The realized capacity of replication is then proportional to $b_i\chi_i$ ($=B_i$). Finally, competition may be introduced in a manner discussed by Eigen (1971) and Eigen and Schuster (1979, p. 30) by assuming a constant level of resources, χ_T, needed for replication (e.g., mononucleotides) so that

$$\sum_i \chi_i = \chi_T$$

These assumptions yield the following dynamics for the replicating molecules (Eigen, 1971; Eigen and Schuster, 1979, p. 30; Michod, 1983a)

$$\frac{\dot{\chi}_i}{\chi_i} = b_i \chi_i - \Psi \tag{5}$$

where

$$\Psi = \frac{\sum_j \chi_j^2 b_j}{\chi_T}$$

is the per capita excess production so that $\sum_i \dot{\chi}_i = 0$. One can see from analysis of Equations (5) that if one replicator is common, no other mutant replicator can increase when rare no matter what its intrinsic capacity for replication, b_i, is. This leads to an evolution law of "survival of the common," instead of "survival of the fittest," since the success of a rare replicator at increasing in frequency is unrelated to its adaptive capacities. Similar dynamical constraints exist for more complicated hypercycles (Eigen, 1971; Eigen and Schuster, 1979). This kind of evolution is more precisely termed "survival of the first," rather than "survival of the common" (Bernstein et al., submitted ms.). Eigen has termed this kind of evolution "once-forever" evolution, and has discussed its role in the evolutionary process (1971; Eigen and Schuster, 1979; Eigen and Winkler, 1981). However, as shown in Michod (1983a, equations 10–14), Eigen's phrase overstates the case somewhat.

This example serves to illustrate the total decoupling of evolutionary success from intrinsic adaptedness. This decoupling would be, perhaps, a mere curiousity confined to the particular hypothetical stage in the history of life just discussed, if it were not for the nonlinearities introduced by sexual reproduction which have similar effects (Bernstein et al., submitted ms.; Hopf and Hopf, submitted ms.).

2.2 Kin Selection

In the previous case of nonencapsulated, protein-mediated molecular replicators, there was no relationship of a replicator's intrinsic adaptedness to its success at increasing when rare. The masking of intrinsic adaptedness for individual survival and reproduction is more subtle in the example of kin selection discussed now.

In many situations, the environment of an organism contains conspecific individuals with whom it shares genes identical by descent. In this case, the presence of relatives constrains the behavior of individuals away from that predicted by considering only adaptedness to an individual's survival and reproduction (Hamilton, 1964; see Michod, 1982, for review). Let c_i be additive effect of i's behavior on its own resulting fitness and a_i the additive effect on the resulting fitness of other individuals whom i encounters. Based on adaptedness for individual survival and reproduction, we would expect genotypes which possess highly positive c_i and highly negative a_i to evolve. Indeed, for a particular model of selection in a nonstructured population the condition for evolutionary success of i over j is found to be $c_i - a_i > c_j - a_j$ (see Wilson, 1980, p. 15 for discussion). Consequently, in a nonstructured population, there is a direct relationship between the adaptedness of a genotype's behavior (for individual survival and reproduction) and its success at increasing over evolutionary time.

However, as is now commonly appreciated, this is not the case in structured populations in which individuals interact with kin. As usual, in a structured population an individual's resulting fitness is a function of its intrinsic capacities and its environment. However, because of sexual reproduction and the presence of kin, a genotype's environment includes relatives that, because they share genes identical by descent, are more likely to have similar intrinsic capacities to behave in the same way as does the individual of interest. For example, genotypes which harm others and consequently have highly negative a_i will be likely to interact with relatives that also have the intrinsic capacity to harm others. Consequently, the resulting fitness of a genotype with negative a_i may actually be lower than a genotype with positive a_i. The intrinsic capacities which are actually selected depend mainly on R, the coefficient of relatedness between interacting individuals. In many such situations (see Michod, 1982, for discussion), the conditions for evolutionary success of genotype i in competition with j is $c_i + Ra_i > c_j + Ra_j$ ($c_i + Ra_i$ is termed the "inclusive fitness effect" of i's behavior). Thus, because of sex and the presence of kin there is no longer any simple relationship between a genotype's intrinsic capacities for individual survival and reproduction and its evolutionary success, as there was in the case of homogeneous populations. This example illustrates the point that the genetic system affects not only the transmission process but may also affect the environment as well.

2.3 Discussion of Environmental Masking

In the case of the nonencapsulated, protein-mediated molecular replicator, the environmental factor, which was the local protein concentration, presents a barrier to further improvement of the adaptive capacity. This barrier may be overcome, presumably by some form of encapsulation of the replicator and its protein. One may argue that this environmental factor is no different from any selective factor and that what I have termed an environmental constraint is no different from any "problem" the environment may "pose" to an evolving system.

The term "constraint" as commonly used in evolutionary discussions refers to some factor that prevents or restricts adaptive evolution. Thus, one use of the term "genetic constraint" refers to a lack of genetic variation for characteristics or capacities that would clearly be adaptive if they existed. Similarly, in the case of the nonencapsulated molecular replicator, it is clearly adaptive to improve the capacity of the protein to replicate the nucleotide sequence. Yet, such evolutionary progress is prevented from occurring until some form of active or passive (Michod, 1983a) encapsulation occurs, for the simple reason that a new mutant replicator is initially rare and unable to encounter its protein.

Another example which illustrates the same type of frequency-dependent barrier or constraint concerns the origin of distastefulness in, for example, butterflies or moths. The distasteful trait is clearly adaptive in a population subjected to predation. However, as first discussed by Fisher (1930), rare distasteful mutants in an otherwise palatable population are at best neutral or may actually be disadvantageous, if there is any cost to producing the distasteful compound. Such frequency dependence presents a barrier, which prevents adaptive evolution by individual selection. However, as commonly hypothesized, this barrier may be overcome by such mechanisms as genetic drift, kin selection, or group selection. Interestingly, the barrier preventing the improvement of protein function in the case of molecular replicators may be overcome by similar mechanisms (Michod, 1983a).

Other ecological variables such as population density or population growth rate may mask a genotype's intrinsic capacities. Consider, for example, density-dependent selection in which a genotype's resulting fitness is some complex function (often logistic) of its intrinsic capacities for birth, b_i, its carrying capacity, k_i, which measures its capacity at efficiently utilizing resources, and the overall density, N; or $W_i = W_i(b_i, k_i, N)$ (see, e.g., Anderson, 1971, or Roughgarden, 1979, Chapter 17). In populations with large N, a genotype's capacity for birth b_i has little influence on its rate of increase which is determined in large part by its carrying capacity, k_i. Conversely, in populations with low N, a genotype's rate of increase is determined mainly by b_i and the low population density masks the genotype's capacity at efficiently utilizing resources (k_i).

In age-structured populations, $l_i(\sigma)$ and $m_i(\sigma)$ measure a genotype's age-specific intrinsic capacities for survivorship and birth, respectively, at age σ (see Charlesworth, 1980, for review of this area). Fitness in populations in demographic and genetic equilibrium growing at rate G is a function of these intrinsic capacities and the environment represented by G the population growth rate, $W_i = \int \exp(-G\sigma)l_i(\sigma)m_i(\sigma)d\sigma$. Consequently, the overall growth rate of the population weights the contribution of the genotype's age-specific intrinsic capacities to fitness. For example, an increasing population masks a genotype's capacity for survivorship and reproduction late in life, whereas the converse is true in decreasing populations. In stable populations (i.e., $G = 0$), the resulting fitness at equilibrium depends only on a genotype's intrinsic capacities, since the environment no longer weights the contribution at different ages.

3 MODIFICATION OF GENETIC CONSTRAINTS ON SOCIAL BEHAVIOR

3.1 Introduction

In the following sections, I consider some genetic constraints on the evolution of social behaviors that result from the processes by which the behaviors are transmitted from parent to offspring. I then ask whether there is selection for an allele that removes these constraints by modifying the genetic system of transmission. A predominant mode of thought concerning the evolution of social behavior involves game theory, the applications of which to evolution have been developed primarily by Maynard Smith and his associates (see Maynard Smith, 1982, for review). The applications of game theory to evolutionary problems is directed at mapping intrinsic capacities (Fig. 1a), such as the probabilities of exhibiting certain behaviors and the effects of those behaviors on fitness (payoffs), onto resulting fitness (Fig. 1b, i.e., the fitness which results when the behaviors of others are taken into account). As discussed at the end of Section 3.2, it is then taken for granted in most of these applications that there is a direct relationship between a phenotype's resulting fitness (Fig. 1b) and its per capita rate of increase (Fig. 1c). Consequently, these applications embrace the general view of "survival of the fittest."

Given that the mode of transmission of many social adaptations is sexual, what is the justification for embracing this view? [Cultural transmission of behavior will be ignored here. However, many forms of cultural transmission lead to masking if not total decoupling of evolutionary success from biological adaptedness (see Cavalli-Sforza and Feldman, 1981)].

In responding to a similar critique of game theory, namely, that it assumes that evolution maximizes resulting fitness, Maynard Smith (1982, p. 5) states that "it can be a simple consequence of the laws of population genetics that, at equilibrium, certain quantities are maximised." In the language of population genetics, social evolution is a kind of frequency-dependent selection. However, it is well known that frequency-dependent selection does not maximize average fitness. As Wright (1969, Chapter 5) has shown, frequency-dependent selection will, in many cases, maximize another quantity (which Wright termed the "fitness function"); however, this quantity bears no obvious relationship to fitness as usually construed. Are there any other principles among the laws of population genetics which can justify the applications of game theory despite the fact that it ignores the constraints imposed by sexual transmission? A likely candidate is evolutionary modification of the genetic system itself.

The genetic system of transmission is not fixed and evolution may modify its basic parameters. The study of the evolution of modifiers of genetic parameters is as old as population genetics itself and has been used to study the evolution of recombination, dominance and linkage relationships, to name a few (see, e.g., Feldman and Krakauer, 1976). It is the modification of dominance which is most relevant to the problem under discussion here.

Fisher (1928a,b, 1929) appealed to the evolution of modifiers of dominance to explain why most mutations are recessive. This application has been extensively criticized by Wright (for review see, 1969, pp. 69–71, 1977, pp. 498–526), who argued that the selection pressures on modifiers of the mutant heterozygote would be too weak in this case (because the heterozygote would exist in too low a frequency by the balance of selection against the prevailing deleterious mutations and recurrent mutation). Fortunately, this controversy is not relevant to the application of the theory of genetic modification used below, since the polymorphic equilibria investigated there may maintain all genotypes in appreciable frequency.

The existence of modifiers of dominance was never questioned by Wright. In fact, Wright (1927) provided early evidence in the guinea pig that dominance at a locus was affected by other loci. Indeed, in 1929, Wright noted that "if for any reason, the proportion of heterozygous mutants reaches the same order as that of type, selection of modifiers approaches the order of direct selection in its effects and might well become of evolutionary importance." The rest of this chapter studies the modification of genetic systems of transmission which prohibit the game-theoretic predictions of social evolution based on resulting fitness from being realized.

3.2 Modifier Model

I will consider the evolution of two social behaviors in a diploid, sexually reproducing, random mating, and nonstructured population. I assume that a single diploid locus, termed the primary locus, with two alleles α_1 and α_2 causes a difference in the expression of social behavior, perhaps by altering the level of a hormone which predisposes the individual to behave in one way or another. A second locus, termed the modifier locus, with two alleles β_1 and β_2 is assumed to modify the expression of the social behavior by the various genotypes at the primary locus (see Fig. 3). The two behaviors are notated X, Y (for example, "Hawk" or "Dove," "Cooperate" or "Defect," etc.; see Maynard Smith, 1981, or Brown et al., 1982) with payoff matrix given in Table 1.

The application of a simple population genetic model to complex adaptations such as social behaviors is not as unrealistic as it first appears. As discussed by Dawkins (1982, Chapter 2), it is never assumed, even in the case of physiological or morphological traits such as eye color, that a single gene completely causes the trait under consideration. Rather, it is assumed that a single gene can cause a *difference* in the trait and this is often a useful modeling assumption if not an assumption that is completely realistic in many cases. Similarly, in the case of complex behaviors, single genes may cause

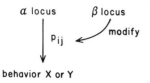

α locus β locus

p_{ij} ← modify

behavior X or Y

Figure 3. Two locus modifier model. The α locus causes a difference in the expression of behaviors X and Y through the intrinsic p_{ij} parameters. The β locus modifies this expression. See text for further explanation.

dramatic differences in behavior as is well known in the case of certain behavioral abnormalities. Indeed, as Dawkins (1982, p. 23) has pointed out, for discussion of a gene "for" a complex behavior (such as reading) to be meaningful, even in humans, it is only necessary to discover a gene for not exhibiting the behavior. [As Dawkins discusses in the case of reading, there may exist a mutant gene causing dyslexia. Such mutants are known to exist (see, e.g., Ehrman and Parsons, 1981, p. 292)]. It is then proper to speak of the "normal" allele at this locus as a gene "for" the behavior. Besides having some degree of realism, a model should be useful. It should lead to understanding and new insights, which can later be refined and subjected to empirical tests. As I have argued and documented elsewhere (Michod, 1981), it is with regard to this heuristic value, that simple genetic models have been most successful in evolutionary biology.

The four haploid gametes $\alpha_1\beta_1$, $\alpha_1\beta_2$, $\alpha_2\beta_1$, and $\alpha_2\beta_2$ exist in frequencies χ_1, χ_2, χ_3 and χ_4, respectively. Let ij denote a diploid genotype formed by the sexual union of gametes i and j. (In this section, it is necessary to use two subscripts to refer to a genotype.) Let p_{ij} be the intrinsic capacity of genotype ij to exhibit behavior X. In other words, genotype ij expresses phenotype X with probability p_{ij} and phenotype Y with probability $1 - p_{ij}$. Table 2 represents the p_{ij} matrix where p_h ($= p_{41} = p_{14} = p_{23} = p_{32}$) is the probability that double heterozygotes exhibit behavior X. Let P without subscripts denote a population frequency of behavior X. The average frequency of X is given by

$$\bar{P} = \sum_{ij} p_{ij}\,\chi_i\chi_j \tag{6}$$

Table 1. Payoff Matrix for Two Behaviors X and Y: The Parameter w_{ij} Is the Fitness of i when Interacting j; $i, j = X$ or Y

Fitness of	When Interacting with	
	X	Y
X	w_{XX}	w_{XY}
Y	w_{YX}	w_{YY}

Table 2. Intrinsic Capacities for Modifier Model: The Parameter p_{ij} Is the Probability That Genotype ij Exhibits Behavior X^a

Modifier Locus	Primary Locus		
	$\alpha_1\alpha_1$	$\alpha_1\alpha_2$	$\alpha_2\alpha_2$
$\beta_1\beta_1$	p_{11}	$p_{13} = p_{31}$	p_{33}
$\beta_1\beta_2$	$p_{12} = p_{21}$	p_h	$p_{43} = p_{34}$
$\beta_2\beta_2$	p_{22}	$p_{24} = p_{42}$	p_{44}

^{a} See text for further explanation.

The resulting fitness of genotype ij is then given by

$$W_{ij} = p_{ij} W_X + (1 - p_{ij})W_Y \qquad (7)$$

where W_X and W_Y are the resulting fitnesses of phenotypes X and Y, respectively, in the population:

$$W_X = \bar{P} w_{XX} + (1 - \bar{P})w_{XY} \qquad (8a)$$

$$W_Y = \bar{P} w_{YX} + (1 - \bar{P})w_{YY} \qquad (8b)$$

Equation (7) maps the intrinsic capacities of a genotype, p_{ij}, onto its resulting fitness W_{ij} (Fig. 1). Note that the resulting fitness of a genotype also depends on the intrinsic capacities of all other genotypes through Equations (8) and (6).

I now present some standard results of evolutionary game theory, which are relevant to the subsequent analysis. The reader is referred to Maynard Smith (1982) for the necessary background and motivation. An ESS, or evolutionary stable strategy (or state), is defined as a strategy (or state of the population) that is stable with regard to the introduction of rare mutant strategies. Let P^* denote the ESS. If $w_{XX} > w_{YX}$ then $P^* = 1$ is an ESS, since $W_X(1) > W_Y(1)$; and if $w_{YY} > w_{XY}$ then $P^* = 0$ is an ESS, since $W_Y(0) > W_X(0)$. Both pure strategies ($P^* = 0, 1$) may exist simultaneously, as they can in the Prisoner's Dilemma game (Axelrod and Hamilton, 1981; Brown et al., 1982). If $w_{XX} < w_{YX}$ and $w_{YY} < w_{XY}$ (equality may exist in one but not both of the previous inequalities), then a mixed ESS exists at

$$P^* = \frac{w_{YY} - w_{XY}}{w_{YY} + w_{XX} - w_{XY} - w_{YX}} \qquad (9)$$

which is found by equilibrating the phenotype's resulting fitnesses given in Equations (8), $W_X = W_Y$, and solving for P^*.

Consequently, the ESS approach to evolution is based on resulting fitness and, as acknowledged by its practitioners, ignores the complications resulting from sexual transmission (Fig. 1). This is more obvious when the dynamic equations underlying the game theory approach are specified (see, e.g., Maynard Smith 1982, Appendix D; Taylor and Jonker, 1978; Zeeman, 1979). These equations assume that phenotypes breed true and are equivalent in form to the asexual haploid equations of population genetics. In Section 3.3.1, the results of Maynard Smith (1981) are presented to show that diploid, sexual transmission of the behaviors may constrain the population from reaching the ESS. In subsequent sections (3.3.2–3.5), I study the evolutionary modification of these genetic constraints.

3.3 Analysis of Model

3.3.1 Fixation of β_1. I will assume that the population is initially fixed for allele β_1 at the modifier locus and that it has come to evolutionary equilibrium, at which point a mutant modifier β_2 is to be introduced as discussed in the next section. Before introduction of the mutant, $\chi_2 = \chi_4 = 0$ and $\chi_3 = 1 - \chi_1$. If β_1 is fixed, the dynamics of the system can be shown to be

$$\Delta\chi_1 = \frac{\chi_1(1 - \chi_1)}{2\bar{W}} (W_X - W_Y) \frac{\partial\bar{P}}{\partial\chi_1} \qquad (10)$$

where the partial of Equation (6) is to be evaluated at $(\chi_1, 0, 1 - \chi_1, 0)$.

From Equation (10) it can be seen that there are three types of evolutionary equilibria for the gamete frequencies:

1. Fixation of α_1 or $\alpha_2 (\hat{\chi}_1 = 1$ or 0 and $\bar{P} = p_{11}$ or p_{13}, respectively).
2. Equilibration of fitness, $W_X = W_Y$, $(\bar{P} = P^*$, the mixed ESS); the genetic equilibria in this case are equidistant from the overdominant equilibrium discussed next by amount

$$\pm \left[\frac{P^*}{(p_{11} - p_{13} + p_{33})} + \frac{(p_{13}^2 - p_{11}p_{33})}{(p_{11} - 2p_{13} + p_{33})} \right]^{1/2}$$

3. Maximization of $\bar{P}$, at $\bar{P}_{max}$ with $\hat{\chi}_1 = (p_{13} - p_{33})/(2p_{13} - p_{11} - p_{33})$. This equilibrium requires overdominance in the p_{ij}'s $(p_{13} > p_{11}, p_{33})$ and is stable if $W_X > W_Y$ at $\hat{\chi}_1$.

Maynard Smith (1981) has discussed these various genetic equilibria and their relations to the ESS equilibria predicted by evolutionary game theory. He made the following points which can be derived from analysis of Equation (10). [The numbering below is the same as in Maynard Smith (1981, Fig. 1), where his p is my χ_1, his F_1 is my $\bar{P}$ and his P^* is my P^*.]

A. No overdominance; $p_{33} > p_{13} > p_{11}$, say.

1 If $P^* > p_{33}$, there is fixation of α_2 ($\hat{\chi}_1 = 0$) and the population is as close to the ESS as the genetic system allows.

2 If $p_{33} > P^* > p_{11}$, there is a mixed ESS.

3 If $P^* < p_{11}$, there is fixation of α_1 ($\hat{\chi}_1 = 1$) and the population is as close to the ESS as the genetic system allows.

B. Overdominance; $p_{13} > p_{33} > p_{11}$, say.

1 If $P^* > \bar{P}_{max}$, population goes to the overdominant equilibrium $\hat{\chi}_1$ where $\bar{P} = \bar{P}_{max}$, which is as close to the ESS as the genetic system allows.

2 If $p_{33} < P^* < \bar{P}_{max}$, there are two genetic equilibria at which the population is at the ESS. Each are equidistant to the (a) left or (b) right of the overdominant equilibrium, $\hat{\chi}_1$.

3 If $p_{11} < P^* < p_{33}$, then (a) if $\chi_1 < \hat{\chi}_1$, population will go away from the ESS to $\chi_1 = 0$, $\bar{P} = p_{33}$, but (b) if $\chi_1 > \hat{\chi}_1$, population goes to P^*.

4 If $P^* < p_{11} < p_{33}$, then (a) if $\chi_1 < \hat{\chi}_1$, then population goes to $\chi_1 = 0$, $\bar{P} = p_{33}$ or (b) if $\chi_1 > \hat{\chi}_1$, population goes to $\chi_1 = 1$ and $\bar{P} = p_{11}$.

Consequently, in cases A.1, A.3, B.1, B.3(a), and B.4 the genetic system constrains the population from reaching the ESS. I now investigate the stability of these equilibria with regard to modification by introducing a rare modifier allele, β_2, and studying the conditions under which it increases in frequency.

3.3.2 Introduction of Mutant Modifier. After introduction of the mutant modifier all four gametic types may exist. In this case, the dynamics are given by the full two-locus recurrence equations (see, e.g., Roughgarden 1979) but with frequency-dependent fitnesses:

$$\chi_1' \bar{W} = \chi_1 W_1 - W_h \rho \Delta$$

$$\chi_2' \bar{W} = \chi_2 W_2 + W_h \rho \Delta$$

$$\chi_3' \bar{W} = \chi_3 W_3 + W_h \rho \Delta \qquad (11)$$

$$\chi_4' \bar{W} = \chi_4 W_4 - W_h \rho \Delta$$

where ρ is the recombination fraction, W_h ($= W_{14} = W_{41} = W_{23} = W_{32}$) is the fitness of double heterozygotes, the linkage disequilibrium variable

$$\Delta = \chi_1 \chi_4 - \chi_2 \chi_3 \qquad (12)$$

the marginal gamete fitnesses

$$W_i = \sum_j \chi_j W_{ij} \tag{13}$$

and the average resulting fitness

$$\bar{W} = \sum_i \chi_i W_i \tag{14}$$

All analyses below involve standard linear (or in some cases nonlinear) stability analyses of Equations (11) at the various equilibria discussed in the previous section.

Monomorphic Equilibria. The analysis of the evolution of modification is considerably simplified in the case of genetically monomorphic equilibria (cases A.1, A.3, B.3(a), and B.4), because at these equilibria only gametes $\alpha_1\beta_1$ or $\alpha_2\beta_1$ exist. The perturbations induced by introduction of a mutant modifier β_2 take place in the $\chi_1 = \chi_2 = 0$ plane [cases A.1, B.3(a), B.4(a)] or in the $\chi_3 = \chi_4 = 0$ plane [cases A.3, B.4(b)]. Consequently in this case, $\Delta = 0$ [Equation (12)] and the analyses are essentially of a single locus with two alleles.

Polymorphic Equilibria. In these cases [cases A.2, B.1, B.2, B.3(b)], all four gametic types exist after introduction of the mutant modifier, β_2, and consequently $\Delta \neq 0$ in general. The linear stability analysis of $(\hat{\chi}_1, 0, 1 - \hat{\chi}_1, 0)$ remains standard but now involves the full two-locus, frequency-dependent system. Of these cases, only case B.1 involves a genetic constraint; in the other cases [A.2, B.2, B.3(b)], the population is at the ESS.

3.4 Results of Modifier Allele

3.4.1 Monomorphic Equilibria

CASES B.3(A) AND B.4(A). In these cases, the population is fixed for the α_2 allele ($\hat{\chi}_1 = 0$) and the frequency of X is above the ESS ($\bar{P} = p_{33} > P^*$). The modified genotype 43 ($\alpha_2\alpha_2\beta_1\beta_2$) will increase when rare if

$$p_{43} < p_{33} \tag{15}$$

CASE A.1. In this case the population is fixed for the α_2 allele ($\hat{\chi}_1 = 0$) and the frequency of X is below the ESS ($\bar{P} = p_{33} < P^*$). The modified genotype 43 ($\alpha_2\alpha_2\beta_1\beta_2$) will increase when rare if

$$p_{43} > p_{33} \tag{16}$$

CASES A.3 AND B.4(B).   In these cases the population is fixed for the α_1 allele ($\hat{\chi}_1 = 1$) and the frequency of X is above the ESS ($\bar{P} = p_{11} > P^*$). The modified genotype 12 ($\alpha_1\alpha_1\beta_1\beta_2$) will increase when rare if

$$p_{12} < p_{11} \tag{17}$$

Consequently, conditions (15)–(17) imply that in all cases of genetically monomorphic equilibria, evolutionary modification of the genetic system will take the population to the ESS. Once the population is at the ESS the eigenvalues are unity.

3.4.2   Polymorphic Equilibria

CASES A.2, B.2 AND B.3(B).   In these cases the population is already at the ESS. The stability of these equilibria with regard to evolutionary modification depends upon the leading eigenvalue, λ, which evaluated at ($\hat{\chi}_1$, 0, $1 - \hat{\chi}_1$, 0) becomes

$$\lambda = \frac{\partial\chi_1'}{\partial\chi_1} = 1 + \frac{\hat{\chi}_1}{W_1}\left(\frac{\partial W_1}{\partial\chi_1} - \frac{\partial\bar{W}}{\partial\chi_1}\right) \tag{18}$$

The variables in Equation (18) are to be evaluated at the various ESS equilibria under consideration (note that in these cases $W_X = W_Y$, so $\bar{W} = W_i = W_{ij}$; $i, j = 1, 2, 3, 4$). For these cases, it can be shown that

$$\frac{\partial W_1}{\partial\chi_1} - \frac{\partial\bar{W}}{\partial\chi_1} = \frac{\partial\bar{P}}{\partial\chi_1}[P_1(w_{YY} + w_{XX} - w_{XY} - w_{YY}) + w_{XY} - w_{YY}] \tag{19}$$

with P_1 given by Equation (20) evaluated at ($\hat{\chi}_1$, 0, $1 - \hat{\chi}_1$, 0):

$$P_i = \sum_j \chi_j p_{ij} \tag{20}$$

In addition for these cases, it can be shown that

$$\frac{\partial\bar{P}}{\partial\chi_1} = 2(P_1 - P_3) \tag{21}$$

By Equation (21) in cases A.2, B.2(b), and B.3(b),

$$P_1 < P_3 \tag{22a}$$

since $\partial\bar{P}/\partial\chi_1 < 0$ (Maynard Smith, 1981, Fig. 1). Likewise for case B.2(a),

$$P_1 > P_3 \tag{22b}$$

since $\partial \bar{P}/\partial \chi_1 > 0$ (Maynard Smith, 1981, Fig. 1). By using Equations (9), (19), and (22), the leading eigenvalue λ in Equation (18) can be shown to be less than 1 in all four cases. Consequently, if the genetic system has allowed the population to reach the ESS, the genetic system is itself an ESS and stable with regards to evolutionary modification.

CASE B.1. In this case, the population is at the overdominant equilibrium $\bar{P}_{max} < P^*$. The linear stability analysis results in the following necessary and sufficient condition for increase of a rare, mutant modifier.
   Either

$$(W_X - W_Y)(2\bar{P} - P_2 - P_4) + W_h\rho < 0, \tag{23}$$

or

$$(W_X - W_Y)(2\bar{P} - P_2 - P_4) + W_h\rho > 0 \tag{24a}$$

and

$$(W_X - W_Y)(\bar{P} - P_2)(\bar{P} - P_4) - W_h\rho \,\bar{\epsilon} < 0 \tag{24b}$$

These conditions are to be evaluated at the overdominant equilibrium $(\hat{\chi}_1, 0, 1 - \hat{\chi}_1, 0)$ and

$$\bar{\epsilon} = \hat{\chi}_1^2 \epsilon_1 + 2\hat{\chi}_1((1 - \hat{\chi}_1)\epsilon_2 + (1 - \hat{\chi}_1)^2\epsilon_3 \tag{25}$$

with

$$p_{12} = p_{11} + \epsilon_1$$
$$p_{14} = p_{13} + \epsilon_2 \tag{26}$$
$$p_{43} = p_{33} + \epsilon_3$$

in Table 2. Consequently, $\bar{\epsilon}$ is the initial change in the frequency of phenotype X among modified genotypes. Conditions (23) and (24) are difficult to interpret and their implications will be discussed in the next section.
   If there is linkage equilibrium or no recombination between the primary and modifier locus [$\Delta = 0$ or $\rho = 0$ in Equation (11)], then the modifier will increase in frequency when rare if either of the following conditions are met:

$$(W_X - W_Y)(P_2 - \bar{P}) > 0 \tag{27a}$$

or

$$(W_X - W_Y)(P_4 - \bar{P}) > 0 \tag{27b}$$

As discussed in the next section, conditions (27a) or (27b), when taken separately, embrace the intuitive result that modifiers increase in frequency if they increase the expression of the most fit phenotype. However, the fact that, for the modifier to increase, only one of these inequalities need hold and not necessarily the other, may allow modifiers to increase which actually decrease the overall frequency of the most fit phenotype.

3.5   Discussion of Modifier Model

As has been shown, if the mutant modifier is introduced into a genetically monomorphic population, the genetic system will evolve to remove the genetic constraints and allow the population to reach the ESS. In addition, polymorphic populations whose genetic system has allowed it to reach the ESS are stable to evolutionary modification.

These results are consistent with the analysis of Slatkin (1978, 1979), which studies modification of the genetic system by taking into account only the fitness of modified genotypes and not the full transmission process (Slatkin, 1979, pp. 390–393). In the present notation, Slatkin's (1979, Equation 17) condition for increase of a modifier allele becomes

$$(W_X - W_Y)\bar{\epsilon} > 0 \qquad (28)$$

Basically, Slatkin's condition [Equation (28)] embodies the intuitive result that modifiers which increase the expression of the most fit phenotype will increase in frequency. (Recall that since the population is not at the ESS, one phenotype must be fitter than the other. Since there is heterozygote superiority in the expression of behavior X, $W_X > W_Y$ at the overdominant equilibrium, if this equilibrium is to be stable in the χ_1 plane.) Condition (28) predicts that modification of the genetic system always takes the population to the ESS. Unfortunately, the results presented above for the case of modification of the overdominant equilibrium do not confirm this in general for the case of a polymorphic equilibrium not at the ESS [Case B.1, Equations (23), (24), and (27)].

It can be seen by comparing conditions (28) with (27), that Slatkin's condition is sufficient but not necessary for evolutionary modification in the case of linkage equilibrium or no recombination. However, condition (28) is neither necessary nor sufficient in general [cf. Equations (28) with (23) and (24)].

4   GENERAL DISCUSSION

Templeton (1982) also discusses the role of adaptedness in the evolutionary process. He considers sickle-cell anemia as a case in point. There are two mutant hemoglobin molecules that are clearly polymorphic in human populations and that figure in Templeton's (1982) analysis: S and C. These mutant hemoglobins differ from the normal, ancestral hemoglobin molecule A by amino acid substitutions in the β chain. Thus, these three hemoglobin molecules correspond to three alleles at a single locus. There are estimates of fitness which suggest that the CC genotype is most adapted to malarial environments (Cavalli-Sforza and Bodmer, 1971, p. 163). Following Templeton (1982), this will be assumed to be the case in what follows; however, there appears to be some confusion on this point (see, e.g., Crozier et al., 1972). Since the C allele is recessive to the predominant A allele, it is not expressed in heterozygous state. Because rare alleles are almost always carried in heterozygous state, the C allele cannot increase when rare in a random mating population. On the contrary, the S allele is advantageous when heterozygous in a malarial environment and so it can increase when rare. The less adapted SS homozygotes, which have severe sickling of the red blood cells, are initially uncommon. For these reasons, the S allele has increased in frequency in malarial environments, even though the C allele is more adapted. Templeton (1982) then considers the role of the mating system in determining whether the more adapted but recessive C allele could increase in frequency. If the ancestral population in which the C mutation occurred were inbred, the frequency of C homozygotes could be high enough for the allele to increase in frequency. Inbreeding would increase the frequency of the CC homozygotes over what it would be in a random mating population with the same gene frequency. Whether the C allele is actually more adapted or not (as pointed out above, there are conflicting data on this point), Templeton's scenario is instructive, for it shows how factors such as the mating system (or in another example which he discusses, linkage relationships) can be the determining factor in whether the more adapted type increases in frequency or not. Viewed within the framework of Section 1.2, these factors would be classed as evolutionary constraints since they mask the evolutionary expression of adaptedness.

An appreciation of evolutionary constraints and their role in evolution is important for understanding general evolutionary questions. For example, Dawkins (1982) has recently argued that the individual is not the unit of selection. Dawkin's argument is based in large part on the many counter-examples which he gives to "the individual as maximizing agent" (Dawkins, 1982, p. 187). In other words, Dawkins argues that if the individual was the unit of selection, then selection should have maximized the individual's adaptedness for survival and reproduction. Since, in many cases, individual adaptedness has not been maximized, Dawkins concludes that the individual is not the unit of selection. However, as pointed out elsewhere (Michod, 1983b), this conclusion does not necessarily follow. As discussed above, there are many factors that mask the evolutionary efficacy of adaptedness. If under individual selection, higher adaptedness always resulted in a higher rate of increase in frequency, then Dawkins' conclusion that individual selection should result in the individual as a "maximized agent" would be correct. However, Dawkins' conclusion is a non sequitur because of the very issues discussed in this chapter. The individual cannot be seen as a "maximizing agent," not because it is not the unit of selection, but because of the reality of evolutionary constraints.

5 SUMMARY

Evolutionary constraints on adaptation are discussed by considering how the intrinsic adaptedness of a trait is related to the trait's per capita rate of increase in frequency. It is suggested that evolutionary constraints be defined as those factors that prohibit a direct, monotonic mapping of the adaptedness of a trait onto its per capita rate of increase. Consequently, evolutionary constraints are said to "mask" intrinsic adaptedness. There are two general considerations involved in mapping the intrinsic adaptedness of a trait onto the trait's rate of increase in frequency. The first is termed environmental and concerns the expression of an organism's intrinsic capacities in an actual environment. The second concerns the process by which an organism's intrinsic capacities are transmitted to its offspring. It is shown by example that genetic constraints may exist at both levels.

A two-locus, population genetic model is then presented to study the modification of genetic systems of transmission which prohibit the game-theoretic predictions of social evolution from being realized. It is shown that if the modifier allele is introduced into a genetically monomorphic population, the genetic system will evolve to remove the genetic constraint and allow the population to reach the ESS. In addition, genetically polymorphic populations whose genetic system has allowed it to reach the ESS are stable to evolutionary modification. However, evolutionary modification of genetically polymorphic populations not at the ESS may take the population farther away from the ESS.

ACKNOWLEDGMENTS

I thank Wyatt Anderson, Harris Bernstein, Henry Byerly, Tom Caraco, Fred Hopf, Conrad Istock, Mike Sanderson, Bill Schaffer, and Art Weiss for discussion and comment.

LITERATURE CITED

Anderson, W. W. 1971. Genetic equilibrium and population growth under density-regulated selection. *Am. Nat.* **105**:489–498.

Axelrod, R., and W. D. Hamilton. 1981. The evolution of cooperation. *Science* **211**:1390–1396.

Bernstein, H., H. C. Byerly, F. A. Hopf, R. E. Michod, and G. K. Vemulapalli. 1983. The Darwinian dynamic. *Q. Rev. Biol.* **58**:185–207.

Bernstein, H., H. C. Byerly, F. A. Hopf, R. E. Michod. Sex and the emergence of species. *Q. Rev. Biol.* Submitted manuscript.

Brown, J., M. Sanderson, and R. E. Michod. 1982. Evolution of social behavior by reciprocation. *J. Theor. Biol.* **99**:319–339.

Cavalli-Sforza, L. L., and W. F. Bodmer. 1971. *The Genetics of Human Populations.* Freeman, San Francisco. 965 pp.

Cavalli-Sforza, L. L., and M. W. Feldman. 1981. *Cultural transmission and evolution: A quantitative approach.* Princeton University Press, Princeton, New Jersey.

Charlesworth, B. 1980. *Evolution in age-structured populations.* Cambridge Univ. Press.

Crozier, R. H., L. A. Briese, M. A. Guerin, T. R. Harris, J. L. McMichael, C. H. Moore, P. R. Ramsey, and S. R. Wheeler. 1972. Population genetics of hemoglobins S, C, and A in Africa: Equilibrium or replacement? *Am. J. Hum. Genet.* **24**:156–167.

Darwin, C. R. 1866. Letter to A. R. Wallace, dated 5 July. In James Marchant (1916), *Alfred Russel Wallace Letters and Reminiscences,* Vol. 1, pp. 174–176. Cassell, London.

Dawkins, R. 1982. *The extended phenotype, the gene as the unit of selection.* Freeman, San Francisco. 307 pp.

Denniston, C. 1978. An incorrect definition of fitness revisited. *Ann. Hum. Genet.* **43**:77–85.

Ehrman, L., and P. A. Parsons. 1981. *Behavior genetics and evolution.* McGraw-Hill, New York.

Eigen, M. 1971. Self-organization of matter and the evolution of biological macromolecules. *Naturwissenschaften* **58**:465–523.

Eigen, M., and P. Schuster. 1979. *The hypercycle, a principle of natural self-organization.* Springer-Verlag, Berlin.

Eigen, M., and R. Winkler. 1981. *Laws of the game: How the principles of nature govern chance.* Knopf, New York. 347 pp.

Feldman, M. W., and J. Krakauer. 1976. Genetic modification and modifier polymorphisms. In S. Karlin and E. Nevo (eds.), *Population genetics and ecology,* pp. 547–584. Academic Press, New York.

Fisher, R. A. 1928a. The possible modification of the response of wild type to recurrent mutations. *Am. Nat.* **62**:115–26.

Fisher, R. A. 1928b. Two further notes on the origin of dominance. *Am. Nat.* **62**:571–74.

Fisher, R. A. 1929. The evolution of dominance. A reply to Professor Sewall Wright. *Am. Nat.* **63**:553–56.

Fisher, R. A. 1930. *The genetical theory of natural selection.* Clarendon Press, Oxford. 2nd rev. ed., 1958. Dover, New York.

Gould, S. J., and R. C. Lewontin. 1979. The spandrels of San Marco and the Panglossian paradigm: A critique of the adaptationist programme. *Proc. R. Soc. London, Ser. B* **205**:581–598.

Hamilton, W. D. 1964. The genetical evolution of social behavior. I and II. *J. Theor. Biol.* **7**:1–51.

Hamilton, W. D. 1972. Altruism and related phenomena, mainly in social insects. *Annu. Rev. Ecol. Syst.* **3**:193–232.

Hopf, F. A., and Hopf. 1983. The role of the Allee effect in species packing. *Theor. Pop. Biol.* Submitted manuscript.

Lohrman, R., P. K. Bridson, and L. E. Orgel. 1980. Efficient metal-ion catalyzed template-directed oligonucleotide synthesis. *Science* **208**:1464–1465.

Lohrman, R., and L. E. Orgel. 1980. Efficient catalysis of polycytidylic acid-directed oligoguanylate formation by Pb^{2+}. *J. Mol. Biol.* **142**:555–567.

Maynard Smith, J. 1981. Will a sexual population evolve to an ESS. *Am. Nat.* **117**:1015–1018.

Maynard Smith, J. 1982. *Game theory and evolution*. Cambridge University Press.

Michod, R. E. 1981. Positive heuristics in evolutionary biology. *Br. J. Philos. Sci.* **32**:1–36.

Michod, R. E. 1982. The theory of kin selection. *Annu. Rev. Ecol. Syst.* **13**:23–55.

Michod, R. E. 1983a. Population biology of the first replicators. *Am. Zool.* **23**:5–14.

Michod, R. E. 1983b. Review of *The extended phenotype—The gene as the unit of selection* by R. Dawkins. *Am. Sc.* 71(5). (in press).

Mills, S. K., and J. H. Beatty. 1979. The propensity interpretation of fitness. *Philos. Sci.* **46**:263–286.

Roughgarden, J. 1979. *Theory of population genetics and evolutionary ecology: An introduction.* Macmillan, New York.

Slatkin, M. 1978. On the equilibration of fitness by natural selection. *Am. Nat.* **112**:845–859.

Slatkin, M. 1979. The evolutionary response to frequency- and density-dependent interactions. *Am. Nat.* **114**:384–398.

Spencer, H. 1864. *The principles of biology*, Vol. 1. Williams and Norgate, London and Edinburgh.

Taylor, P. D., and L. B. Jonker. 1978. Evolutionary stable strategies and game dynamics. *Math. Biosci.* **40**:145–156.

Templeton, A. R. 1982. Adaptation and the integration of evolutionary forces. In R. Milkman (ed.), *Perspectives on evolution*, pp. 15–31. Sinauer, Sunderland, Massachusetts.

Wallace, A. R. 1866. Letter to Charles Darwin dated 2 July. In J. Marchant (1916) *Alfred Russel Wallace letters and reminiscences*, Vol. 1, pp. 170–174. Cassell, London.

Wilson, D. S. 1980. *The natural selection of populations and communities.* Benjamin/Cummings. Menlo Park, California.

Wright, S. 1927. The effects in combination of the major color factors of the guinea pig. *Genetics* **12**:530–569.

Wright, S. 1929. The evolution of dominance. *Am. Nat.* **63**:556–561.

Wright, S. 1969. *Evolution and the genetics of populations*, Vol. 1, *The theory of gene frequencies.* University of Chicago Press, Chicago.

Wright, S. 1977. *Evolution and the genetics of populations*, Vol. 3, *Experimental results and evolutionary deductions.* University of Chicago Press, Chicago.

Zeeman, E. C. 1979. Population dynamics from game theory. *Proc. Int. Conf. Global Theory Dynamical Syst.* Northwestern, Evanston, Illinois.

CHAPTER 10

Sociality and Survivorship in Animals Exposed to Predation

THOMAS CARACO
Department of Biology
University of Rochester
Rochester, New York

H. RONALD PULLIAM
Department of Biological Sciences
State University of New York
Albany, New York

CONTENTS

1 INTRODUCTION

Ecologists generally believe that predation influences social organization in a number of prey species. The potentially dire consequences for the prey individual justify a strong theoretical emphasis on the selective aspects of predation. An individual's failure to avoid predation often results in death, whereas less than maximally efficient foraging, for example, need not induce such a drastic cost.

Alexander (1974) suggests that sociality arose in many species, particularly primates, as a response to predation. However, a hypothetical distinction between adaptations that initially favored sociality and those subsequently maintaining group living (Hoogland, 1981) is most often impossible to test empirically. Therefore, we prefer to integrate predation with other selective forces influencing survivorship and reproductive success. We view social organization generally as a set of coadapted traits, so that antipredator, foraging, and mating strategies covary dependently (e.g., Crook, 1972; Pulliam, 1973; Altmann, 1974; Hoogland and Sherman, 1976; Caraco, 1979a; Brown, 1982).

Historically, ecological analyses of social organization have depended a good deal on the comparative method (e.g., Clutton-Brock and Harvey, 1977). Different species, occupying different habitats, are compared to establish correlations between social attributes and the intensity of predation or other ecological variables. The method possesses clear heuristic value, and has generated important hypotheses (e.g., Eisenberg et al., 1972; Jarman, 1974). However, in many comparative studies it is difficult to disentangle effects of predation from other influences on sociality, and no correlation demonstrates causality (Altmann, 1974). More generally, evolutionary problems may admit more than one adaptive solution under similar environmental conditions, and not every social trait need be a direct consequence of strong adaptation (Maynard Smith, 1978; Lewontin, 1979). But the value of the comparative approach and the importance of an ecology of social organization lie in the generation of falsifiable hypotheses concerning benefits and costs to individuals, and we develop our discussion in this context.

We begin by reviewing current hypotheses that attribute advantages or disadvantages to individuals in groups exposed to predation. We then discuss game theory and suggest how this technique might be applied to some questions regarding sociality. Game theoretical analyses of animal social behavior have a short, but rich history (e.g., Maynard Smith, 1974; Maynard Smith and Parker, 1976; Oster and Wilson, 1978). Depending on the particular question, we explore both Pareto ("cooperative") and Nash (apparently more "selfish") solutions to a game. Most models predicting evolutionarily stable strategies (ESSs) are developed in terms of Nash solutions. The idea that individual selection might produce an ESS embracing an element of cooperation only recently has been advanced (e.g., Axelrod and Hamilton, 1981; Pulliam et al., 1982).

2 PREY SOCIALITY: BENEFITS AND COSTS

Some prey individuals are constrained to the proximity of a critical resource, so that aggregation follows. Hamadryas baboons sleep on cliffs to deter nocturnal predation, and the cliffs are rare in space (Kummer, 1968). Therefore, Hamadryas sleep in large herds, though other activities are conducted in much smaller groups. Sleeping in a group, per se, might be less beneficial than the security provided by the rocky ledges.

Prey individuals might aggregate in an attempt to interpose conspecifics between themselves and an attacking predator (Williams, 1966; Hamilton, 1971). When "selfish herding" induces the formation of foraging groups, nesting colonies, or communal roosts, central locations could offer greater safety than peripheral positions. Horn (1968) and Tenaza (1971) find that predation can be more intense at the perimeter of avian nesting colonies, and in many groups social dominants occupy geometrically central positions (e.g., Moore, 1972; but see Altmann, 1979). An interesting consequence of Hamilton's (1971) model is that some individuals may enhance their probability of avoiding predation even though the formation of the group can increase the predator's efficiency.

Pulliam (1973) points out a difficulty encountered in invoking the selfish herding hypothesis as an *unitary* explanation of the benefits of prey sociality. If susceptibility to predation increases toward the group's periphery, the group's outermost individuals might leave. Their departure creates a new set of peripherals, and so on, until the group has disbanded. Either solitary living is especially precarious, or other factors must help maintain the persistence of grouped prey. Selfish herding could interact with other benefits of sociality, or the spatial distribution of food, mating opportunities, and other critical resources might keep individuals in groups. When a peripheral individual decides to remain in a particular group, it may allocate more time to watching for approaching predators than do more centrally positioned group members. Inglis and Lazarus (1981) document this geometric effect on vigilance behavior in flocks of foraging geese. Hoogland (1981) obtains a qualitatively similar result for black-tailed prairie dogs.

The "early warning" function of socialization in prey animals represents one of the most thoroughly documented benefits of grouping. Earlier detection of an approaching predator enhances the probability of escaping successfully. Compared to a solitary existence, this benefit of sociality could provide a significant advantage if group members give an alarm when a predator is sighted. A good deal of empirical evidence indicates that groups of birds detect approaching predators sooner than a solitary does (Powell,

1974; Siegfried and Underhill, 1975; Hoogland and Sherman, 1976; Kenward, 1978; Lazarus, 1979; Caraco, 1979b). Hoogland (1981) reports the same benefit of sociality in two species of prairie dogs; the phenomenon is probably very common.

Under certain circumstances, the dilution effect (Bertram, 1978) can reduce an individual's chance of falling to a predator. Suppose that attack rate does not vary significantly with prey group size, and that predators kill only a single individual per successful attack. If a predator's chance of succeeding also is independent of group size, then the individual's probability of becoming a victim is inversely proportional to the size of its group. The advantage remains (though weaker) when predators are either more persistent or more successful against larger groups, as long as the predator's probability of making a kill during a given time interval increases sufficiently slowly as group size increases. The possibility of multiple kills also can reduce the dilution effect, but solitary predators seldom take more than one victim per successful attack when the prey are mobile.

Aggregated prey might confuse an attacking predator (Bertram, 1978). If prey individuals aggregate tightly, a predator could encounter difficulty focusing its efforts on a particular target. Tinbergen (1951) suggests that prey might coalesce into a tight group to threaten injury to a predator approaching at high velocity. If a falcon should collide with a flock of birds, the predator itself might be hurt seriously. However, once an attacking predator has been detected, selfish herding might be a more parsimonious explanation of tightly aggregated groups of fleeing prey (Pulliam and Millikan, 1982). Additionally, confusion effects may protect prey from only inexperienced predators. Milinski (1979) demonstrated that predatory fish were initially confused by aggregated prey, but soon learned to handle highly clumped food items.

When attempting to escape a predator's attack, individuals in groups might be able to attain a greater speed than can a solitary. This hypothesis presupposes an aerodynamic or hydrodynamic advantage to group members (Lissaman and Shoolenberger, 1970). Individuals in groups may be able to interactively reduce costs of maintaining certain velocities (Gould and Heppner, 1974; Higdon and Corrsin, 1978). However, no evidence suggests that grouping increases escape speed sufficiently to enhance the antipredator benefits of sociality. In fact, empirical investigations of the geometry of fish schools indicate that individuals are not positioned so that they could increase their speed through hydrodynamic efficiency (Weihs, 1973; Partridge and Pitcher, 1979).

Females might benefit by grouping births of dependent offspring in both space and time. A pulse of new individuals entering a prey population might temporarily satiate predators. Consequently, an individual female's *a priori* probability of keeping her offspring alive until independence might be greater when births are synchronized than when births are aseasonal (Estes, 1976). Hoogland and Sherman (1976) portray this potential benefit of sociality as a corollary of the selfish herding hypothesis, though it appears more closely related to dilution. The benefits and costs to an individual female, accured as a result of giving birth at a particular time, should depend on the times when other females give birth. Therefore, the hypothesis suggests application of game theory (see Section 3.2).

Individual survivorship may be increased when prey groups can mob a predator (Kruuk, 1964; Hoogland and Sherman, 1976; Curio et al., 1978). Further, some prey aggressively and capably defend themselves against attacking predators (Crook, 1972; Altmann, 1974; Jarman, 1974). The benefit of mobbing should, at least initially, accelerate with increasing prey group size (Pulliam and Caraco, in press). Benefits of active defense should increase with group size, but they also depend on body size (relative to the predator's size), weaponry (e.g., antlers), and fighting ability. Mobbing or aggressive defense might be more advantageous in species where young offspring have essentially no chance of evading an attack if their parents flee. Adults choosing to defend against a predator may increase their offspring's survival at the expense of their own survivorship. An individual's choice between mobbing or fleeing a detected predator can be viewed as a decision to cooperate or defect, so we investigate the choice with a game-theoretical approach in Section 3.2.

Group members also may incur costs in terms of predation. Aggregated prey may be more conspicuous to a searching predator than is a solitary prey animal. Crypticity and a near solitary existence might reduce the rate at which an individual is encountered by a predator in some species (Jarman, 1974; Pulliam and Mills, 1977), particularly if polyphagous predators temporarily restrict searching to the proximity of their latest meal (Tinbergen et al., 1967). Whether grouping increases or decreases encounter rate with predators can depend on a variety of factors, including the mobility of predators, the sensory modality used by predators, and sensory acuity (e.g., Vine, 1971; Treisman, 1975; Taylor, 1976, 1979).

Attack rate also could depend on group size if a predator's capture efficiency varies with degree of prey aggregation. Solitaries might be easier to capture when a predator is detected more quickly by a group (e.g., Page and Whitacre, 1975; Kenward, 1978). Alternatively, a predator might be more likely to find a weakened or diseased target, and an easy meal, in a larger group.

Whether due to increased conspicuousness or predator preference, attack rate does sometimes increase with the level of aggregation in the prey. Krebs (1971) and Andersson and Wicklund (1978) experimentally demonstrated that egg predators attack clumped birds' nests at a greater frequency than they attack dispersed nests. In the second study, egg losses were lower in clumped nests, despite the attack rate difference, since the adult birds vigorously mobbed any predators they detected. On balance, the benefit of sociality apparently outweighed the cost, at least in terms of predation.

As indicated above, social organization in most group-living species will be influenced simultaneously by predation, feeding requirements, etc. Analyses of time budgets have demonstrated, at least qualitatively, interactions between avoiding predation and averting starvation in social groups of birds (e.g., Pulliam, 1976; Barnard, 1980; Caraco and Pulliam, 1980). It may be a fairly common advantage of sociality that groups detect predators sooner than solitaries, yet individual group members allocate less time to vigilance and more time to feeding than do solitaries.

Our discussion of advantages and disadvantages of prey sociality has been phrased in terms of direct fitness of individuals. Brown and Brown (1981) include parental care, quite logically, as a means of increasing direct fitness, though Maynard Smith (1964) considers parental care as part of kin selection. Outside of obvious parental care, the only common antipredator mechanisms inspiring explanations more complicated than individual survivorship are alarm signals. Rump patches, white tail feathers, and alarm calls are very common in nature and usually associated (though not always solely) with the detection of a predator. Dunford (1977) and Sherman (1977), among others, view such signals as kin-directed altruism, selected because they enhance the indirect component of inclusive fitness. However, certain alarm signals also have been interpreted as examples of selfish manipulation (Charnov and Krebs, 1975), standard parental care (Shields, 1980), and cooperation (Trivers, 1971). All of these mechanisms could enhance an individual's direct fitness. Envisioning a unitary explanation of alarm signals is probably unrealistic.

Most of the arguments we develop in the remainder of the chapter focus on individual benefits and costs. Although we emphasize direct fitness, we recognize that some of our methods may require consideration of kinship in application (e.g., Mirmirani and Oster, 1978; Grafen, 1979).

The infusion of evolutionary theory into behavioral ecology has focused increasing attention on ultimate, rather than proximate, causation. Models and tests of "optimal strategies" abound. We feel that game theory and the concept of evolutionarily stable strategies will assume increasingly important roles in at least the near future of behavioral ecology. Therefore, we decided to write an elementary introduction to the topic and suggest that game theory might be usefully employed in examining certain aspects of social organization.

3 MODELS FOR PREY SOCIALITY

An ecological model takes a simplified view of nature; quantitative models sometimes do not achieve the richness of strictly verbal approaches. However, quantification entails a clear statement of assumptions, and so avoids the vaguery of some verbal arguments (Maynard Smith, 1978).

Optimality principles form the basis of much of the predictive theory in behavioral ecology. Properly interpreted, optimization models do not envision utopia via adaptation. A model of optimal behavior intially assumes a plausible relationship between natural selection and some behavior of interest. For a given environment the model then deduces hypotheses concerning behavioral strategies and attainable "fitness." If observation and experimentation at least qualitatively uphold the model's testable hypotheses, we may have learned something about adaptation. If the model fails, we must evaluate its assumptions, but we need not reject optimization as a reasonable basis for generating falsifiable hypotheses concerning adaptation.

Many optimality models involve a single objective function that maps an individual's behavior into only the same individual's benefits and costs, in terms of a currency of fitness (Schoener, 1971). Strengths and weaknesses, as well as the mathematical techniques of this type of optimization, have been discussed by a number of authors (e.g., Maynard Smith, 1978; Oster and Wilson, 1978; Lewontin, 1979). The apparent success of optimization models with a single objective function in areas such as foraging behavior indicates that the philosophy will remain an essential element of behavioral ecology. However, many problems involve interactions among two or more individuals, so that a model must incorporate more than one objective function. Such situations can be analyzed with game theory (Owen, 1968; Vincent and Grantham, 1981).

Consider two competing individuals whose payoffs (in the currency of fitness) show mutual dependence. F_i is the payoff to player i; $i = 1, 2$. Let u_1 represent the behavior of player 1, while u_2 represents the behavior of player 2. The payoff functions are $F_1(u_1 \mid u_2)$ and $F_2(u_2 \mid u_1)$. Since each player's payoff is conditioned on the other's strategy, we may not have a single function to maximize in order to predict an individual's behavior. However, game theory may allow us to solve simultaneously for equilibrial "best" strategies for the two players. Hence, the application of game theory to animal behavior does not represent a conceptual departure from the philosophy of optimality.

A particular game may admit more than one type of solution. Among the various solution concepts developed by game theorists (see Vincent and Grantham, 1981), only the Nash and the Pareto solutions currently find favor as possible consequences of natural selection. The former concept assumes strictly selfish behavior, and so arises immediately as a reasonable outcome of selection. The latter solution assumes an element of cooperation and may provide some insight into the evolution of altruistic behavior through natural selection. Formal definitions of the solutions are given below.

Suppose we locate a Nash solution to a game. Then we identify a strategy for each player (the strategies may be indentical), such that no player will be tempted to change its strategy as long as every other player continues to use its Nash strategy. Hence, the solution has an equilibrial character. Altering strategies at a strong Nash solution (Oster et al., 1977) ordinarily should be avoided, since the one player deviating from the Nash equilibrium receives a decreased payoff. Therefore, a Nash solution can show certain stability properties.

Nash solutions assume selfish behavior in a competitive context. Therefore, they make reasonable candidates for strategies favored by natural selection, when the game abstracts the essence of the biology. However, some nontrivial games need not have a solution qualifying as a Nash ESS (Haigh, 1975), and not every Nash solution satisfies conditions for an ESS. Additionally, the evolutionary stability of a Nash solution might depend on population size (Riley, 1979). One application of game theory identifies Nash solutions with the equilibrium frequencies of different phenotypes in a population (e.g., Krebs and Davies, 1981). The analogy is suggested when the expected fitness of an individual (of any particular phenotype) depends on the frequencies at which various phenotypes in the population are encountered. The typical approach considers a linear discrete game between two identical players (players with the same payoff matrix, see below). Maynard Smith (1974, 1978) suggests that the Nash solution to such a game represents an ESS where each extant phenotype has the same fitness. The ESS may specify a single behavior (a pure ESS) or a mixture of different behaviors. A mixture can be interpreted as different pure phenotypes occurring in frequencies where fitnesses equilibrate, or a population where all individuals play the same mixed strategy and select each behavior with a probability equaling the associated equilibrium frequency.

Applications of game theory to evolution often equate phenotype with genotype, since this allows a very simple evolutionary interpretation of a game's solution. Given this assumption, attainability and stability of the equilibrium can depend on the manner of genetic transmission. In a sufficiently large asexual population composed of randomly interacting individuals, a Nash ESS will be locally (and perhaps globally) stable to perturbation. A mutant (in the sense of strategy, not in the sense of an entirely new behavior) so rare that it interacts only with individuals playing the equilibrium Nash strategy will not have an expected payoff greater than that of its competitors. Furthermore, in a large, sexually recombining population, the ESS may be attained or closely approximated (Maynard Smith, 1981; Michod, Chapter 9).

Concern over intergenerational transmission of strategies appears less important when we adopt a perhaps more realistic, but more complicated, mapping from genotype to phenotype. We have argued (Pulliam and Caraco, in press) that selection should often favor learning and fitness enhancing decision-making abilities in higher organisms involved in variable social interactions. The consequent phenotypic plasticity increases the attainability of any particular mixed strategy.

If learning is important to some Nash strategies, it is probably even more significant when we consider Pareto solutions, which assume that players can cooperate. At the Pareto solution, no player can increase its payoff without reducing the payoff to another player. We might hesitate to expect that natural selection somehow could constrain individual selfishness under these conditions, at least in games involving nonrelated individuals. For example, when two individuals engage in a single play of the discrete game known as the Prisoner's Dilemma, the Pareto solution is not stable against defection to a Nash strategy. However, in this game, each of two cooperators playing each other receives a greater payoff than does each of two Nash strategists playing against each other. Therefore, selection can favor cooperation in certain forms of repeated play if the same two individuals continue to interact or if noncooperators can be distinguished in a randomly interacting population (Brown et al., 1983).

The ESS associated with Pareto solutions involves conditional cooperation, labeled the tit-for-tat or judge strategy (Axelrod and Hamilton, 1981; Pulliam et al., 1982). The judge cooperates initially and continues to cooperate as long as the other player cooperates. If the opponent defects to a Nash strategy, the judge also defects until the other player reinitiates cooperation. Each of two interacting judges cooperates and fares better than each of two interacting Nash strategists. Yet a judge playing a defector also will defect and thereafter fare no worse than the guilty player. Hence, the judge strategy can be considered as an ESS.

In this section we have described some basic qualitative concepts. The next section demonstrates simple quantitative aspects of behavioral games.

3.1   Static Games

Most game-theoretical models in behavioral ecology involve discrete games, where a player selects one of a finite number of behaviors every time the game is played. In a continuous game a player chooses over an infinite number of possible behaviors at every play of the game. The ecological assumptions underlying a particular type of solution are essentially the same for both discrete and continuous games.

Consider a linear discrete game between two players (1 and 2) with identical payoff matrices. Each player chooses one of behaviors A and B on every play of the game. The common payoff matrix is M_1:

$$
\begin{array}{c}
 \\
A \\
B
\end{array}
\begin{array}{cc}
A & B \\
\left[\begin{array}{cc}
m_{11} & m_{12} \\
m_{21} & m_{21}
\end{array}\right]
\end{array}
$$

Allowing player 1 to control the rows, we can show payoffs to both players in a simple form:

$$
\begin{array}{cc}
 & \text{player 2} \\
 & \begin{array}{cc} A & \qquad\qquad B \end{array}
\end{array}
$$

$$
\text{player 1} \;
\begin{array}{c}
A \\
B
\end{array}
\left[\begin{array}{cc}
(m_{11}, m_{11}) & (m_{12}, m_{21}) \\
(m_{21}, m_{12}) & (m_{22}, m_{22})
\end{array}\right]
$$

For the pairs (x, y) player 1's payoff is x and player 2's payoff is y.

A strategy is designated by the column vector p (or q), which assigns probability π to behavior A and probability $1 - \pi$ to behavior B ($0 \leq \pi \leq 1$). With the superscript T representing transposition, the expected payoff to a player adopting strategy p, when its opponent adopts q, is $p^T M_1 q$, abbreviated $E(p, q)$. If player 1's strategy is $p = (\pi_1, 1 - \pi_1)^T$, and player 2's strategy is $q = [\pi_2, 1 - \pi_2]^T$, then

$$E(p, q) = m_{22} + \pi_1(m_{12} - m_{22}) + \pi_2(m_{21} - m_{22})$$
$$+ \pi_1 \pi_2(m_{11} - m_{21} - m_{12} + m_{22}) \tag{1a}$$

$$E(q, p) = m_{22} + \pi_2(m_{12} - m_{22}) + \pi_1(m_{21} - m_{22})$$
$$+ \pi_1 \pi_2(m_{11} - m_{21} - m_{12} + m_{22}) \tag{1b}$$

Every nontrivial 2×2 payoff matrix has at least one Nash ESS (though this need not be true for matrices of greater dimension; see Haigh, 1975), so we confidently analyze M_1. Suppose p^* is a Nash ESS. Then

$$E(p^*, p^*) \geq E(q, p^*) \tag{2a}$$

for all q, and for any $q \neq p^*$ where $E(p^*, p^*) = E(q, p^*)$,

$$E(p^*, q) > E(q, q) \tag{2b}$$

There is no better response to p^* than p^* itself, and for any other strategy q that does as well against p^*, p^* does better against q than q does against itself.

With *identical* players and any $k \times k$ payoff matrix, the simplest way to locate a *pure* Nash strategy qualifying as an ESS is to identify diagonally dominant columns. If $m_{jj} > m_{ij}$ ($1 \leq i \leq k; i \neq j$), the jth column exhibits diagonal dominance. Therefore, the jth pure strategy is a Nash ESS. There may be as many such pure strategies as there are columns. More importantly, when two players share the same payoff matrix where column j is diagonally dominant, no mixed strategy including use of the behavior associated with the jth column can qualify as a Nash ESS (Haigh, 1975). Hence, no 2×2 payoff matrix with one or two diagonally dominant columns admits a mixed Nash ESS, though larger $k \times k$ ($k \geq 3$) matrices may have both pure and mixed solutions.

To clarify pure and mixed solutions, we consider several possibilites for m_1. First assume $m_{11} > m_{21}$ and $m_{12} > m_{22}$. Diagonal dominance indicates that $p^* = [1, 0]^T$ (pure A) is the Nash ESS. $E(p^*, p^*)$ is, of course, m_{11}. Suppose player 1 plays p^* and player 2's strategy is $q = [\pi_2, 1 - \pi_2]^T$ with $0 < \pi_2 < 1$. Then $E(q, p^*) = m_{21} + \pi_2(m_{11} - m_{21})$. Comparing expected payoffs, we have $m_{11} > m_{21}(1 - \pi_2) + m_{11} \pi_2$, since $m_{11} > m_{21}$. Then $E(p^*, p^*) > E(q, p^*)$ for all $q \neq p^*$. Condition (2a) is satisfied as a strict inequality and p^* is the Nash ESS. Behavior A is the better response to either behavior A or B. The pure A strategy (p^*) cannot be beaten (invaded evolutionarily) by pure B or by a mixture of A and B played aginst it.

Now suppose $m_{11} < m_{21}$ and $m_{12} < m_{22}$. Diagonal dominance indicates that $p^* = [0, 1]^T$ is the Nash ESS; $E(p^*p^*) = m_{22}$. If player 2 uses $q \neq p^*$, $E(q, p^*) = m_{22} + \pi_2(m_{12} - m_{22})$. Since $m_{12} < m_{22}$, $E(p^*, p^*) > E(q, p^*)$ for all $q \neq p^*$. Then pure B is the Nash ESS, which cannot be beaten by any alternative strategy played against it.

If $m_{11} > m_{21}$ and $m_{12} < m_{22}$, both columns show diagonal dominance. Either pure strategy qualifies as a Nash ESS; $p^* = [1, 0]^T$ or $p^* = [0, 1]^T$. Chance or initial conditions might govern which equilibrium is attained evolutionarily when phenotype and genotype are coupled strongly. Suppose $m_{11} < m_{22}$ so that $E(p_A^*, p_A^*) < E(p_B^*, p_B^*)$. Allowing the behavioral flexibility envisioned in our comments on learning and decision making, two players in a population using only behavior A might agree that each use B. As long as two such players interact only with each other (nonrandom encounter), the increased payoff would favor *mutual* switching to the second Nash solution (Pulliam and Caraco, in press; see Fagen, 1980). Note that such behavior does not constitute cooperation in the sense of a Pareto solution to a game (discussed below).

For identical players a pure Nash ESS also may occur without strict diagonal dominance. Suppose $m_{11} = m_{21}$ and $m_{12} > m_{22}$. The Nash ESS is $p^* = [1, 0]^T$ (pure A). Let player 1's strategy be p^* and player 2's strategy be $q = [\pi_2, 1 - \pi_2]^T$, where $0 < \pi_2 < 1$. $E(p^*, p^*) = m_{11}$ and $E(q, p^*) = m_{11} \pi_2 + m_{21}(1 - \pi_2)$. Since $m_{11} = m_{21}$, $E(p^*, p^*) = E(q, p^*)$. Therefore, if p^* is the Nash ESS, it must satisfy condition (2b). $E(p^*, q) = m_{11} \pi_2 + m_{12}(1 - \pi_2)$. $E(q, q) = m_{11} \pi_2^2 + m_{12} \pi_2(1 - \pi_2) + m_{21} \pi_2(1 - \pi_2) + m_{22}(1 - \pi_2)^2$, which is $\pi_2 E(p^*, q) + m_{21} \pi_2(1 - \pi_2) + m_{22}(1 - \pi_2)^2$. Calculating the difference in expected payoffs, we have $E(p^*, q) - E(q, q) = (m_{12} - m_{22})(1 - \pi_2)^2$. Since $m_{12} > m_{22}$, $E(p^*, q) > E(q, q)$, and p^* is the Nash ESS by condition (2b).

For 2×2 payoff matrices diagonal dominance eliminates the possibility of a mixed Nash ESS and, therefore, stable behavioral polymorphisms. But suppose $m_{11} < m_{21}$ and $m_{12} > m_{22}$, so neither column exhibits diagonal dominance. The first inequality indicates that B can invade pure A; the second inequality indicates that A can invade pure B. The Nash ESS must be a mixture. Recognizing that π_1 and π_2 in conditions (1a) and (1b) are continuous variables, though the behaviors are discrete, the Nash ESS is (Haigh, 1975):

$$\pi_1^* = \pi_2^* = \frac{m_{12} - m_{22}}{m_{12} + m_{21} - m_{11} - m_{22}} \tag{3}$$

Then $p^* = [\pi_i^*, 1 - \pi_i^*]^T$, and the strategies are the same for the identical players. When a mixed strategy is the Nash ESS,

$$\frac{\partial E(p, q)}{\partial \pi_1}\bigg|_{\pi_2^*} = \frac{\partial E(q, p)}{\partial \pi_2}\bigg|_{\pi_1^*} = 0 \tag{4a}$$

$$\frac{\partial^2 E(p, q)}{\partial \pi_1 \, \partial \pi_2} = \frac{\partial^2 E(q, p)}{\partial \pi_2 \, \partial \pi_1} < 0 \tag{4b}$$

When M_1 lacks diagonal dominance, Equation (4a) gives the best mixed strategy. We noted that either pure strategy is invasible by the other, and it is easy to show that p^* is stable against a pure strategy. Consider $q = [0, 1]^T$, a strategy of pure B. From (1a) and (1b),

$$E(p^*, p^*) = E(q, p^*) = m_{22} + \frac{(m_{12} - m_{22})(m_{21} - m_{22})}{m_{12} + m_{21} - m_{11} - m_{22}}$$

Since $E(p^*, p^*) = E(q, p^*)$, we compare $E(p^*, q)$ and $E(q, q)$. $E(q, q) = m_{22}$, and

$$E(p^*, q) = m_{22} + \frac{(m_{12} - m_{22})^2}{m_{12} + m_{21} - m_{11} - m_{22}}$$

Then $E(p^*, q) > E(q, q)$ and p^* is stable by criterion (2b).

For identical players, the mixed Nash ESS can be obtained simply by equating expected payoffs across the rows of M_1. This reflects the biological assumption that any individual encounters each phenotype in proportion to its frequency in the population. Of course, the mixed ESS also can be interpreted as a population of individuals playing the same mixed strategy. For a discussion of equilibrial mixed strategies in identical player games with general $k \times k$ payoff matrices, see Haigh (1975).

In an asymmetric game (e.g., Maynard Smith and Parker, 1976) we drop the assumption of identical players, so that the two players' payoff matrices differ. Each payoff matrix may be associated with a particular role (e.g., dominant and subordinate roles, or intruder and defender), but the set of possible discrete behaviors is the same for each player. Neither the Nash equilibrium strategies nor the expected payoffs need be identical for the two roles. The difference in expected payoffs need not imply evolutionary instability. As discussed by Maynard Smith and Parker (1976), the two Nash strategies can represent an ESS when individuals assume the different roles at different times. When in the role of player 1, an individual plays the Nash strategy associated with the applicable payoff matrix. When in the role of player 2, the same individual plays the Nash strategy associated with player 2's payoff matrix. The paired, role dependent Nash strategies then correspond to a noncooperative ESS. Of course, if certain players usually (or always) find themselves associated with the role accruing the greater expected payoff, then selection will favor the characteristic(s) providing this advantage.

Maynard Smith and Parker (1976) also point out that pure strategies are more likely for certain asymmetric games than for games with identical players. Suppose one player (role) necessarily has a pure strategy. Then the other player will respond rationally and select the pure strategy providing the greatest payoff, conditional on the first player's behavior, unless different responses give the same payoff.

To demonstrate this point, suppose player 1's payoff matrix is

$$\begin{bmatrix} 3 & 5 \\ 4 & 6 \end{bmatrix}$$

Let player 2's payoffs be

$$\begin{bmatrix} 1 & 4 \\ 2 & 3 \end{bmatrix}$$

By diagonal dominance player 1 will use only the second behavior, $p^* = [0, 1]^T$. Then player 2's best strategy is also pure, and player 2 will use only the first behavior, $q^* = [1, 0]^T$, even though player 2's payoff matrix lacks diagonal dominance. If neither payoff matrix in an asymmetric game exhibits diagonal dominance, the Nash ESS may be a pair of mixed strategies (Oster and Wilson, 1978).

We deal only briefly with the Prisoner's Dilemma, since cooperation in this particular discrete game is discussed elsewhere (Owen, 1968; Axelrod and Hamilton, 1981; Brown et al., 1983; Michod, Chapter 9; Slobodchikoff, Chapter 8). Assume that two identical players may cooperate (C) or play guilty (G). The Prisoner's Dilemma has the form of payoff matrix M_2:

$$\begin{array}{c} \quad\quad C \quad G \\ \begin{array}{c} C \\ G \end{array} \begin{bmatrix} R & S \\ T & P \end{bmatrix} \end{array}$$

In M_2, $T > R > P > S$ and $R > (T + S)/2$. R rewards mutual cooperation, whereas T tempts a player to defect against a cooperator. S is the sucker's payoff for cooperating when the opponent defects. P is the punishment received by each of two guilty players.

A pure guilty strategy is the Nash ESS (diagonal dominance). As indicated above, if a player chooses the pure guilty strategy, he cannot be beaten by any other strategy. Pure cooperation by both players is the Pareto solution, since either player can improve its payoff only be reducing the other's payoff. Mutual cooperation can be attractive since $R > P$, but a cooperator risks being cheated (Trivers, 1971).

Under seemingly realistic assumptions concerning memory and/or population structure, natural selection may achieve a cooperative ESS for this game. That is, the wary cooperator of the judge (tit-for-tat) strategy (Axelrod and Hamilton, 1981; Pulliam et al., 1982) may be favored under certain rules for repeated play. We caution, however, that not all forms of repeated play will favor cooperation (see Owen, 1968, p. 153).

Now we consider continuous static games, where payoff functions replace payoff matrices. Let $F_1(u_1, u_2)$ be player 1's payoff function, while $F_2(u_1, u_2)$ is player 2's payoff function. The u_i are decision (control) variables, representing behaviors quantified on a continuous scale. Sufficient conditions for a Nash solution are (Oster and Wilson, 1978)

$$\frac{\partial F_1}{\partial u_1}\bigg|_{u_2^*} = \frac{\partial F_2}{\partial u_2}\bigg|_{u_1^*} = 0 \tag{5a}$$

$$\frac{\partial^2 F_i}{\partial u_i^2}\bigg|_{(u_1^*, u_2^*)} < 0; \quad i = 1, 2 \tag{5b}$$

At the equilibrium we have a Nash ESS if (1) $F_1(u_1^*, u_2^*) = F_2(u_1^*, u_2^*)$, or (2) if players adopt u_1^* when in the role of player 1 and adopt u_2^* when in the role of player 2, that is, in an asymmetric game where $F_1(u_1^*, u_2^*) \neq F_2(u_1^*, u_2^*)$. For the case of identical players, where $F_1(u_1, u_2) = F_2(u_2, u_1)$, (u_1^*, u_2^*) is a Nash ESS and $u_1^* = u_2^*$. Then the payoff functions satisfy the conditions $\partial^2 F_i/\partial u_i \partial u_j \leq 0$ $(i \neq j)$.

Sometimes the payoff functions may depend on a "state" variable, as well as the control variables. Further, the players' strategies and the state variable may be collectively constrained by a functional relationship. Let $F_i(x, u_1, u_2)$ be the payoff functions $(i = 1, 2)$. The state variable is x. Let an equality constraint be given by the function $g(x, u_1, u_2) = 0$. The Nash equilibrium is (u_1^*, u_2^*), and let x^* be the value of the state variable satisfying the constraint at the equilibrium. That is, $g(x^*, u_1^*, u_2^*) = 0$. We form the functions L_i:

$$L_1 = F_1(x, u_1, u_2) + \lambda_1 g(x, u_1, u_2)$$

$$L_2 = F_2(x, u_1, u_2) + \lambda_2 g(x, u_1, u_2)$$

where the λ_i are Lagrange multipliers. Necessary conditions for a Nash equilibrium are (Vincent and Grantham 1981)

$$\frac{\partial L_1}{\partial x}\bigg|_{(u_1^*, u_2^*)} = \frac{\partial L_2}{\partial x}\bigg|_{(u_1^*, u_2^*)} = 0 \tag{6a}$$

$$\frac{\partial L_1}{\partial u_1}\bigg|_{(x^*, u_2^*)} = \frac{\partial L_2}{\partial u_2}\bigg|_{(x^*, u_1^*)} = 0 \tag{6b}$$

$$g(x^*, u_1^*, u_2^*) = 0 \tag{6c}$$

For identical players we are assured that $F_1(x^*, u_1^*, u_2^*) = F_2(x^*, u_1^*, u_2^*)$, a Nash ESS.

Vincent and Grantham (1981) discuss rational reaction sets (RRS) as a technique of solving games. The method can be particularly useful when the payoff functions lack the continuity assumed in Equations (5) and (6). Suppose player 1 knows u_2 (player 2's behavior) before the game is played. Then player 1 could select u_1 to achieve the highest attainable payoff to itself, given the value of u_2. For all possible values of u_2, identify these values of u_1 as player 1's RRS (R_1). Using the same logic, we can identify player 2's RRS (R_2). The intersection of the two sets, $R_1 \cap R_2$, gives the Nash solution to the game. If the sets do not intersect, the game has no Nash solution. In the next section we demonstrate the method in locating the Nash ESS for a simple game.

Now suppose that mutual cooperation provides each player with a greater payoff than that acquired at the Nash ESS. The judge (tit-for-tat) strategy might then qualify as a reasonable ESS, and the Pareto solution might predict the players' behavior. As in the discrete case, cooperation invites cheating, so the cooperative ESS for a continuous game requires certain forms of repeated play and assumptions concerning population structure or memory for evolutionary stability.

Pareto solutions satisfy

$$\nabla F_1 = -c \, \nabla F_2 \tag{7}$$

where c is a positive constant and ∇F_i is the gradient $< \partial F_i/\partial u_1, \partial F_i/\partial u_2 >$. The "Pareto optimal front" is defined by Equation (7). To locate a Pareto solution we combine the payoff functions $F_1(u_1, u_2)$ and $F_2(u_1, u_2)$ into a single function (F). As an example, we assume that addition is biologically realistic (see Oster and Wilson, 1978) and write $F = \theta F_1 + (1 - \theta)F_2$, where $0 < \theta < 1$. Then the conditions for a Pareto solution (u_1°, u_2°) are

$$\left. \frac{\partial F}{\partial u_1} \right|_{u_2^\circ} = \left. \frac{\partial F}{\partial u_2} \right|_{u_1^\circ} = 0 \tag{8a}$$

$$\left. \frac{\partial^2 F}{\partial u_i^2} \right|_{(u_1^\circ, u_2^\circ)} < 0; \qquad i = 1, 2 \tag{8b}$$

The optimal strategies depend on the value of θ. Since we seek a solution consistent with natural selection, it is realistic to take $\theta = \frac{1}{2}$ and weight the players equally (or $\theta = 1/n$ in a n-person game). Selection ordinarily should not weight the players' fitnesses unequally. For nonidentical players fitnesses may equilibrate on a single play of the game only for $\theta \neq \frac{1}{2}$. If the strategies are role dependent, as in Maynard Smith and Parker's (1976) asymmetric games, then $\theta = \frac{1}{2}$ is still a reasonable weighting. However, an additional possibility exists for Pareto solutions. Nonidentical players might "bargain" to select a value of θ, if both do better at the resulting Pareto solution than they would do at a Nash equilibrium.

In many applications the players are identical (e.g., Pulliam et al., 1982). In this case the solution obtained from Equation (8), with $\theta = \frac{1}{2}$ for two players, is found merely by setting $u_1 = u_2 = u$ and maximizing $F_i(u)$ with respect to u. This approach works for identical and equally weighted players, since $u_1^o = u_2^o$ at the solution associated with the cooperative ESS.

As indicated in our discussion of Nash strategies, state variables and constraint equations (or inequalities) can be incorporated as necessary into the analysis of Pareto solutions (see Vincent and Grantham, 1981). In passing we note that our discussion deals with only static games. Some questions are formulated best as differential games and require time dependent solutions (Maynard Smith, 1978).

We end this section with a simple example. For two identical players we have

$$F_1 = 10 + 2u_1 - u_1^2 - u_2^2$$

$$F_2 = 10 + 2u_2 - u_2^2 - u_1^2$$

(9)

The Nash solution $u_1^* = u_2^* = 1$ satisfies condition (5). $F_1(u_1^*, u_2^*) = F_2(u_1^*, u_2^*) = 10$, so we have found the noncooperative ESS. As an aside, suppose player 1 plays u_1^* and player 2 chooses $u_2 = 1.1$. Then $F_1(u_1^*, 1.1) = 9.79$ and $F_2(u_1^*, 1.1) = 9.99$. Player 2 does better than player 1, but player 2 has decreased its own payoff (slightly) by deviating from the Nash solution. Player 2 has a lower payoff than each member of every Nash strategist pair in the population, and "spite" would not then be favored over the Nash ESS.

Mutual cooperation can increase each payoff. We take $F = (F_1 + F_2)/2 = 10 + u_1 + u_2 - u_1^2 - u_2^2$. Then $\partial F/\partial u_i = 1 - 2u_i$, and $\partial^2 F/\partial u_i^2 < 0$. The Pareto solution $(u_1^o, u_2^o) = (\frac{1}{2}, \frac{1}{2})$. $\nabla F_1 = <2 - 2u_1, -2u_2>$, and $\nabla F_2 = <-2u_1, 2 - 2u_2>$. At (u_1^o, u_2^o), $\nabla F_1 = -c \nabla F_2$, with $c = 1$. $F_1(u_1^o, u_2^o) = F_2(u_2^o, u_2^o) = 10.5$, which exceeds $F_i(u_1^*, u_2^*)$. Therefore, two cooperators fare better against each other than do two guilty players against each other, so that cooperation might become the dominant behavior. However, the wary cooperator should defect against a guilty strategist. To amplify this point, suppose player 1 is a pure cooperator (never defects) and that player 2 is a guilty strategist. Then $u_1 = \frac{1}{2}$ and $u_2 = 1$ in our example. Consequently, $F_1(\frac{1}{2}, 1) = 9.75$ and $F_2(\frac{1}{2}, 1) = 10.75$. Again, the Pareto solution is unstable against cheating on a single play of the game, but cooperation (i.e., the judge strategy) can be evolutionarily stable under certain forms of repeated play.

3.2 Applications: Conflict and Cooperation

In this section we apply game theory to some questions concerning social behavior in animals subject to predation. Earlier we mentioned that females might synchronize their production of young in response to predation on their offspring. Although seasonal factors may constrain the timing of births, reproductive synchrony remains striking in some colonial birds and social mammals (e.g., Emlen and Demong, 1975; Estes, 1976). A pulse of reproduction might quickly satiate predators. Consequently, a particular offspring's *a priori* probability of surviving neonatal susceptibility to predation could exceed the survivorship attained when births occur more uniformly through time. Lloyd and Dybas (1966) adopt a conceptually similar approach in discussing the 13- and 17-year life cycles of certain *Cicada* species. We first consider the timing of births with a very simple model of a strictly competitive game between females. Analysis suggests that synchrony could be a Nash ESS when offspring survivorship depends only on avoiding predation. We then consider a slightly more sophisticated model where the timing of births involves an antagonistic interaction between exposure to predation and growth prior to migration.

The first example considers discontinuous payoff functions. Let female 1 choose to produce her offspring at time t_1. Female 2 choose t_2; t_1, t_2, $\epsilon(0, T)$. We place an upper bound on the birthing season, since the females might reinitiate the reproductive process, or the young may have to be prepared to migrate in response to seasonal resource patterns. Each payoff function gives the probability that the female's offspring survives predation during the birthing season:

$$F_1(t_1, t_2) = \begin{cases} \alpha, & \text{if } t_1 < t_2 \\ (1 + \alpha)/2, & \text{if } t_1 = t_2 \\ 1 - \beta, & \text{if } t_1 > t_2 \end{cases}$$

where $1 + \alpha < 2(1 - \beta)$. By reversing the order of the two inequalities involving t_1 and t_2, we obtain $F_2(t_1, t_2)$. The first offspring born (when $t_1 \neq t_2$) has the lower survivorship value; its predation probability is $1 - \alpha$. With synchrony ($t_1 = t_2$) the probability of a single kill is still $1 - \alpha$, but the predation risk is shared equally by the two offspring. That is, we assume a single kill reduces the intensity of predation. The second offspring born (when $t_1 \neq t_2$) has the greater survivorship value.

Suppose female 1 knows t_2. Then female 1's rational reaction set, R_1, must be $R_1 = (t_1 \mid t_1 > t_2 \text{ for } t_2 < T; t_1 = T \text{ for } t_2 = T)$. Similarly, if female 2 knows t_1, $R_2 = (t_2 \mid t_2 > t_1 \text{ for } t_1 < T; \quad t_2 = T \text{ for } t_1 = T)$. The intersection of these sets gives the Nash solution:

$$R_1 \cap R_2 \Rightarrow t_1^* = t_2^* = T$$

Since $F_1(t_1^*, t_2^*) = F_2(t_1^*, t_2^*)$, $t_i^* = T$ qualifies as a selfish ESS. Payoffs are equal whenever $t_1 = t_2$, but only $t_i^* = T$ is stable against deviations in strategy. Given the simplistic assumptions, selection should favor synchronous births, delayed as long as is feasible.

The above model demonstrates an application of rational reaction sets, but oversimplifies the biology of predation. Therefore, we examine the timing of births in a more realistic context and consider both strictly selfish and cooperative strategies.

Consider two female birds, each of which will produce a single egg. If her egg escapes predation, female 1 will incubate for a fixed amount of time, as will female 2. During incubation, an egg may suffer predation with probability $\leq \theta$; the exact value will depend on the ordering of births. At time t_1 (t_2 for female 2) the offspring hatches if incubation is completed; $t_1, t_2 \in (0, T)$. For simplicity, we assume that an offspring, once hatched, can dilute predation on either egg or nestling of the other female, but an egg exerts no dilution effect.

The probabilistic rate of predation during (t_1, T) is α per unit time if $t_1 < t_2 = T$. The same force of mortality applies to female 2's offspring during (t_2, T) if $t_2 < t_1 = T$. For all t such that $t_1, t_2 < t \leq T$, the two offspring equally share the probability of predation. Therefore, the first offspring born (when $t_1 \neq t_2$) is exposed to a greater risk of predation. The *a priori* probability that female 1 loses her offspring to predation is d_p^1,

$$d_p^1 = \theta + \alpha (T - t_1) - \frac{\alpha}{2} (T - t_2)$$

where $0 < \Theta - \alpha T/2 < \Theta + \alpha T < 1$. For the second female's offspring, d_p^2 is obtained by interchanging t_1 and t_2 in d_p^1.

At time T the reproductive season ends, and migration begins. We assume that death during migration can result from starvation. During (t_1, T), if female 1's offspring avoids predation, the offspring will grow, learn to recognize food, and so on. To accommodate this circumstance, we assume that the probability that female 1's offspring dies from starvation during migration (d_s^1) declines over (t_1, T),

$$d_s^1 = 1 - \gamma (T - t_1)$$

d_s^2 is the same, after t_2 is substituted for t_1.

We assume that death due to predation and death due to starvation are independent events. Consequently, female 1's payoff, the *a priori* probability that her offspring survives both sources of mortality, is $F_1 = (1 - d_p^1) (1 - d_s^1)$. Then

$$F_1 = \left[1 - \Theta - \alpha (T - t_1) + \frac{\alpha}{2} (T - t_2) \right] [\gamma (T - t_1)] \qquad (10)$$

F_2 is obtained by interchanging t_1 and t_2. Using Equations (5a) and (5b), we locate the Nash ESS:

$$t_1^* = t_2^* = T - \frac{2(1 - \Theta)}{3\alpha} \tag{11}$$

Since the payoff functions are symmetric, synchrony is the ESS; reproductive synchrony need not have this equilibrial property if the payoff functions differ.

Note that $\partial t_i^*/\partial \alpha > 0$ and $\partial t_i^*/\partial (1 - \Theta) < 0$. As the intensity of predation (α) increases, the equilibrial hatching dates occur later in the reproductive season. As the probability of surviving egg predation $(1 - \Theta)$ increases, eggs will be produced earlier. At the Nash ESS, the payoff to each female is

$$F_i (t_1^*, t_2^*) = \frac{4\gamma(1 - \Theta)^2}{9\alpha}$$

Now we consider the cooperative solution to the same game. We take $F = (F_1 + F_2)/2$, seeking the Pareto solution where the fitnesses of the identical players are equal. Using Equation (8), we find the cooperative solution:

$$t_1^\circ = t_2^\circ = T - \frac{1 - \Theta}{\alpha} \tag{12}$$

Cooperation also suggests reproductive synchrony for identical players. For given values of α and Θ, $t_i^\circ < t_i^*$. Cooperation implies earlier breeding than does a Nash strategy, and the naive cooperator can be cheated by a female delaying reproduction.

As was the case for the Nash solution, $\partial t_i^\circ/\partial \alpha > 0$ and $\partial t_i^\circ/\partial(1 - \Theta) < 0$. However, the gradients are steeper about the Pareto solution. The payoff for mutual cooperation is $F_i (t_1^\circ, t_2^\circ) = \gamma (1 - \Theta)^2/2\alpha$, and $F_i (t_1^\circ, t_2^\circ) > F_i (t_1^*, t_2^*)$. Cooperation in timing reproduction may not be as probable a consequence of natural selection as it can be in games played more often during an individual's lifetime (e.g., Pulliam et al., 1982). Hence, the judge strategy might encounter difficulty invading a population of strictly selfish breeders. Again, environmental or physiological constraints may in some cases strongly limit the timing of reproduction. Breeding synchrony, inducing synchronous births, also may be a female strategy deterring male desertion (e.g., Evans, 1982).

The two models above predict reproductive synchrony because we assumed that the females were identical players. To demonstrate this assumption's significance and the consequent limitations of the prediction, we consider nonidentical players.

Suppose female 1 possesses a better nesting site, so that $\Theta_1 < \Theta_2$, but the predation process if otherwise unchanged. Then we have

$$F_1(t_1, t_2) = \left[1 - \Theta_1 - \alpha(T - t_1) + \frac{\alpha}{2}(T - t_2)\right] \gamma (T - t_1)$$

$$F_2(t_1, t_2) = \left[1 - \Theta_2 - \alpha(T - t_2) + \frac{\alpha}{2}(T - t_1)\right] \gamma (T - t_2)$$

(13)

At the Nash solution,

$$t_1^* = T - \frac{2}{15\alpha}(5 - 4\Theta_1 - \Theta_2)$$

$$t_2^* = T - \frac{2}{15\alpha}(5 - 4\Theta_2 - \Theta_1)$$

(14)

Since $5 - 4\Theta_2 - \Theta_1 < 5 - 4\Theta_1 - \Theta_2, t_1^* < t_2^*$. The female with the safer nesting site (female 1) reproduces first, but still has a greater payoff:

$$F_1(t_1^*, t_2^*) = 4\gamma \frac{[5 - 3(\Theta_1 + \Theta_2)] [5 - 4\Theta_1 - \Theta_2]}{15\alpha}$$

$$F_2(t_1^*, t_2^*) = 4\gamma \frac{[5 - 3(\Theta_1 + \Theta_2)] [5 - 4\Theta_2 - \Theta_1]}{15\alpha}$$

Since $\Theta_1 < \Theta_2$, $F_1(t_1^*, t_2^*) > F_2(t_1^*, t_2^*)$. If strategies are role dependent (Maynard Smith and Parker, 1976), (t_1^*, t_2^*) is the Nash ESS. This assumes that over her lifetime a female is likely to hold both better and worse nest sites and that she behaves appropriately in each circumstance. Of course, if some intrinsic property of female 1 ensures her access to better nest sites, that property will be favored by selection. In any case, variation in nest site quality could tend to desynchronize reproduction. In fact, any differences in predation probabilities have this consequence.

Now suppose the two nonidentical females choose to cooperate. Again, we weight the players equally and take $F = (F_1 + F_2)/2$. The Pareto solution is

$$t_1^\circ = T - \frac{1}{3\alpha}(3 - \Theta_2 - 2\Theta_1)$$

$$t_2^\circ = T - \frac{1}{3\alpha}(3 - 2\Theta_2 - \Theta_1)$$

(15)

Since $3 - 2\Theta_2 - \Theta_1 < 3 - \Theta_2 - 2\Theta_1, t_1^\circ < t_2^\circ$ and female 1 reproduces first. $F_1(t_1^\circ, t_2^\circ) > F_2(t_1^\circ, t_2^\circ)$, as at the Nash solution to the same problem. The interpretation of the asymmetry in payoffs is the same as that for the Nash solution.

If our analysis sufficiently abstracts the biological essence of the problem, we can make four qualitative predictions. As the intensity of predation on offspring increases, births should be delayed at the expense of offspring growth (unless renesting, or its equivalent, is possible). Cooperative strategies favor births earlier in the season than do strictly selfish strategies. Individuals whose offspring have greater *a priori* survivorship probabilities should tend to produce offspring earlier, whether they play cooperative or strictly selfish strategies. Finally, and most importantly, the degree of reproductive synchrony should be inversely related to the variation in *a priori* offspring survivorship probabilities.

Prey individuals sometimes mob a predator or actively defend themselves by fighting. Even when the adults themselves are immune to a particular predator's efforts, so that the real or potential costs of mobbing are accepted primarily to protect dependent offspring, these responses to predation suggest prey cooperation. To advance this notion we consider a game where each of two prey may mob or flee (i.e., cooperate or defect).

The two prey forage together. When a predator is sighted nearby, we assume that each prey animal must decide whether to flee or mob the predator. For simplicity, we assume that the predator's strategy is probabilistically governed by the behavior of the prey. The probability that the predator initiates an attack is:

$$Pr(\text{attack}) = \begin{cases} \alpha_2, & \text{if both prey mob} \\ \alpha_1, & \text{if one prey mobs} \\ \alpha_0, & \text{if both prey flee} \end{cases}$$

The conditional probability that the predator makes a kill, given an attack, is

$$Pr(\text{kill} \mid \text{attack}) = \begin{cases} k_2, & \text{if both prey mob} \\ k_1, & \text{if one prey mobs} \\ k_0, & \text{if both prey flee} \end{cases}$$

If the predator attacks, it can kill at most only one of the prey. The probability of a kill is simply $\alpha_i k_i$; $i = 0, 1, 2$. If one prey individual mobs and the second flees, we assume the second is never attacked.

The payoff matrix, M, for each of the identical prey gives survivorship probabilities:

	mob	flee
mob	$1 - \dfrac{\alpha_2 k_2}{2}$	$1 - \alpha_1 k_1$
flee	1	$1 - \dfrac{\alpha_0 k_0}{2}$

To obtain the form of a Prisoner's Dilemma, we assume

$$1 > 1 - \frac{\alpha_2 k_2}{2} > 1 - \frac{\alpha_0 k_0}{2} > 1 - \alpha_1 k_1 \qquad (16)$$

Mutual cooperation in mobbing provides a greater payoff than mutual defection. But in deciding to mob, the individual risks the possibility that the other player will flee. Hence, the dilemma, since mobbing alone provides the lowest payoff.

By diagonal dominance, flee is the Nash ESS. Most social prey flee upon detecting a predator, and we do not require an explanation via game theory. However, mobbing or active defense occurs in invertebrates, birds and mammals, so we examine conditions for a cooperative ESS. From condition (16) this example of a Prisoner's Dilemma is defined by

$$\frac{\alpha_2 k_2}{2} < \frac{\alpha_0 k_0}{2} < \alpha_1 k_1 \qquad (17)$$

Natural selection might favor cooperative defense if $\alpha_2 k_2$ is relatively small. The probability of a kill when both prey mob will be small if the prey can threaten the predator with aggressive signals, deterring an attack, or if group defense renders an attacking predator relatively ineffective. Note that if the dilution effect is removed from our assumptions, so that the predator might kill both of two mobbing prey, cooperation seems less probable. Of course, if $\alpha_2 k_2 > \alpha_0 k_0$, there is no advantage from cooperation, and the prey should always flee.

A potential difficulty for a cooperative ESS involves the necessary conditions concerning repeated play (Owen, 1968). If the prey rarely encounter predators, they might treat each interaction as a game played once. In this case, defection to the guilty strategy (flee) always would be predicted. If attacks occur quite frequently and many prey are killed, a given individual might often interact with strangers. In this circumstance also it can be difficult for the conditional cooperation of the judge strategy to displace the guilty strategy (Brown et al., 1983). However, cooperation could be favored if attacks are fairly frequent, but the predator only rarely succeeds in making a kill.

When mobbing is both associated with the defense of offspring only, and extracts only negligible costs from parents, mobbing could arise without cooperation (and the above model does not apply). When both parents and their offspring can be captured by the same predator, the game increases in complexity. A parent's payoffs should trade off the value of the present offspring against expected future reproductive success.

Scanning for approaching predators and time spent foraging must interactively govern survivorship in many social prey. The interaction will be particularly important when the total time for vigilance and feeding is constrained, as in certain flocks of overwintering birds (Pulliam, 1973; Caraco et al., 1980). Pulliam et al. (1982) develop a detailed model of vigilance behavior for flocks of yellow-eyed juncos. The rate at which an individual scans for predators is taken as the decision variable in an n-person game, and survivorship is the currency of fitness. For any particular flock size, increased scanning increases the probability that the individual will survive a single predatory attack. However, increased vigilance also increases the time to obtain a required amount of food, so that the individual is exposed to a greater expected number of such attacks.

The game was solved numerically for both the Nash and Pareto solutions, for groups ranging in size from two to 10 birds. The cooperative solution provided a far better fit to observed scanning rates than did the Nash solution. The result is tentative, since the payoff functions required a series of assumptions regarding the predator–prey interaction. We find the result intriguing, especially since most game theoretical models in behavior have been constructed around the assumption of a Nash ESS. Cooperation may be far more common in nature than we have suspected previously.

4   MODELS OF PREDATOR SOCIALITY

In Section 2 we considered the costs and benefits of prey sociality, and in Section 3 we analyzed some specific games involving prey sociality. In this section, we consider some costs and benefits of sociality to predators and briefly discuss how predator sociality too can be modeled as a game. The advantages of sociality to obtaining food have been reviewed recently by Bertram (1978) and Pulliam and Caraco (in press). In this chapter we emphasize one particular advantage of predator sociality: the reduction in search time required to locate prey.

Krebs et al. (1972) demonstrated that captive great tits (*Parus major*) were more successful at locating hidden food when foraging in small groups than when foraging alone. In particular, when one individual located a food patch, others were attracted to the proximity of the food source. This sharing of information about the location of food patches reduces the variance in the rate of individual food intake and may provide one of the primary benefits of social foraging. The results of Krebs et al. (1972) as well as those of Murton (1971) demonstrate that social interactions between individuals also influence the costs and benefits of social foraging. In particular, dominant individuals may take a larger proportion of the food located by groups but subordinates may gain by observing the food choices of more experienced dominant individuals. Clearly then, any general theory of predator sociality must take into account the role of dominance status in determining costs and benefits to group participants.

4.1   The Mean Rate of Food Intake

We envision a group of n foragers exploiting food that occurs in widely dispersed patches. The time required to locate a patch of food can be taken to be

$$T_s(n) = T_s(1)/n^\beta \tag{18}$$

where $0 \le \beta \le 1$. When search paths do not overlap and foragers do not otherwise reduce one another's searching efficiency, β may approach 1. On the other hand, when all individuals search the very same area, β approaches 0 or may even by negative. In this case there is no advantage to group foraging in terms of reduced search time. We analyze here only the case where $\beta = 1.0$. This is the situation most favorable to group foraging and can thus be used as a standard of comparison for situations involving interference.

We assume that once one individual has located a food patch, the other members of the foraging group immediately congregate at the patch and start feeding. If the initial food abundance in the patch is X_0 and patch depletion is given by the expression

$$dX(t)/dt = -\alpha n X(t) \tag{19}$$

the amount of food remaining after T_p seconds is

$$X(T_p) = X_0 e^{-\alpha n T_p} \tag{20}$$

(Hereafter, T_p is referred to as the time in the patch and will be treated as a decision variable under control of the birds in the group.)

The *mean* amount of food consumed by individuals in the group is given by

$$X_0/n(1 - e^{-\alpha n T_p}) \tag{21}$$

Since this amount is consumed in T_p seconds and T_s/n seconds were spent searching for the patch, the mean rate of food intake is given by

$$r_n(T_p) = \frac{X_0/n(1 - e^{-\alpha n T_p})}{T_s/n + T_p} \tag{22}$$

The optimal length of time to spend in the patch can be found by setting the derivative $\partial r_n(T_p)/\partial T_p$ equal to zero and solving for T_p. An approximate solution is found by taking a Taylor's series expansion of r_n around the point

$T_p = 0$. This yields the following approximation

$$r_n(Tp) = \frac{\alpha X_0 n T_p}{T_s} - \frac{\alpha^2 X_0 n^2 T_p^2}{2T_s} - \frac{2\alpha T_p^2 X_0 n^2}{T_s^2} \tag{23}$$

The approximation represents only the first three terms in the Taylor's series expansion and is good only for relatively small values of T_p.

Taking the derivative of Equation (23) and setting it equal to zero, we find that optimal time to spend in a patch is

$$\hat{T}_p = \frac{T_s}{\alpha n \, T_s + 4n} \tag{24}$$

Note that

$$\hat{T}_p(n) = \hat{T}(1)/n \tag{25}$$

and

$$r_1[\hat{T}_p(1)] = r_n[\hat{T}_p(n)] \tag{26}$$

[Relationships (25) and (26) also hold when the fourth term of the Taylor series is included in the expansion.] Relationship (26) means that group foraging neither increases nor decreases the mean rate of food intake. Since the case analyzed represents that most favorable to group foraging ($\beta = 1$), any interference, no matter how slight, between foragers results in a lower overall rate of food intake.

Though the above analysis argues against any general advantage to predators from group foraging, at least two other factors promote group foraging in a patchy environment. First group foraging may allow predators to exploit prey items that cannot otherwise be utilized. Whether or not this mechanism actually results in a higher mean rate of food intake remains to be established. A second factor promoting group foraging in a patchy environment is the reduction in the variance in food intake. Since we think this likely to be an important factor favoring predator sociality, we treat it in some detail.

4.2 The Variance in Food Intake

Thompson et al. (1974) developed a computer model of social foraging which demonstrated that predators in groups may have less risk of going long periods of time without finding any food. Pulliam and Millikan (1982) and Caraco (1981) develop more analytical models of the advantage of variance reduction resulting from group foraging. Here we briefly illustrate this approach by describing a model according to which group foragers and solitary foragers have the same mean daily food intake but group foragers have a lower variance in daily food intake.

If each patch exploited is treated as an independent event and foragers use many patches each day, the law of large numbers can be invoked to show that total daily food intake has an approximately normal distribution (see Cox, 1962). The mean of this distribution is given by

$$\bar{I}(n) = r_n(T_p) \cdot \tau \tag{27}$$

where $r_n(T_p)$ is given by Equation (22) and τ is the available foraging time in a day. As previously discussed [Equation (26)], the expected daily intake is independent of foraging group size, when pursuit time is taken to be at the optimal value for each group size. On the other hand, the variance in daily food intake is given by

$$\sigma_I^2 = \text{Var } [I(n)] = \frac{\text{Var } [I(1)]}{n} \tag{28}$$

where $I(n)$ is the daily food intake of an individual in a group of size n. In other words, solitary and group foragers have the same mean daily intake but the variance in intake is n times as great for solitary foragers.

This result is illustrated in Figure 1a where the probability distribution of daily food intake is compared for a solitary individual and one in a group of size two. Figure 1b compares the probability of starvation for the same two birds. The curves in Figure 1b represent the integrated area

$$\int_0^r N(\bar{I}, \sigma_I^2) dx \tag{29}$$

under the normal curves in Figure 1a, and thus give the probability of having a daily food intake of less than r. If r is taken to be the required daily intake, an individual starves if it consumes less food. Thus, expression (29) can be taken as the probability of starvation as a function of required food intake, r.

For the situation shown in Figure 1, group foraging is favored when expected daily food ($\bar{I}$) intake is greater than required intake (r). However, in circumstances where the required intake is greater than the expected intake, solitary foraging is favored. The prediction is then that flocks should break up and individuals forage alone when food supply declines to the point that starvation is likely.

The graphs in Figure 1 assume that individuals in a group share all food found equally. In reality, dominant individuals will be likely to have greater rates of food intake than do subordinates. The result is a game situation, in which group foraging may be advantageous to dominant individuals and disadvantageous to subordinates. The resolution to such social games can

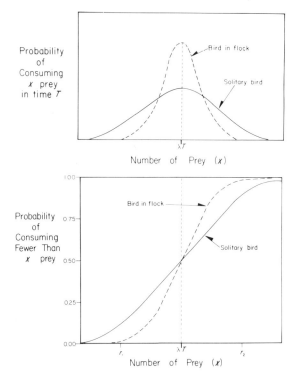

Figure 1. The probability distribution of daily food intake differs for solitary and group foraging individuals. As shown in the upper graph, if solitary and flocking birds have the same expected daily food intake (λT), we expect the flocking individuals to have lower variance in daily food intake. The probability of consuming less than a given quantity (x) of prey is the integral of the curves of the upper graph evaluated from zero to the quantity x. As indicated in the lower graph, if the number (r) of prey required to avert starvation is less than the expected number of prey consumed daily, individuals in a group have a lower probability of starving (i.e., a lower probability of finding fewer than the required number). If, however, the required number of prey exceeds the expected daily intake, solitary individuals have a greater survival probability than individuals in a group.

be found by application of game theory techniques discussed in Section 3 and in Pulliam and Caraco (in press).

5 SUMMARY

Although not all animals subject to predation exhibit true social behavior, predation may have exerted a potent influence on sociality in many species (Alexander, 1974; Hoogland, 1981). We have reviewed major hypotheses linking predation and sociality. We have pointed out some of the simpler aspects of game theory, since many ideas concerning social responses to predators are most effectively developed in that theoretical context. We have further suggested that cooperation might form the biological basis for certain antipredator strategies in social animals.

Strategies deterring predation merit ecological interest in their own right. Further, they sometimes may offer insight into interspecific relationships. Although their functional link to the dynamics of predator–prey interactions is obvious, a thorough knowledge of antipredator strategies sometimes can enhance our understanding of ecological relationships among coexisting prey species.

In Section 4 we took the viewpoint of the predator. Foraging theory is one of the most rapidly growing areas of behavioral ecology and predators, as well as their prey, sometimes can benefit from sociality.

ACKNOWLEDGMENTS

We thank the editors, P. Price, C. N. Slobodchikoff, and W. S. Gaud, for their efforts and consideration. We particularly appreciate the comments and discussion provided by A. Ives and S. Lima. C. Cicero and D. Riley helped prepare the manuscript. Financial support was provided by NSF grant BNS-8020717.

LITERATURE CITED

Alexander, R. D. 1974. The evolution of social behavior. *Annu. Rev. Ecol. Syst.* **5**:325–383.

Altmann, S. A. 1974. Baboons, space, time and energy. *Am. Zool.* **14**:221–248.

Altmann, S. A. 1979. Baboon progressions: Order or chaos? A study of one-dimensional group geometry. *Anim. Behav.* **27**:46–80.

Andersson, M., and C. G. Wicklund. 1978. Clumping versus spacing out: Experiments on nest predation in fieldfare (*Turdus pilaris*). *Anim. Behav.* **26**:1207–1212.

Axelrod, R., and W. D. Hamilton. 1981. The evolution of cooperation. *Science* **211**:1390–1396.

Barnard, C. J. 1980. Flock feeding and time budgets in the house sparrow, *Passer domesticus*, L. *Anim. Behav.* **28**:295–309.

Bertram, B. C. R. 1978. Living in groups: Predators and prey. In J. R. Krebs and N. B. Davies (eds.), *Behavioural ecology: An evolutionary approach*, pp. 64–96. Blackwell, Oxford.

Brown, J. L. 1982. Optimal group size in territorial animals. *J. Theor. Biol.* **95**:793–810.

Brown, J. L., and E. R. Brown. 1981. Kin selection and individual selection in babblers. In R. D. Alexander and D. Tinkle (eds.), *Natural selection and social behavior: recent results and new theory*, pp. 417–420. Chiron Press, New York.

Brown, J. S., M. J. Sanderson and R. E. Michod. 1983. Evolution of social behavior by reciprocation. *J. Theor. Biol.*

Caraco, T. 1979a. Time budgeting and group size: A theory. *Ecology* **60**:611–617.

Caraco, T. 1979b. Time budgeting and group size: A test of theory. *Ecology* **60**:618–627.

Caraco, T. 1981. Risk-sensitivity and foraging groups. *Ecology* **62**:527–531.

Caraco, T., and H. R. Pulliam. 1980. Time budgets and flocking dynamics. *Proc. Int. Ornith. Congr.* **17**:807–812.

Caraco, T., S. Martindale, and H. R. Pulliam. 1980. Avian flocking in the presence of a predator. *Nature (London)* **285**:400–401.

Charnov, E. L., and J. R. Krebs. 1975. The evolution of alarm calls: Altruism or manipulation? *Am. Nat.* **109**:107–112.

Clutton-Brock, T. H., and P. H. Harvey. 1977. Primate ecology and social organization. *J. Zool. London* **183**:1–39.

Cox, D. R. 1962. *Renewal theory.* Methuen, London.

Crook, J. H. 1972. Sexual selection, dimorphism, and social organization in primates. In B. Campbell (ed.), *Sexual selection and the descent of man,* pp. 231–281. Aldine Press, Chicago.

Curio, E., U. Ernst, and W. Vieth. 1978. The adaptive significance of avian mobbing. *Z. Tierpsych.* **48**:184–202.

Dunford, C. 1977. Kin selection for ground squirrel alarm calls. *Am. Nat.* **111**:782–785.

Eisenberg, J. F., N. A. Mackenhirn, and R. Rudran. 1972. The relation between ecology and social structure in primates. *Science* **176**:863–874.

Emlen, S., and N. Demong. 1975. Adaptive significance of synchronized breeding in a colonial bird: a new hypothesis. *Science* **188**:1029–1031.

Estes, R. D. 1976. The significance of breeding synchrony in the wildebeest. *E. Afr. Wildl. J.* **14**:135–152.

Evans, R. M. 1982. Colony desertion and reproductive synchrony of black-billed gulls *Larus bulleri. Ibis* **124**:491–501.

Fagen, R. M. 1980. When doves conspire: evolution of nondamaging fighting tactics in a nonrandom encounter animal conflict model. *Am. Nat.* **115**:858–869.

Gould, L. L., and F. Heppner. 1974. The vee formation of Canada Geese. *Auk* **91**:494–506.

Grafen, A. 1979. The hawk-dove game played between relatives. *Anim. Behav.* **27**:905–907.

Haigh, J. 1975. Game theory and evolution. *Adv. Appl. Prob.* **7**:9–12.

Hamilton, W. D. 1971. Geometry for the selfish herd. *J. Theor. Biol.* **31**:295–311.

Higdon, J. J. L., and S. Corrsin. 1978. Induced drag of a bird flock. *Am. Nat.* **112**:727–744.

Hoogland, J. L. 1981. The evolution of coloniality in white-tailed and black-tailed prairie dogs (Sciuridae: *Cynomys leucurus* and *C. ludovicianus*). *Ecology* **62**:252–272.

Hoogland, J. L., and P. W. Sherman. 1976. Advantages and disadvantages of bank swallow (*Riporia riporis*) coloniality. *Ecol. Mon.* **46**:33–58.

Horn, H. S. 1968. The adaptive significance of colonial nesting in the Brewer's Blackbird (*Euphagus cyanocephalus*). *Ecology* **49**:682–694.

Inglis, I. R., and J. Lazarus. 1981. Vigilance and flock size in brent geese: The edge effect. *Z. Tierpsych.* **57**:193–200.

Jarman, P. J. 1974. The social organization of antelope in relation to their ecology. *Behaviour* **58**:215–267.

Kenward, R. E. 1978. Hawks and doves: Attack success and selection in goshawk flights at woodpigeons. *J. Anim. Ecol.* **47**:449–460.

Krebs, J. R. 1971. Territory and breeding density in the great tit, *Parus major. Ecology* **52**:2–22.

Krebs, J. R., and N. B. Davies. 1981. *An introduction to behavioural ecology.* Blackwell, Oxford.

Krebs, J. R., M. H. MacRoberts, and J. M. Cullen. 1972. Flocking and feeding in the Great Tit, *Parus major* L.—an experimental study. *Ibis* **114**:507–530.

Kruuk, H. 1964. Predators and anti-predator behaviour of the black-headed gull (*Larus ridibandus*). *Behav. Suppl.* **11**:1–129.

Kummer, H. 1968. *Social organization of hamadryas baboons: A field study.* Univ. Chicago Press, Chicago.

Lazarus, J. 1979. The early warning function of flocking in birds: an experimental study. *Anim. Behav.* **27**:855–865.

Lewontin, R. C. 1979. Fitness, survival, and optimality. In D. J. Horn, G. R. Stairs, and R. D. Mitchell (eds.), *Analysis of ecological systems*, pp. 3–22. Ohio State Univ. Press. Columbus.

Lissaman, P. B. S., and C. A. Shollenberger. 1970. Flight formation of birds. *Science* **168**:1003–1005.

Lloyd, M., and H. S. Dybas. 1966. The periodical cicada problem: II. Evolution. *Evolution* **20**:466–505.

Maynard Smith, J. 1964. Group selection and kin selection. *Nature (London)* **201**:1145–1147.

Maynard Smith, J. 1974. The theory of games and the evolution of animal conflicts. *J. Theor. Biol.* **47**:209–221.

Maynard Smith, J. 1978. Optimization theory in evolution. *Annu. Rev. Ecol. Syst.* **9**:31–56.

Maynard Smith, J. 1981. Will a sexual population evolve to an ESS? *Am. Nat.* **117**:1015–1018.

Maynard Smith, J., and G. A. Parker. 1976. The logic of asymmetric contests. *Anim. Behav.* **24**:159–175.

Milinski, M. 1979. Can an experienced predator overcome the confusion of swarming prey more easily? *Anim. Behav.* **27**:1122–1126.

Mirmirani, M., and G. Oster. 1978. Competition, kin selection and evolutionary stable strategies. *J. Theor. Biol.* **83**:489–501.

Moore, N. J. 1972. *The ethology of the Mexican junco (Junco phaeonotus palliatus).* Ph.D. thesis University of Arizona, Tucson.

Murton, R. K. 1971. Why do some bird species feed in flocks? *Ibis* **113**:534–536.

Oster, G., I. Eshel, and D. Cohen. 1977. Worker-queen conflict and the evolution of social insects. *Theor. Pop. Biol.* **12**:49–85.

Oster, G., and E. O. Wilson. 1978. *Caste and ecology in the social insects.* Princeton University Press, Princeton, New Jersey.

Owen, G. 1968. *Game theory.* Saunders, Philadelphia.

Page, G., and D. F. Whitacre. 1975. Raptor predation on wintering shorebirds. *Condor* **77**:73–83.

Partridge, B. L., and T. J. Pitcher. 1979. Evidence against a hydrodynamic function for fish schools. *Nature (London)* **279**:418–419.

Powell, G. V. N. 1974. Experimental analysis of the social value of flocking by starlings (*Sturnus vulgaris*) in relation to predation and foraging. *Anim. Behav.* **22**:501–505.

Pulliam, H. R. 1973. On the advantages of flocking. *J. Theor. Biol.* **38**:419–422.

Pulliam, H. R. 1976. The principle of optimal behavior and the theory of communities. In P. P. G. Bateson and P. H. Klopfer (eds.), *Perspectives in ethology*, pp. 311–333. Plenum Press, New York.

Pulliam, H. R., and T. Caraco. 1983. Living in groups: Is there an optimal group size? In J. R. Krebs and N. B. Davies (eds.), *Behavioural ecology: An evolutionary approach*, 2nd ed. Blackwell, Oxford (in press).

Pulliam, H. R., and G. C. Millikan. 1982. Social organization in the non-reproductive season. In D. S. Farner and J. R. King (eds.). *Avian biology*, Vol. 6, pp. 45–87. Academic Press, New York.

Pulliam, H. R., and G. S. Mills. 1977. The use of space by sparrows. *Ecology* **58**:1393–1399.

Pulliam, H. R., G. H. Pyke, and T. Caraco. 1982. The scanning behavior of juncos: A game-theoretical approach. *J. Theor. Biol.* **95**:89–103.

Riley, J. G. 1979. Evolutionary equilibrium strategies. *J. Theor. Biol.* **76**:109–123.

Schoener, T. W. 1971. Theory of feeding strategies. *Annu. Rev. Ecol. Syst.* **2**:369–404.

Sherman, P. 1977. Nepotism and the evolution of alarm calls. *Science* **197**:1246–1253.

Shields, W. M. 1980. Ground squirrel alarm calls: nepotism or parental care? *Am. Nat.* **116**:599–603.

Siegfried, W. R., and L. G. Underhill. 1975. Flocking as an antipredator strategy in doves. *Anim. Behav.* **23**:504–508.

Taylor, R. J. 1976. Value of clumping to prey and the evolutionary response of ambush predators. *Am. Nat.* **110**:13–29.

Taylor, R. J. 1979. The value of clumping to prey when detectability increases with group size. *Am. Nat.* **113**:299–301.

Tenaza, R. 1971. Behavior and nesting success relative to nest location in Adelie penguins (*Rigoscelis adeliae*). *Condor* **73**:81–92.

Thompson, W. A., J. Vertinsky, and J. R. Krebs. 1974. The survival value of flocking in birds: A simulation model. *J. Anim. Ecol.* **43**:785–820.

Tinbergen, N. 1951. *The study of instinct.* Clarendon Press, Oxford.

Tinbergen, N., M. Impekoven, and D. Franck. 1967. An experiment on spacing out as a defense against predators. *Behaviour* **28**:307–321.

Treisman, M. 1975. Predation and the evolution of gregariousness. I. Models for concealment and evasion. *Anim. Behav.* **23**:779–800.

Trivers, R. L. 1971. The evolution of reciprocal altruism. *Q. Rev. Biol.* **46**:35–57.

Vincent, T. L., and W. J. Grantham. 1981. *Optimality in parametric systems.* Wiley, New York.

Vine, I. 1971. Risk of visual detection and pursuit by a predator and the selective advantage of flocking behaviour. *J. Theor. Biol.* **30**:405–422.

Weihs, D. 1973. Hydromechanics of fish schooling. *Nature (London)* **241**:290–291.

Williams, G. C. 1966. *Adaptation and natural selection.* Princeton University Press, Princeton, New Jersey.

PART IV

Organization
of Communities

CHAPTER **11**

Density-Vague Ecology and Liberal Population Regulation in Insects

DONALD R. STRONG
Department of Biological Science
Florida State University
Tallahassee, Florida

CONTENTS

1 INTRODUCTION

This chapter draws attention to the high variance in many natural density relationships and makes a plea for rekindled interest by academic ecologists in empirical studies of population dynamics. The scatter around lines of best fit for the effects of density on population parameters is frequently great, and parameters that one might expect to vary with density often can be seen to do so only with some imagination, especially in natural populations; density vagueness is common. With the shift in focus by much of ecology to communities and ecosystems during the last several decades, many researchers do not work closely with real population change, and the diversity of influences other than density on population parameters may be underappreciated. However, such processes must be studied if the mechanisms involved with community organization are to be understood.

Of course, the interest of theoreticians in populations dynamics has remained high, with such popular topics as stability analysis (May, 1974) and food webs (Pimm, 1982); population models are the basis of many community models. The effects of density in many population models are deterministic; they are density explicit. Stochastic models can, in principle, accommodate variance in density relationships of population parameters and explore the implications of density vagueness on long-term dynamics.

I approach population dynamics as an empiricist who places stock in actual measurements of real organisms. Variance in density relationships and the absence of density effects are real and interesting and an important part of how nature works. Density vagueness, though inconvenient to some theory, is not merely noise or irrelevant detail that obscures to our eyes some ideal regulatory process. If density relationships for birth or death rates are vague, weak, inconsistent, highly discontinuous, or totally absent, then the potential for population regulation is much different from that of life histories with strictly consistent and explicit density effects.

One of the most fruitful empirical approaches to density vagueness is mechanistic and deterministic. It seeks to explain the variance in terms of variables other than density. For example, age, weather, and migration may have great effects on mortality rates independently of density, and birthrates can be greatly influenced by season, the condition of food, and many other variables that have no relationship to density. These effects are very obvious to ecologists who actually work with natural populations, but the variance produced by these additional variables is not often a part of general population and community models. Investigating density vagueness by means of additional variables is contingency analysis rather than stochastic modeling (Chesson, 1978). The stochastic modeler accepts variance and explores its implications, whereas the deterministic empiricist attempts to understand or explain the variance. Stochastic modeling and empiricism are not working at cross purposes but rather toward answers to complementary aspects of the same questions.

I am not making the argument of the classical density-independent school of population dynamics, and density vagueness is consistent with most widely accepted general evidence of population regulation in nature. This general evidence is the disparity between the dynamics of real populations and those of hypothetical populations that are not regulated at all. Hypothetical "random walks" climb to infinity and fall to extinction far too frequently to mimic the behavior of real populations (Cole, 1957; Slobodkin, 1976). It is important to realize that this general evidence does not necessarily imply the sort of explicit, consistent, and continuous regulation around a central set point or equilibrium that is assumed by much mathematical population theory. Regulation at extremes of density can involve very different mechanisms from regulation around central or average densities. Real populations that usually avoid extinction and always avoid infinity may have density-vague or even density-independent dynamics at central population sizes. Regulation in many real populations may be much more liberal than that assumed by equilibrium mathematics.

2 SOME DEFINITIONS

A brief historical catalogue of population terminology sets the stage for the notion of density vagueness (Table 1). Many terms combine the word "density," and even today different usage of these terms can cause substantial disagreement (Murray, 1982; Charlesworth, 1981; Hassell, 1981). Table 1 classifies population jargon two ways, in a–d by the kind of feedback that a density relationship imparts to a population (positive, negative, or mixed) and in I, II, and III by the amount of variance in density relationships. Parts I and II (density explicit and density vague) can have the various sorts of feedback in a–d. In density explicit relationships feedback is essentially deterministic. In density-vague relationships feedback is greatly affected by variance about the density trend. Milne (1957) discriminated between density relationships with high and low variance by means of the terms "imperfect" and "perfect." Unfortunately, the ecological literature has not paid much attention to this aspect of Milne's work.

Either positive or negative feedback can be generated from population parameters that are related to density (Table 1a–c). Negative feedback resists change in density by means of decreasing birth, survivorship, immigration, and/or net growth rate with increasing density and by the opposite changes in rates with decreasing density. This pattern is now usually called "density dependence," but note the variety of terms that have been applied in the past to parameters with an influence of negative feedback on population density. Positive feedback (= inverse density dependence) is the opposite of density dependence (Table 1, Ic). The Allee effect (Table 1, Ic') is an example of positive feedback in population parameters that takes place at the extreme low end of the density spectrum, when increasing sparseness impairs basic social and sexual functions of population. In very sparse populations, females might not find mates or groups might be too small to provide protections from predators. Thus, increased density could result in increased birth or survivorship rate.

Explicit density relationships do not have to be linear or even monotonic.

Table 1. A Brief Historical Catalogue of Terminology Used to Describe Density Relationships[a]

I. Density Explicit: "perfect" (Milne, 1957)
- a. Density Feedback, Either Positive or Negative "density related" (Solomon, 1949), "reactive" (Nicholson, 1954), "individualized" (Thompson, 1928).
- b. Negative Feedback in Density
  "density dependent" (Krebs, 1978; Smith, 1935[b]), "compensatory" (Neave, 1953[c]), "controlling" (Nicholson, 1933), "facultative" (Howard and Fiske, 1911), "density governing" (Nicholson, 1954), "concurrent" (Solomon, 1949), "negatively density dependent" (Haldane, 1953), directly density dependent (Allee, 1931), population "regulation," or "control" (many authors; Varley et al., 1973).
- c. Positive Feedback in Density
  "inversely density dependent" (Krebs, 1978; Allee, 1931), "inverse" (Howard and Fiske, 1911), "positively density dependent" (Haldane, 1953), "depensatory" (Neave, 1953).
- c′. "Allee effect" (Allee, 1931).
- d. Delayed Feedback in Density "time lag" (Varley, 1949; Royama, 1971; Oster, 1976; May, 1976).

II. Density Vague[d] "imperfect" (Milne, 1957)

III. Density Effects Absent or Invisible:
  "catastrophic" (Howard and Fiske, 1911), "density independent" (Smith, 1935[b]), "general" (Thompson, 1928), "nonreactive" (Nicholson, 1954), "extrapensatory" (Neave, 1953[c])

[a] Different kinds of feedback appear under "a" through "d." I, II, and III discriminate among relationships with low variance (density explicit), with high variance (density vague), and with so much variance that any trend is totally obscured (density effects absent or invisible).

[b] Smith (1935) meant "density dependent" to refer to negative feedback only (Solomon, 1958).

[c] Neave (1953) used his terms only in reference to mortality rates.

[d] The relationships a–d in Part I can all occur in a density-vague fashion as well.

One likely form of nonlinear density dependence is caused by time delay or time lag (Table 1, Id). Delays or time lags in the effects of density can cause convex functions of birth, survivorship, and/or net population growth with increasing density and concave functions of these parameters with decreasing density. One simple sort of nonmonotonic density relationship combines an Allee effect at low densities with negative feedback at high densities to yield a humped birth, survivorship, or net growth rate over a wide range of densities (Krebs, 1978). Of course, variance makes the details of nonlinear relationships much more difficult to discern for the empiricist.

3 DENSITY VAGUENESS

Density vagueness describes parameters of birth and death rate that are only weakly explained by density (Fig. 1), ranging from low but statistically significant correlations to relationships in which the influence of density can be inferred only with imagination. Thus, the borderlines of density vagueness are not sharp, and a broad spectrum of possible relationships fall into its category. Density-vague population parameters are most likely in natural situations at realistic densities, in which factors such as the weather, natural enemies, habitat heterogeneity, and phenology may have a variable influence on demographic rates. Density-explicit relationships are most likely at artificially high densities under controlled and experimentally homogeneous conditions, where the influences of variables that are normally important in nature are kept low. Density-explicit relationships are common in homogeneous, crowded laboratory cultures (e.g., Thomas et al., 1980; Mueller and Ayala, 1981).

The empirical literature on density relationships is fairly large. Insects contribute most studies, but vertebrates and invertebrates other than insects have been studied in this regard as well (Podoler and Rogers, 1975; Dempster, 1975). Many empirical population studies are based on repeated measurements of survivorship and reproduction, over a sequence of generations. Data from these repeated life table studies are often summarized with a "k factor analysis" (Varley and Gradwell, 1960). k refers to "killing power," and the analysis compares the relative population influence of mortalities that operate serially during the life history of an organism. Each distinct life history stage has a k value, which is calculated as the $\log_{10}$ difference between population density entering and leaving that stage. For example, the k value for second instar nymphs is the difference between log densities of first and third instar nymphs. k values can be calculated for ecological factors, such as predation, parasitism, or starvation, that are responsible for portions of mortality during a life history stage. Reproduction is taken into account various ways, as the stage between adults and newly born offspring. Overlapping generations make the calculation of k values more complicated, but the method can be used for iteroparous species (Southwood, 1978). More sophisticated versions of life table analysis deal with the variance in k values

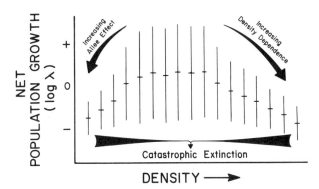

Figure 1.   Density vagueness in population parameters means that factors other than density have great influence.

within generations (e.g., Pielou, 1974) and the statistical interactions of mortality factors (Royama, 1981b).

Patterns of k values among generations show how various factors influence population dynamics. The "key factor" is usually defined as that responsible for the largest fraction of population change between generations. In practice, however, no single factor of life history stage may stand out as a key factor (e.g., McNeill, 1973), and correlation techniques used to identify key factors are not well founded statistically (Smith, 1973). The original conception of key factors, as density-dependent influences that were largely responsible for population dynamics because they operated against a background of constant, density-independent mortality (Morris, 1959), often does not apply. Density-independent mortality is not constant for many species, especially in natural populations, and the life history stage or factor that is responsible for the greatest fraction of change between generations is often not highly correlated with density, if it is correlated at all (Dempster, 1975; Varley et al., 1973).

The incidence of density vagueness is not readily apparent in the literature, mainly because many studies ignore variance (Stubbs, 1977) or treat density-explicit relationships exclusively (Bellows, 1981). The commonness of high variance in population functions can be crudely seen in the list of repeated life table studies in Podoler and Rogers (1975). All of these studies ran for at least five generations, and some ran for more than 15 generations. I have summarized some density patterns from the 26 of these studies that were done in natural populations (Table 2). Since most of the studies were not experimental, tendencies of density dependence usually mean an observed negative correlation between density and birth or survivorship rate. The incidence of inverse density dependence was not consistently recorded in Podoler and Rogers (1975).

The gross tally suggests that many vertebrate populations have at least one life history stage that the authors of the study interpret as density dependent, whereas a bit less than half of insect populations are reported as having one apparently density-dependent stage. In the second row of Table 2, the gross tally suggests that about half of the vertebrate stages and factors were interpreted as having some tendency of density dependence while about $\frac{1}{10}$ of insect stages and factors were so interpreted. The numerators in table

Table 2.   The Crude Incidence of Density Dependence as Discovered in Some Life History Studies[a]

	Insects	Vertebrates
Studies	8/20	5/6
Life history stages and factors	9/102	8/20

[a] Numerators are the incidences of density dependence and denominators are the total number of studies, or stages and factors, as reported in Podoler and Rogers (1975).

two include factors and stages that range from density explicit to density vague, and the denominators include factors that range from density vague through density independent and inversely density dependent.

The main lesson of the raw tallies in Table 2 is the high frequency of studies and stages that were not detected to be density dependent. Density vagueness is the gray area between explicit density dependence and density independence. Thus, a high incidence of apparently density-independent relationships indicates a large potential gray area. If most relationships were clearly density dependent, then density vagueness would be relatively uncommon.

Raw tallies such as those in Table 2 give only the most preliminary indications of the incidence of different density relationships because variance makes the relationships hard to discern. The denominators in the tallies may be underestimates of density dependence for at least two reasons. First, Podoler and Rogers (1975) were more concerned with finding key factors than with ferreting out vague density dependence. Second, the sample size of many of the studies is small (quite justifiably, given that the number of points is equal to the number of years in most cases!), and the power of correlation tests with low sample size is low. A closer look at the data and steadfast Malthusian faith might turn up a slightly larger fraction of density dependence. But the fact remains that the influence of density in many of the cases is at best weak or inconsistent.

The raw tallies in Table 2 might be biased in the opposite direction, toward "pseudo-density dependence," with more stages and factors appearing to be density dependent than really are. Royama (1977, 1981a) has studied this possibility in depth and has found at least three sources of pseudo-density dependence. First, the estimated correlation coefficients can be biased upward toward density dependence by small sample size and inaccurate measurement of density [this is part and parcel of the well-known problems with purely statistical detection of density relationships (Maelzer, 1970; St. Amant, 1970; Pielou, 1974; Slade, 1977)]. Second, correlations can be "spurious," a result of independent cause by a third factor rather than birth or death rate changed by density, "such as the correlation between human birth rate and the stork population of Holland (Yule, 1926)" (Royama, 1981a, p. 479). Third, apparent density correlations could be the false product of temporal autocorrelations.

4   SOME INDIVIDUAL CASE HISTORIES

A close look at individual studies tells us more about density vagueness. In Table 3, I have listed some highlights of density effects for the cinnabar moth with its host plant ragwort and with its natural enemies. This system is among the most studied of all insect–plant associations (Dempster, 1982), and similar measurements of demographic parameters made in different environ-

Table 3. Some Potentially Density Vague Population Functions for the Cinnabar Moth[a]

Organism	Parameter	Other Important Factors	Reference
Cinnabar moth on ragwort	(DD) caterpillar survivorship	1. Density of host plants (compare Wheeting and Monk's Wood Populations)	Dempster (1971, 1975)
		2. Food plant patchiness	van der Meijden (1979)
		3. Soil drainage	Dempster (1971, 1975)
		4. Nitrogen content of soil and host plants	Myers and Post (1981)
	(DI) pupal predation by rodents	5. Rodent's autecology	Dempster (1975)
	(INV DD) caterpillar parasitism by *Apanteles*	6. Nutritional state of caterpillar	van der Meijden (1980)
		7. Depressed dispersal of parasitoids at high host density	
		8. Hyperparasitism of *Apanteles*	

[a] DI, density independent; INV DD, inversely density dependent; DD, density dependent.

ments give a good idea of how variable and contingent these parameters can be. High environmental contingency is one cause of density vagueness.

From Table 3, we see that caterpillar survival for the cinnabar moth is density dependent at Wheeting, although there is sufficient variance in the relationship that 8 years of data are not enough to show the relationship with the Varley–Gradwell statistical test (Dempster, 1975, p. 71). Some of the variance comes from the discontinuity or threshold of starvation rate as a function of density. At low ratios of caterpillar to plant biomass, there is no appreciable starvation. Only in years of high ratios does defoliation occur, and defoliation years are starvation years. A host of factors in addition to density influence caterpillar survival. Some of the more important are the density and patchiness of host plants. This patchiness causes a lack of co-ordination in grazing rates at Wheeting, allowing a few individuals to mature before defoliation causes mass starvation and mortality. Density of host plants must be high for caterpillar densities to build to levels that defoliate host plants. The cinnabar moth population at Monk's Wood, where ragwort plants are sparser, did not defoliate its host plants and did not show the density dependence in caterpillar survivorship that was evident at Wheeting. Soil moisture greatly affects ragwort density in Britain, which in turn affects cinnabar moth dynamics. Also, the fluctuation of cinnabar moth densities is greater where plants have higher nitrogen content.

Mortality to cinnabar moth caterpillars from parasites and predators is normally inversely density dependent and density independent, respectively (Table 3). Although both sources of mortality can be very heavy (predation on pupae was recorded at over 99% in 1 year at Wheeting), they do not have a regulatory capacity for the population. Additional factors, other than density, have been best studied for caterpillar parasitism by *Apanteles*. The population density ratio of caterpillars to host plants greatly affects mortality rate from *Apanteles*. Starved caterpillars do not fledge many parasitoids. As well, the dispersal of parasitoids is greatly affected by the nutritional state of caterpillars. Starved caterpillars do not disperse as far as well fed ones, and at high caterpillar densities parasitism rates fall. Hyperparasitism also affects density relationships for mortality of cinnabar moth.

I do not believe that resource depletion, as shown by the defoliation of ragwort by some populations of cinnabar moth, is the norm for phytophagous insects. Examples of species that do not severely deplete their resources are common in the literature (e.g., Hayes, 1981; Strong, 1982; Lawton and Strong, 1981), but Dempster and Pollard (1981) have ideas opposite to mine on this subject. Suffice it to say that some insect species under some conditions do deplete resources. Other species do not deplete resources and do not suffer density dependent mortality from this source.

For mortality from natural enemies, an instructive study for understanding density vagueness is that of the biocontrol of the olive scale by two species of parasitoids (Huffaker and Kennett, 1966). The olive scale appeared in central California in the early 1930s and spread to a wide area within the next 20 years to become a major pest. Two parasitoid species were introduced. *Aphytis maculicornis*, which attacks in early spring, brought scale densities down in many areas. *Coccophagoides utilis* further suppressed some scale populations by causing mortality from late spring through early fall. The pair of parasitoids buffer biological control of the olive scale because they are adversely affected by different abiotic conditions. Bad years for one are not so for the other species. Although some interspecific competition occurs between the two parasitoids, their phenological differences are more than sufficient for separation from much mutual negative effect.

Even in this case of successful biological control, however, population parameters are density vague. Life table analysis did not show a density-dependent trend among generations, even though scale mortality caused by the parasitoids was judged to be the key factor. One reason that density effects are vague is that *A. maculicornis* is so efficient. It causes mortality as high as 90% even at quite low scale densities. This situation is analogous to that with starvation in cinnabar moth in the Wheeting population (Table 3). Both have discontinuous density effects, with a low threshold above which mortality is usually high and only vaguely related to density.

So, we have two sorts of apparently inconsistent evidence from the olive scale study. On the one hand, the biocontrol indicates density-dependent influence from the parasitoids. This is to say that parasitoids greatly reduced olive scale density but do not cause the scale to become extinct; at some low density the parasitoids lose their influence and cause no further density decrease in the scale. On the other hand, higher resolution with data comparing mortalities and densities between years does not reveal any distinct regulatory tendencies; "several sets of data suggest this relationship (density dependence), but they are all inconclusive, possibly because of the many interrelated and confusing variables involved" (Huffaker and Kennett, 1966, p. 326). Other studies of insect parasitoids that show density-vague and density-independent effects include Waloff and Thompson (1980), Morrison and Strong (1980), Stiling and Strong (1982), and Weis (1982).

A qualitative model of density relationships for the olive scale that includes variance is shown in Figure 2. The shaded horizontal stripe is a confidence band for birthrate of the scale, which apparently does not vary markedly with density (Huffaker and Kennett, 1966). The diagonal clouds of dots cover what might be confidence regions for mortality rates with and without the parasitoids. Intersection of birth and death confidence regions indicate the range of densities that olive scale would be limited to under the two conditions, where net growth rate of the population would average at zero over a series of generations. Within the overlap of birth and mortalty regions one would not expect to observe density-dependent change, because the model is not based on equilibrium or stability around some distinct carrying capacity. Any net regulation would only occur outside the regions of overlap. This is a model of liberal population regulation that occurs at density extremes. It sets indistinct boundaries for excursion but does not assume regulation within the boundaries.

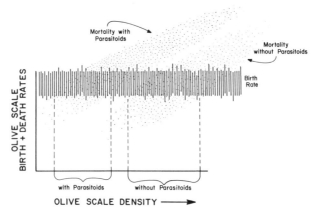

Figure 2. A liberal model of regulation for population density of the olive scale by its parasitoids. Density dependence is not expected within the overlap of confidence regions for birth and death rates. Only outside these regions are birth and death rate differences sufficiently consistent to regulate population density.

5   LIBERAL POPULATION REGULATION

Density-vague population dynamics is consistent with other ideas of liberal population regulation, the mathematical models of stochastic boundedness (Chesson, 1978; Murdoch, 1979) and the conceptual model of "envelopist" population dynamics (Ehrlich et al., 1972). All three of these ideas make a point of the variance in density relationships of population parameters and suggest that forces of population change are not tightly linked to density at intermediate densities. Liberal population regulation can be seen in a model such as Figure 3, which does not have precise bilateral attraction toward some central equilibrium density. I have drawn Figure 3 with density vagueness decreasing at extreme densities, where single factors can become so pronounced as to override the variable and compound influences of other factors acting on the population. The width and distinctness of the regions in the figure will vary with conditions and species. Some populations may have a vary narrow region of Allee effect; others may have a broader region of density dependence, and so on. At all densities there is some probability of catastrophic extinction, and this should be higher at the lowest densities, as illustrated by the width of the line at the base of the figure. There might also be an increased probability of catastrophic extinction at the highest densities; if very crowded populations were weakened by starvation, a second factor, such as bad weather or disease, could have a higher probability of wiping out the population entirely.

One major attraction of liberal models of regulation, of regulation away from extremes rather than around the more conventional and mathematically convenient set point or equilibrium, is that crucial features of the model are operational in terms of empirical field ecology. Average densities and confidence limits of densities can really be measured in the field. As put by Murdoch (1979, p. 306), "it is extremely difficult for a field ecologist to define or recognize a real population's equilibrium state, or even to determine if such a state exists, or to measure the stability properties of a real population," but it is possible to deal with fluctuations, variances and averages of densities.

I would like to end with what has become a refrain among many ecologists: experiments in natural populations provide much of the real promise for understanding populations and communities. The bounded stochasticity no-

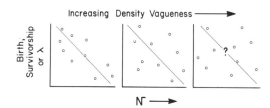

Figure 3.   A density vague model of population change.

tions and other realistic, operational, yet theoretical approaches to population dynamics (e.g., Bigger, 1973; Kareiva, 1982) give us a conceptual framework within which to work, and the wealth of descriptive studies of population dynamics that I have referred to gives us a solid biological base to design experiments from. Murdoch (1970) made an earlier plea for experiments with population dynamics that was, perhaps, ahead of its time.

6 SUMMARY

Demographic parameters of birth, survivorship, and net growth rate in natural populations have a great variety of relationships with density. Rates may increase, decrease, or vary in even more complex ways as the numbers and concentrations of individuals increase in populations; the simple Malthusian assumption of dominant and monotonic density dependence is probably rare for many organisms. In virtually all natural circumstances, density relationships have associated variance. For insects, the variance is normally high, and the influence of density on population parameters is often weak, intermittent, or discontinuous, even though it may be statistically significant. As a result, the cybernetic or regulatory influences acting within and among populations are not as clear as assumed by theory that ignores such variance. Density vague relationships, albeit messy, are the basis of communities in nature. Theories such as "bounded stochasticity" (Chesson, 1978), which take density vagueness seriously, are the most promising avenue to a valid general understanding of the dynamics and integration of natural communities.

ACKNOWLEDGMENTS

My ideas on these subjects have been greatly influenced by Gerold Morrison, Daniel Simberloff, Joel Trexler, William Murdoch, and especially by Joseph Travis.

LITERATURE CITED

Allee, W. C. 1931. *Animal aggregations: A study in sociology*. University of Chicago Press, Chicago.

Bellows, T. S. 1981. The decriptive properties of some models for density dependence. *J. Anim. Ecol.* **50**:139–156.

Bigger, M. 1973. An investigation into the interaction between coffee leaf-miners and their larval parasites. *J. Anim. Ecol.* **42**:417–434.

Charlesworth, B. 1981. Ecological models (Review of Murray). *Ecology* **64**:1401–1402.

Chesson, P. 1978. Predator-prey theory and variability. *Annu. Rev. Ecol. Syst.* **9**:323–347.

Cole, L. C. 1957. Sketches of general and comparative demography. In *Cold Spring Harbor Symp. Quant. Biol.* **22**:1–16.

Dempster, J. P. 1971. The population ecology of the cinnabar moth, *Tyria jacobaeae* L. (Lepidoptera, Arctiidae). *Oecologia* **7**:26–67.

Dempster, J. P. 1975. *Animal population ecology.* Academic Press, New York.

Dempster, J. P. 1982. The ecology of the cinnabar moth, *Tyria jacobaeae* L. (Lepidoptera: Arctiidae). *Adv. Ecol. Res.* **12**:1–36.

Dempster, J. P., and E. Pollard. 1981. Fluctuations in resource availability and insect populations. *Oecologia* **50**:412–416.

Ehrlich, P. P., D. E. Breedlove, P. F. Brussard, and M. A. Sharp. 1972. Weather and the "regulation" of subalpine populations. *Ecology* **53**:243–247.

Haldane, J. B. S. 1953. *New Biology*, Vol. 15:9. Penguin, London.

Hassell, M. P. 1981. Review of Murray. *J. Anim. Ecol.* **50**:642–643.

Hayes, J. L. 1981. The population ecology of a natural population of the pierid butterfly *Colias alexandra. Oecologia* **49**:188–200.

Howard, L. O., and W. F. Fiske. 1911. The importation into the United States of the parasites of the gipsy-moth and the brown-tail moth. *U.S. Dept. Agric., Bur. Entomol., Bull.* 91. (Reprint edition, 1977. Arno Press, NY, 344 pp.)

Huffaker, C. B., and C. E. Kennett. 1966. The biological control of *Parlartoria oleae* (Colvee) through the compensatory action of two introduced parasites. *Hilgardia* **37**:283–334.

Kareiva, P. 1982. Experimental and mathematical analyses of herbivore movement: Quantifying the influence of plant species and quality on foraging discrimination. *Ecol. Monog.* **52**:261–282.

Krebs, C. J. 1978. *Ecology: The experimental analysis of distribution and abundance*, 2nd ed. Harper & Row, New York.

Lawton, J. H., and D. R. Strong. 1981. Community patterns and competition in folivorous insects. *Am. Nat.* **118**:317–338.

McNeill, S. 1973. The dynamics of a population of *Leptoterna dolabrata* (Heteroptera: Miridae) in relation to its food resources. *J. Anim. Ecol.* **42**:495–507.

Maelzer, D. A. 1970. The regression of log N_{n+1} on log N as a test of density dependence: An exercise with computer-constructed density-independent populations. *Ecology* **51**:810–822.

May, R. M. 1974. *Stability and complexity in model ecosystems.* Princeton University Press, Princeton, New Jersey.

May, R. M. 1976. Models for two interacting populations. In R. M. May (ed), *Theoretical ecology.* Blackwell, Oxford.

Milne, A. 1957. The natural control of insect populations. *Can. Entomol.* **89**:193–213.

Morris, R. F. 1959. Single factor analysis in population dynamics. *Ecology* **40**:580–588.

Morrison, G., and D. R. Strong. 1980. Spatial variations in host density and the intensity of parasitism: Some empirical examples. *Environ. Entomol.* **9**:149–152.

Mueller, L. D., and F. J. Ayala. 1981. Dynamics of single-species population growth: Stability or chaos? *Ecology* **62**:1148–1154.

Murdoch, W. W. 1970. Population regulation and population inertia. *Ecology* **51**:497–502.

Murdoch, W. W. 1979. Predation and the dynamics of prey populations. *Fortschr. Zool.* **25**:295–310.

Murray, B. G. 1982. On the meaning of density dependence. *Oecologia* **53**:370–373.

Myers, J. H., and B. J. Post. 1981. Plant nitrogen and fluctuations of insect populations: A test with the cinnabar moth-tansy ragwort system. *Oecologia* **48**:151–156.

Neave, F. 1953. Principles affecting the size of pink and chum salmon populations in British Columbia. *J. Fish. Res. Bd. Can.* **9**:450–491.

Nicholson, A. J. 1933. The balance of animal populations. *J. Anim. Ecol.* **2**:132–178.

Nicholson, A. J. 1954. An outline of the dynamics of animal populations. *Aust. J. Zool.* **2**:9–65.

Oster, G. 1976. Internal variables in population dynamics. In S. A. Levin (ed.), *Lectures on mathematics in the life sciences*, Vol. 8, pp. 37–68. American Mathematical Society, Providence, Rhode Island.

Pielou, E. C. 1974. *Population and community ecology: Principles and methods.* Gordon and Breach, New York.

Pimm, S. L. 1982. *Food webs.* Chapman and Hall, London.

Podoler, H., and D. Rogers. 1975. A new method for the identification of key factors from life-table data. *J. Anim. Ecol.* **44**:85–114.

Royama, T. 1971. A comparative study of models of predation and parasitism. *Res. Pop. Ecol. Suppl.* **1**:1–91.

Royama, T. 1977. Population persistence and density dependence. *Ecol. Monogr.* **47**:1–35.

Royama, T. 1981a. Fundamental concepts and methodology for the analysis of animal population dynamics, with particular reference to univoltine species. *Ecol. Monogr.* **51**:473–493.

Royama, T. 1981b. Evaluation of mortality factors in insect life table analysis. *Ecol. Monogr.* **51**:495–505.

St. Amant, J. L. S. 1970. The detection of regulation in animal populations. *Ecology* **51**:823–828.

Slade, N. A. 1977. Statistical detection of density dependence from a series of sequential censuses. *Ecology* **58**:1094–1102.

Slobodkin, L. B. 1976. Comments from a biologist to a mathematician. In S. A. Levin (ed.), *Ecosystem analysis and prediction*, pp. 318–329. SIAM SIMS, Alta, Utah.

Smith, H. S. 1935. The role of biotic factors in the determination of population densities. *J. Econ. Entomol.* **28**:873–898.

Smith, R. H. 1973. The analysis of intra-generation change in animal populations. *J. Anim. Ecol.* **42**:611–622.

Solomon, M. E. 1949. The natural control of animal populations. *J. Anim. Ecol.* **18**:1–35.

Solomon, M. E. 1958. Measuring of density-dependence and related terms in population dynamics. *Nature (London)* **181**:1778–1780.

Southwood, T. R. E. 1978. *Ecological methods.* Chapman and Hall, London.

Stiling, P. D., and D. R. Strong. 1982. Egg density and the intensity of parasitism in *Prokelisia marginata* (Homoptera: Delphacidae). *Ecology* **63**:1630–1635.

Strong, D. R. 1983. Exorcising the ghost of competition past from insect communities. In D. R. Strong, L. G. Abele, D. Simberloff and A. B. Thistle (eds.), *Ecological communities: Conceptual issues and the evidence.* Princeton University Press, Princeton, New Jersey (in press).

Stubbs, M. 1977. Density dependence in the life-cycles of animals and its importance in K and r strategies. *J. Anim. Ecol.* **46**:667–688.

Thomas, W. R., M. J. Pomerantz, and M. E. Gilpin. 1980. Chaos, asymmetric growth and group selection for dynamical stability. *Ecology* **61**:1312–1320.

Thompson, W. R. 1928. A contribution to the study of biological control and parasitic introduction in continental areas. *Parasitology* **20**:90–112.

van der Meijden, E. 1979. Herbivore exploitation of a fugitive plant species: Local survival and extinction of the cinnabar moth and ragwort in a heterogeneous environment. *Oecologia* **42**:307–323.

van der Meijen, E. 1980. Can parasitoids escape from their parasitoids? The effects of food shortage on the braconid parasitoid *Apanteles popularis* on its host *Tyria jacobaeae*. *Neth. J. Zool.* **30**:382–392.

Varley, G. C. 1949. Population changes in German forest pests. *J. Anim. Ecol.* **18**:117–122.

Varley, G. C., and G. R. Gradwell. 1960. Key factors in population studies. *J. Anim. Ecol.* **29**:399–401.

Varley, G. C., G. R. Gradwell, and M. P. Hassell. 1973. *Insect population ecology: An analytical approach.* Blackwell, Oxford.

Waloff, N., and P. Thompson. 1980. Census data of populations of some leafhoppers (Auchenorrhyncha, Homoptera) of acidic grassland. *J. Anim. Ecol.* **49**:395–416.

Weis, A. 1982. Resource utilization patterns in a community of gall attacking parasitoids. *Environ. Entomol* **11**:809–815.

Yule, U. 1926. Why do we sometimes get nonsense-correlations between time-series? *J. R. Stat. Soc.* **89**:1–64.

CHAPTER 12

Herbivore Community Organization: General Models and Specific Tests with Phytophagous Insects

J. H. LAWTON
Department of Biology
University of York
Heslington, York, England

CONTENTS

---

1   INTRODUCTION

"Herbivore" is a general term for a bewildering variety of animals, as different as ungulates, slugs, tortoises, and *Daphnia*. Accordingly, it is neither sensible nor practicable, in one short chapter, to consider "herbivore communities" in general. To prevent discussion from becoming hopelessly diffuse I propose to focus on communities of one of the most important groups of herbivores, namely, phytophagous insects, defined (again to reduce the problem to manageable proportions) as insects that exploit the living tissues of higher land plants.

So defined, phytophagous insects are virtually confined to nine extant orders: Coleoptera, Collembola, Diptera, Hemiptera, Hymenoptera, Lepidoptera, Orthoptera, Phasmida, and Thysanoptera. These are not an obscure subset of the world's biota. At least one-third of a million species of phytophagous insects have been described (Strong et al., 1983). Contrast this with a paltry 8500 species of birds and 4500 species of mammals. Herbivorous ungulate mammals muster little more than 200 species.

Aspects of the community organization of phytophagous insects have recently been reviewed by several authors (e.g., Denno et al., 1981; Gilbert, 1979; Gilbert and Smiley, 1978; Lawton, 1978a; Lawton and Strong, 1981; Price, 1980, 1983; Root, 1973; Southwood, 1978; Strong et al., 1983). What are required now are detailed studies of a range of systems, designed to disentangle the many possibilities that can, in theory, determine their organization. As a contribution to such an endeavor, this chapter focuses on one system, the insect community on bracken fern *Pteridium aquilinum*.

The chapter is organized as follows. First, I describe the host plant and its associated insects, and in so doing touch on the problem of "coevolution" in phytophagous insect communities. Then, I describe five patterns in the organization of bracken–herbivore systems that imply determinism in their structure; in other words, there is some organization worth explaining. Finally, I look at the possible mechanisms that might contribute to these patterns, and describe experiments, most of them as yet unfinished, to test which are important in this particular system, and which are not.

2 BRACKEN AND ITS HERBIVOROUS INSECTS

Bracken grows naturally on all continents except Antarctica, and may well be one of the five commonest plants in the world (Harper, 1977). Contrary to popular opinion, it supports a varied fauna of herbivorous insects (Auerbach and Hendrix, 1980; Kirk, 1977, 1980; Lawton, 1976, 1978a,b, 1982; Lawton and Eastop, 1975; Rigby and Lawton, 1981; Smith and Lawton, 1980). The British fauna is best known and provides the focus for this chapter.

Thirty-five species of insects definitely, probably or possibly feed on the aboveground parts of the plant in Britain, of which the most important are a "core group" of 27 species. [These figures are from Lawton (1982), and update earlier summaries in Lawton (1976). The eight species not included in the core group either each have less than five records on the host in 10 years collecting, or are recorded as feeding on bracken in the literature but I have never collected them myself.] The 27 core species are drawn from five of the nine orders listed above (Table 1); Coleoptera, in particular, are conspicuous by their absence. Table 1 also summarizes the feeding specificities of these insects, and those of their close relatives.

Several things are noteworthy about the data. First, the British species are an almost equal mixture of strict monophages (8 or 9 species), oligophages (9 or 10 species) and polyphages (9 species). Second, these species are of mixed evolutionary pedigree. Only the Selandriinae sawflies as a group are primarily associated with ferns (Benson, 1952). *Chirosia* Diptera (a relatively primitive group of Anthomyiidae; Collins, 1955) have also radiated among this group of plants. Comparable close associations between particular groups of insects and plants are provided by cynipid gall wasps on *Rosa* and *Quercus* (e.g., Cornell and Washburn, 1979), heliconiine butterflies and the Passifloraceae (Benson et al., 1976; Gilbert and Smiley, 1978), and the "*machaon*" complex within the butterfly genus *Papilio* on Umbelliferae (Berenbaum, 1981; Slansky, 1973). However, there is nothing unusual about the hodgepodge of other taxa that constitutes the majority of species recruited onto bracken over evolutionary time. Studies of many insects reveal similar, idiosynchratic patterns of host use. Examples will be found among the sawflies (Benson, 1950), macrolepidoptera (Holloway and Hebert, 1979), microlepidoptera (Powell, 1980), Lycaenidae (Atsatt, 1981), and aphids and psyllids (Eastop, 1979; Hille Ris Lambers, 1979). In brief, the herbivorous insect fauna of most host plants, including bracken, is a potpourri of long-standing evolutionary associations, isolated and apparently opportunistic colonizations, and every shade in between, in varied and at present unpredictable proportions.

"Coevolution" is a popular theme in studies of herbivore community organization. However, I am not at all clear what the differences ought to be between coevolved communities, and other communities, in such things as species richness, species relative abundancies, and so on (cf. Kuris et

Table 1. Feeding Habits of the "Core Species" of Insects Found on Bracken in Britain[a]

Species	Feeding Specificity			Feeding Behavior of Relatives	
	Monophagous Feeds only on bracken	*Oligophagous* Feeds on bracken and other ferns	*Polyphagous* Feeds on angiosperms as well as on bracken	Are other British species in same genus mainly bracken and fern specialists?	Predominant host plants of species in same family
Collembola					
Bourletiella viridescens			√	No	Sminthuridae Angiosperms
Diptera					
Chirosia albifrons	√ (probably)			Yes	Anthomyiidae Very varied, including fungi, angiosperms, and seaweed
C. albitarsis	√ (probably)			Yes	
C. histricina	√ (probably)			Yes	
C. parvicornis		√		Yes	
Dasineura filicina	√ (probably)			No	Cecidomyiidae Angiosperms
D. pteridicola		√ (probably)		No	Cecidomyiidae Angiosperms
Phytoliriomyza hilarella = *Pteridomyza hilarella*		√		Rarely, or placed in monospecific genus	Agromyzidae Angiosperms
Phytoliriomyza pteridii	√			Rarely (one species above), or no	Agromyzidae Angiosperms
Hemiptera					
Ditropis pteridis	√			Monospecific genus	Delphacidae Angiosperms

332

Taxon				
Macrosiphum ptericolens	√ (probably) (overwintering host unknown)	√ (possibly)	Not in Britain	Aphididae Angiosperms
Monalocoris filicis		√	Monospecific genus	Miridae Angiosperms
Philaenus spumarius	√		Monospecific genus	Cercopidae Angiosperms
Hymenoptera				
Aneugmenus padi	√		Yes	Tenthredinidae Angiosperms
A. fürstenbergensis	√		Yes	Members of subfamily Selandriinae (including *Aneugmenus*, *Stromboceros*, and *Stronglyogaster*) are mainly on ferns
A. temporalis	√		Yes	
Stromboceros delicatulus	√		Monospecific genus	
Strongylogaster lineata	√		Yes	
Tenthredo sp. and *T. ferruginea*	√		No	
Lepidoptera				
Ceramica pisi	√		Monospecific genus	Noctuidae Angiosperms
Euplexia lucipara	√		Monospecific genus	
Lacanobia oleracea	√		No	
Phlogophera meticulosa	√		Monospecific genus	
Petrophora chlorosata		√	Monospecific genus	Geometridae Angiosperms
Olethreutes lacunana	√		No	Tortricidae Angiosperms
Paltodora cytisella		√	Monospecific genus	Gelechiidae Angiosperms

[a] Subterranean feeders, and those of uncertain status (see Lawton, 1982, for details) have been excluded from the analysis.

al., 1980). Moreover, I share with Fox (1981) the view that species-rich communities, such as those on bracken, are unlikely to be coevolved in the sense originally envisaged by Ehrlich and Raven (1964). There are too many species for any one herbivore to have had an obvious, overriding impact on the evolution of the plant, and too many of the insects (or their close relatives) feed on host plants other than bracken, for an intimate association with this one host to have been of paramount importance in organizing the community over evolutionary time. This is not to say that the plant has not evolved chemical and other defenses against herbivory; it has (Jones, 1983; Lawton, 1976). Or that insects are not adapted in varying degrees to their host(s); they are. But mundane evolution does not automatically constitute coevolution.

I think contemporary patterns in herbivore community structure are best understood by ignoring coevolution as an explanation until the data absolutely force us to do otherwise. Even then we ought to think very hard about how coevolved systems might differ, if at all, in their organization from the majority of communities, or parts of communities, that are not conspicuously coevolved.

3   PATTERNS IN CONTEMPORARY COMMUNITIES

The fact that species making up communities of insects on bracken are of mixed evolutionary pedigree does not imply that contemporary communities have no structure or pattern. Chance may have played a large part in determining which species colonized the plant over evolutionary time; nonetheless, the communities so formed are not random and unpredictable in their structure. In this section, I summarize five patterns shown by bracken herbivore communities. Gathering the patterns together in this way does not necessarily imply that they are all produced by the same mechanism(s).

3.1   Species Composition and Relative Abundance

Except for one year (1979) I have sampled a 2600 m^2 patch of bracken at Skipwith Common, Yorkshire, every year since 1972. The site is described fully in Lawton (1982). The phytophagous insects at this site are a very predictable assemblage (Table 2).

In several (unfortunately not all) years I not only established species' presence and absence, but also sampled carefully enough to estimate mean maximum abundances per frond. On average, some of the species at this site are much commoner than others (e.g., Lawton and McNeill, 1979; Caughley and Lawton, 1981). How stable is this species abundance distribution? The data are summarized in Table 3 and Fig. 1. With three exceptions, species rank abundancies in one year are significantly correlated with their rank abundancies in all other years. Moreover, there is no indication that the correlation between years gets worse with the passage of time (Fig.

Table 2. Annual Occurrence of Species at Skipwith[a]

Species	Open Site									Woodland Site	
	1972	1973	1974	1975	1976	1977	1978	1980	1981	1980	1981
Bourletiella viridescens	√	√	√	√	?	√	√	√	√	0	√
Chirosia albifrons	√	?	√	√	?	√	√	√	√	√	0
C. histricina	√	√	√	√	?	√	√	√	√	√	√
C. parvicornis	√	√	√	√	?	√	√	√	√	√	√
Dasineura filicina	√	√	√	√	?	√	√	√	√	√	√
D. pteridicola	√	√	√	√	?	√	√	√	√	√	0
Phytoliriomyza hilarella	0	?	0	√	?	√	0	0	0	0	0
Ditropis pteridis	√	√	√	√	√	√	√	√	√	√	√
Macrosiphum pteicolens	√	√	√	√	√	√	0	√	√	√	√
Monalocoris filicis	0	√	0	0	√	√	√	√	√	0	√
Philaenus spumarius	√	?	√	√	?	0	0	√	√	√	√
Aneugmenus spp. (two species)	√	?	√	√	√	√	√	√	√	√	√
Stromboceros delicatulus	√	√	√	√	?	0	0	√	√	0	0
Strongylogaster lineata	√	?	√	√	√	√	√	√	√	√	√
Tenthredo spp. (two species not distinguished until 1980)	√	?	√	√	?	√	√	√	√	√	√
Euplexia leucipara	0	?	√	0	?	0	0	√	0	0	0
Olethreutes lacunana	0	?	√	√	√	√	√	√	√	√	√
Paltodora cytisella	√	√	√	√	?	√	√	√	√	√	√
Petrophora chlorosata	√	?	0	0	?	0	0	√	√	√	√
Phlogophora meticulosa	0	?	0	0	?	0	0	√	√	0	0
Number of taxa recorded each year[b]	15	(10)	16	16	(6)	15	13	19	18	14	14

[a] Woodland sampling was not started until 1980. The open site was not sampled in 1979. In 1973 and 1976 sampling was not intensive, so failure to record a species (indicated by ?) is probably not indicative of absence. In all years, a positive record is indicated by √; failure to find a species despite sampling by 0. *Aneugmenus* caterpillars (known from adult records to belong to two species) cannot be distinguished. Caterpillars of the two species of *Tenthredo* were only distinguished in 1980. (A zero does not necessarily imply extinction, and should not be construed as such.)

[b] *Aneugmenus* spp. throughout, and *Tenthredo* spp. until 1978, counted as one each.

Table 3. Mean Maximum Numbers per Frond for Feeding Stages of Skipwith Species in Each of 6 Years[a]

Species	1972	1974	1975	1977	1980	1981
Bourletiella	4.79	5.22	3.43	3.30	3.19	3.05
Chirosia albifrons	0.01	0.085	0.06	0.05	0.11	0.01
C. histricina	1.20	0.47	0.30	10.00	17.81	8.20
C. parvicornis	0.22	3.88	2.26	8.00	9.20	7.25
Dasineura filicina	14.73	5.80	6.83	7.15	25.07	11.35
D. pteridicola	7.30	1.64	3.70	6.45	16.67	10.50
Phytoliriomyza	0	0	0.03	0.05	0	0
Ditropis	12.10	1.21	6.25	8.03	1.17	5.30
Macrosiphum	133.4	0.84	0.73	0	1.44	19.9
Monalocoris	5.40	0.17	14.23	0.03	0.20	0.60
Philaenus	0	0	0	0	0.01	0
Sawflies (all species)	1.04	0.78	0.48	0.40	0.70	0.50
Euplexia	0	0	0	0	0.037	0
Olethreutes	0	0.01	0.03	0.067	0.015	<0.01
Paltodora	0.08	0.07	0.33	0.10	0.01	0.03
Petrophora	0.05	0.04	0.01	0.05	1.00	0.95
Phlogophora	0	0	0	0	0	<0.01

[a] Sawfly caterpillars have been combined because abundance (rather than presence and absence—Table 2) was not determined for each species separately until 1980.

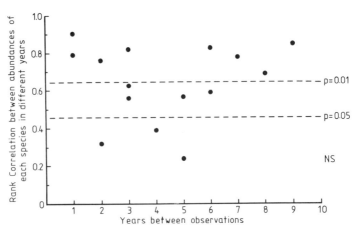

Figure 1. Spearman rank correlation coefficients between species abundances at Skipwith for all possible pairs of years, taken from Table 3. The maximum number of years between observations is 9 (1972 with 1981), the minimum 1 (1974 with 1975 and 1980 with 1981). The three nonsignificant correlations all involve 1977 (see text). Three species that have each only been recorded in 1 year (*Philaenus, Euplexia,* and *Phlogophora*) were omitted from the calculations to minimize tied ranks, and to prevent them giving undue weight to the analysis.

1). In other words, there is no evidence of long-term changes in the relative abundance of species. I conclude that this community has a predictable structure that has not changed significantly over 9 years.

As an aside, note that all three nonsignificant comparisons in Fig. 1 involve 1977 (1977 with 1981, 1977 with 1975, and 1977 with 1972). I have no idea why. In other words, in most years this community has a reasonably predictable structure, but it is not absolutely predictable.

3.2 Species–Area Relationships

The Skipwith study site is a large bracken patch. Smaller patches on the North York Moors support fewer species (Rigby and Lawton, 1981). This species–area relationship is generated by most species becoming progressively rarer *per frond* on smaller patches. The rate of decline in average abundance per frond with patch size is different for the different species; hence, each declines to zero at a characteristic, average size of island. The result is a species (S)–area (A) relationship of the form:

$$S = 1.84 \log_{10} A + 5.77$$

with area of the bracken patch measured in square meters.

3.3 Altitudinal Gradients

Data from the North York Moors (J. H. Lawton, M. MacGarvin, and P. A. Heads, unpublished) show that the diversity of bracken herbivores, and individual species abundancies, are also strongly influenced by altitude (Figs. 2 and 3). Again, the patterns are very clear.

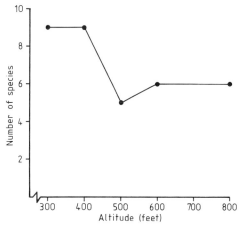

Figure 2. The number of species feeding on bracken recorded on between 36 and 38 fronds at five altitudes on the North York Moors. Samples at 300–600 feet were all taken in the same huge patch of bracken; the sample at 800 feet was in an isolated patch of over 1000 m^2 (8 June 1982).

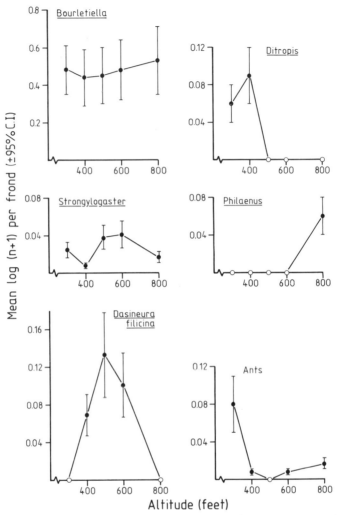

Figure 3. Mean numbers per frond [as $\log_{10}(n + 1)$] ± 95% CI, for representative species along the same altitudinal transect as Figure 2.

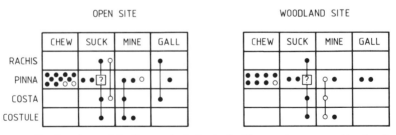

Figure 4. Feeding sites and feeding methods for the herbivorous insects attacking bracken in the open and in the woodland at Skipwith. Feeding sites of species exploiting more than one part of the frond are joined by lines. Open circles are for species recorded in 50%, or less, of the years of study (after Lawton, 1982). Rachis, main stem; pinna, "leaves"; costa, main stalks of pinnae, arising from rachis; costules, main "leaf veins" of pinnae).

3.4   Open Ground to Woodland

A second, less striking distributional gradient is that between bracken growing in the open, and under trees, at Skipwith. Six species that occur on the
open site have never been recorded feeding on the plant just a few meters
away under trees (Lawton, 1982) (Table 2). Particularly noticeable by their
absence are *Paltodora* and *Strongylogaster*. Examples of other changes in
abundance across this gradient are given in Table 4. Similar patterns have
been documented for bracken insects on a transect from open ground to
woodland at Imperial College Field Station on the outskirts of London (V.

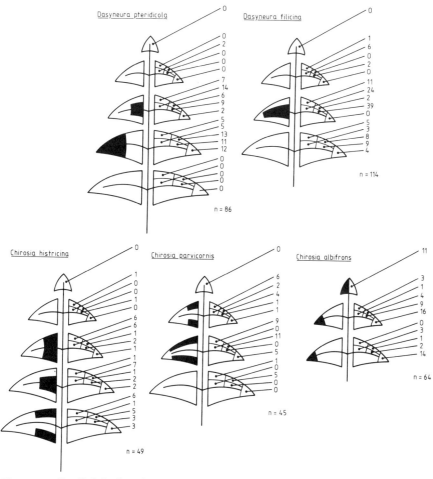

Figure 5.   Detailed feeding sites of Diptera in two genera exploiting bracken at Skipwith.
Fronds are shown diagramatically, with increasing numbers of pinnae as the season progresses.
The total number of records for each species at each position on the frond is shown; the shaded
areas shows those parts of the frond that are most heavily used (50% of records). Species differ
from close relatives in their preferred feeding sites.

Table 4. Abundance of Representative Species on a Transect Approximately 25 m Long, from Bracken Growing in the Open, to Bracken Growing under Trees at Skipwith (May and June 1982)[a]

Species and Date	Abundance Category	Number of Fronds in Each Category			χ^2 under Null Hypothesis That Species Abundance Is Independent of Position on Transect
		Wood	Edge	Open	
Tenthredo ferruginea egg-masses 14 July 1982	Fronds with 0	14	14	15	$\chi^2_2 = 0.16$
	Fronds with > 0	6	6	6	N.S.
Monalocoris adults 26 May 1982	Fronds with 0	155	225	47	$\chi^2_2 = 18.25$
	Fronds with > 0	26	10	1	$P < 0.001$
Dasineura filicina galls 14 July 1982	Fronds with 0–9	8	11	18	$\chi^2_4 = 18.21$
	Fronds with 10–19	9	6	2	$P \simeq 0.001$
	Fronds with > 20	3	3	0	
Ditropis nymphs 26 May 1982	Fronds with 0	179	196	29	$\chi^2_4 = 60.30$
	Fronds with 1	2	26	7	$P < 0.001$
	Fronds with > 1	0	13	12	

[a] Because most fronds contain no individuals of most species, analysis is based on the proportion of fronds with more than, or less than, a certain number of individuals, rather than on means. *Tenthredo* sawflies are equally abundant in the wood and in the open. *Monalocoris* and *Dasineura filicina* are commoner in the woodland than in the open. *Ditropis* is more common in the open than in woodland. (J. H. Lawton, M. MacGarvin, and P. A. Heads, unpublished data.)

K. Brown and G. McGavin, personal communication) and for bracken insects in New Mexico (Lawton, 1982).

3.5 Feeding Niches

Patterns of a qualitatively quite different nature are generated by the feeding niches of each species on this host. Patterns of niche differentiation are summarized in Figures 4 and 5. On present evidence, each species' method and position of feeding on the frond is relatively stereotyped, and insensitive to the presence or absence of other species.

4 COMPARISONS WITH OTHER STUDIES

Patterns like these are commonplace in community ecology (e.g., Cody and Diamond, 1975). However, the data do have some features that are worth discussing; briefly they are these.

I know of only one other study (Grossman, 1982) that tests whether species relative abundancies are stable over time using the recipe summarized in Figure 1, although Grossman's study spanned only some two years, not nine, as here. Even nine years is not very long. Of course, to many field naturalists, such statistical tests are an irritating irrelevance; they know that in many habitats some species are almost always rare, and others almost always common. Nonetheless, the problem is not as simple as this, because there are types of communities where the relative abundances of species change markedly over time, for example, successional systems (Connell and Slatyer, 1977; Horn, 1975) and other "nonequilibrium" assemblages (Connell, 1979; Price, 1980; Washburn and Cornell, 1981). Hence, it would be useful to have data of the type summarized in Figure 1 for a variety of communities, from habitats varying in their spatial and temporal predictability (Southwood, 1977). Bracken is a long-lived plant, with individual clones that persist for hundreds of years (Oinonen, 1967). We must not expect data from bracken to be typical of phytophagous insect communities associated with ephemeral weeds.

Year to year stability in relative species abundancies aside, the other patterns documented here have all been reported in at least one other phythagous insect community, and often in several.

Species–area relationships on patches or clumps of plants of various sizes are well documented (see Lawton, 1978a; MacGarvin, 1982; Rey, 1981; Strong, 1979). Altitudinal gradients have been less well studied, but examples will be found in Claridge and Singhrao (1978) and Herbert (1980). (In both cases, unlike the bracken data, food plants also changed with altitude.) In contrast, numerous papers yield results similar to the woodland–open ground comparison by showing that an insect may use a host plant in some habitats but fail to exploit it in others. Examples include Bale (1981), Benson

(1978), Dempster and Hall (1980), Heath (1959), Robertson (1981), and Wiklund (1974). Finally, many phytophages sharing the same host show classical "niche differentiation" (Askew, 1962; Benson, 1978; Nielsen, 1978; Root and Chaplin, 1976; Skuhravy, 1978; Waloff, 1979; and many others).

It is easy to believe that each of these patterns, and others like them (Lawton and Strong 1981), are generated in whole or in significant part by interspecific competition between members of the community for limiting resources. For example, the rare species in the community may be kept rare by superior competitors (Hebert et al., 1974); competition may limit species coexistence on small islands (Diamond, 1975); harsh, unpredictable climates at higher altitudes may favor broader niches and reduce species richness (May and MacArthur, 1972; J. W. MacArthur, 1975); competition coefficients and "carrying capacities" change along habitat gradients and may lead to changes in dominance and to exclusion (Pielou, 1975; Tilman, 1977); and competition favors niche differentiation (Hutchinson, 1978). But, each and every one of these patterns can also be explained by alternative hypotheses (e.g., Lawton and Strong, 1981). It is to these competing explanations that we must now turn.

5 MECHANISMS

It is impossible to think seriously about community organization without considering population dynamics. Depending on the form and intensity of density-dependent population control, a series of theoretically ideal communities exist, as follows:

1. The simplest possibility is for density dependence to have no, or only very feeble, sporadic effects on most of the component species. The result is a shambles of randomly fluctuating, independent populations that many might argue do not constitute a "community" in the normal sense of the word. Local extinctions should be commonplace in this ill-controlled world.

2. Next are communities constructed from populations each controlled by density dependent events, but with the component populations operating more or less independently of all other populations in the same trophic level. This might happen for two reasons. Populations may be resource limited; but for one of several possibilities (e.g., isolation, lack of time for species to evolve), the system is nowhere near saturated with occupants. Alternatively, natural enemies (predation, parasitism, and disease) may regulate each population well below the point where resource limitation becomes important. In either case, the result is a community predictable in composition and species abundancies, but one that is not influenced to any significant degree by interactions between species on the same trophic level.

3. Finally, component populations may experience density-dependent control, *and* the structure of the community may be influenced significantly

by interspecific competition. These interspecific competitive interactions may be contemporary and amendable to study by experimental manipulation; or "ghosts" (Connell, 1980), detectable because species' ecologies and morphologies have evolved to minimize competitive interactions; or they may lie somewhere between these two extremities.

Moreoever, competitive interactions need not simply be for food. One interesting possibility is that species compete for "enemy free space" (Askew, 1961; Atsatt, 1981; Lawton, 1978a; Lawton and Strong, 1981; Price, 1983; and references therein). The enemy free space hypothesis notes that victims vulnerable to one species or guild of natural enemies may be relatively or completely safe from others, by virtue of their color, mode of feeding, or what have you. But few refuges are absolute. In consequence, there are limits to the numbers of herbivore species able to coexist under the influence of a given set of natural enemies. A formal theoretical exploration of this verbal argument will be found in Holt (1977). The result is a stable community, structured by interspecific interactions propagated via polyphagous natural enemies, as species jostle to avoid their neighbors' predators and parasitoids.

Possibilities 1–3 represent no more than high points on a multidimensional continuum, and most ecologists now accept that different sorts of organisms and habitat templets (Southwood, 1977) combine to generate real communities throughout this continuum. Cogent descriptions of various types of communities will be found in Cody (1974), Connell (1975), Price (1980), and Zaret (1980).

The available evidence suggests that many phytophagous insect communities are not structured to any major extent by interspecific competition (Lawton and Hassell, 1983; Lawton and Strong, 1981). But a proportion may be (e.g., Benson, 1978; Denno et al., 1981). On the other hand, only a handful seem totally unregulated (examples in Price, 1980; see also Washburn and Cornell, 1981). These fuzzy and tentative generalizations are at least consistent with data on phytophagous insect life tables (Caughley and Lawton, 1981; Lawton and McNeill, 1979), in which natural enemies are the commonest source of density-dependent (or delayed density-dependent) control (12 out of 20 life tables—60%). In only three populations (15%) is there absolutely no evidence of any density dependence. Intraspecific competition, a necessary but not sufficient condition for interspecific competition, appears to control the remaining 25% of populations.

Where does the bracken community fit in this scheme of things?

5.1 Density Dependence in Bracken Herbivores

Unfortunately, detailed life tables do not exist for any of the bracken herbivores. However, long-term population minima and maxima lying within an order of magnitude of one another, as is the case with seven of the fourteen populations summarized in Table 3, imply effective density dependent control (e.g., Hassell et al., 1976; Williamson, 1972). At the other extreme, it would be difficult to argue that the *Macrosiphum* population at Skipwith is effectively regulated by anything, and *Chirosia histricina* and *Monalocoris* populations are erratic. On balance, however, and with some interesting exceptions, most of the populations at Skipwith are relatively stable, indicative of density dependent control. The result is a community with a reasonably predictable structure.

5.2 Intraspecific Resource Limitation

The species on bracken are usually so rare (Table 4), even at peak densities (Table 3), that it is difficult to believe they are food limited. On the other hand, it could be that the low protein content of the fronds over most of the growing season, and the large number of allelochemics which they contain (Jones, 1983; Lawton, 1976, 1978a), mean that most parts of most fronds are inedible for most of the species most of the time. Experiments to examine this possibility are under way, but it is too early to say what the results will be.

5.3 Interspecific Competition

Logically, a search for interspecific competition should follow demonstrations of intraspecific resource limitation (Reynoldson and Bellamy, 1971). Revealing impatience, I have already looked for signs of interspecific competition in the contemporary population dynamics of congeners and/or species sharing similar feeding habits on bracken, but found none. (Lawton, 1982, 1983). These species show no evidence of depressing each others abundancies when they co-occur on the same fronds at Skipwith; no evidence of excluding one another from sites at different localities; and do not show density compensation or niche expansion at species-poor sites.

5.4 Natural Enemy Regulation

Experiments to test for the effects of natural enemies have been started. They include bird-exclusion experiments (e.g., Holmes et al., 1979), parasitoid and arthropod predator exclusion experiments, and insecticide treatments. Again, it is too early to say what most reveal, except in two cases.

In 1980, the sawfly *Aneugmenus padi* colonized unaided, transplanted, herbivore-free bracken on the University of York campus. The result has been a population explosion (Table 5), both outside and inside fine-screen cages constructed around some of the bracken after it had been colonized, to exclude parasitoids. The maximum abundance of *Aneugmenus* caterpillars at Skipwith (actually two different species, namely, *A. fürstenbergensis* and

Table 5. A Population Outbreak of *Aneugmenus* Eggs and Caterpillars on Bracken Planted on the University of York Campus, Compared with Maximum Numbers on the Skipwith Study Site[a]

Site	Date	Mean Numbers per Frond	Range (minimum–maximum numbers per frond)	Number of Fronds Examined
Skipwith Common	29 July 1980	0.35	0–2	20
Skipwith Common	4 Aug. 1981	0.15	0–1	20
Campus, unenclosed bracken	July 1981	2.50	0–6	20
Campus, unenclosed bracken	14 June 1982	30.6	4–76	10
Campus, unenclosed bracken	13 July 1982	35.3	4–52	20
Campus, bracken inside fine screen cages	29 June 1982	8.88	3–24	8
Campus, bracken inside fine screen cages	13 July 1982	9.75	0–26	12

[a] See text for details (J. H. Lawton, M. MacGarvin, and P. A. Heads, unpublished observations).

A. temporalis) has never exceeded 0.35 per frond, two orders of magnitude less than densities achieved on the campus plots.

These data are characteristic of an insect herbivore relieved of the effects of its natural enemies (e.g., Beddington et al., 1978; Caughley and Lawton, 1981; Lawton and McNeill, 1979). What these enemies are remains to be seen. Obviously, other explanations are possible, but I find them less compelling than natural enemy release.

The second experiment is an attempt to exclude a known group of predators, ants, from the fronds. Bracken has extrafloral nectaries attractive to ants (Lawton, 1976). Several species are involved (primarily in the genera *Formica* and *Myrmica*) but no species are obligate exploiters, so a tight coevolutionary relationship between these particular ants and bracken seems unlikely. This aside, there are now several studies demonstrating the effects of such ant–plant mutualisms on phytophagous insects (e.g., Bentley, 1977; Schemske, 1980). Accordingly, I confidently expected that ant predation would have an effect on bracken herbivores. I investigated the effect of ants at Skipwith in three ways:

1. Introduction of nonbracken feeding caterpillars onto fronds, to follow their subsequent removal by ants.
2. Exclusion of ants from fronds using tree-banding grease.
3. Comparison of the bracken fauna on fronds with a naturally low density of ants with the fauna on areas where ants are common.

Introduced caterpillars are rapidly removed by ants; Nicholsonian attack rates (Varley et al., 1973) for the ants, estimated from rates of caterpillar removal, ant numbers, and frond area are summarized in Table 6. These are fierce rates of predation.

Despite these high attack rates on nonbracken insects, excluding ants from fronds had no obvious, consistent effect on the bracken fauna (Table 7), and the differences between areas with naturally high and low numbers of ants are not consistent with a hypothesis that ant predation markedly reduces the abundance of bracken herbivores (Table 7). Figure 2 makes much the same point. Species abundances and the number of species are not obviously reduced by larger numbers of ants at low altitudes.

5.5 A Dilemma: The Ghosts of Predation and Competition Past

I expected that removing ants would have a major impact on the community, but found no significant effects at all. In other words, it would appear that bracken herbivores have evolved, or by coincidence possess, characteristics that make them largely immune to ant predation. Possibilities include methods of feeding (the larvae of leaf-tiers, e.g., *Olethreutus*, and miners and gall

Table 6. Nicholsonian Attack Rates for Ants (*Myrmica* and *Formica*) on Bracken Fronds at Skipwith, Calculated from Experiments Done in Late May and Early June 1980[a]

Date	Mean Number of Ants Present per Frond at Any One Time (P)	Duration of Experiment (hr)	Initial and Final Numbers of *Plodia* Caterpillars (the latter corrected for losses from controls)	Attack rate, a[b] ($cm^2\ hr^{-1}\ ant^{-1}$)
18 June, exp. 1	0.013	5.00	20–15	60.0
18 June, exp. 2	0.57	1.00	15–1	199.7
25 June	0.30	1.25	15–1	279.4
3 July, exp. 1	0.26	3.50	17–2	166.2
3 July, exp. 2	0.96	2.50	17–1	41.7
16 July, exp. 1	0.57	1.50	17–1	222.8
16 July, exp. 2	0.18	1.67	17–9	195.5
Mean a	—	—	—	166.5

[a] The prey were caterpillars of the moth *Plodia interpunctella* which have no means of defense against ants and do not feed on bracken (J. H. Lawton, unpublished observations).

[b] Attack rates, a, were calculated from $a = -ma/P$, where m is slope of regression of $\log_e$ (number of *Plodia* caterpillars surviving) against time (in hours); A is total frond area (cm^2), P is average number of ants per frond (see Varley et al., 1973). Losses of *Plodia* from control fronds, from which ants were excluded, were negligible, but are allowed for in these calculations.

Table 7. Effect of Excluding Ants from Fronds (Using Tree-Banding Grease) for 3 Weeks (27 May 1981–17 June 1981) on the Bracken Fauna[a]

	Mean $\log_{10}(n+1)$ per Frond ($\pm 95\%$ CI)		
Species and Date	Fronds Banded to Exclude Ants (sample size 30)	Control (Unbanded) Fronds (sample size 30)	Unbanded Fronds on an Adjacent Area with Naturally Low Ant Activity (sample size 40)
Ants, 27 May 1981	0	0.28 ± 0.11	—
Ants, 8 June 1981	0.02 ± 0.007	0.16 ± 0.06	0
Ants, 17 June 1981	0.05 ± 0.019	0.11 ± 0.04	0.008 ± 0.015
Bourletiella, 17 June 1981	0.32 ± 0.15	0.36 ± 0.17	0.52 ± 0.11
D. filicina, 17 June 1981	0.46 ± 0.20	0.28 ± 0.15	0.09 ± 0.09
D. pteridicola, 17 June 1981	0.13 ± 0.10	0.08 ± 0.07	0.03 ± 0.04
Sawfly eggs, 17 June 1981	0.076 ± 0.061	0.187 ± 0.083	0.053 ± 0.043
Total number of species (all other species too rare to make sensible comparisons)	11	7	8

[a] The grease was applied in a narrow band, 2–3 cm long, on the rachis (stalk) of the plant just above ground level. A small number of ants were able to gain access to these fronds from other vegetation, later in the experiment, but ant numbers remained significantly lower on experimental fronds, compared with controls. Also shown are insect numbers on another part of the Skipwith study area where ant activity was naturally lower than on the experimental area 20 m away.

formers, e.g., *Paltodora* and *Chirosia*, are totally immune), distastefulness (sawfly caterpillars), behavioral escape reactions (*Ditropis*), and so on. So, unless my logic or my experiments are faulty, it is difficult to avoid the conclusion that part, at least, of the structure of the contemporary community must be due to the ghost of ant-predation past. Nonadapted species (Table 6) suffer intense mortality.

The dilemma is obvious. It has, so far, also proved impossible to detect significant interspecific competition between any of the bracken herbivores (see above). But, if I am honest, I have to admit that this does not preclude evolution to minimize interspecific competition in the past. If the ghost of ant predation past has left its mark on the structure of the present community, why not the ghost of competition past?

Part of the answer probably lies with the abundance of bracken herbivores; most are too rare most of the time for significant interspecific competition for a limited food supply to seem vaguely likely. Also, there is no evidence that local communities saturate with species in a way that suggests competition has seriously constrained community evolution (Lawton, 1982, 1983). But these arguments are very speculative.

I see the task of establishing the roles of "competition and predation past" as one of the most difficult and challenging problems in contemporary community ecology.

6 SUMMARY

Drawing these arguments together, my best guess is that bracken communities conform most closely to the idealized model (2) in Section 5. Most of the component populations are reasonably well regulated by density dependent processes, generating a community that is predictable in structure (Fig. 1) without strong, contemporary interspecific competitive interactions between species. Most of the density dependence is probably due to natural enemies, but some populations may be resource limited.

These are sufficient conditions to explain each of the patterns discussed in Sections 3.1–3.5, as one, or more, of each population's vital rates (birth, death, immigration, and emigration) change along environmental gradients. Perhaps the most outstanding question to be answered is whether niche differentiation (Section 3.5) is greater than one might expect under some reasonable null hypothesis (see Lawton, 1983; Lawton and Strong, 1981, for further discussions on this problem).

Two things remain to be done. The first is to test this scenario properly. The second is to assemble comparable data for other phytophagous insect communities on different species of plants. Only then can be hope to develop general theories of herbivore community organization from a sound foundation of specific tests.

ACKNOWLEDGMENTS

Work on bracken is supported by grant GR3/4383 from NERC. Permission to use Skipwith Common is given by the Yorkshire Naturalists' Trust.

LITERATURE CITED

Askew, R. R. 1961. On the biology of the inhabitants of oak galls of Cynipidae (Hymenoptera) in Britain. *Trans. Soc. Br. Entomol.* **14**:237–268.

Askew, R. R. 1962. The distribution of galls of *Neuroterus* (Hym: Cynipidae) on oak. *J. Anim. Ecol.* **31**:439–455.

Atsatt, P. R. 1981. Lycaenid butterflies and ants: Selection for enemy-free space. *Am. Nat.* 638–654.

Auerbach, M. J., and S. D. Hendrix. 1980. Insect-fern interactions: Macrolepidopteran utilization and species-area association. *Ecol. Entomol.* **5**:99–104.

Bale, J. S. 1981. Seasonal distribution and migratory behaviour of the beech leaf mining weevil, *Rhynchaenus fagi* L. *Ecol. Entomol.* **6:**109–118.

Beddington, J. R., C. A. Free, and J. H. Lawton. 1978. Characteristics of successful natural enemies in models of biological control of insect pests. *Nature (London)* **273:**513–519.

Benson, R. B. 1950. An introduction to the natural history of British sawflies (Hymenoptera, Symphyta). *Trans. Soc. Br. Entomol.* **10:**46–142.

Benson, R. B. 1952. *Handbooks for the identification of British insects. Hymenoptera 2. Symphyta.* Section (b). Royal Entomological Society, London.

Benson, W. W. 1978. Resource partitioning in passion vine butterflies. *Evolution* **32:**493–518.

Benson, W. W., K. S. Brown, Jr., and L. E. Gilbert. 1976. Coevolution of plants and herbivores: Passion flower butterflies. *Evolution* **29:**659–680.

Bentley, B. L. 1977. Extrafloral nectaries and protection by pugnacious bodyguards. *Annu. Rev. Ecol. Syst.* **8:**407–427.

Berenbaum, M. 1981. Effects of linear furanocoumarins on an adapted specialist insect (*Papilio polyxenes*). *Ecol. Entomol.* **6:**345–351.

Caughley, G., and J. H. Lawton. 1981. Plant-herbivore systems. In R. M. May (ed.), *Theoretical ecology. Principles and applications,* pp. 132–166. Blackwell, Oxford.

Claridge, M. F., and J. S. Singhrao. 1978. Diversity and altitudinal distribution of grasshoppers (Acridoidea) on a Mediterranean mountain. *J. Biogeog.* **5:**239–250.

Cody, M. L. 1974. *Competition and the structure of bird communities.* Princeton University Press, Princeton, New Jersey.

Cody, M. L., and J. M. Diamond (eds.). 1975. *Ecology and evolution of communities.* Harvard University Press, Cambridge, Massachusetts.

Collins, J. E. 1955. Genera and species of Anthomyiidae allied to *Chirosia* (Diptera). *J. Soc. Br. Entomol.* **4:**94–100.

Connell, J. H. 1975. Some mechanisms producing structure in natural communities: A model and evidence from field experiments. In M. L. Cody and J. M. Diamond (eds.), *Ecology and evolution of communities,* pp. 460–490. Harvard University Press, Cambridge, Massachusetts.

Connell, J. H. 1979. Tropical rain forests and coral reefs as open non-equilibrium systems. *Symp. Br. Ecol. Soc.* **120:**141–163. Blackwell, Oxford.

Connell, J. H. 1980. Diversity and the coevolution of competitors, or the ghost of competition past. *Oikos* **35:**131–138.

Connell, J. H., and R. O. Slatyer. 1977. Mechanisms of succession in natural communities and their role in community stability and organization. *Am. Nat.* **111:**1119–1144.

Cornell, H. V., and J. O. Washburn. 1979. Evolution of the richness-area correlation for cynipid gall wasps on oak trees: A comparison of two geographical areas. *Evolution* **33:**257–274.

Dempster, J. P., and M. L. Hall. 1980. An attempt at re-establishing the swallowtail butterfly at Wicken Fen. *Ecol. Entomol.* **5:**327–334.

Denno, R. F., M. J. Raup, and D. W. Tallamy. 1981. Organization of a guild of sap-feeding insects: Equilibrium vs non-equilibrium coexistence. In R. F. Denno and H. Dingle (eds.), *Insect life history patterns: Habitat and geographic variation,* pp. 151–181. Springer-Verlag, New York.

Diamond, J. M. 1975. Assembly of species communities. In M. L. Cody and J. M. Diamond (eds.), *Ecology and evolution of communities,* pp. 342–444. Harvard University Press, Cambridge, Massachusetts.

Eastop, V. 1979. Sternorrhyncha as Angiosperm taxonomists. *Synb. Bot. Upsal.* **22:**120–134.

Ehrlich, R., and P. Raven. 1964. Butterflies and plants: A study in coevolution. *Evolution* **18:**586–608.

Fox, L. R. 1981. Defense and dynamics in plant-herbivore systems. *Am. Zool.* **21**:853–864.

Gilbert, L. E. 1979. Development of theory in the analysis of insect-plant interactions. In D. J. Horn, R. D. Mitchell, and G. R. Stairs (eds.), *Analysis of ecological systems*, pp. 117–154. Ohio State University Press, Columbus.

Gilbert, L. E., and J. H. Smiley. 1978. Determinants of local diversity in phytophagous insects: Host specialists in tropical environments. *Symp. R. Entomol. Soc. London* **9**:89–104.

Grossman, G. D. 1982. Dynamics and organization of a rocky intertidal fish assemblage: The persistence and resilience of taxocene structure. *Am. Nat.* **119**:611–637.

Harper, J. L. 1977. *Population biology of plants*. Academic Press, London.

Hassell, M. P., J. H. Lawton, and R. M. May. 1976. Patterns of dynamical behaviour in single species populations. *J. Anim. Ecol.* **45**:471–486.

Heath, J. 1959. The autecology of *Eustroma reticulata* Schiff. in the Lake District with notes on its protection. *J. Soc. Br. Entomol.* **6**:45–51.

Hebert, P. D. N. 1980. Moth communities in montane Papua New Guinea. *J. Anim. Ecol.* **49**:593–602.

Hebert, P. D. N., P. S. Ward, and R. Harmsen. 1974. Diffuse competition in Lepidoptera. *Nature (London)* **252**:389–391.

Hille Ris Lambers, D. 1979. Aphids as botanists? *Symb. Bot. Upsal.* **22**:114–119.

Holloway, J. D., and P. D. N. Hebert. 1979. Ecological and taxonomic trends in macrolepidopteran host plant selection. *Biol. J. Linn. Soc.* **11**:229–251.

Holmes, R. T., J. C. Schultz, and P. Nothnagle. 1979. Bird predation on forest insects: An exclosure experiment. *Science* **206**:462–463.

Holt, R. D. 1977. Predation, apparent competition, and the structure of prey communities. *Theor. Pop. Biol.* **12**:197–229.

Horn, H. 1975. Markovian properties of forest succession. In M. L. Cody and J. M. Diamond (eds.), *Ecology and evolution of communities*, pp. 196–211. Harvard University Press, Cambridge, Massachusetts.

Hutchinson, G. E. 1978. *An introduction to population ecology*. Yale University Press, New Haven, Connecticut.

Jones, C. G. 1983. Phytochemical variation, colonization, and insect communities: The case of bracken fern, (*Pteridium aquilinum*). In R. F. Denno and M. S. McClure (eds.), *Variable plants and herbivores in natural and managed systems*. pp. 513–558. Academic Press, New York.

Kirk, A. A. 1977. The insect fauna of the weed *Pteridium aquilinum* (L.) Kuhn (Polypodiaceae) in Papua New Guinea: A potential source of biological control agents. *J. Aust. Entomol. Soc.* **16**:403–409.

Kirk, A. A. 1980. *The Insects Associated With Bracken Fern*, Pteridium aquilinum *(L.) Kuhn in Papua New Guinea and Their Possible Use in Biological Control*. Unpublished M.Phil. Thesis, University of London.

Kuris, A. M., A. R. Blaustein, and J. J. Alio. 1980. Hosts as islands. *Am. Nat.* **116**:570–586.

Lawton, J. H. 1976. The structure of the arthropod community on bracken. *Bot. J. Linn. Soc.* **73**:187–216.

Lawton, J. H. 1978a. Host plant influences on insect diversity: The effects of space and time. *Symp. R. Entomol. Soc. London* **9**:105–125.

Lawton, J. H. 1978b. *Olethreutes lacunana* (Lepidoptera: Tortricidae) feeding on bracken. *Entomol. Gaz.* **29**:131–134.

Lawton, J. H. 1982. Vacant niches and unsaturated communities: A comparison of bracken herbivores at sites on two continents. *J. Anim. Ecol.* **51**:573–595.

Lawton, J. H. 1983. Non-competitive populations, non-convergent communities, and vacant niches: The herbivores of bracken. In L. G. Abele, D. Simberloff, and D. R. Strong, Jr. (eds.), *Ecological communities: Conceptual issues and the evidence.* Princeton University Press, Princeton, New Jersey (in press).

Lawton, J. H., and V. F. Eastop. 1975. A bracken feeding *Macrosiphum* (Hem. Aphididae) new to Britain. *Entomol. Gaz.* **26**:135–138.

Lawton, J. H., and M. P. Hassell. 1983. Interspecific competition in insects. In C. B. Huffaker and R. L. Rabb (eds.), *Ecological entomology.* Wiley, New York (in press).

Lawton, J. H., and S. McNeill. 1979. Between the devil and the deep blue sea: On the problem of being a herbivore. *Symp. Br. Ecol. Soc.* **20**:223–244.

Lawton, J. H., and D. R. Strong, Jr. 1981. Community patterns and competition in folivorous insects. *Am. Nat.* **118**:317–338.

MacArthur, J. W. 1975. Environmental fluctuations and species diversity. In M. L. Cody and J. M. Diamond (eds.), *Ecology and Evolution of Communities*, pp. 74–80. Harvard University Press, Cambridge, Massachusetts.

MacGarvin, M. 1982. Species-area relationships of insects on host plants: Herbivores on rosebay willowherb. *J. Anim. Ecol.* **51**:207–223.

May, R. M., and R. H. MacArthur. 1972. Niche overlap as a function of environmental variability. *Proc. Natl. Acad. Sci. USA* **69**:1109–1113.

Nielsen, B. O. 1978. Food resource partitioning in the beech leaf-feeding guild. *Ecol. Entomol.* **3**:193–201.

Oinonen, E. 1967. The correlation between the size of Finnish bracken (*Pteridium aquilinum* (L.) Kuhn) clones and certain periods of site history. *Acta For. Fenn.* **83**:1–51.

Pielou, E. C. 1975. *Ecological diversity.* Wiley, New York.

Powell, J. A. 1980. Evolution of larval food preferences in microlepidoptera. *Annu. Rev. Entomol.* **25**:133–159.

Price, P. W. 1980. *Evolutionary biology of parasites.* Princeton Univ. Press, Princeton, New Jersey.

Price, P. W. 1983. Hypotheses on organization and evolution in herbivore communities. In R. F. Denno and M. S. McClure (eds.), *Variable plants and herbivores in natural and managed systems.* pp. 559–596. Academic Press, New York.

Rey, J. R. 1981. Ecological biogeography of arthropods on *Spartina* islands in northwest Florida. *Ecol. Monogr.* **51**:237–265.

Reynoldson, T. B., and L. S. Bellamy. 1971. The establishment of interspecific competition in field populations, with an example of competition between *Polycelis nigra* (Mull.) and *P. tenuis* (Ijima) (Turbellaria, Tricladida). *Proc. Adv. Study Inst. Dynamics Numbers Popul. (Oosterbeek, 1970)*, 282–297 PUDOC, Wageningen, Netherlands.

Rigby, C., and J. H. Lawton. 1981. Species-area relationships of arthropods on host plants: Herbivores on bracken. *J. Biogeog.* **8**:125–133.

Robertson, T. S. 1981. The decline of *Carterocephalus palaemon* (Pallas) and *Maculinea arion* (L.) in Great Britain. *Entomol. Gaz.* **32**:5–12.

Root, R. B. 1973. Organization of a plant-arthropod association in simple and diverse habitats: The fauna of collards (*Brassia oleracea*). *Ecol. Monogr.* **43**:95–124.

Root, R. B., and S. J. Chaplin. 1976. The life-styles of tropical milkweed bugs, *Oncopeltus* (Hemiptera: Lygaeidae) utilising the same host. *Ecology* **57**:132–140.

Schemske, D. W. 1980. The evolutionary significance of extrafloral nectar production by *Costus woodsonii* (Zingiberaceae): An experimental analysis of ant protection. *J. Ecol.* **68**:959–967.

Skuhravy, V. 1978. Invertebrates: destroyers of common reed. In D. Dykyjova and J. Kvet (eds.), *Ecological Studies*, Vol. 28, pp. 376–88. Springer-Verlag, Berlin.

Slansky, F. Jr. 1973. Latitudinal gradients in species diversity of the New World swallowtail butterflies. *J. Res. Lep.* **11**:201–217.

Smith, D. R., and J. H. Lawton. 1980. Review of the sawfly genus *Eriocampidea* (Hymenoptera: Tenthredinidae). *Proc. Entomol. Soc. Wash.* **82**:447–453.

Southwood, T. R. E. 1977. Habitat, the templet for ecological strategies? *J. Anim. Ecol.* **46**:337–365.

Southwood, T. R. E. 1978. The components of diversity. *Symp. R. Entomol. Soc. London* **9**:19–40.

Strong, D. R., Jr. 1979. Biogeographic dynamics of insect-host plant communities. *Annu. Rev. Entomol.* **24**:89–119.

Strong, D. R., Jr., J. H. Lawton, and T. R. E. Southwood. 1983. *Insects on plants: Community patterns and mechanisms*. Blackwell, Oxford.

Tilman, D. 1977. Resource competition between planktonic algae: an experimental and theoretical approach. *Ecology* **58**:338–348.

Varley, G. C., G. R. Gradwell, and M. P. Hassell. 1973. *Insect population ecology. An analytical approach*. Blackwell, Oxford.

Washburn, J. O., and H. V. Cornell. 1981. Parasitoids, patches, and phenology: Their possible role in the local extinction of a cynipid gall wasp population. *Ecology* **62**:1597–1607.

Waloff, N. 1979. Partitioning of resources by grassland leafhoppers (Auchenorrhyncha, Homoptera). *Ecol. Entomol.* **4**:379–385.

Wiklund, C. 1974. The concept of oligophagy and the natural habitats and host plants of *Papilio machaon* L. in Fennoscandia. *Entomol. Scand.* **5**:151–160.

Williamson, M. 1972. *The analysis of biological population*. Edward Arnold, London.

Zaret, T. M. 1980. *Predation and freshwater communities*. Yale University Press, New Haven, Connecticut.

CHAPTER **13**

Alternative Paradigms in Community Ecology

PETER W. PRICE
Department of Biological Sciences
Northern Arizona University
Flagstaff, Arizona

CONTENTS

1   THE STATE OF COMMUNITY ECOLOGY

The conceptual basis of community ecology is in a state of flux. On the one
hand, widely read textbooks in ecology (e.g., Odum, 1971; Pianka, 1974;
Ricklefs, 1979) develop the theme of interspecific competition as a potent
organizing force in communities and present a well-developed theory as if
it had been verified in nature. On the other hand, it has been claimed that
tests for competition have been inadequate (Connell, 1980), other mechanisms may result in patterns in nature attributed to the results of competition
(Lawton and Strong, 1981), evidence for competition is lacking in many
communities (e.g., Rathcke, 1976a,b; Strong et al., 1979; Strong, 1981; Price,
1980), and the nonequilibrium state of many assemblages greatly weakens
the force of competition if it exists at all (e.g., Grime, 1979; Connell, 1979;
Dayton, 1971; Wiens, 1974; Sale, 1977; Price, 1980). The new ecology of
the 1980s must face these discrepancies and bring them together in a unified
theory of community organization in ecological systems.

In fact competition theory lives in a dreamworld where everything can
be explained, but the validity of these explanations has not been adequately
established in the real world. As an example, it is widely held that in the
tropics many species are packed together and interspecific competition has
forced species to specialize in their niche exploitation patterns, whereas in
northern latitudes communities are loosely packed, competition is less severe, and species are more generalized in their exploitation patterns (e.g.,
MacArthur and Wilson, 1967; MacArthur, 1972; Pianka, 1974). However,
the larger number of species in the tropics can be attributed to more kinds
of resources in the tropics (Karr, 1975), and in the few cases where adequate
data are available species are no less specialized in their feeding habits in
temperate regions than in the tropics (Rohde, 1978a; Beaver, 1979; Price,
1980). Bark and ambrosia beetles are actually more specialized in host plant
utilization in temperate regions than in the tropics (Table 1), and butterflies
seem to be equally specialized along the latitudinal gradient (Table 2), as do
marine monogenean parasites on fishes (Table 3). Clearly, the real world
differs substantially from theory on community organization.

One major problem with the theory on community organization is its
global application. Although conditions for any organism differ dramatically
within any region the general theory has largely ignored the effects of dif-

Table 1. Specificity of Some Tropical and Temperate Bark and Ambrosia Beetles Showing the Percentage of Species in a Fauna Which are Confined to a Particular Species, Genus, or Family of Host Plants or Which Are More Polyphagous[a]

	Tropical		Temperate	
	West Malaysia	Fiji	France	California
Species	0	2	2	12
Genus	4	14	51	58
Family	35	16	38	24
Polyphagous	61	68	9	6
Number of species	275	56	132	173

[a] After Beaver (1979).

fering resource states, generation times of organisms, length of time over which resources are available, dispersal ability, and so on. A theory explaining global patterns in community ecology is doomed to failure because the array of resources and the organisms that exploit them is so diverse, even in one geographic region, that this diversity must be taken into account.

One approach to resolving this problem is to concentrate on resources,

Table 2. Specificity of Some Tropical and Temperate Butterfly Species, Showing the Percentage of Species with Known Food Plants Which Fall into Each Class of Number of Host Species Utilized[a]

Number of Hosts	Tropical	Temperate
1	24	21
2	14	24
3	16	13
4	4	4
5	6	16
6	2	9
7	2	3
8	4	4
9	2	0
10–19	12	4
20–29	8	0
30–39	8	0
Number of species	51	67

[a] After Price (1980).

Table 3. Specificity of Monogenean Parasites on Marine Fish
Hosts on a Latitudinal Gradient[a]

Degrees Latitude	Percent of Parasite Species			
	1 Host	2 Hosts	3 Hosts	>3 Hosts
8°	100			
13°–22°	91	6	1	
15°	71	18	4	7
23°	82	16	1	1
17°–27°	83	12	5	
20°	71	14	7.5	7.5
30°	89	9.5	1.5	
45°	83	17		
63°–64°	83	13		4
65°	80	20		
70°	86	14		

[a] Data from Rohde (1978a).

and the response of individuals and populations to these resources. Resources will always play an organizing role in communities whereas other
factors, such as weather or enemies, although important in some cases, are
not inevitably significant. Thus, in the search for patterns in community
organization, the resources on which they are based provide a central theme
and a foundation for building a general theory. The recognition of substantially different kinds of resources is an elementary step in the development
of a conceptual basis for viewing community organization. This is treated
in Section 2. Discussion of other factors, such as the role of enemies, will
be treated later in the chapter.

2 TYPES OF RESOURCES AND THEIR EXPLOITATION

The resource base on which communities are built must be understood in
detail if the communities themselves are to be understood. No study can
pretend to portray community phenomena without establishing clearly the
nature of the resource base. This is a major failing in many studies (e.g.,
Cody, 1974; Diamond, 1975). If resources are unmeasured it cannot be established that they are in limited supply, and the mechanisms involved with
community organization cannot be identified, including the force of interspecific competition. The new community ecology is concentrating on a
precise definition of relevant resources and how they are distributed in space
and change through time. Southwood (1977) has made an important contribution to this end.

2.1   Resources

Resource types may be broadly categorized into the temporal patterns: (a) rapidly increasing, (b) pulsing or ephemeral, (c) steadily renewed, (d) constant, and (e) rapidly decreasing (Table 4). Each type depicts the condition of the resource most of the time and will be explained in the following paragraphs. After this the population response of organisms to these resources will be considered as will the effects of uniformity and patchiness in resource distribution. Finally, the consequences for important mechanisms in community organization will be considered.

Rapidly increasing resources are those that do so over much of the active season of the exploiting populations, and then decline rapidly at the end of this season. A good example is foliage on temperate deciduous trees. Leaves are absent at the beginning of the warm season, a flush of leaves is produced, and the number of leaves increases rapidly through the season. Herbaceous perennial plants and many annuals produce leaves in a similar way, for example, soybean (Price, 1976), stinging nettle (Davis, 1973), and bracken fern (Lawton, 1978). Foliage then declines rapidly with leaf fall as the probability of cold conditions increases. Changes in leaf quality may modify this pattern.

Pulsing or *ephemeral resources* increase rapidly, then decline rapidly. Only a short pulse of resources is displayed in a relatively long period of

Table 4.   Types of Resources, Examples of These Types, and the Organisms That Exploit Them

Resource Type		Resource	Exploiters
(a)	Rapidly increasing	Temperate deciduous foliage	Herbivores
		Flowers	Generalist pollinators
		Insects in temperate regions	Insectivors
(b)	Pulsing (ephemeral)	Phytoplankton	Zooplankton
		Herbaceous plant parts	Specialist herbivores
		Flowers in temperate regions	Specialist pollinators
(c)	Steadily renewed	Gut contents, blood	Internal parasites
		Marine plankton	Intertidal filter feeders
		Flowers in wet tropics	Pollinators
		Foliage in wet tropics	Herbivores
(d)	Constant	Space	Intertidal organisms
		Space	Plants
		Nesting sites	Hole nesters
(e)	Rapidly decreasing	Seeds in deserts	Granivores
		Insect stages	Specialized parasitoids

otherwise favorable conditions. For a pollinator specialized on one plant, blooms may be available for only 1 month of the year, or herbivores ex-, ploiting specific plant parts such as flower buds may have resources available for even less time (e.g., Thompson and Price, 1977).

Steadily renewed resources are those that are produced for prolonged periods of time and are not readily overexploited, for example, internal parasites of vertebrates living in a constant flow of blood, or food in the alimentary canal of the host, with the host continually replenishing resources. Marine plankton drifts into the intertidal zone becoming available for filter feeders, and is produced largely at sites independent of the feeding community. Leaves of wet tropical plants are likely to be produced in this way also, as in the rolled leaves of *Heliconia* utilized by hispine chrysomelid beetles (Strong, 1981, 1982), where each leaf unfurls becoming unsuitable after a few days.

Constant resources are physical in nature, largely uninfluenced by seasonal change, exploitation or other factors. Surfaces on which to settle for marine intertidal organisms are likely to remain stable over long periods of time. Holes in wood used by nesting birds or insects may change little through time.

Rapidly decreasing resources are produced in a short period of time each season and subsequently decline through the remainder of the season. An example is seed for granivores produced largely soon after the single rainy season in the Chihuahuan and Mohave deserts, or after the two wet seasons in the Sonoran desert. The seed resource declines through exploitation, burial, and ultimately germination in the next rainy season. Specialized insectivores may face a similar situation, for one exploiting only the cocooned stage that overwinters exists in a sudden flush of resources and these decline as they are exploited and ultimately emerge as adults.

These temporal resource patterns are summarized in Figure 1. The important point is that resource patterns differ profoundly and will have divergent effects on community ecology. This is clearly an effort to categorize a continuum of resource types, so we should not spend much time debating the validity of this classification in comparison with others, or the appropriateness of the examples given. Depending on the length of the life cycle of the insect resource in relation to an exploiting population it could be classified as a rapidly decreasing resource, or a pulsing resource. Constant resources and steadily renewed resources could be grouped. But if we accept that resources do differ in ways similar to those described here, it then becomes interesting to discuss the population responses of exploiters to these resources.

2.2 Population Responses

Response of exploiter populations to the five kinds of resources, while that resource persists, in any one season can be regarded as rapid or slow in

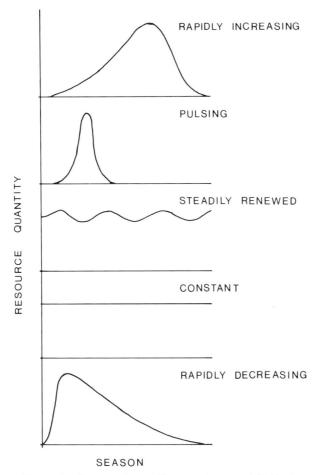

Figure 1. Resource types. For examples consult Table 4.

order to depict the extremes in the range of possibilities. In general, if generation time of the exploiter population is short relative to availability of the resource then the population response can be rapid, whereas populations with long generation times can show only a slow response or none to resource change. Then it is possible to predict the probability that the resource will be in short supply in some period during its availability, and we can predict the relative importance of interspecific competition under these conditions if two or more species exploit that resource.

For example, herbivores feeding on tree leaves in temperate regions (a rapidly increasing resource) may have an annual life cycle or several generations per year. Those with an annual life cycle cannot respond rapidly to the increasing resource, and interspecific competition is unlikely to develop. On the other hand, species with several generations per year can track the increasing resource and interspecific competition may develop. Leaf hoppers in the genus *Erythroneura* feeding on sycamore leaves may pass through four generations per year, resulting in a strong positive response to leaf availability (McClure, 1974). Among these species, interspecific competition can be significant at the end of the season (McClure and Price, 1975). These predictions are entered into the general summary of probabilities of occurrence of interspecific competition relevant to the five resource types (Table 5). Even in the case of *Erythroneura* leaves disappear completely during the winter, and adults enter a period of high mortality because of poor overwintering sites, spring floods, and perhaps inadequate fat storage. Thus, numbers of adults starting in the next season are very low in relation to resources, and even after four generations, populations may not be exploiting a large proportion of the resources (McClure and Price, 1975). Therefore, although the *Erythroneura* example is classed as leading to a high probability of interspecific competition, this may be short and local in any one season.

Pulsing or ephemeral resources exist so briefly that it is unlikely that any species can respond rapidly to such a resource. Numbers will be dictated by other factors affecting the population at other times of the year and population size will be usually independent of the amount of resources present. For example, *Depressaria* moths oviposit on unopened umbels of wild parsnip which are available for only 13 days per season in some patches (Thompson and Price, 1977) because the plants develop so rapidly from overwintered rosette to senescent plant. The life cycle of the moth is annual and much of the year is spent as an overwintering adult and much of the mortality is probably incurred at this time. Therefore it is unlikely that resources will be largely exploited because of high mortality of adults, shortage of time for them to find suitable oviposition sites, and probably asynchrony in some years between moth flight and resource availability. If another insect species exploited umbels of wild parsnip in a similar manner interspecific compe-

Table 5. **Probability of Interspecific Competition Occurring under Different Sets of Resource Rate of Change, Distribution, and Population Response of the Exploiters**

Spatial Distribution:	Uniform		Patchy	
Population Response: Rate of Resource Change	Rapid	Slow	Rapid	Slow
(a) Rapidly increasing	High	Low	Low	Low
(b) Pulsing (ephemeral)	[High]	Low	Low	Low
(c) Steadily renewed	High	Moderate	Low	Low
(d) Constant	High	High	Moderate	Low
(e) Rapidly decreasing	High	High	Moderate	Low

tition would have a low probability. Species with a rapid population response to an ephemeral resource probably do not exist so the high probability of competition in this category is placed in brackets in Table 5.

Similar kinds of reasoning can be used to arrive at the remainder of the predictions in Table 5. A rapid population response to steadily renewed resources, such as in blood parasites like *Plasmodium* and *Trypanosoma*, is likely to lead to interspecific competition if two species coexist. But in many gut parasites existing in replenished resources numbers are dictated solely by the number of colonists and not by multiplication within the host, so in the absence of very effective colonization, competition is likely to be absent or of low intensity. In constant and rapidly decreasing resources, interspecific competition is likely to be important ultimately whether the population responds rapidly or slowly.

2.3 Spatial Distributions

In a uniform environment, with resources spread evenly, interspecific competition is predicted to be important over much of the matrix of resources (Table 5), although it must be recognized that resources that are rapidly increasing or pulsing are exploited by a very large number of species with a relatively slow population response (Table 4), in which the probability of interspecific competition is low.

However, uniformity of resource distribution is most unusual: patchiness is the common dispersion pattern. Because patches frequently have a low probability of colonization by any one species, this greatly reduces the probability of co-occurrence of any two potential competitors, making the probability of interspecific competition low in most cases. Such conditions have been widely recognized in plant communities (e.g., Benzing, 1978a,b; Connell, 1979; Grime, 1977, 1979), invertebrate communities (e.g., Dayton, 1971; Rathcke, 1976a,b; Beaver, 1977, 1979; Connell, 1979; Price, 1980), in communities of vertebrates, both birds (Wiens, 1973, 1974) and fishes (Sale, 1977), and in theoretical studies (e.g., Shorrocks et al., 1979).

Where resources are patchy, disturbance frequently disrupts biotic interactions leading to short residence times of species, and resources that tend to be available briefly. Thus, many resources, although they are persistent on a global scale, tend to become available for colonization in local pulses, and availability is ephemeral (e.g., Dayton, 1971; Beaver, 1977, 1979; Connell, 1979; Price, 1980). Such conditions result in nonequilibrium conditions in both populations and communities.

Consideration of spatial distribution of resources and recognition of their generally patchy nature dramatically change predictions on the probability of interspecific competition (Table 5). Such competition is likely only where the probability is high that two or more potential competitors will occur in the same patch, with population responses rapid enough to exploit much of the resources while they are available.

2.4   Experimental Evidence

Two factors forced a consideration of resources as the basis for understanding pattern in the occurrence of competition. One was the convincing work of Brown and his colleagues on organization of granivorous desert rodent and ant communities, showing experimentally that competition for seeds played a significant role (reviewed in Brown et al., 1979; see also Davidson et al., 1980; Munger and Brown, 1981; Bowers and Brown, 1982). Seeds are produced synchronously and availability declines through time representing a rapidly decreasing resource, as discussed in Section 2.1, where competition is likely to occur (cf. Table 5). This body of work stood in contrast to my own perception of the role of competition in communities of specialists like parasites (Price, 1980, 1983b). The second factor was John Thompson's challenge, after reading my monograph on parasites, to explain why competiton played such an important role in the parasitic wasps I had studied (Price, 1970, 1971) whereas I largely discounted its role in endoparasitic helminth communities (Price, 1980). The answer is that resources are very different. Cocoons as a resource for parasitic wasps are like seeds in the desert. Cocoons are produced once a year and must sustain the community of parasitic wasps throughout the year, even after the resource has been drastically depleted by emergence of sawflies in early summer (Price and Tripp, 1972). Cocoons are a rapidly decreasing resource. For a helminth parasite community developing in the alimentary canal of a host, resources are either constant, in terms of space available, or steadily renewed, in terms of nutrients flowing down the canal. Given that hosts are patchy and population responses of helminth parasites are likely to be slow because they do not multiply within the host, we should predict relatively low probabilities of interspecific competition (Table 5). Experimental work on helminth parasites shows clearly that competition can occur in some communities (Holmes, 1961, 1962a) and not in others (Holmes, 1962b; Wertheim, 1970). Even the same parasite species in different hosts may show competitive (Holmes, 1961, 1962a) and noncompetitive (Holmes, 1962b) relationships. More experimental work is desirable on helminth community organization, but the results so far tend to fit predictions in Table 5.

The rare cases of experimental work on competition in birds tend to fit predictions in Table 5, although it is difficult to characterize the resource type in some cases, leaving them open to subjective interpretation. In winter the bark-foraging guild of birds face a declining resource of insects and seeds in temperate regions and competition is evident and important (Williams and Batzli, 1979). Given a constant overall nectar resource, Pimm (1978) showed competition in hummingbirds if the resource was predictable, with declining interference competition as resources became more unpredictable. Högstedt (1980) showed competition between jackdaws and magpies during the breeding season when relatively large insects would be an important prey item. The resource base was not described by Högstedt, but like seeds and cocoons, it is likely to be a decreasing resource. Although not an experimental study, Lister (1980) provides good evidence that competition between insectivorous birds increases progressively through winter when insects represent a declining resource.

Among herbivorous insects several types of resources are experienced. On rapidly increasing resources competition has been shown to be unimportant (Lawton, Chapter 12) except where population responses to these resources are very rapid (McClure and Price, 1975; Addicott, 1978; Stiling, 1980). In the tropics where resources may be more constant but very patchy or ephemeral, and population responses are slow, experiments indicate lack of competition (Seifert and Seifert, 1976, 1979; Strong, 1981, 1982). I know of no experimental data on competing insects on pulsing, ephemeral plant resources so common in temperate regions.

As in Pimm's (1978) study of hummingbirds discussed above, pollinators commonly colonize new resources rapidly. This can result in competition even on rapidly increasing resources, as Inouye (1978) found among bumblebees. He also found competition on the same flower species as resources declined later in the year. Significantly, Inouye's (1977) example of morphological divergence between these bees is one of the few that stood up to tests for significant differences between competitors conducted by Simberloff and Boecklen (1981).

The very rapid growth rates of populations of microorganisms and the frequency of antibiotic products predispose them to interspecific competition on all kinds of resources. The body of literature documenting experimental evidence for this is considerable (e.g., Alexander, 1971; Lynch and Poole, 1979; Cole, 1982; Tilman, 1982; Tilman et al., 1982). Rapid population growth is also possible in zooplankters, among which competition has been documented with experimental studies (e.g., Neill, 1975; Lynch, 1978).

Plants compete for sunlight, a constant resource, nutrients, which in most systems are more or less constant or declining, and in many cases water supply is seasonal with declining supplies between these seasons (e.g., Harper, 1977).

Unfortunately, ecologists have not found it profitable in the past to report negative results on competition. This situation must change before the predictions in Table 5 can be fully assessed objectively. In developing an understanding of patterns in nature it will be just as important to report lack of interaction as to report interaction. In a new ecology this requires a new sensitivity among editors and peer reviewers to this need. But in the context of the very biased literature available the predictions in Table 5 appear to be valid.

2.5 Juxtaposition of Species

The final point relevant to a prediction on probability of interspecific competition is whether potential interspecific competitors exist that could colonize the same patch of resources. The question is rarely asked because the answer has been assumed to be positive without an objective check on the real world. My position is that on many resource gradients exploited by specialized organisms, of which there are a multitude, many vacant niches exist (Price, 1980, 1983a,b). Therefore, co-occurrence of species does not lead to interspecific competition because their exploitation patterns do not overlap significantly.

Some examples will clarify this assertion. Tropical fish species can harbor at least five monogenean parasites in discrete niches on the complex gill surface (Rohde, 1978b, 1979). However, in the Atlantic Ocean the mean number of parasite species per host species ranges from 0.3 species per host in the coldest seas to only about 1.5 species per host in the warmest seas. In the Pacific Ocean the mean numbers range from 1.75 to 2.5 parasite species per host species in cold and warm seas respectively (Rohde, 1980). On average, half or more of the ecological niches physically defined for monogenean parasites are unoccupied on any one fish species. In cold seas such as the Black Sea, White Sea, and Barents Sea 66–82% of fish species have no monogeneans on the gills (Rohde, 1978b). Reality shows that many more ecological niches remain on the majority of fish species than are presently utilized. There is no question of species being packed into a restricted set of resources, a situation usually envisaged in competition theory. Rather, individual species appear as specks in a vast matrix of resources which could support a number of coexisting species at least one order of magnitude higher [e.g., in the Black Sea, of 34 fish species 28 have no monogeneans, and 6 have only 1 monogenean. If we assume that each fish species could support five monogenean species the total possible would be 170 monogenean species in the Black Sea, assuming all are specific to a single host. In fact about 80% of monogeneans are this specific (see Rohde, 1978a,b, for data)]. Under these circumstances competition is likely to be uncommon, even if every fish is infected in a population. In support of this, Rohde (1978a,b, 1979) has shown that specificity of niche occupation is very high and independent of the number of coexisting species in the community.

Another example concerns a similarly specific group of terrestrial parasites: the leaf mining flies in the family Agromyzidae. In Canada there is only one fly species for every 20 plant species approximately (Price, 1977), when we know that up to nine species do exist on one host, and 12 to 15 probably could exist on one host (Lawton and Price, 1979). For the British flora and agromyzid fauna there is no family with more fly species than plant species. Attempts to explain this small fauna relative to the resources available by looking for preempting species from other taxa have failed (Lawton and Price, 1979). Again, many ecological niches remain vacant, and many resources remain unexploited (see also Lawton, 1982).

To conclude this section, when specific resource types are considered together with the population responses of the exploiters and the resource distribution, it is evident that interspecific competition is probably unimportant in the majority of real situations, but is predictably important under circumscribed conditions. In contrast we have a general theory of community ecology in which competition plays a central organizing role, whatever the kind of resource or organism. Where the theory becomes vaguely specific, for example, by making predictions about tropical vs. temperate communities, reality does not provide the supporting evidence.

Thus, a major paradigm of community ecology is applicable to a minority of natural situations. At present in the literature there seems to be no alternative theme that has the structuring power of interspecific competition. We must therefore ask if communities are really unstructured of if alternative paradigms will take the place of competition theory. This question will be addressed in the next section.

3   ALTERNATIVE PARADIGMS

In advocating alternatives to competition as an organizing force in communities, my intention is not to expunge consideration of competition, but rather to foster a phase in ecological research in which several alternative hypotheses are tested simultaneously. Only then will competition be placed in its proper perspective and the relative importance of it and alternatives adequately assessed.

Competition theory dealt with problems central to community ecology. "Which factors influence the presence or absence of species in a community?" "What forces limit the number of species on a particular range of resources?" "What selective pressures result in the divergence of exploitation patterns between closely related species?" In any alternatives to competition these questions must be addressed.

Here I use the term paradigm in its general sense, for an example or model, without any connotation of it playing a dominant role in the conceptual development of a science. In this sense the word becomes interchangeable with the term hypothesis.

3.1   The Individualistic Response Paradigm

The null hypothesis in community ecology is that species respond individualistically to selective pressures and resources independently of other species, and simply live wherever autecological conditions are suitable. In evolutionary time species diverge in their resource exploitation patterns because of idiosynchratic responses to selective pressures imposed by the physical environment or the resource base that is exploited. In ecological time, species colonize a patch simply because the conditions are adequate for survival, uninfluenced by the presence of other species on the same trophic level, or by those on levels above. The individualistic response paradigm was proposed by Gleason (1926) and Ramensky (1926) (see also Whittaker, 1967), so it has deep roots in ecological literature. However, it has not been given the prominence in the literature that a null hypothesis deserves. Clearly, any thesis invoking structuring and organizing forces in a community should consider this null hypothesis as an alternative and attempt its falsification.

Whereas character displacement (Brown and Wilson, 1956) gained considerable attention during the 1960s as evidence of the role of competition in structuring communities in evolutionary time, unequivocal examples have actually declined in number since. The classic case of character displacement in the rock nut hatches was undermined by Grant (1972, 1975), and in Darwin's finches by Strong et al. (1979), and Simberloff and Boecklen (1981). In these assemblages differences between species may well result from individualistic responses to physical factors or resources. For example, Boag and Grant (1981) documented evolutionary change of size in one of Darwin's finches, *Geospiza fortis*, in response to change in seed size, without invoking any interspecifically induced selective forces. However, the case against competition in Darwin's finches remains debatable (e.g., Grant and Abbott, 1980), and some competition may be important (Grant and Grant, 1982).

In ecological time it appears that each species responds in an idiosynchratic way to one resource or a set of resources. Whittaker (1952, 1956) demonstrated this in the Great Smoky Mountains for foliage insects and plants. Cromartie (1975) stated that the diversity of the community of insects on crucifer plants was maintained by each species showing a distinct pattern of colonization and subsequent population change. Flea beetles responded strongly to concentrated resources both in their colonization and reproduction (Root, 1973), but were poor colonizers of small isolated patches of resources. But cabbage butterflies responded weakly to concentrated resources and laid more eggs per plant on isolated plants (Cromartie, 1975). Such differences in colonization behavior most likely result from selective pressures through intraspecific competition independent of interaction between species. The small flea beetles must reach high densities before intraspecific competition becomes important, and even then individualistic differences between species, such as phenology and plant colonization, coupled with resource density effects, greatly reduce the potential impact of competition (Kareiva, 1982). But the large caterpillars that have evolved on small wild crucifers are likely to compete should two occur on the same plant. Hence, ovipositing females inspect plants carefully and avoid those with eggs already present (e.g., Shapiro, 1981).

Idiosyncratic responses of species to resources and the patchy nature of resources is likely to result in underexploitation and lack of interspecific competition. Rathcke (1976a,b) found that species of stem boring insects in prairie plants best fitted a random distribution. Usually less than 20% of stems were attacked by any species, and less than 10% of any stem was utilized. She concluded that interspecific competition has been unimportant in both evolutionary and ecological time. This kind of study supports at the local level the inferences made from geographic regions in Section 2.5 above. So many resources exist with so few species to exploit them that each species can evolve and behave independently of others.

If we were to select at random communities for study, without knowing about the species present or the distribution of resources, we may well find that the individualistic response paradigm is relevant in many cases. In the past it has seemed to be more relevant and interesting to document cases of interspecific competition than to report on its absence. Thus, cases have been selected in which the probability of competition is high without any attempt to estimate how frequent such cases are in the community at large. The preferable approach would be to identify a resource or area, study all the species present, attempt to estimate which are interacting and which are not, and apply tests that will discriminate among several hypotheses, including the individualistic response hypothesis and the competition hypothesis.

The individualistic response paradigm is likely to be relevant wherever time is short in a patch, as on pulsing or ephemeral resources, and where

Table 6. Summary of Predictions in Which Paradigms Are Likely to Be Most Important in Relation to Resources, and Population Response to These Resources[a]

Spatial Distribution:	Uniform		Patchy	
Population Response: Rate of Resource Change	Rapid	Slow	Rapid	Slow
(a) Rapidly increasing		3.1	3.1	3.1
	3.3c	3.3c	3.3c	3.3c
	3.4b	3.4a	3.4a,c	3.4a.c
	3.5			
(b) Pulsing (ephemeral)		3.1	3.1	3.1
	3.4c	3.4c	3.4c	3.4c
(c) Steadily renewed	3.2	3.2		
	3.3a,b,c,	3.3a,b,c	3.3c	3.3c
	3.4b	3.4b	3.4a,c	3.4a,c
	3.5	3.5		
(d) Constant	3.2	3.2	3.2	
	3.3a,b,c	3.3a,b,c	3.3a,b,c	3.3c
	3.4b	3.4b	3.4b	3.4a
	3.5	3.5	3.5	
(e) Rapidly decreasing	3.2	3.2	3.2	
	3.3a,b,c	3.3a,b,c	3.3a,b,c	3.3c
	3.4b	3.4b	3.4b	3.4a
	3.5	3.5	3.5	

[a] Numbers refer to the section number in which a paradigm is considered and letters to hypotheses considered as subunits of the general paradigm (cf. Table 5).

resources increase rapidly, leaving large amounts per capita and a condition of plenty for any colonizing species (Table 6).

3.2   The Resource Heterogeneity Paradigm

This paradigm asserts that the number of species present in a community is positively related to the number of resources present, the limit to species number is set by the number of resources present, and the relative abundance of species is dictated by the relative abundance of the resources they need.

As environmental heterogeneity increases more species are likely to co-exist in a community because more ecological niches exist, and because no resources are abundant enough to enable a few species to become dominant and exclude others through interspecific competition. Therefore, in general, if the total quantity of resources in a community stays constant and we contrast one community with a narrow range of resources, but large supplies of them, against another with a wide range of resources, but relatively small quantities of each, we would predict large numbers of a few species in the former, and small numbers of many species in the latter (Fig. 2).

Heterogeneity of resources may take many forms. If our focus is on herbivore communities on specific host plant species then the relevant heterogeneity may be generated by genotypic differences within a plant population (see Price, 1983a, for a consideration of the genotypic heterogeneity hypothesis). For example, coexistence of two galling cecidomyiid flies in the genus *Semudobia* on birches in Holland seems to depend on the genetic heterogeneity of the plants, with one species specialized to exploit trees with a largely *Betula pendula* genotype, and the other on trees with predominantly a *B. pubescens* genotype (Roskam and van Uffelen, 1981). Since the birch species hybridize freely there is a spectrum of genotype mixtures in a birch population and the presence and absence of the cecidomyiids and their relative abundance are dictated by the birch genotypes present. Phenotypic heterogeneity may also be important in herbivore communities, involving

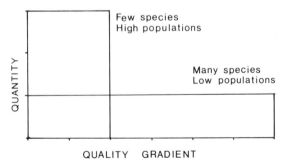

Figure 2.   Relationships between range of resources on a quality gradient and quantity of resource in each resource state, comparing two communities with the same amount of resources but distributed in contrasting ways.

environmentally induced differences in chemical defense (e.g., Berenbaum, 1978, 1981), nutrition, architecture of plants that differ according to plant density (Thompson and Price, 1977) or patch size (Rigby and Lawton, 1981), or unknown differences between plants growing in sun and shade (Lawton, 1982, 1983). Thus, habitat heterogeneity may also play a role in community organization, as proposed by Williams (1964), and supported by Strong (1979), Connor and McCoy (1979), Lawton and Strong (1981), and Fowler and Lawton (1982). Karr (1975) accounted for the larger number of bird species in tropical vs. temperate forests largely in the basis of resources being present in the tropics, but absent in temperate regions, on which species specialized, such as large insects and fruits. There seems to be a simple positive relationship between the number of resources and the number of species present.

The resource concentration hypothesis proposed by Root (1973) is also relevant to the resource heterogeneity hypothesis, and is basically its converse. Large bodies of a resource, such as a pure stand of a plant species for herbivores, provides a concentrated resource for specialists which increases the attraction and accumulation of specialist species, the time they spend on the resource, and their reproductive success. A few species will be abundant under these conditions because their requirements are fully provided (cf. Fig. 2). Evidence indicating that these mechanisms account for a major part of community organization has been provided by Root (1973), Ralph (1977a,b), Bach (1980), and Risch (1981).

The resource heterogeneity paradigm is likely to be relevant and play its part in all resource conditions discussed in Section 2. However, its organizing role may be observed most clearly where the potential for competition is high or moderate, on steadily renewed, constant, and especially rapidly decreasing resources (cf. Tables 5 and 6).

3.3 The Island or Patch Size Paradigm

This paradigm accounts for differences in the number of species in a community through the effect of area alone (MacArthur and Wilson, 1963, 1967). Area alone has a direct and positive effect on the number of species in a community because a larger area has a higher probability of receiving colonists, and they will persist longer because larger populations on larger islands are less likely to go extinct through the influences of interspecific competition, impact of natural enemies, or stochastic events. Although the pattern predicted by the theory of island biogeography has been observed repeatedly (e.g., MacArthur and Wilson, 1967; Connor and McCoy, 1979; Strong, 1979; Price, 1980) simpler mechanisms may be involved. Area is likely to be correlated with resource diversity in which case the resource heterogeneity paradigm is relevant (Section 3.2). Alternatively, larger areas sample larger numbers of colonizing species, and thus accumulate more species per unit time. This passive sampling hypothesis, defined by Connor and McCoy (1979), invokes no interaction between species, and acts as the null hypothesis of the species–area relationship. The maximum number of species on an island is defined by the number of species in the vicinity capable of colonizing that island or patch—the species exhaustion hypothesis (Lawton and Strong, 1981)—and not through interspecific interaction. Thus, we have three hypotheses relating to island or patch size (Connor and McCoy, 1979): (a) the area per se hypothesis, (b) the habitat or resource diversity hypothesis, and (c) the passive sampling hypothesis. Studies are needed which test these hypotheses simultaneously.

Experimental tests of the area per se hypothesis by Wilson and Simberloff (Simberloff, 1978) and Rey (1981) are supportive. The habitat or resource diversity hypothesis is supported by studies discussed in Section 3.2, whereas the passive sampling hypothesis is akin to the individualistic response hypothesis discussed in Section 3.1.

This paradigm also makes predictions on the kinds of species present in a community. Species differ in their colonizing abilities such that small islands or patches may be found by only the best colonizers, while large patches may have residents with poor colonizing ability. For example, Opler (1974) showed that such differences created patterns in the leaf miner communities on oak trees in California. An oak species with a small geographic distribution (a small island for colonization), like *Quercus dunnii*, was exploited only by the best colonizers, in the genera *Cameraria* and *Lithocolletis*. With wider geographic distribution of oak species number of miner species increased in the communities, with some species of poor colonizing ability in the genera *Neurobathra* and *Careospina* found only on the most common oak species, *Quercus agrifolia*. This recognition that the character of species can differ is an important development in community ecology that needs more attention.

Making predictions on the conditions under which this paradigm is likely to apply depends on the mechanisms underlying the observed pattern. The area per se hypothesis (3.3a) is likely to apply where interaction between species is most likely: on steadily renewed, constant, or rapidly decreasing resources (Table 6). The habitat or resource diversity hypothesis (3.3b) will apply where the resource heterogeneity paradigm applies. The passive sampling hypothesis (3.3c) will be relevant wherever there is enough time for sampling to run to completion, that is, on all but pulsing or ephemeral resources.

3.4 The Time Paradigm

This paradigm accounts for differences in species numbers in communities through the effect of time, which results in more species on long-established sites or resources, and fewer species on relatively young sites or resources. The paradigm has been applied most frequently to evolutionary time, starting with Southwood's (1961) study showing that species of tree in Britain with a more extensive fossil record supported more species of insects than those with a lesser record. Wilson (1969) also invoked evolutionary time as a necessary component of increasing species number in a community arguing that the equilibrium number of species would increase through coevolution of species which adapts them to more efficient coexistence. In Southwood's scenario no limit to accumulation of species on a resource is acknowledged, so presumably more time would continue to result in more species on a resource; species colonize independently of others. This is a noncompetitive view of the time paradigm (3.4a). Wilson's view regards interspecific competition as one factor setting an upper limit to the number of species in a community. Here the time paradigm invokes competition as an important organizing force (3.4b).

Strong (1974a,b) criticized the time hypothesis by showing that in the data used by Southwood present geographic area over which the trees are distributed correlates better with herbivore species number than evolutionary time. But Birks (1980) substantiated Southwood's view using better estimates of residence times of trees in Britain. Also, as argued in Section 2.5, in so many cases species are so depauperate relative to the resources available that evolutionary time may well be an important component in the richness of communities (e.g., Lawton and Price, 1979).

Time is also likely to be a significant factor, in a contemporary or ecological sense, where resources are pulsing or ephemeral. Andrewartha and Birch (1954) made a strong case for regarding many species as living under conditions in which time is short for successful reproduction. It also takes time to accumulate species in a community, as in the colonization studies by Simberloff and Wilson (1969, 1970), Davis (1973), Price (1976), and Lawton (1978). This period of accumulation of species has been called the noninteractive phase by Wilson (1969) and I have argued that many communities of specialists seldom if ever reach beyond this stage of individualistic colonization unhampered by interaction from resident species (Price, 1980, 1983b). Recent studies of insect communities on plants are consistent with this view (Lawton, 1978; Strong, 1979; Lawton and Strong, 1981). If many communities exist in a nonequilibrium state (Section 2.3), then time since disruption is likely to be an important factor. The ecological time hypothesis (3.4c) should be distinguished from the evolutionary time hypothesis (3.4a and b).

A prediction on where the time paradigm is likely to operate must again depend on the mechanisms invoked. Hypothesis 3.4a, which assumes lack of competition, is likely to be important where resources are rapidly increasing most of the time, where population responses are low, or where resources are patchy, that is, where competition is unlikely to be important (Tables 5 and 6). Hypothesis 3.4b regards competition as a strong organizing force that is likely to apply in an array of situations complementary to hypothesis 3.4a. The ecological time hypothesis (3.4c) will apply most frequently where resources are available briefly, where they are pulsing or ephemeral, but also in other situations where the probability of colonizing a resource is low, as is the case in many communities of specialists on patchily distributed living resources that rapidly increase or steadily renew resources (Table 6).

3.5 The Enemy Impact Paradigm

The enemy impact paradigm develops the model that enemies such as predators and parasites play an organizing role in communities by keeping populations below those at which resources become limiting, by limiting the number of species in a community to those that can be distinct enough not to be exploited by a common enemy, and by selecting for divergence of sympatric species such that enemies are not shared. Thus, the patterns resulting from enemy impact may bear a striking resemblance to patterns assumed to result from competition (Lawton and Strong, 1981).

The paradigm seems to have its roots in Brower's (1958) suggestion that predators can limit the similarity between prey species and diets of prey will be restricted to those resources on which protection from enemies is most effective. Support was provided by Ricklefs and O'Rourke (1975) and Otte and Joern (1977). An independent source came through Askew (1961) who reasoned that in closely related coexisting gall wasps divergence of gall form was selected for by parasitic wasps, with the result that overlap of host exploitation by the parasitoids was minimized. But Park (1948) provided one of the earliest prominent examples of enemy effects by demonstrating the role of the parasitic protozoan *Adelina tribolii* in modifying the outcome of competition between *Tribolium castaneum* and *T. confusum*. I have argued that such biological warfare is common in nature and is in need of much more serious consideration (Price, 1980). The term "keystone species" was applied by Paine (1966, 1969) to signify an organism that played a central role in community organization: the top predator in his studies of the marine intertidal community. Risch and Carroll (1982) provide data on the predaceous ant *Solenopsis geminata* that can act in a similar way in insect communities.

Divergence of form in gall shape and texture and leaf shape cannot be explained easily by competition theory, but enemies have an obviously important role to play. Galls of closely related coexisting species are frequently widely divergent in shape and texture (Askew, 1961; cf. Felt, 1940; Darlington, 1975), but such differences in no way reduce the impact of competition for the plant's resources. In a similar way, the discrete differences in leaf shapes between coexisting *Passiflora* species are best explained by the impact of visually hunting herbivores among the heliconiine butterflies (Gilbert, 1975). The number of coexisting plant species and the kinds of leaf shapes that can coexist seem to be dictated by the higher trophic level.

The widely diversified food plants utilized by butterflies in the speciose lycaenid subfamilies Theclinae and Polyommatinae seems to result from opportunistic association with ants that provide protection from enemies (Atsatt, 1981a,b; Pierce and Mead, 1981). Thus, the evolution of these largest of lycaenid butterfly subfamilies seems to have resulted through the interaction of mutualistic ants and their interference with enemy impact.

This paradigm has received considerable recent attention and support (e.g., Lawton, 1978; Holloway and Herbert, 1979; Lawton and McNeill, 1979; Price et al., 1980; Lawton and Strong, 1981), but its importance relative to other organizing influences in communities has not been adequately assessed.

This paradigm is likely to be applicable in situations where enemies can colonize readily and when resources are stable enough to permit prolonged interaction, and where, in the absence of enemy impact, competition might play an important role (Table 6).

4 THE FUTURE OF COMMUNITY ECOLOGY

4.1 Testing among Paradigms

Several paradigms discussed here have their roots deep in the ecological literature. Darwin (1859) was clearly the major force in generating a preoccupation with competition in ecological systems (even though he employed the term "competitors" loosely, Grime 1977). Gleason (1926) and Ramensky (1926) independently emphasized the importance of individualistic responses of species to physical factors in determining where species live and with which other species they associate. Andrewartha and Birch (1954) stressed time as one of the most limiting resources, with shortage of time preventing biotic interactions from running their course.

Although so many hypotheses have been extant in the literature, we have tended to test one at a time: it has not been traditional to test among alternative hypotheses in community ecology. Yet there is clearly a need to devise studies to test among alternatives so that we can determine the relative merits of each, and their relative importance in communities of differing resource type. Only then will community ecology advance significantly beyond the stage where different schools of thought are espousing contrasting mechanisms in the organization of natural communities, in the absence of critical studies testing each school's concepts simultaneously.

The challenge is not only to evaluate the relative merits of each paradigm, but to determine the mechanisms involved, for example, to distinguish among subhypotheses 3.3a, 3.3b, and 3.3c, or 3.4a, 3.4b, and 3.4c. More than one paradigm seems to be applicable at any one time or on any one kind of resource (Table 6, Fig. 3). So the question should never be is this or that paradigm relevant, but which of all the paradigms plays the most important role under a defined set of resource conditions.

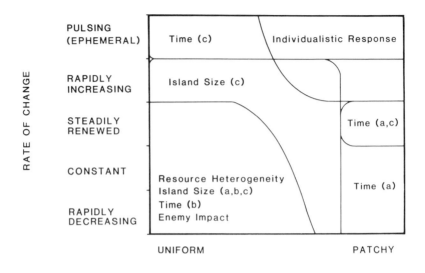

Figure 3. Predictions on where each paradigm is likely to be important on a matrix of resource rate of change and distribution of resources from uniform to patchy.

Communities on which critical tests are to be made need to have certain qualities:

1. Members of the community must be easily defined and recognized.
2. Resources must be easily measured at a level of detail relevant to the exploiting organisms: that is, assessed in much more detail than has been usually achieved in the past.
3. The pool of potential colonists must be clearly defined.
4. The community and resources must be amenable to extensive experimental manipulation. Good tests among paradigms and the underlying mechanisms will require a stronger experimental element than is generally found in the literature.

Selection of such communities may take two forms. Selections might be made almost at random without any knowledge of the specific organisms present or how resources are arrayed, so that no preconceived ideas hamper the objective testing among hypotheses. Alternatively, coexisting phylogenetically related species in a community enable a particularly cogent test of competition theory, and the other hypotheses also, because the evolutionary divergence of species from a common stock can be measured, and the mechanisms involved with this divergence assessed. Connell's (1980) "Ghost of competition past" becomes more earthly and testable.

As an example of the kind of community that is amenable to detailed testing, I will reveal my own interests and biases. Six species of sawfly

(Hymenoptera: Tenthredinidae: Nematinae: Nematini) occur on one species of willow, *Salix lasiolepis*, throughout its range in the western United States (Fig. 4). They show a clearly defined phylogenetic line with the free feeders as the basic stock from which have been derived the leaf folder, leaf galler, petiole galler, stem galler, and bud galler, probably in that order (Price, 1984). The resources they utilize are clearly defined, easily measured, and they are adjacent on the plant, such that competition could play a role in community organization. Willow can be propagated readily, and stems from different source plants and locations can be planted in common conditions in a willow garden. Plants potted in a uniform soil can be treated under different regimes of water, nutrients, and light to test their effects on sawfly choice and survival. Enemies are usually abundant in galling species, as in this case, so data on the third trophic level can be readily collected, enabling a test of the enemy impact paradigm. Desirable community qualities listed above are met.

4.2 Kinds of Species

Community ecology in the past has paid some attention to the kinds of species in communities: strong competitors and weak competitors; herbi-

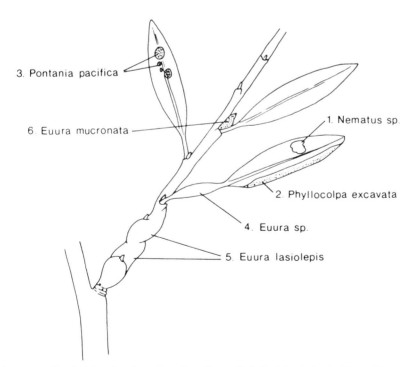

Figure 4. Six closely related species of sawfly on *Salix lasiolepsis* in the Flagstaff area, and the resources each species exploits.

vores, predators, parasites, and so on; keystone species (Paine, 1966, 1969); organizer species (Price, 1971); specialists and generalists (Southwood, 1960; Berenbaum, 1981); high incidence species; and A–D type tramps and supertramps (Diamond, 1975). In the future, as well as considerations of species number and relative abundance, the kinds of species present or absent must be given central importance. We all acknowledge that an elephant and a mouse have different requirements and respond in different ways to a given environment, but such differences are not commonly accounted for in community studies.

Of particular importance is the recognition that communities largely composed of different kinds of organisms may come under the influence of different paradigms. Communities of specialists may fit the individualistic response paradigm whereas generalists are more likely to fit the resource heterogeneity paradigm. For example, communities of parasites and predators contrast in their modes of community organization. Communities composed of year-round residents (e.g., desert rodents) are likely to be organized in ways different from those in which resources are colonized anew each year (e.g., migratory birds). Thus, more consideration must be given, not only to the different kinds of resources that lead to different kinds of community organization, but to the different kinds of species that exploit these resources.

4.3   Kinds of Interaction

Ecologists have recognized for a long time that many kinds of interaction occur between living organisms in nature: mutualism, parasitism and predation, commensalism, competition and amensalism (Burkholder, 1952; Odum, 1959), plus the condition of neutralism or noninteraction. Yet our literature is dominated by selected antagonistic relationships—predation and competition—probably resulting from Darwin's (1859) early influence. We have yet to assess the relative importance of all kinds of relationships.

Community ecology should consider in much more detail the neutralistic, mutualistic, amensalistic, and parasitic relationships, while continuing to assess the importance of predation and competition. There is evidence that amensalism is more common than competition in nature in certain taxa (Lawton and Hassell, 1981), and parasitism undoubtedly plays an important role in natural communities (Price, 1980). Mutualism certainly is also important and yet it has hardly been studied as an organizing influence in natural communities, a point that is emphasized in Chapter 16.

One of the challenges in the future of community ecology is to adequately assess the roles of all types of interaction between species.

5   SUMMARY

In the consideration of organizing influences in natural communities it is essential to recognize major differences between resource patterns in time:

whether they are rapidly increasing, pulsing or ephemeral, steadily renewed, constant, or rapidly decreasing. By recognizing differences in population response to such resources, predictions are made on when competition is most likely to be an important organizing influence in communities, it being most intense when the response is rapid and the resource is steadily renewed, constant, or rapidly declining. However, patchiness of resources, their nonequilibrium state in many cases, and the commonness of vacant niches in natural communities lessen the probability of competition playing an important role.

Alternative paradigms are reviewed that are likely to play a role in community organization, and predictions are made in relation to resource type and population response on the conditions under which each paradigm will apply. The paradigms considered are the individualistic response, resource heterogeneity, island or patch size, time, and enemy impact paradigms.

In a discussion of the future of community ecology an emphasis is placed on testing simultaneously among paradigms using communities in which experimental manipulation is possible, focusing more attention on the kinds of species coexisting in communities, and evaluating the importance of interactions other than competition and predation, in particular, neutralism amensalism, parasitism, and mutualism.

ACKNOWLEDGMENTS

I am grateful to my colleagues John Lawton, Donald Strong, and John Thompson for their reviews of this manuscript. Financial support was provided through National Science Foundation grants DEB 80-21754 and DEB 82-05904.

LITERATURE CITED

Addicott, J. F. 1978. Niche relationships among species of aphids feeding on fireweed. *Can. J. Zool.* **56**:1837–1841.

Alexander, M. 1971. *Microbial ecology.* Wiley, New York.

Andrewartha, H. G., and L. C. Birch. 1954. *The distribution and abundance of animals.* University of Chicago Press, Chicago.

Askew, R. R. 1961. On the biology of the inhabitants of oak galls of Cynipidae (Hymenoptera) in Britain. *Trans. Soc. Br. Entomol.* **14**:237–268.

Atsatt, P. R. 1981a. Lycaenid butterflies and ants: Selection for enemy-free space. *Am. Nat.* **118**:638–654.

Atsatt, P. R. 1981b. Ant dependent oviposition by the mistletoe butterfly *Ogyris amaryllis.* *Oecologia* **48**:60–63.

Bach, C. E. 1980. Effects of plant density and diversity on the population dynamics of a specialist herbivore, the striped cucumber beetle, *Acalymma vittata* (Fab.). *Ecology* **61**:1515–1530.

Beaver, R. A. 1977. Non-equilibrium "island" communities: Diptera breeding in dead snails. *J. Anim. Ecol.* **46**:783–798.

Beaver, R. A. 1979. Host specificity of temperate and tropical animals. *Nature (London)* **281**:139–141.

Benzing, D. H. 1978a. The life history profile of *Tillandsia circinnata* (Bromeliacae) and the rarity of extreme epiphytism among the angiosperms. *Selbyana* **2**:325–337.

Benzing, D. H. 1978b. Germination and early establishment of *Tillandsia circinnata* Schlecht. (Bromeliaceae) on some of its hosts and other supports in southern Florida. *Selbyana* **5**:95–106.

Berenbaum, M. 1978. Toxicity of a furanocoumarin to armyworms: A case of biosynthetic escape from insect herbivores. *Science* **201**:532–534.

Berenbaum, M. 1981. Patterns of furanocoumarin distribution and insect herbivory in the Umbelliferae: Plant chemistry and community structure. *Ecology* **62**:1254–1266.

Birks, H. J. B. 1980. British trees and insects: A test of the time hypothesis over the last 13,000 years. *Am. Nat.* **115**:600–605.

Boag, P. T., and P. R. Grant. 1981. Intense natural selection in a population of Darwin's finches (Geospizinae) in the Galápagos. *Science* **214**:82–85.

Bowers, M. A., and J. H. Brown. 1982. Body size and coexistence in desert rodents: Chance or community structure. *Ecology* **63**:391–400.

Brower, L. P. 1958. Bird predation and food plant specificity in closely related procryptic insects. *Am. Nat.* **92**:183–187.

Brown, J. H., O. J. Reichman, and D. W. Davidson. 1979. Granivory in desert ecosystems. *Annu. Rev. Ecol. Syst.* **10**:201–227.

Brown, W. L., Jr., and E. O. Wilson. 1956. Character displacement. *Syst. Zool.* **5**:49–64.

Burkholder, P. R. 1952. Cooperation and conflict among primitive organisms. *Am. Sci.* **40**:601–631.

Cody, M. L. 1974. *Competition and the structure of bird communities*. Princeton University Press, Princeton, New Jersey.

Cole, J. J. 1982. Interactions between bacteria and algae in aquatic ecosystems. *Annu. Rev. Ecol. Syst.* **13**:291–314.

Connell, J. H. 1979. Tropical rain forests and coral reefs as open non-equilibrium systems. *Symp. Br. Ecol. Soc.* **20**:141–163.

Connell, J. H. 1980. Diversity and the coevolution of competitors, or the ghost of competition past. *Oikos* **35**:131–138.

Connor, E. F., and E. D. McCoy. 1979. The statistics and biology of the species-area relationship. *Am. Nat.* **113**:791–833.

Cromartie, W. J. 1975. The effect of stand size and vegetational background on the colonization of cruciferous plants by herbivorous insects. *J. Appl. Ecol.* **12**:517–533.

Darlington, A. 1975. *A pocket encyclopaedia of plant galls in colour*, rev. ed. Blandford Press, Poole, England.

Darwin, C. 1859. *The origin of species by means of natural selection*. Murray, London.

Davidson, D. W., J. H. Brown, and R. S. Inouye. 1980. Competition and the structure of granivore communities. *Bioscience* **30**:233–238.

Davis, B. N. K. 1973. The Hemiptera and Coleoptera of stinging nettle (*Urtica dioica L.*) in East Anglia. *J. Appl. Ecol.* **10**:213–237.

Dayton, P. K. 1971. Competition, disturbance, and community organization: The provision and subsequent utilization of space in a rocky intertidal community. *Ecol. Monogr.* **41**:351–389.

Diamond, J. M. 1975. Assembly of species communities. In M. L. Cody and J. M. Diamond (eds.), *Ecology and evolution of communities*, pp. 342–444. Belknap Press, Cambridge, Massachusetts.

Felt, E. P. 1940. *Plant galls and gall makers*. Comstock, Ithaca, New York.

Fowler, S. V., and J. H. Lawton. 1982. The effects of host-plant distribution and local abundance on the species richness of agromyzid flies attacking British umbellifers. *Ecol. Entomol.* 7:257–265.

Gilbert, L. E. 1975. Ecological consequences of a coevolved mutualism between butterflies and plants. In L. E. Gilbert and P. H. Raven (eds.), *Coevolution of animals and plants*, pp. 210–240. University of Texas Press, Austin.

Gleason, H. A. 1926. The individualistic concept of the plant association. *Bull. Torrey Bot. Club* 53:7–26.

Grant, B. R., and P. R. Grant. 1982. Niche shifts and competition in Darwin's finches: *Geospiza conirostris* and congeners. *Evolution* 36:637–657.

Grant, P. R. 1972. Convergent and divergent character displacement. *Biol. J. Linn. Soc.* 4:39–68.

Grant, P. R. 1975. The classical case of character displacement. *Evol. Biol.* 8:237–337.

Grant, P. R., and I. Abbott. 1980. Interspecific competition, island biogeography and null hypotheses. *Evolution* 34:332–341.

Grime, J. P. 1977. Evidence for the existence of three primary strategies in plants and its relevance to ecological and evolutionary theory. *Am. Nat.* 111:1169–1194.

Grime, J. P. 1979. *Plant strategies and vegetation processes*. Wiley, London.

Harper, J. L. 1977. *Population biology of plants*. Academic Press, New York.

Högstedt, G. 1980. Prediction and test of the effects of interspecific competition. *Nature (London)* 283:64–66.

Holloway, J. D., and P. D. N. Herbert. 1979. Ecological and taxonomic trends in macrolepidopteran host plant selection. *Biol. J. Linn. Soc.* 11:229–251.

Holmes, J. C. 1961. Effects of concurrent infections on *Hymenolepis diminuta* (Cestoda) and *Moniliformis dubius* (Acanthocephala). I. General effects and comparison with crowding. *J. Parasitol.* 47:209–216.

Holmes, J. C. 1962a. Effects of concurrent infections on *Hymenolepis diminuta* (Cestoda) and *Moniliformis dubius* (Acanthocephala). II. Effects on growth. *J. Parasitol.* 48:87–96.

Holmes, J. C. 1962b. Effects of concurrent infections on *Hymenolepis diminuta* (Cestoda) and *Moniliformis dubius* (Acanthocephala). III. Effects in hamsters. *J. Parasitol.* 48:97–100.

Inouye, D. 1977. Species structure of bumblebee communities in North America and Europe. In W. J. Mattson (ed.), *The role of arthropods in forest ecosystems*, pp. 35–46. Springer-Verlag, New York.

Inouye, D. W. 1978. Resource partitioning in bumblebees: Experimental studies of foraging behavior. *Ecology* 59:672–678.

Kareiva, P. 1982. Exclusion experiments and the competitive release of insects feeding on collards. *Ecology* 63:696–704.

Karr, J. R. 1975. Production, energy pathways, and community diversity in forest birds. In F. B. Golley and E. Medina (eds.), *Tropical ecological systems: Trends in terrestrial and aquatic research*, pp. 161–176. Springer-Verlag, New York.

Lawton, J. H. 1978. Host-plant influences on insect diversity: The effects of space and time. In L. A. Mound and N. Waloff (eds.), *Diversity of insect faunas*, pp. 105–125. Symp. Roy. Entomol. Soc. London 9. Blackwell Scientific, Oxford.

Lawton, J. H. 1982. Vacant niches and unsaturated communities: A comparison of bracken herbivores at sites on two continents. *J. Anim. Ecol.* 51:573–595.

Lawton, J. H. 1983. Non-competitive populations, non-convergent communities, and vacant niches: The herbivores on bracken. In D. R. Strong, D. S. Simberloff, and L. G. Abele (eds.), *Ecological communities: Conceptual issues and the evidence*. Princeton University Press, Princeton, New Jersey. In press.

Lawton, J. H., and M. P. Hassell. 1981. Asymmetrical competition in insects. *Nature (London)* **289**:793–795.

Lawton, J. H., and S. McNeill. 1979. Between the devil and the deep blue sea: On the problem of being a herbivore. In R. M. Anderson, B. D. Turner, and L. R. Taylor (eds.), *Population dynamics*, pp. 223–244. Symp. Brit. Ecol. Soc. 20.

Lawton, J. H., and P. W. Price. 1979. Species richness of parasites on hosts: Agromyzid flies on the British Umbelliferae. *J. Anim. Ecol.* **48**:619–637.

Lawton, J. H., and D. R. Strong, Jr. 1981. Community patterns and competition in folivorous insects. *Am. Nat.* **118**:317–338.

Lister, B. C. 1980. Resource variation and the structure of British bird communities. *lProc. Natl. Acad. Sci. USA* **77**:4185–4187.

Lynch, J. M., and N. J. Poole (eds.). 1979. *Microbial ecology: A conceptual approach*. Blackwell, Oxford.

Lynch, M. 1978. Complex interactions between natural coexploiters—*Daphnia* and *Ceriodaphnia*. *Ecology* **59**:552–564.

MacArthur, R. H. 1972. *Geographical ecology: Patterns in the distribution of species*. Harper and Row, New York.

MacArthur, R. H., and E. O. Wilson. 1963. An equilibrium theory of insular zoogeography. *Evolution* **17**:373–387.

MacArthur, R. H., and E. O. Wilson, 1967. *The theory of island biogeography*. Princeton University Press, Princeton, New Jersey.

McClure, M. S. 1974. Biology of *Erythroneura lawsoni* (Homoptera: Cicadellidae) and coexistence in the sycamore leaf-feeding guild. *Environ. Entomol.* **3**:59–68.

McClure, M. S., and P. W. Price. 1975. Competition among sympatric *Erythroneura* leafhoppers (Homoptera: Cicadellidae) on American Sycamore. *Ecology* **56**:1388–1397.

Munger, J. C., and J. H. Brown. 1981. Competition in desert rodents: An experiment with semipermeable exclosures. *Science* **211**:510–512.

Neill, W. E. 1975. Experimental studies of microcrustacean competition, community composition and efficiency of resource utilization. *Ecology* **56**:809–826.

Odum, E. P. 1959. *Fundamentals of ecology*, 2nd ed. Saunders, Philadelphia.

Odum, E. P. 1971. *Fundamentals of ecology*, 3rd ed. Saunders, Philadelphia.

Opler, P. A. 1974. Oaks as evolutionary islands for leaf-mining insects. *Am. Sci.* **62**:67–73.

Otte, D., and A. Joern. 1977. On feeding patterns in desert grasshoppers and the evolution of specialized diets. *Proc. Acad. Nat. Sci., Philadelphia* **128**:89–126.

Paine, R. T. 1966. Food web complexity and species diversity. *Am. Nat.* **100**:66–75.

Paine, R. T. 1969. The *Pisaster-Tegula* interaction: Prey patches, predator food preference, and intertidal community structure. *Ecology* **50**:950–961.

Park, T. 1948. Experimental studies of interspecies competition. I. Competition between populations of the flour beetles, *Tribolium confusum* Duval and *Tribolium castaneum* Herbst. *Ecol. Monogr.* **18**:265–308.

Pianka, E. R. 1974. *Evolutionary ecology*. Harper and Row, New York.

Pierce, N. E., and P. S. Mead. 1981. Parasitoids as selective agents in the symbiosis between lycaenids and ants. *Science* **211**:1185–1186.

Pimm, S. L. 1978. An experimental approach to the effects of predictability on community structure. *Am. Zool.* **18**:797–808.

Price, P. W. 1970. Characteristics permitting coexistence among parasitoids of a sawfly in Quebec. *Ecology* **51**:445–454.

Price, P. W. 1971. Niche breadth and dominance of parasitic insects sharing the same host species. *Ecology* **52**:587–596.

Price, P. W. 1976. Colonization of crops by arthropods: Non-equilibrium communities in soybean fields. *Environ. Entomol.* **5**:605–611.

Price, P. W. 1977. General concepts on the evolutionary biology of parasites. *Evolution* **31**:405–420.

Price, P. W. 1980. *Evolutionary biology of parasites.* Princeton University Press, Princeton, New Jersey.

Price, P. W. 1983a. Hypotheses on organization and evolution in herbivorous insect communities. In R. F. Denno and M. S. McClure (eds.), Variable plants and herbivores in natural and managed systems, pp. 559–596. Academic Press, New York.

Price, P. W. 1983b. Communities of specialists: Vacant niches in ecological and evolutionary time. In D. R. Strong, D. S. Simberloff, and L. G. Abele (eds.), *Ecological Communities: Conceptual issues and the evidence.* Princeton University Press, Princeton, New Jersey. In press.

Price, P. W. 1984. The gall-inducing Tenthredinidae. In J. D. Shorthouse and O. Rohfritsch (eds.), *Biology of insect and Acarina induced galls.* Praeger, New York. In press.

Price, P. W., and H. A. Tripp. 1972. Activity patterns of parasitoids on the Swaine jack pine sawfly, *Neodiprion swainei* (Hymenoptera: Diprionidae), and parasitoid impact on the host. *Can. Entomol.* **104**:1003–1016.

Price, P. W., C. E. Bouton, P. Gross, B. A. McPheron, J. N. Thompson, and A. E. Weis. 1980. Interactions among three trophic levels: Influence of plants on interactions between insect herbivores and natural enemies. *Annu. Rev. Ecol. Syst.* **11**:41–65.

Ralph, C. P. 1977a. Effect of host plant density on populations of a specialized, seed-sucking bug *Oncopeltus fasciatus. Ecology* **58**:799–809.

Ralph, C. P. 1977b. Search behavior of the milkweed bug, *Oncopeltus fasciatus* (Hemiptera: Lygaedae). *Ann. Entomol. Soc. Am.* **70**:337–342.

Ramensky, L. G. 1926. Die Grundgesetzmässigkeiten im Aufbau der Vegetationsdecke. *Bot. Centralbl. N. F.* **7**:453–455.

Rathcke, B. J. 1976a. Insect plant patterns and relationships in the stem-boring guild. *Am. Midl. Nat.* **99**:98–117.

Rathcke, B. J. 1976b. Competition and coexistence within a guild of herbivorous insects. *Ecology* **57**:76–87.

Rey, J. R. 1981. Ecological biogeography of arthropods on *Spartina* islands in northwest Florida. *Ecol. Monogr.* **51**:237–265.

Ricklefs, R. E. 1979. *Ecology*, 2nd ed. Chiron Press, Newton, Massachusetts.

Ricklefs, R. E., and K. O'Rourke. 1975. Aspect diversity in moths: A temperate-tropical comparison. *Evolution* **29**:313–324.

Rigby, C., and J. H. Lawton. 1981. Species-area relationships of arthropods on host plants: Herbivores on bracken. *J. Biogeog.* **8**:125–133.

Risch, S. J. 1981. Insect herbivore abundance in tropical monocultures and polylcultures: An experimental test of two hypotheses. *Ecology* **62**:1325–1340.

Risch, S. J., and C. R. Carroll. 1982. Effect of keystone predaceous ant, *Solenopsis geminata*, on arthropods in a tropical agroecosystem. *Ecology* **63**:1979–1983.

Rohde, K. 1978a. Latitudinal differences in host-specificity of marine Monogenea and Digenea. *Marine Biol.* **47**:125–134.

Rohde, K. 1978b. Latitudinal gradients in species diversity and their causes. II. Marine parasitological evidence for a time hypothesis. *Biol. Zentrabl.* **97**:405–418.

Rohde, K. 1979. A critical evaluation of intrinsic and extrinsic factors responsible for niche restriction in parasites. *Am. Nat.* **114**:648–671.

Rohde, K. 1980. Diversity gradients of marine Monogenea in the Atlantic and Pacific Oceans. *Experientia* **36**:1368–1369.

Root, R. B. 1973. Organization of a plant-arthropod association in simple and diverse habitats: The fauna of collards (*Brassica oleracea*). *Ecol. Monogr.* **43**:95–124.

Roskam, J. C., and G. A. van Uffelen. 1981. Biosystematics of insects living in female birch catkins. III. Plant-insect relation between white birches, *Betula* L., Section Excelsae (Koch) and gall midges of the genus *Semudobia* Kieffer (Diptera, Cecidomyiidae). *Neth. J. Zool.* **31**:533–553.

Sale, P. F. 1977. Maintenance of high diversity in coral reef fish communities. *Am. Nat.* **111**:337–359.

Seifert, R. P., and F. H. Seifert. 1976. A community matrix analysis of *Heliconia* insect communities. *Am. Nat.* **110**:461–483.

Seifert, R. P., and F. H. Seifert. 1979. A *Heliconia* insect community in a Venezuelan cloud forest. *Ecology* **60**:462–467.

Shapiro, A. M. 1981. The pierid red-egg syndrome. *Am. Nat.* **117**:276–294.

Shorrocks, B., W. Atkinson, and P. Charlesworth. 1979. Competition on a divided and ephemeral resource. *J. Anim. Ecol.* **48**:899–908.

Simberloff, D. S. 1978. Colonization of islands by insects: Immigration, extinction, and diversity. In L. A. Mound and N. Waloff (eds.), *Diversity of insect faunas*, pp. 139–153. Symp. Roy. Entomol. Soc. London 9.

Simberloff, D. S., and W. Boecklen. 1981. Santa Rosalia reconsidered: Size ratios and competition. *Evolution* **35**:1206–1228.

Simberloff, D. S., and E. O. Wilson. 1969. Experimental zoogeography of islands: The colonization of empty islands. *Ecology* **50**:278–296.

Simberloff, D. S., and E. O. Wilson. 1970. Experimental zoogeography of islands: A two-year record of colonization. *Ecology* **51**:934–937.

Southwood, T. R. E. 1960. The abundance of the Hawaiian trees and the number of their associated insect species. *Proc. Hawaiian Entomol. Soc.* **17**:299–303.

Southwood, T. R. E. 1961. The number of species of insect associated with various trees. *J. Anim. Ecol.* **30**:1–8.

Southwood, T. R. E. 1977. Habitat, the templet for ecological strategies? *J. Anim. Ecol.* **46**:337–365.

Stiling, P. D. 1980. Competition and coexistence among *Eupteryx* leafhoppers (Hemiptera: Cicadellidae) occurring on stinging nettles (*Urtica dioica* L.). *J. Anim. Ecol.* **49**:793–805.

Strong, D. R., Jr. 1974a. Nonsymptotic species richness models and the insects of British trees. *Proc. Natl. Acad. Sci. USA* **71**:2766–2769.

Strong, D. R., Jr. 1974b. The insects of British trees: Community equilibrium in ecological time. *Ann. Missouri Bot. Gdn.* **61**:692–701.

Strong, D. R., Jr. 1979. Biogeographic dynamics of insect-host plant communities. *Annu. Rev. Entomol.* **24**:89–119.

Strong, D. R., Jr. 1981. The possibility of insect communities without competition: Hispine beetles on *Heliconia*. In R. F. Denno and H. Dingle (eds.), *Insect life history patterns: Habitat and geographic variation*, pp. 183–194. Springer-Verlag, New York.

Strong, D. R., Jr. 1982. Harmonious coexistence of hispine beetles on *Heliconia* in experimental and natural communities. *Ecology* **63**:1039–1049.

Strong, D. R., Jr., L. A. Szyska, and D. S. Simberloff. 1979. Tests of community-wide character displacement against null hypotheses. *Evolution* **33**:897–913.

Thompson, J. N., and P. W. Price. 1977. Plant plasticity, phenology and herbivore dispersion: Wild parsnip and the parsnip webworm. *Ecology* **58**:1112–1119.

Tilman, D. 1982. *Resource competition and community structure*. Princeton University Press, Princeton, New Jersey.

Tilman, D., S. S. Kilham, and P. Kilham. 1982. Phytoplankton community ecology: The role of limiting nutrients. *Annu. Rev. Ecol. Syst.* **13**:349–372.

Wertheim, G. 1970. Experimental concurrent infections with *Strongyloides ratti* and *S. venezuelensis* in laboratory rats. *Parasitology* **61**:389–395.

Whittaker, R. H. 1952. A study of summer foliage insect communities in the Great Smoky Mountains. *Ecol. Monogr.* **22**:1–44.

Whittaker, R. H. 1956. Vegetation of the Great Smoky Mountains. *Ecol. Monogr.* **26**:1–80.

Whittaker, R. H. 1967. Gradient analysis of vegetation. *Biol. Rev.* **42**:207–264.

Wiens, J. A. 1973. Pattern and process in grassland bird communities. *Ecol. Monogr.* **43**:237–270.

Wiens, J. A. 1974. Habitat heterogeneity and avian community structure in North American grasslands. *Am. Midl. Nat.* **91**:195–213.

Williams, C. B. 1964. *Patterns in the balance of nature*. Academic Press, London.

Williams, J. B., and G. O. Batzli. 1979. Competition among bark-foraging birds in central Illinois: Experimental evidence. *Condor* **81**:122–132.

Wilson, E. O. 1969. The species equilibrium. In G. M. Woodwell and H. H. Smith (eds.), *Diversity and stability in ecological systems*, pp. 38–47. Brookhaven Symp. Biol. 22.

PART V

Synthesis

CHAPTER **14**

What's New?
Community Ecology
Discovers Biology

ROBERT K. COLWELL
Department of Zoology
University of California
Berkeley, California

CONTENTS

1 COOKING-POT ECOLOGY

Traditionally, we are told that a community is no more than a set of interacting or potentially interacting populations that coexist in a habitat. Like
some fragrant Indian curry, you just sauté the right proportions of a great

number of species together, then simmer gently, stirring periodically. If you measure carefully, and the ingredients are pure, the flavor and consistency should be identical each time you cook up that particular community.

Although it may well be that no one ever truly believed that this cooking-pot theory of community ecology was generally applicable to nature, a rather great proportion of the mathematical theory of communities over the past 20 years relies on cooking-pot assumptions. These assumptions ultimately boil down to four.

1. *Homogeneity of Ingredients.* A gram of coriander is a gram of coriander; a vireo is a vireo and an oak leaf is an oak leaf, regardless of genotype, sex, age, condition, or geographical origin.

2. *Continuity of Ingredients.* The cinnamon this year tastes the same as it did last year; this year's spider mites are just like last year's in spite of the passage of the gene pool through 30 generations of the tribulations of Moses, and in spite of year-to-year vagaries of environment.

3. *Homogeneity of Interactions.* A serving of really well-stirred curry tastes the same from anywhere in the pot; species interactions in a cooking-pot community occur everywhere simultaneously with the same distribution of probabilities.

4. *Continuity of Interactions.* The curry turns out the same regardless of the exact stirring schedule; the role of unique events, and episodic or catastrophic disturbance in shaping community structure is not part of the world depicted by cooking-pot community models, because they are deterministic models. The curry is also supposed to taste the same forever, once cooked; the role of pulsed or intrinsically declining resources is not part of cooking-pot models, because they are equilibrium models.

There is absolutely nothing new, to ecologists in 1982, about the observations just made, although I have tried to state them in a fresh way. In fact, they have not been new for a long time. After drafting my own analysis, I found that, in one of his last papers, MacArthur (1972a) listed four shortcomings of ecological theory that map almost exactly on my own four categories. He appealed to everyone to work toward a more realistic theory. In my view, much progress has been made in the intervening 10 years, both in uncovering the kinds of reality that need to be incorporated, and in building new and better theoretical mousetraps. It is puzzling, but I guess expected, that some people would rather complain about the old mousetraps than try the new ones.

2   TOWARD A COMMUNITY BIOLOGY

Levins (1966) pointed out long ago that no model can simultaneously emphasize realism, generality, and precision. I doubt that community models will ever be very precise, in Levins' sense, so we are left with trying to optimize the mix of realism and generality.

Cooking-pot community theory was intended to capture the essentials of community structure, regardless of biological detail; it was intended to be very general. In fact, the assumptions I outlined earlier make this body of theory very specific. It works best for cases in which the four assumptions are either roughly true (vireo gene pools probably change very little on ecological time scales) or else false in detail but approximately true for averages (as in the neglect of age structure and physiological condition, for species in fairly stable conditions).

When one or more of the assumptions of homogeneity and continuity of ingredients and interactions are grossly violated, traditional theory is likely to be downright wrong in its predictions. The message of the past 10 years of ecological research is clear: those pesky "biological details" matter a lot, because they reflect the violation of orthodox cooking-pot assumptions in the daily lives of real organisms.

In fact, I have become so impressed with the necessity for an intimate understanding of lives of particular organisms in the study of communities that I have all but abandoned the term "community ecology" in favor of "community biology." The new term certainly embraces the traditional pursuits of community ecologists, for example, ecological biogeography, patterns of resource use, patterns of species interaction, relative abundance, and species diversity. But community biology approaches questions in these areas by the study of particular *kinds* of organisms. The focus is on features that affect species interactions, recognizing that these important features vary enormously among living things.

I will give a few examples. To understand fully the role of plants as resources for herbivores and the role of herbivores in the lives of plants, it will be necessary to study the mechanical and biochemical properties of the many kinds of plant tissue that are food to herbivores, the physiological and genetic responses of plants to herbivory and of herbivores to plants, the life history patterns of both, the cost of defense to plants and the cost of processing to herbivores, the cost of herbivory to plants, the way that herbivores detect food plants and discriminate among them, and the patterns of variability in all of these phenomena. On the other hand, with vertebrate predators and vertebrate prey, we may more often ignore biochemical defenses than we can with plants, but must certainly look more closely at behavioral defenses.

The investigation of competitive relationships among plants may require the study of growth rates, plant architecture, leaf morphology, photosynthetic response curves, nutrient requirements and uptake rates, water relations, mechanisms of reproduction and dispersal, and relations with mutualists and herbivores. Investigations of competition between animal species demand a comparable list of considerations, but with rather different emphasis, and very different techniques.

The special features of mutualistic interactions are simply too diverse even to attempt a sketch. Many mutualisms illustrate clearly that the nature of the interaction depends strongly on quirks of each species: thorns on acacias, leaky cell membranes for lichens, fecal licking behavior in termites, to name a few (see also Chapter 16). An understanding of the complex relations between competitive and trophic interactions, and the way that direct and indirect mutualistic relations affect competition and predation makes even greater demands on our grasp of the biology of the species involved.

3 A ROSE IS NOT A ROSE

By the term "community structure," we usually mean patterns of resource use and species interactions, as well as community composition. Cooking-pot theories of community structure failed as generalities precisely because not only population structure, but organismal features were ignored.

The other contributors to this volume ·have presented a great variety of examples of the need to integrate individual and population phenomena into our consideration of communities. It may be useful to categorize some of these examples, along with others, according to which of the four cooking-pot assumptions they violate.

1. *Homogeneity of Ingredients.*   Witham et al. (Chapter 2) and Frankie and Morgan (Chapter 4) warn against assuming that a leaf is a leaf, even on the same tree. Edmonds and Alstad (1978) showed that herbivores of the same species may differ genetically in consistent ways on adjacent trees. Dingle (Chapter 6) discusses ecological consequences of genotypic differences for diapause and migration. (A milkweed bug is not a milkweed bug.) Istock (Chapter 5) documents the surprising degree of heritable variation for life history traits, and points out the importance of age structure to resource tracking.

Wilbur (Chapter 7) makes a convincing case that organisms with complex life cycles must be treated realistically to understand their role in communities: a tadpole is not a frog. Age/size structures of populations often play a crucial role in species interactions, even for species that lack metamorphosis (Murdoch and Oaten, 1979). In plants and other modular organisms (Harper, 1981) the age structure, not only of "individuals" but of organs (e.g., leaves), must often be considered. Geographic variation among populations of widespread species, though recognized for centuries for morphological characters and now known for many biochemical and behavioral traits, has been largely ignored in ecological models, in spite of its pertinence to species interactions. Arnold's work with geographic variation in the innate prey preferences of garter snakes is a striking example of the integrative approach that is needed (Arnold, 1981). Lawton (Chapter 12) shows that, to an ant on bracken, a caterpillar is not a caterpillar: if the caterpillar lives on Yorkshire bracken, it is distasteful to the ant, but might be delicious elsewhere.

2. *Continuity of Ingredients.* The long-held belief of ecologists that evolution may be safely ignored on "ecological" time scales is increasingly discredited. The fact that even "fitness" traits turn out to have a considerable degree of heritable variation in natural populations (Istock, Chapter 5; Dingle, Chapter 6) means that short-term evolution of a meaningful sort is not only possible but commonplace. The experience of plant breeders with loss of "resistance" to pathogens and the genetic response of arthropods to pesticides cannot be ignored in community ecology, if our models are to have any relevance to human affairs.

3. *Homogeneity of Interactions.* The assumption that individual organisms of two species have the same probability of interacting, and the same consequences of interacting, throughout the physical extent of a community, is so rarely true that it brings a blush to the face of any god-fearing ecological modeler nowadays, although some population geneticists still behave quite shamelessly in this regard.

Patchiness, or graininess, exists on every scale in the natural world, and has profound consequences for species interactions. Plants and other sessile organisms spend most of their lives with the same neighbors. Trophic links are dependent not only on the density but on the spatial pattern of the species involved. Coevolution depends critically on the structuring of species interactions, as Wilson (1980) and others have shown.

Slobodchikoff (Chapter 8) discusses the role of patchiness in animal social systems, and Caraco and Pulliam (Chapter 10) show how social interactions can affect trophic links. Lawton's bracken ferns (Chapter 12) show the classic relation between patch size and species diversity, a nearly universal pattern that makes a prima facie case against the homogeneity of interactions in space. Price's grand scheme for categorizing the forces that affect species diversity in general has patchiness as one coordinate (Chapter 13).

Homogeneity of species interactions may itself vary in intriguing ways on a local or geographic scale. Inger and I showed that the microhabitat distribution of some tropical amphibians and reptiles is more uniform in assemblages from climatically unpredictable habitats than in assemblages from contiguous, but more predictable habitats (Inger and Colwell, 1977). Among hummingbirds, conspecific individuals within higher elevation tropical communities tend to have rather similar foraging repertories, whereas individuals within lowland tropical communities vary markedly among themselves in the plant species they visit, judging from the pattern of occurrence of host-plant-specific flower mites found on birds (Colwell, unpublished data).

There is a growing body of evidence that the distribution of the importance of interactions in natural communities is not homogeneous, and not even unimodal: species interact strongly within subsets of the species present, and weakly between subsets (Inger and Colwell, 1977, and subsequent references reviewed in May 1981, pp. 221–225).

4. *Continuity of Interactions.* The importance of regular or episodic disturbance to community structure is now well established for intertidal organisms and terrestrial plants (Connell, 1978; Sousa, 1983). Less well documented is the role of rare catastrophic events in the evolution of interactions and the structure of communities. In the winter of 1972 in Berkeley, California, where any frost is extraordinary, there were 11 consecutive nights of freezing weather (Nora, 1973). The hills were soon a graveyard of century-old introduced *Eucalyptus* trees, though many resprouted. The University Botanic Garden, where plants from virtually any climate will normally grow (if not reproduce), was a ghastly morgue of sagging *Saguaros*, frozen fuchsias, and other exotic casualties. Yet native plants were scarcely affected, beyond the loss of fruits and flowers. They had seen it all before. On the other hand, the effect on certain wintering birds was disastrous, as, for example, in the case of frugivores and hummingbirds. Knowlton and colleagues (1981) followed the fate of staghorn corals in Jamaican waters after a devastating hurricane. They found that mortality from predation during a 5-month period after the hurricane far exceeded mortality directly attributable to the storm. The most likely explanation appears to be that the hurricane affected the corals far more than their predators, who lived on long enough to decimate the corals. We can only surmise that an unknown but perhaps surprisingly large proportion of organismal adaptations and demographic patterns have been molded by events that are exceedingly infrequent on a human time scale, but nonetheless recurrent.

Price's classification (Chapter 13) of patterns of resource renewal and decline emphasizes the importance of time-varying interactions between resources and consumer, and effects transmitted to higher trophic levels.

4 DEATH OF THE "OLD ECOLOGY"?: EPITAPH FOR A STRAW MAN

It may sound as if I am saying that nature is so complex and variable, and the products of evolution and biogeography so special that useful community theory is not possible. In fact, I am quite convinced that community theory is by no means at the catastrophic bifurcation where some seem to see it (e.g., Simberloff, 1980; Price, Chapter 13): an old paradigm discredited, and no new one to replace it, or a new one that somehow claims to have explained a phenomenon once it is reduced to chance.

In fact, the integration of realistic substitutes for one or more of the four cooking-pot assumptions has been steadily under way in ecological theory on several fronts (May, 1981; Schoener, 1982). Meanwhile, the useful work of pointing out the shortcomings of the old theories, and, especially, exposing the excesses of naive adherents to the old orthodoxy has been carried on by others.

To a degree, these two activities appear to have been independent, at least until recently. In the 1970s, experimental field ecology came of age. In 1975, Connell (1975) and Colwell and Fuentes (1975) were able to cover the field of experimental studies of species interactions with (intentionally) complementary reviews, citing a combined total of some 300 references. Today the task of reviewing the same scope of experimental ecology would be staggering [although Schoener (1983) covers competition]. Some journals will now accept *non*experimental papers only under unusual circumstances. The experimental evidence does not often reveal the well-ordered, equilibrium world of MacArthur and Levins (1967), but rather, a sometimes turbulent and bewildering place where disturbance, natural enemies, biochemistry, life histories, and behavior play leading roles, along with the original cast of competitors.

The response of theorists was to begin incorporating these new roles into the script. A comparison of MacArthur's *Geographical Ecology* (1972b), the first edition of May's *Theoretical Ecology* (1976), and the second edition of the latter (May, 1981) reveals the progress that has been made. The shift in emphasis from competition to predation (in the broad sense), the incorporation of stochastic effects in models, and the development of increasingly *biological* models for species interactions are all apparent. In a recent review of the second edition of May's book (May, 1981) for *The Auk* (Colwell, 1983), I was obliged to tell the journal's ornithological readers that birds were mentioned on only 18 pages out of 500. Theoretical ecology is no longer the ecology of birds.

A parallel response to the mounting experimental evidence for the limited domain of deterministic, equilibrium models of competitive interactions has been ardent skepticism regarding inferences from nonexperimental data (Strong et al., 1983; Strong, Chapter 11). The effect on the field will surely prove to have been quite positive in the long run, whatever divisiveness may be engendered in the short run. An insistence on statistical and biological evidence and on the disproof of alternative hypotheses ("null" or otherwise) is a thoroughly healthy development. The expulsion of resource competition, within or between species, from the pantheon of forces affecting populations and communities because it does not act strongly, consistently, and ubiquitously would not be healthy (Arthur, 1982; Schoener, 1982). Gravity is a weak force throughout the vast interstellar spaces of the universe, yet it would not be wise to ignore its critical role in the tiny regions where it is strong, or to expect to discover it at all by dropping apples at random points in the universe.

Interpretation of natural systems inevitably requires an interplay between the particular and the general, and people differ among themselves in the mix they find most interesting and explanatory. The special features (adaptive or not) of particular kinds of organisms constrain the ecological generalities they can support. In the other direction, generalities define the expected, and thereby permit us to identify and explore the exceptional. When the balance between these two processes swings too far in one direction or the other, the explanatory power of a science suffers: a compendium of case histories is unenlightening if recurrent patterns are not sought and interpreted, and a generality is of little use if the exceptions far outweigh the cases that approximate the expected. It is my view that ecology, particularly as it concerns species interactions, is undergoing not a revolution in paradigms (Simberloff, 1980), but a salubrious readjustment in the balance between our increasingly detailed appreciation of nature and the domain of our theories and models. [Schoener (1972) predicted just such a readjustment.] I see no call for despair that the direction of this shift is (currently) toward models of more limited domain. After all, classical and medieval philosophers proposed grand schemes that embraced the form and function of every entity in the universe, living and nonliving. We find their efforts merely quaint nowadays, from a scientific perspective, yet those models are the intellectual ancestors of our own. That ecology now has its very own quaint, overgeneral models to look back on, we may take as a sign of the maturity and vigor of our science.

To conclude, I can do no better than to quote a pertinent passage from another optimist:

I predict there will be erected a two- or three-way classification of organisms and their geometrical and temporal environments, this classification consuming most of the creative energy of ecologists. The future principles of the ecology of coexistence will then be of the form "for organisms of type A, in environments of structure B, such and such relations will hold." This is only a change in emphasis from present ecology. All successful theories, for instance in physics, have initial conditions; with different initial conditions, different things will happen. But I think initial conditions and their classification in ecology will prove to have vastly more effect on outcomes than they do in physics. . . . Bird censuses in a habitat in successive years or in similar habitats in one year are usually very similar, while insect censuses (to the extent they can be taken) seem often to differ dramatically from place to place and year to year. Thus, plausibly in our classification, insects, at least of some kinds, will go into a non-equilibrium category and birds into an equilibrium category. But the classification will be more pervasive than these examples suggest; many morphological, behavioral, and genetic parameters will probably be included. There has been a biological tradition of searching for the best organism to solve a problem—like Drosophila for chromosome genetics and viruses and bacteria for aspects of molecular genetics. The ecologist should resist this temptation. . . . Anyone familar with the history of science knows it is done in the most astonishing ways by the most improbable people and that its only real rules are honesty and validity of logic, and that even these are open to public scrutiny and correction.

The author was Robert MacArthur, more than 10 years ago (MacArthur,

1972a, p. 257–259). Given the content of this quotation, it is ironic that MacArthur is so frequently held responsible for the supposed intellectual tyranny of the concept of competition, and for the continuing prevalence of deterministic, equilibrium models in ecological theory. It is even more ironic that not one person in the large, well-read audience at the symposium this volume records was able to identify the author of the preceding passage when I presented it.

The last sentence of this quotation brings up one final point I wish to make. "The only real rules in science are honesty and validity of logic." *Experimental* falsifiability is not a rule, it is a tool, like mathematical modeling, statistical inference, a pair of binoculars, or an electron microscope. Observation, logical inference, and plausibility arguments are sometimes as capable of scientific revelation as experiments and statistics. Realistic experiments with primate or bird communities are not very much more feasible than experiments in astrophysics, but our curosity about stars and starlings is not thereby lessened.

ACKNOWLEDGMENTS

I am grateful to Robin Chazdon, Lloyd Goldwasser, and Peter Price for their useful comments on the manuscript.

LITERATURE CITED

Arnold, S. J. 1981. Behavioral variation in natural populations. II. The inheritance of a feeding response in crosses between geographic races of the garter snake, *Thamnophis elegans*. *Evolution* **35**:510–515.

Arthur, W. 1982. The evolutionary consequences of interspecific competition. *Adv. Ecol. Res.* **12**:127–187.

Colwell, R. K. 1983. Review of *Theoretical Ecology: Principles and applications*, 2nd edition. *Auk* **100**:261–262.

Colwell, R. K., and E. R. Fuentes. 1975. Experimental studies of the niche. *Annu. Rev. Ecol. Syst.* **6**:281–310.

Connell, J. H. 1975. Some mechanisms producing structure in natural communities. In M. L. Cody and J. M. Diamond (eds.), *Ecology and evolution of communities*, pp. 460–490. Belknap Press, Cambridge, Massachusetts.

Connell, J. H. 1978. Diversity in tropical rain forests and coral reefs. *Science* **199**:1302–1310.

Edmonds, G. F., and D. N. Alstad. 1978. Coevolution in insect herbivores and conifers. *Science* **199**:941–945.

Harper, J. L. 1981. The concept of population in modular organisms. In R. M. May (ed.), *Theoretical ecology*, 2nd ed., pp. 53–77. Sinauer, Sunderland, Massachusetts.

Inger, R. F., and R. K. Colwell. 1977. Organization of contiguous communities of amphibians and reptiles in Thailand. *Ecol. Monogr.* **47**:229–253.

Knowlton, N., J. C. Lang, M. C. Rooney, and P. Clifford. 1981. Evidence for delayed mortality in hurricane-damaged Jamaican staghorn corals. *Nature (London)* **294**:251–252.

Levins, R. 1966. The strategy of model building in population biology. *Am. Sci.* **54**:421–431.

MacArthur, R. H. 1972a. Coexistence of species. In J. Behnke (ed.), *Challenging biological problems*, pp.253–259. Oxford University Press, New York.

MacArthur, R. H. 1972b. *Geographical ecology.* Harper and Row, New York.

MacArthur, R. H., and R. Levins. 1967. The limiting similarity, convergence, and divergence of coexisting species. *Am. Nat.* **101**:377–385.

May, R. M. 1976. *Theoretical ecology*, 1st ed. Sinauer, Sunderland, Massachusetts.

May, R. M. 1981. *Theoretical ecology*, 2nd ed. Sinauer, Sunderland, Massachusetts.

Murdoch, W. W., and A. Oaten. 1979. Predation and population stability. *Adv. Ecol. Res.* **9**:1–131.

Nora, F. 1973. Effects of the December, 1972, freeze on University of California Botanic Garden. *Calif. Hort. J.* **34**:106–108.

Shoener, T. W. 1972. Mathematical ecology and its place among the sciences. *Science* **178**:389–391.

Schoener, T. W. 1982. The controversey over interspecific competition. *Am. Sci.* **70**:586–595.

Schoener, T. W. 1983. Field experiments on interspecific competition. *Am. Nat.* In press.

Simberloff, D. 1980. A succession of paradigms in ecology: Essentialism to materialism to probabilism. *Synthese* **43**:3–39.

Sousa, W. P. 1983. The role of disturbance in terrestrial and marine communities. *Annu. Rev. Ecol. Syst.* **14** (in press).

Strong, D. R., D. S. Simberloff, L. G. Abele, and A. B. Thistle (eds.). 1983. *Ecological communities: Conceptual Issues and the evidence.* Princeton University Press, Princeton, New Jersey.

Wilson, D. S. 1980. *The natural selection of populations and communities.* Benjamin/Cummings, Menlo Park, California.

CHAPTER **15**

Resource Systems, Populations, and Communities

JOHN A WIENS
Department of Biology
University of New Mexico
Albuquerque, New Mexico

CONTENTS

Resources are central to all considerations of ecological and evolutionary phenomena. As the contributions to this volume attest, such diverse topics as life history strategies, physiological allocation patterns, population structure and dynamics, mating systems, community organization, social behavior, individual spacing patterns, nutrient cycles, habitat selection, foraging behavior, or the genetic structure of populations are related in one way or another to underlying patterns of resource supply or variability. We talk incessantly, in both our empirical and theoretical endeavors, of "resources." Despite this, we really known very little of what resources actually are, how they vary, and how they influence individuals, populations, or communities. Detailed, intensive, quantitative studies that explicitly consider resources and link cause with effect are excrutiatingly difficult, whereas, at the opposite extreme, global theoretical approaches to resources and their effects (e.g., Hairston et al., 1960; Fretwell, 1977; Owen and Wiegert, 1976) provide only ethereal hypotheses that hopelessly simplify the complexities of biological reality.

In this chapter I develop a framework for viewing the operations of *resource systems*: series of components (see Fig.1) and the processes that affect the linkages between them. I then discuss these components and processes and suggest some priorities and directions for future studies. The topic of this chapter is impossibly broad, and my treatment will thus represent a personal viewpoint rather than a comprehensive review. Trophic resources (food, nutrients, energy) will be emphasized, and a bias toward animals (especially vertebrates) will be apparent. This mirrors my own research interests, which have forced me to think about resource systems often.

1   THE DEFINITION OF RESOURCES

"Resource" is a term that is rarely defined carefully and explicitly in the ecological literature, perhaps because it is used so ubiquitously that everyone is sure that he knows what it means and that everyone else shares the same conception. This definitional looseness, however, fosters a variety of operational approaches to resources, not all of which have a firm biological or logical foundation. It is commonplace, for example, to measure several environmental features that seem possibly important and/or are easily measured, subject the values to correlation analyses with the variable of interest (e.g., individual performance, population size, community structure), and conclude that those features exhibiting significant correlations are in fact important resources. In more elaborate multivariate analyses of such information, synthetic gradients or dimensions may be derived that have no apparent reality in nature; yet, if these dimensions produce interesting patterns in population or community variables, they are often considered as "resource dimensions" with little hesitation.

The inherent circularity of such approaches is especially clear in studies of "resource partitioning" in communities. The competitive exclusion principle states that species sharing resources in limited supply cannot stably coexist. Coexisting species, therefore, must differ in their use of resources. By searching among a series of environmental variables or species attributes (or multivariate axes), clear ecological differences between the species may be detected. These variables are thus identified as resources or "resource dimensions," the partitioning of which permits the coexistence of the species. Schoener (1974) surveyed many such studies, identifying food size, macrohabitat, microhabitat, food type, time of day, and time of year as the major resource dimensions on which species are partitioned. In a similar fashion, Cody (1974) concluded that the only important resource dimensions for birds were habitat type, foraging height and site, and feeding behavior, simply because he was able to differentiate coexisting species in a manner consistent with competition theory by using these parameters. Certainly some of the confusion and controversy over such resource partitioning studies (see Wiens, 1983a) may be a consequence of the failure to adopt a clear and rigorous definition of resources. Instead, a resource is implicitly defined as that which is partitioned among coexisting species; in turn, it is the partitioning of that resource that explains their coexistence. The circle is unbroken.

David Tilman (1982) has recently considered the problem of resource definition in considerable detail. He defines a resource as "any substance or factor which can lead to increased growth rates as its availability in the environment is increased, and which is consumed by an organism" (1982, p. 11). Tilman's focus is on resource competition and community structure, and he suggests (1982, p. 11) that "it is through depression of resource levels caused by consumption that species may compete with each other, and thus that resources may influence the structure of communities." To be a resource a factor thus must (1) be consumed; (2) be limiting; and (3) have a direct effect on fitness (growth rate or, as Tilman considers it, the average, long-term rate of population increase). Approaching resources from the perspective of their acquisition and physiological allocation to various individual functions, Townsend and Calow (1981) place similar emphasis on consumption, limitation, and (at times via competition) effects on fitness.

What sorts of factors qualify as resources under these definitional criteria? Obviously, food, and the essential nutrients, energy, and materials that it contains, is a resource. Space is clearly a resource for sessile organisms, as occupancy "consumes" it, it is often limiting, and its availability can have profound effects on fitness, perhaps involving intense competition. Mates may be limiting and have a clear effect on fitness, and are "consumed," at least temporarily, during mating. In a similar sense, pollinators may be resources to plants, as the plants are in turn to their pollinators (e.g., Feinsinger, 1978); indeed, participants in any mutualistic system are reciprocal resources for one another. There may, of course, be disagreement over exactly what constitutes the resource in these examples. Is it the food or a specific limiting nutrient such as nitrogen (White, 1978) that is the resource? Are individual plants or the nectar or pollen rewards that they provide pollinators the resource? The resolution of resource definition in such instances must rest on a thorough knowledge of the attributes of the organisms: while entire plants may represent resources to a large grazing herbivore, for example, only specific portions of the same plants may be resources for stem-boring insects (Rathcke, 1976), and a cottonwood tree may appear to an aphid as a heterogeneous assemblage of resources of vastly differing suitabilities (Whitham et al., Chapter 2).

Other factors are clearly not resources under Tilman's definition. Temperature, for example, is not a resource, despite its importance to organisms and its quantitative similarities to resources such as food (Magnuson et al., 1979), because it cannot be consumed. Foraging behavior is not a resource dimension, but rather represents details of the use of resources. Time is probably not a resource either, but a factor that influences the availability of resources (e.g., seasonal phenology) or that constrains resource use (e.g., prey-handling time). "Patches" in a heterogeneous environment are not resources, but instead represent aspects of the spatial dispersion of factors that are resources, such as food or a particular type of space. "Habitat" is a more difficult factor to consider. Some components of habitats or their use, such as foliage-height diversity or foraging height, are clearly not themselves resources but factors influencing resource distribution or accessability. Other components (e.g., nest sites, intertidal crevices that provide shelter from desiccation) may be limiting, influence fitness, and be "consumed" by being occupied, and thus qualify as resources.

Much of the confusion and uncertainty in the definition of resources arises from a failure to distinguish aspects of resource use (e.g., foraging position) or its consequences (e.g., resource partitioning) from actual resources as defined by strict criteria, such as Tilman's. Tilman's definition, however, is not without problems. Tilman clearly restricts resources to factors that are actually limiting to individuals or populations. Thus, "It is imperative that ecological studies of resource competition focus on those resources which actually limit species in their natural habitats. . . . Although studies of competition for resources which are never limiting in nature have some short-term heuristic value, in the long term such studies may confuse much more than they clarify" (1982, p. 96). Quite aside from the very real operational difficulty of documenting a limiting role for a resource under field conditions, however, this insistence on limitation becomes difficult to apply in a fluctuating environment. If food supply varies in abundance, for example, it may be limiting at times and superabundant at others (Wiens, 1977a). Does food cease to be a resource when it is superabundant, and should we then not study it closely? The assumption of limitation may be necessary to Tilman's models, but it may be unrealistically restrictive in dealing with real world situations.

The criterion of consumption, although generally helpful, may also pose difficulties in some situations. Taitt et al. (1981), for example, provided *Microtus* populations with additional cover and recorded greater survival and more successful breeding, which they attributed to reduced bird predation. In this situation the response to the experimental treatment indicates both prior limitation and a clear effect on fitness, although cover is not consumed. In his studies of desert rodents, Thompson (1982a,b) has indicated the importance of desert shrubs in providing high concentrations of seeds and cover from predators, and has demonstrated that experimental manipulation of the distance between shelters can have major effects on habitat use and rodent community composition. But shelter or shrubs are not consumed by the rodents, although they may have obvious effects on individual fitness.

These considerations suggest that Tilman's definition may be unnecessarily restrictive and that some important relations of organisms and populations to their environments and to one another may be ignored if one focuses too strongly on consumption, limitation, and fitness effects. Instead, I believe that resources should be defined by their use by and *potential* effects on individuals: a resource is an environmental factor that is directly used by an organism and that may potentially influence individual fitness. Limitation of either individual physiological performance or population dynamics may be a consequence of resource levels or their utilization, but it is not an essential ingredient of a definition of resources. Resource limitation will be discussed more fully later in this chapter.

2 RESOURCE SYSTEMS

It is apparent that definitions of resources are complicated by the variety of factors involved: elements of the resources themselves, their distribution in the environment, the ways in which they are used by consumers, how individuals allocate them to various functions, and the population and community consequences of these consumption/allocation patterns. I have used a general definition of resources so that the details and dynamics of resource processing by individuals and the individual and populational consequences of this resource use may be explored more fully in the framework of resource systems (Fig. 1). "Resources," as defined above, exist in the environment in a particular abundance and distributional configuration. Not all of these resources will be available to a consumer at a given time, however; various factors of the environment, the resources, or the consumer ("translators") act to determine resource availability. The actual use of some subset of the available resources by an individual is, in turn, influenced by the resource acquisition characteristics of the organism: in animals, chiefly components of foraging behavior. Once taken in or used, resources may be allocated in various ways by the physiological processing systems of the individual. It

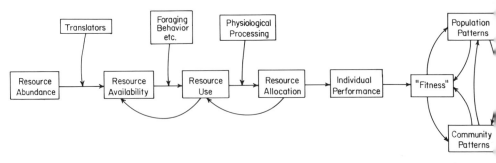

Figure 1. Components of "resource systems." These components and the factors influencing their relationships or transitions are the framework for Sections 2–4 of the text.

is through such allocation systems that the effects of resources on survivorship and reproductive output, and thus individual fitness, are determined.

Others (e.g., Ellis et al., 1976; Johnson, 1980; Dunning and Brown, 1982) have noted the importance of some of the components of resource systems in the ecological dynamics of individuals or populations, but the more conventional approach to resources in ecology glosses over many of these details. The conceptualization depicted in Fig. 1, however, suggests that a comprehensive understanding of the ways in which resources act to influence features of populations and communities must rest on knowledge of the dynamics of the components of resource systems at the individual level. Thus, despite the fact that individuals do not exist in an ecological vacuum and many feedbacks from features of population and community structuring (e.g., social dominance, territorial exclusion, interspecific competition, predation pressure, mutualism) may influence individual resource relations, consideration of resource systems prompts a somewhat more reductionist approach to resource effects on populations and communities than has been fashionable in the recent past.

Let us, then, examine the components of such resource systems (Fig. 1) and the ways they have been considered in ecological studies in greater detail.

2.1 Resource Abundance

The abundance of resources in the environment lies at the foundation of all considerations of the resource relations of organisms, and thus of populations and communities. Resource abundance is simply the quantity of a resource in the environment, independent of the consumer (Johnson, 1980). Measuring resource abundance, however, is by no means straightforward. Tepedino and Stanton (1981), for example, examined the relationship between the abundance of bees and the abundance of their floral resources, which was measured as the number of open flowers. Zimmerman and Pleasants (1982) challenged this measure, noting that resources should be measured in a manner that reflects their use by consumers and that nectar and pollen, rather than the flowers themselves, served as resources for the bees. They suggested that the absence of a close association between the number of open flowers and nectar and pollen availability invalidated Tepedino and Stanton's measure, and thus their conclusions. In their response, however, Tepedino and Stanton (1982) criticized Zimmerman and Pleasants' analysis and reaffirmed (with some reservations) their original measure. Other studies of bees and their flowers have used still other definitions and measures of resource abundance: corolla tube length, nectar concentration, flower size, and so on. There is little doubt that something about flowers represents an important resource to bees. The disagreement is over exactly what is the resource and how it should be measured. Similar problems characterize attempts to define and measure resources in other systems (Brown, 1975; Lemen, 1978); Case (1981) has considered such measurement definition problems and their consequences in detail.

Faced with the difficulties of defining resources and measuring their abundance, many investigators have resorted to the use of indirect estimates of resource levels. Thus, Courtney and Blokpoel (1980) assessed food availability to terns by measuring the percentage of fish brought to chicks that was accepted, foraging time by the adults, and chick growth rates; they inferred from the apparently nonlimiting values of these parameters that food was not a limiting factor to the reproductive success of the birds. In a similar fashion, Thorman (1982) used temporal changes in the stomach contents of fish to infer changes in food resource abundance, which were then used to interpret patterns in fish community structure. Thorman rationalized this indirect approach on the basis that the real food supply was impossible to measure. This may often be true, but it does not justify the practice of using estimates of resource *use* as inferential measures of resource *abundance*, or of deriving resource partitioning patterns in communities from features of resource use in the absence of any measures of resource abundance (e.g., Brown and Lieberman, 1973; Cody, 1974; Pacala and Roughgarden, 1982; see Wiens, 1983a). Without some knowledge of the factors influencing these components of resource systems (Fig. 1), attempts to equate use with abundance in defining or measuring resources are simply invalid.

When attempts are made to measure levels of resource abundance, they usually produce quantitative estimates that bear an unknown relationship to actual resource abundance. "Sticky boards" (small squares coated with tanglefoot) have been used in several studies (e.g., Ballinger, 1977; Cody, 1981; Hutto, 1981) to measure the abundance of arthropod prey of birds or lizards. Temporal variations in these measures have then been taken as indications of changes in absolute resource abundance. Others have used sweep-net sampling (e.g., Stamps et al., 1981; Folse, 1982; see also Fleischer et al., 1982), pitfall traplines (e.g., Brooke, 1979), suction traps (e.g., Bryant and Westerterp, 1980), vacuum (D-Vac) samples (e.g., Dunham, 1980; Christian, 1982), or various other procedures to estimate the abundance of arthropod resources. Each of these methods provides an index of resource abundance, but each has its own inherent biases (see Southwood, 1978). This has led some investigators to use several different measurement procedures. Case (1979), for example, sampled insect abundances by using sticky boards, sweep netting, sifting soil beneath shrubs and litter, beating bushes, and collecting insects under rocks and plant debris. He recognized that each of these procedures was biased, but he nonetheless proceeded to lump the measures from all years and all methods to derive a frequency distribution of insect sizes for his habitats, which he then used as a measure of the prey available to lizards.

In general, such approaches to measuring resource abundance have suffered from a lack of any variance measures (e.g., Ballinger, 1977; Cody, 1981; but see Dunham, 1980; Hutto, 1981), from an ignorance of how the measures calibrate the actual resource levels, and from an insensitivity to the biases of the methods. I have come to appreciate the importance of these problems through the progression of methods we have employed in our own attempts to measure resource abundance for insectivorous shrubsteppe birds. Initially, we used a combination of sweep netting and pitfall trapping, but we abandoned the netting when it became apparent that it suffered from insurmountable biases. The pitfall trapping has been continued for over 5 years but will provide only a coarse index of temporal changes in general resource levels, not a measure of resource abundance. More recently (Wiens, Cates, and Rotenberry, unpublished), we have used D-Vac sampling to determine the arthropod faunas occupying the sagebrush (*Artemisia*) shrubs in which the birds forage. Comparisons with calibration samples that have completely enumerated the insects occurring on these shrubs, however, have indicated that D-Vacing is only 55% efficient in sampling arthropods on the plants, and that different taxa are sampled with differing effectiveness. We now measure arthropod abundance by fumigating entire shrubs with methyl ethyl ketone; calibration tests indicate that this procedure is over 95% effective in estimating the actual abundance of arthropods on the shrubs. Achieving this level of sampling accuracy requires substantial effort, however, and it is easy to understand why many studies have employed simpler procedures. These usually provide only general measures, however, and, although they may be useful in qualitative comparisons as indices of resource abundance, they justify only general and conservative interpretations and conclusions. Unfortunately, such restraint has characterized rather few resource studies.

Even careful and intensive efforts to measure resource abundance generally record only levels of the standing crop, and thus do not consider the input–output dynamics of resources. Such temporal dynamics are important, however. Where these changes are obvious, as between seasons, they are generally not ignored, although they are usually recorded by measuring the standing crop at different times rather than through direct consideration of input and output dynamics. This may be a reasonable procedure if it is conducted at the appropriate temporal scale, given the logistic constraints on actually documenting resource turnover dynamics. But the importance of these dynamics relative to the standing crop may differ dramatically between different resources. Consider, for example, seeds vs. nectar. Seeds often are abundant in the soil "seed bank" (Harper, 1977), and erratic flushes of inputs through production or outputs through consumption, germination, and so on, may represent only a small fraction of the standing crop (Reichman, 1975; Brown et al., 1979; Templeton and Levin, 1979; Rabinowitz, 1981; but see Reichman, 1979). Measures of seed abundance in the soil may thus provide a reasonable estimate of resource conditions, even if inputs and outputs are not considered. Nectar, on the other hand, normally exists in small reservoirs in flowers. Production inputs and consumption outputs may far exceed the size of the standing crop over even a short time period. To consider nectar resource abundance by occasionally measuring standing crop without also considering the input–output dynamics would give an erroneous impression of resource levels. Renewal or turnover rates are thus an important constituent of resource abundance, especially for resources that normally exist in relatively small standing crops. Levinton and Lopez (1977), in fact, suggested that renewal rates rather than abundance per se may often be the limiting feature of resources to populations, and Waser and Wiley (1980) have noted the profound effects that differences in resource renewal rates may have on the expression of sociality in populations.

Resources exhibit spatial as well as temporal dynamics in their abundance patterns. Probably all resources are heterogeneous or patchy in their spatial distribution, at least to a degree, and these variations in abundance may have diverse effects on individuals, populations, and communities (Wiens, 1976). Reichman (1975, 1981), for example, has documented the extreme patchiness of seed abundances in desert soils, which may span an order of magnitude over a distance of a meter. Granivorous rodents may respond to this patchiness in different ways. Laboratory experiments (Reichman and Oberstein, 1977; Price, 1978; Hutto, 1978) and field observations (Bowers, 1982; Thompson, 1982a) suggest that some species of *Dipodomys* select and preferentially forage on clumped seeds, while *Perognathus* occurring in the same habitat may forage more efficiently on scattered seeds (but see Frye and Rosenzweig, 1980; Trombulak and Kenagy, 1980). The importance of seed dispersion to these species, however, is influenced by microhabitat. Reichman (1975) and Bowers (1982) proposed that *Dipodomys* may forage more often in open areas between shrubs, but Thompson's more intensive observations (1982a) and experiments (1982b) indicate that both *Dipodomys* and *Perognathus* concentrate their activities beneath the shrub canopies, perhaps because predation risk is greater in the open. The influence of resource dispersion pattern on these species may thus vary as a function of the microhabitat in which the resources occur. M'Closkey's (1981) observations suggest that variations in population densities of the rodents may also affect these relationships.

The pattern of resource heterogeneity in space (or time) is clearly a function of the scale on which the resources are considered and measurements made (Wiens, 1981), and what is a proper scale depends on attributes of the organisms being studied. It is obvious that a resource distribution that is clumped on a scale appropriate for large grazing herbivores may be quite different on a scale relevant to a phytophagous insect, for example, but it is not clear that the dispersion patterns of desert seeds are necessarily viewed on the same scale by a *Perognathus* and a *Dipodomys*, even though they have been measured as if they were. Tilman (1982) is one of the few to have considered such problems of resource scale carefully. He suggests that spatial heterogeneity may be considered as the variance in resource levels between randomly sampled areas of a size in which one individual may obtain resources during one reproductive bout, averaged over the reproductive lifetime of the individual. Patchiness is thus considered in the context of the average long-term reproductive success of individuals; although this approach may be suitable to Tilman's model analyses of resource competition among plants, it is less useful to investigations of resource heterogeneity in the context of foraging behavior. Still, the emphasis on evaluating scale in terms of the area in which an individual obtains resources during some suitable time period is quite appropriate.

2.2 Resource Availability

The abundance of resources in an environment at a particular time may or may not be of direct importance to organisms using the resources. What *is* important is the *availability* of the resources, the direct accessibility of the resources to the consumer (Johnson, 1980). Generally, only a restricted subset of the resources actually present in an area is available to an organism; a variety of factors act as "translators" (Fig. 1) to determine this subset (Menge, 1972; Ellis et al., 1976; Myers et al., 1980). Thus, the ability of a shorebird foraging in intertidal mudflats to encounter prey that are actually present may be influenced by the depth of the prey in the mud, the penetrability of the substrate by the bird's bill, or the activity of the prey (Goss-Custard, 1980; Myers et al., 1980). Horned lizards (*Phrynosoma*) forage almost exclusively on ants in desert environments, but the actual availability of ants to the lizards is strongly affected by the aggressiveness of different ant species (Rissing, 1981). The distinction between resource abundance and resource availability has been recognized at times, although in manners that range from "rigorous applications to rather loose handwaving" (Myers et al., 1980, p. 1573), and often the two measures of resource levels are simply confounded (e.g., Reichman, 1975).

Features of the abiotic environment, the resources themselves, or the organisms using the resources may all contribute to the translation of resource abundance to resource availability. (Aspects of resource use by consumers are considered in the next section.) Abiotic factors may have a clear direct effect on resource availability, as by snow cover limiting the accessibility of browse herbage to deer, but other influences of such factors may be more complex. The foraging activity of heteromyids such as *Dipodomys*, for example, is reduced on moonlit nights, when animals shift their foraging to shaded microhabitats, presumably in response to increased risk of predation (Kaufman and Kaufman, 1982); this effectively alters the availability of seed resources to the rodents. Jaeger (1980) has demonstrated how some prey in leaf litter become limited in their availability to *Plethodon* during dry periods, when the salamanders do not leave moist microhabitats to forage in the dry litter.

Resource attributes act as translators of resource abundance in a variety of ways. Perhaps least obvious but most dramatic are the effects that variations in the concentrations of secondary chemical compounds in the tissues of different plant parts, individuals, or species may have in restricting their suitability (and thus availability) to phytophagous animals (Atsatt and O'Dowd, 1976; Feeny, 1976; Rhoades and Cates, 1976; Rhoades, 1979; Rosenthal and Janzen, 1979; Bryant and Kuropat, 1980). Toxins have also been implicated as a factor causing foraging sparrows to use plant seeds nonrandomly (Pulliam, 1980). Perhaps the most elegant demonstration of such effects is provided by the intensive studies of Whitham (1978, 1980, 1981; Whitham et al., Chapter 2) on *Pemphigus* aphids colonizing leaves of cottonwood (*Populus*). Leaves vary in availability and suitability to aphids as a function of their size and shoot position at bud burst, and various portions of an individual leaf also vary in suitability. Zucker (1982) has shown that these patterns are associated with variations in the concentrations of phenolic compounds, and Whitham (1981) has suggested that this within-plant variation is an adaptive response by the plant to herbivory.

Other features of resources may also be important. Prey, for example may occur in habitat refuges where they are not vulnerable to predation (this, in fact, was a primary treatment in the early "bottle experiments" on competition and predation; e.g., Crombie, 1946). If a predator is size selective, some prey individuals may be unavailable to the predator by virtue of being too small or becoming too large (e.g., tadpoles; Wilbur, Chapter 7). Aspects of spatial distribution may interact with other features of resources to determine the overall availability of the resources: Schluter (1982a) has shown how the preference of finches for certain seeds on the basis of their size and efficiency of use is altered by seed clumping, and Thompson (1982a) has demonstrated how interactions among seed quality, abundance, and clumping influence seed availability to desert rodents. Moreover, variations in the temporal or phenological characteristics of resources in relation to those of their consumers may produce patterns of availability that differ from those expected on the basis of abundance alone (e.g., Hutto, 1981; Holdren and Ehrlich, 1982). Differences in the behavior of prey resources may also influence the abundance-availability translation, as in Rissing's (1981) *Phrynosoma*–ant system. Some indication of the complexity of the interactions underlying such translations is provided by studies of the effects of parasitism on individuals (e.g., Holmes and Bethel, 1972; Bethel and Holmes, 1973, 1974, 1977; Camp and Huizinga, 1979; Hamilton and Zuk, 1982). Moore (1981, 1983), for example, found that infected individuals of the isopod *Armadillidium vulgare*, the intermediate host of the acanthocephalan *Plagiorhynchus cylindraceus*, differed in their behavior from uninfected individuals, exhibiting a greater affinity for light-colored substrates and open areas and a reduced preference for humid microhabitats. These attributes would seem to render infected individuals more vulnerable to predation, and, indeed, Starlings (*Sturnus vulgaris*) fed infected isopods to their young far more frequently than would be expected on the basis of the frequency of infected individuals in the isopod population as a whole. Parasitism thus markedly altered the availability of different individuals in the population of isopods to avian predators, presumably through the changes in behavior that it produced.

Some appreciation of the complexity of the interactions among variables that influence resource availability may be gained by considering one additional example in some detail. Many species of hummingbirds feed largely on the nectar contained in flowers. Nectar is thus a critical resource to hummingbirds, and variations in its availability may have profound effects on the foraging behavior, social systems, and distributions of the birds. Nectar availability, in turn, is determined by a web of interacting factors (Fig. 2). Hummingbirds must not only obtain energy from the nectar, for example, but they must also maintain osmotic homeostasis by regulating salt and fluid balance. As a consequence, the concentration of the nectar contributes importantly to its availability (Calder, 1979). Ambient weather conditions may affect the concentration of the nectar directly, but they may also influence the water balance of the birds and thus modify their nectar-concentration requirements. In communities containing several hummingbird species,

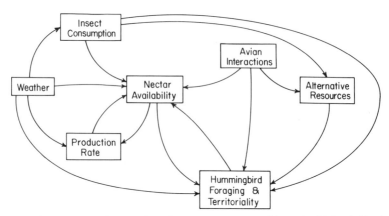

Figure 2. Components of nectar-based resource systems of hummingbirds, indicating the major influences upon nectar availability to the birds.

competitive interactions between the birds may be severe, and it is not unusual for a competitively dominant species to restrict access to high-quality
floral resources by more subordinate species, forcing the latter to use more
unpredictable, lower-quality flowers in a more generalist fashion or to aban-
don the local community altogether (e.g., Feinsinger, 1976; Pimm, 1978).
The availability of nectar to a given species is thus strongly influenced by
the size and aggressiveness (competitive ability) of the species and by the
community context.

Because plants are rarely uniformly distributed, nectar is usually distinctly patchy in its distribution in an area (Pleasants and Zimmerman, 1979;
Gass and Montgomerie, 1981). This distribution prompts area-restricted foraging by the birds (as well as by nectarivorous insects), and the resulting
concentration of foraging in rich patches may lead to localized resource
depletion, further altering the patch structure of nectar distribution (Zimmerman, 1982). The effects of such depletion are a function of nectar production (renewal) rates, which are sensitive to proximate weather conditions.

Nectar dispersion and availability further act to influence the expression
of resource-related territorial behavior in the birds. Territory sizes may be
adjusted as a function of the floral resources they contain, so that a similar
level of daily productivity within the territory is maintained (Gass et al.,
1976). If nectar availability falls below some critical level, however, the
territorial system may dissolve as individuals move about more widely, opportunistically exploiting flowers and other resources. Gass (1978), for example, has documented the differences between meadows containing extensive swards of typical hummingbird flowers, where the birds defended
clearly defined territories quite vigorously, and areas containing only discrete and widely separated *Rhamnus* shrubs, in which the birds showed only
a faint affinity to their territories and foraged widely during much of the day.
Nectar availability to the birds in the *Rhamnus* areas was further reduced
by the activity of nectarivorous insects, which exploited *Rhamnus* more
heavily and continuously than typical hummingbird flowers. Insects such as
bees may in fact have a profound influence on nectar availability to hummingbirds in many systems, both through exploitation of the nectar and
through direct interference with foraging birds (Brown et al., 1981; Gill et
al., 1982). Because bees are more sensitive to ambient temperature than are
hummingbirds, however, nectar resources are effectively unavailable to
them when temperatures are low, permitting the hummingbirds to dominate
resource use and establish resource-based territories (Brown et al., 1981;
Heinemann, unpublished). Such weather-driven systems can be quite dynamic; Heinemann (unpublished), for example, has documented how hummingbird species that had abandoned a montane meadow containing *Scro-
phularia* flowers, apparently as a consequence of nectar exploitation by bees
and other insects, rapidly returned to the meadow and established territories
with the onset of a period of unusually cool, rainy weather. With the return

of warmer conditions a day later, insect activity increased and the hummingbirds once again abandoned their territories and the meadow.

These examples indicate the wide array of features of environments, resources, and consumers that may influence the translation of resource abundance into resource availability. It is *availability* that determines the individual, population, and community patterns that are founded on resources—resources that are present in an area but, for one reason or another, are not available to an organism cannot contribute to such patterns, at least in an immediate manner. Resources are therefore sometimes defined in terms of their apparent availability rather than their abundance. To do so, however,, ignores the important influences of factors acting as translators (Fig. 1), and may obscure the details of the temporal and spatial dynamics of resource systems.

2.3 Resource Use

Organisms use a portion of the resources that are available to them to meet their needs for living space, breeding sites, energy and nutrient demands, and the like. A variety of factors acts to constrain the use of available resources or determine resource preferences (Fig. 1). These are perhaps most apparent among herbivores and secondary consumers, but many of these features apply generally to a wide range of organisms or resource types.

Animals generally exhibit morphological traits that can be considered adaptations to various forms of resource use (e.g., gill raker spacing in fishes, bird bill shapes, lizard jaw sizes, herbivore tooth structure), and in a broad sense these may constrain the ability of a consumer to use the resources potentially available to it. Smith (1981), for example, has noted the differences in jaw structure and musculature between two squirrel (*Tamiasciurus*) species, and has suggested that these differences limit the ability of one species to harvest cones of certain conifer species, thereby determining the distributional niche relationships of the two species. The assumption that the details of morphology are closely matched to the details of resource use has often been employed to justify using morphology as an indirect measure of resource use in ecological studies (Schoener, 1974; Brown, 1975; Ricklefs and Cox, 1977; but see Pyke, 1980; Wiens and Rotenberry, 1980; Wiens, 1982). Certainly one can infer a *general* relationship between morphology and resource use (e.g., Karr and James, 1975), but in more detailed studies resource use must be assessed directly.

The mode of foraging behavior adopted by individuals has several clear effects on resource use, as well as on other attributes of individuals and populations (Huey and Pianka, 1981). Most organisms exhibit clear preferences among resources, and a thorough understanding of resource use patterns entails some consideration of the components of resource preference (Ellis et al., 1976). In most models of foraging behavior, preferences for food types are associated with their relative energy value (e.g., Schoener,

1971), although some models introduce nutrient constraints as well (Pulliam, 1975). Stamps et al. (1981) used such a model to conclude that juvenile *Anolis* lizards selected prey so as to maintain a nutritionally balanced diet. This conclusion was based solely on the apparent agreement of stomach contents data with the predictions of the model, however, not on actual determination of the nutritional value of the prey consumed in relation to those available to the lizards. Other studies have shown that organisms may select habitats or foods that contain relatively high levels of nitrogen or other nutrients (e.g., Moss, 1972; White, 1978; Cole and Batzli, 1979; Cockburn, 1981), or that the selection of leaf tissues by herbivores is closely related to the age of the leaves (Savory, 1978; Coley, 1980). Foster (1977) has documented a particularly enlightening example of the nutritional consequences of selection of different resources. The manakins (*Chiroxiphia linearis*) that she studied switched from eating ripe fruits to green fruits when the ripe fruits became unavailable. Foster calculated that the birds needed to consume 6.6, 4.5, 8.5, and 18.6 times as many green fruits as ripe fruits of *Ardisia revoluta* to obtain equivalent amounts of metabolizable energy, water, protein, and trichloroacetic acid-soluble carbohydrates, respectively. The birds maintained normal weights under these conditions, so they apparently were able to alter their foraging behavior and time budget to compensate for these substantial differences in resource quality.

The energetic or nutritional quality of food resources does not inevitably dominate resource selection and use by consumers, however. Chew (1980), for example, found that larvae of *Pieris* butterflies were behaviorally flexible in their resource preferences, even to the point of not rejecting plants that do not support larval growth, and Sorensen (1981) was unable to find any correlations between the feeding preferences of woodland birds among fruits and energy content, total nitrogen, total protein, total carbohydrates, or total fat content of the fruits. Milton (1979), however, noted how howler monkey leaf choice is influenced by a variety of factors of leaf quality, including protein and fiber content and secondary compounds, and Bryant and Kuropat (1980) concluded that nutritional quality of vegetation was a major positive factor influencing resource use by browsing vertebrates, whereas secondary compounds played a major (and more important) negative role in determining preferences.

Other features of the resources, such as size, shape, texture, or the efficiency with which they are handled, may also contribute to preferences of consumers among the resources available to them (Holling, 1965; Abbott et al., 1977; Pearson and Stemberger, 1980; Lawhon and Hafner, 1981; Sherry and McDade, 1982). There is a rich literature in behavior dealing with such components of food selection; Morse (1980) provides a good entry point to this literature. Two additional aspects of food resource use merit brief mention, however, as they clearly indicate how the different components of resource systems may interact. Resources are usually patchily distributed, and this patchiness both affects and is affected by the resource use patterns of consumers, as in the nectar-based systems discussed previously. Thus, Hart (1981) has experimentally demonstrated how larval caddisfly foraging patterns are strongly influenced by the spatial patchiness of their periphyton prey, but in such a manner that the return rate of larvae to previously grazed patches is low, permitting time for the resource to recover. Oystercatchers (*Haematopus ostralegus*) concentrate their foraging in areas of high density of cockles (*Cerastoderma*), locally depleting resource levels; this forces some birds to move elsewhere to forage, whereas others relax their prey-size selection criteria, thus changing the relative value of other patches in the cockle beds that were previously unused because prey sizes were only marginally suitable (O'Connor and Brown, 1977). For heteromyid rodents, the value of a seed as a resource is apparently a function not only of its size and energy content, but of its distance from a plant canopy (Hay and Fuller, 1981) and its occurrence in a clump of seeds or by itself (Reichman and Oberstein, 1977; Hutto, 1978). Seed distribution patterns, however, are also influenced by the resource use characteristics of harvester ants, which may forage beneath plant canopies and transport seeds some distance from their source (O'Dowd and Hay, 1980) or, by storing seeds in their mounds, create large seed clumps that may be especially important to heteromyids in early spring (Wiens, 1976).

Resources are rarely encountered alone in nature, but occur in concert with other resources, and the context in which a given resource is found affects its use. The pattern of functional response to variations in prey density differs for a given prey type depending on the presence and relative quality of alternate foods (Holling, 1965). Under some conditions this may produce "switching," the consumer suddenly shifting its use from one resource to another when some threshold value of their relative availabilities is passed (Murdoch, 1969; Murdoch and Oaten, 1975). Resources thus interact in a variety of ways (Tilman, 1982). The use of different portions of the habitat by some Namib Desert rodents thus changes in relation to water availability (Christian, 1980), and the tendency of Redshanks (*Tringa totanus*) to feed on mudflat amphipods is a function not only of the abundance of amphipods but of the relative availability of intertidal worms as well (Goss-Custard, 1977).

Theoretical treatments of resource use are dominated by formulations that predict how organisms should forage optimally so as to maximize some criterion (e.g., energy intake) that is presumably related to fitness (Pyke et al., 1977; Krebs et al., 1981). The value of such optimization approaches has been the subject of much debate. Despite the caveats of Maynard Smith (1978) and Oster and Wilson (1978), however, optimality in resource use is often regarded not just as a simplified and idealized model against which we can test reality, but as an expectation of how nature *should* be. Organisms are seemingly expected to be optimal or at least close to it, and the challenge of research thus becomes the identification of the proper constraints and currency of the models that will show them to be so. Optimization models that generate point solutions are satisfactory because we expect more or less perfect adaptation in organisms (Cain, 1964; Gould and Lewontin, 1979).

The real world, however, presents complications for simple optimization approaches. Models that predict that resources should be used in accordance with their quality ranking (e.g., Schoener, 1971; Pyke et al., 1977), for example, fail to deal with situations that involve resource complementarity (Rapport, 1980, 1981; Tilman, 1982). Optimal patch use theory predicts that an animal should leave a given resource patch to search for another when the rate of energy intake in the patch falls to a value equal to the average net rate for the habitat as a whole (the marginal value theorem; Charnov, 1976). Patch-dependent foraging should thus act to deplete all patches in an environment to the same resource availability level: the habitat tends toward greater homogeneity. Area-restricted searching, on the other hand, may act to concentrate resource use, leading to localized resource depletion well below the general level for the habitat as a whole: patchiness is accentuated. Temporal variation in resource levels that occurs more rapidly than the response time of the consumers may also erode the match between optimality expectations and reality (Wiens, 1977a).

Part of the difficulty in optimality thinking may relate to a misconception of how natural selection operates and what its consequences should be. Selection is not a process that continuously acts to favor the "best" (i.e., optimal) phenotype in a population or the most efficient use of resources or time, but rather acts negatively, against individuals that fall beneath some minimal standard of performance. Charles Elton (1927) observed that "animals . . . spend an unexpectedly large amount of their time doing nothing at all, or at any rate nothing in particular . . . [they] are not always struggling for existence." Thus, precisely what organisms do or how they use resources may not matter much of the time. Variations in resource use, or "doing nothing," may be of no consequence to the variations in individual fitness on which selection acts. The "strategy" of individuals is to avoid doing poorly or to be adequate in their use of resources (Schluter, 1982a), not necessarily to match some ideal "best." Rather than there being a single optimal pattern of resource use under a given set of conditions, there is an array of equivalent, alternative ways of using resources, an "optimality pla-

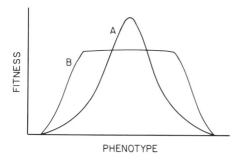

Figure 3. Two contrasting views of optimality in ecological systems. Curve A represents the optimal point pattern predicted by most current optimality theory: there is a single "best" or most fit phenotype for a given set of environmental circumstances. Curve B presents an alternative that suggests the existence of a wide range of equivalently fit phenotypes, an "optimality plateau."

teau'' (Fig. 3). In a given environment at a particular time, many individuals may pay no penalties in fitness by departing somewhat from the theoretically optimal phenotype (Wiens, 1977a).

2.4   Resource Allocation

The resources that are acquired by organisms are allocated to a variety of physiological functions within the organism (Fig. 1): metabolism, growth, reproduction, molt, defense, and so on. Energy or nutrients allocated to one function are thus unavailable for other functions. If the components and processes of resource systems limit the availability of resources to con-sumers or restrict their use, differences in allocation patterns may affect individual performance, and perhaps fitness. The book edited by Townsend and Calow (1981) provides a good perspective on the details of physiological allocation and their consequences, but the entire book is founded on the optimization premise that there is a direct relationship between resource allocation and fitness, that resources are limited and their "wise" allocation is thus essential. This is an assumption of resource allocation theory, how-ever, not an empirically supported conclusion. If resources are not limiting, variations among organisms in precisely how they are allocated to physio-logical functions may not matter much, for example, and even under con-ditions of limitation, differences in allocation pathways may not necessarily produce differences in individual fitness.

Such arguments rest on considerations of resource limitation. Limitation, however, can operate at two distinct levels. In the context of resource al-location, limitation has its effects on individuals, potentially altering their metabolic performance, growth, reproduction, and so on. These effects may or may not be translated into expressions of resource limitation at the pop-ulation level, recognized as variations in population size, age structure, and so on. Although most discussions of resource limitation in the ecological literature consider the latter level, the distinction between the two levels of limitation is not always clearly recognized. In some situations (e.g., Whi-tham's cottonwood–aphid system) limitation can be assessed at both levels, but this is rarely done, and often is not possible. Sinclair et al. (1982) have suggested an indirect approach to detecting resource limitation. Individuals within a population may respond to decreasing food availability by (1) main-taining a constant rate of food intake by including more poor-quality food in their diet, (2) increasing the quantity of food eaten to compensate for poor quality, or (3) eating only high-quality foods and thus decreasing the quantity eaten. Experiments with snowshoe hares (*Lepus americanus*) indicated that the animals used the first option; mean diet quality may thus be used to predict whether or not animals can maintain body weight under natural con-ditions, as there is a threshold level (given constant intake) below which diet quality can no longer maintain body weight. Sinclair et al. used fecal crude protein to infer diet quality, and found that if this measure fell below 7.5%, animals in the field did indeed lose weight, suggesting food limitation.

A more direct documentation of the physiological and population consequences of apparent resource limitation emerges from Jones and Ward's (1979) study of a breeding colony of an African ploceid weaver-finch (*Quelea quelea*) under conditions of a severe drought that depressed food supplies. Birds attempting to breed at the colony contained far fewer insects in their crops than did birds at a successful colony not influenced by the drought. Inadequate energy intake apparently caused males to leave before completing the nests; the females that remained thus deposited thousands of eggs through bottomless nests onto the ground. Jones and Ward had previously (1976) shown that breeding performance in quelea is closely governed by reserve protein levels. In females at the drought-affected colony the levels of flight muscle protein and body lipids were lower than those of females breeding in the successful colony, and the rates of decline during the laying cycle were more rapid. Jones and Ward (1979, p. 16) summarized the situation as follows: "as long as a female was able to maintain her protein reserve at the required level she continued egg formation normally, even though this was achieved by feeding intensively 'for protein', and thereby neglecting the high-energy food (seeds). This situation could not persist, however, once fat reserves became dangerously low. Then it became necessary for the birds to switch to seeds. The consequent reduction of protein intake led in turn to atresia of yolks, and eventually to abandonment of the colony. For a few this came too late and they died." Such clear documentation of food limitation and its physiological effects is rare.

The physiological functions that relate to resource allocation also have feedback effects on resource use and foraging behavior, as metabolism, growth, and so on, influence energy and nutrient demands, and thus the amount and types of resources required by an individual. In *Plethodon*, for example, increasing ambient temperature increases metabolic demands. When this effect is combined with the restrictions on microhabitat use imposed by moisture stresses, it means that resources are much more likely to be limiting to the salamanders on dry, warm days (when energy demands are high and resource availability low) than on cool, wet days (Jaeger, 1980). More indirect approaches to estimating the energy demands of animals and their effects on resource use have relied on bioenergetic models (e.g., Wiens and Innis, 1974; Collier et al., 1975; Ferns, 1980; Puttick, 1980, 1981). By coupling such model projections of energy demands with information on population features and food habits, estimations of overall resource demands placed on the pool of available resources and of the spatial distribution patterns of members of the consumer population have been generated for several systems (e.g., Wiens and Nussbaum, 1975; Wiens, 1977b; Ford et al., 1982). Unfortunately, such indirect approaches require field verification, which will be difficult. Bryant and Westerterp (1980), however, have used doubly labeled water ($D_2{}^{18}O$) techniques to obtain direct measures of energy demands of house martins (*Delichon urbica*). They combined this information with suction trap estimates of insect abundance and observations of the foraging behavior and breeding activity of the birds to construct a detailed budget of the energy allocation pattern of the birds and to predict the conditions under which resource limitation was likely to occur. Studies such as this are laborious, but they hold great promise.

In order to derive meaningful speculations about the ultimate consequences of variations in resource allocation patterns and individual performance, the operations of resource systems must be linked with some measure of individual fitness (Fig. 1). In a few systems, such as Whitham et al.'s (Chapter 2) aphids using cottonwood leaves, meaningful measures of individual fitness can indeed be derived and related to the details of resource systems. Boag and Grant (1981) were able to document nonrandom survival of individuals in a population of *Geospiza fortis* during a period of reduced seed abundance caused by a severe drought in the Galapagos Islands. Overall, the finch population declined by 85%, but large individuals predominated among the survivors, primarily because of their ability to crack the large, hard seeds that were available during the drought. Such documentations are rare, however, and more often the fitness consequences of resource use and allocation patterns are simply asserted or inferred from the foundations of adaptationist or optimality logic. My concerns with this approach have been expressed above.

3 POPULATION CONSEQUENCES

Populations of organisms relate to resources through the effects of the components of resource systems on individual performance and, ultimately, fitness (Fig. 1). It is in this way that resource patterns and their variations influence such population attributes as density, distribution, age structure, territorial spacing patterns, dominance relationships, cooperative breeding, group size and composition, genetic structure, morphological variation, and the like. It is not possible to consider such resource effects here; some perspective may be obtained from the works of Wilson (1975), Krebs and Davies (1978), and Morse (1980). In this section I will instead focus on some aspects of resource limitation of populations and on how the temporal dynamics of resource systems and populations relate to the equilibrium status of ecological systems.

3.1 Resources and Population Limitation

The view that populations are limited or regulated by resource supplies has dominated population ecology over the past 25 years (e.g., Lack, 1954, 1966), and forms the foundation of most current work (e.g., Townsend and Calow,

1981; May, 1981; Tilman, 1982). Resource limitation occurs when the resource demand of a population coincides with resource supply (availability);
an increase or decrease in resource levels thus generates a parallel increase
or decrease in population size through its effects on the components of population change (reproduction, mortality, emigration, and immigration).
These considerations are theoretically straightforward, but there is surprisingly little direct evidence of clear resource limitation in natural populations
(see Newton, 1980). Certainly many populations seem to vary in the degree
to which resources directly determine their dynamics. Most obvious are
populations occupying highly seasonal environments, in which resource levels may vary between superabundant in summer and scarce in winter (Fretwell, 1972; Schoener, 1982). Jansson et al. (1981) have experimentally demonstrated that food abundance during winter may be critical to survival of
Willow Tits (*Parus montanus*) and Crested Tits (*P. cristatus*), thereby limiting the size of the breeding populations. In many arid and semiarid environments, resource abundance varies dramatically between years, and apparently limiting conditions may occur only once every few years (Wiens,
1977a; Pulliam and Parker, 1979; Newton, 1980; Dunham, 1980). Singer and
Ehrlich (1979) found that some populations of *Euphydryas* butterflies varied
in relation to their food supplies in a clear density-dependent manner,
whereas other populations fluctuated in a more complex manner that was
unrelated to density but attuned to the interactions between climatic variations and food plant conditions.

Unfortunately, much of the evidence relating to resource limitation of
populations relies heavily on inference. For example, Puttick (1981) suggested on the basis of bioenergetic calculations that food was not limiting
to wintering populations of Curlew Sandpipers (*Calidris ferruginea*) at a
South African estuary. He then suggested that food must be limiting on the
breeding grounds; this was not investigated, however, but only inferred from
the morphological differences between the sexes. The observation that carrion beetles are attracted to carrion bait was used by Anderson (1982, p.
1323) as evidence that "food is in high demand and thus can be treated as
a factor limiting population growth." Dunning and Brown (1982) found that
densities of wintering sparrows in southeastern Arizona were positively correlated with rainfall during the previous summer; they suggested that this
provided strong support for the proposition that the local abundance and
distribution of the sparrows were regulated primarily by the availability of
food resources (seeds).

Newton (1980) undertook a detailed evaluation of the evidence bearing
on the role of food in limiting bird populations. Much of the evidence relies
on correlations, as between population density and primary productivity or
fluctuations in bird numbers and variations in cone or mast crops. Other
arguments use information on the birds themselves (e.g., starvation, low
weights, altered feeding rates, food-related aggression) to support conclusions of food limitation. Newton notes that such patterns, although sugges-

tive of food limitation, can be explained in other ways. This kind of evidence is strengthened if direct measures of the food supply are also made. If the birds then are found to remove a "considerable proportion" of the standing crop of food over some time period, one may argue that they are more likely to be limited by food supply than if only a small proportion is removed. But even this does not actually demonstrate food *limitation*; it merely defines the periods when food is scarcest.

Newton concludes that a rigorous and satisfying documentation of food limitation is possible only if such circumstantial evidence is coupled with experiments. Indeed, experimental investigations of population and community dynamics have been gaining force in recent years, and several workers have manipulated food and other resources of populations to determine whether or not resource limitation occurs. Jansson et al. (1981), for example, provided supplemental food to wintering populations of Willow and Crested Tits in western Sweden, and recorded improved overwinter survival of marked individuals as well as considerable immigration into the supplemented area. Breeding population sizes were consequently doubled over those in control areas, although there was a subsequent density-dependent reduction of breeding success in the Willow Tit population. The supplementation of food supplies of breeding populations of these species (Brömssem and Jansson, 1980) produced changes in breeding phenology but not in population sizes, clutch sizes, or fledging success. From such experiments and other evidence, Alatalo (1982) concluded that food resources may generally be limiting to tit populations in northern Europe. Similar food supplementation experiments with various small mammal populations (e.g., Smith, 1971; Flowerdew, 1972; Cole and Batzli, 1978; Andrzejewski, 1975; Taitt, 1981; Taitt and Krebs, 1981) generally produced changes in overwinter survival, reproductive activity, and population density.

Not all experiments have produced such clear results, however. Källander (1981) provided supplemental food to Great and Blue Tits (*Parus major* and *P. caeruleus*) during two winters in Sweden. Following the first winter Great Tits increased on the experimental plots while decreasing in the control areas, but they increased in both experimental and control plots following the second winter. Blue Tits exhibited no response to the treatment in either year [*contra* Krebs (1971), whose food supplement experiments in England affected *caeruleus* but not *major*]. The first winter, however, was severe, and much of the beech mast on the ground was unavailable because of snow cover. The following winter was average, and mast was readily available, apparently in nonlimiting quantities. Hansen and Batzli (1979) likewise attributed the failure of their food supplements to produce clear demographic responses in a *Peromyscus* population to an abundance of natural foods during the experiment. Other experiments have produced equivocal results for other reasons, such as differences in responses to different experimental food types (Gilbert and Krebs, 1981), food additions at the wrong stage of the breeding cycle (Franzblau and Collins, 1980), or confounding influences of immigration (Gilbert and Krebs, 1981; Taitt and Krebs, 1981). Even carefully designed experiments may thus sometimes complicate more than resolve our understanding of resource effects.

A second approach to experimental testing of the resource relations of populations has involved manipulation of the access of consumers to resources rather than the resources themselvs. By excluding birds from deciduous woodland foliage, for example, Holmes et al. (1979) demonstrated that birds had a significant depressing effect upon lepidopteran larvae and (to a lesser extent) coleopterans. Such results, however, indicate only that a consumer has an impact on its resources, *not* that the resources are necessarily limiting to the consumer populations. Similar restrictions apply to investigations that have demonstrated consumer impacts by using bioenergetic models to calculate the percentage of resource productivity used by the consumers (e.g., Nielsen, 1978; Puttick, 1980).

Other experiments have involved manipulations of the consumer populations themselves, generally to determine the competitive relationships between populations that are presumed to be resource limited (e.g., Dunham, 1980; Strong, 1982; Schroder and Rosenzweig, 1975). Finding that annual fluctuations in Great Tit breeding densities in a British woods were inversely related to Blue Tit densities, Minot (1981) experimentally removed Blue Tit young from one area and added them to another. Great Tit fledging weights were greatest in the removal area and least in the addition area, leading Minot to conclude that the species did indeed normally complete, Blue Tits depleting food resource availability sufficiently to affect Great Tit reproductive ability. Alatalo (1982) has carefully reviewed the results of such experimental studies of *Parus* populations, finding considerable (but not universal) evidence for resource limitation and competition.

3.2 Equilibrium Assumptions

Experimental investigations such as those of Källander (1981) or Dunham (1980), which have found evidence of resource limitation effects in some years but not in others, cast doubt on the view that resources are more or less continuously limiting to populations. A basic assumption of much of the theory of population dynamics and community structuring, however, is that populations and communities are in equilibrium with their resources (MacArthur, 1972); if resources exhibit temporal dynamics, these changes are presumed to be closely tracked by the consumer populations (Cody, 1981; Lister, 1981; but see Smith, 1982). Tilman's (1982, p. 5) theory of resources and communities, for example, is explicitly equilibrial, not because equilibrium is necessarily the most likely state of nature, but because "it is most appropriate to consider the simpler, equilibrium explanations first." This may be reasonable in theory [Real (1975) and others have noted the formidable difficulties of mathematical treatment of nonequilibrium systems], but it turns out that the predictions of Tilman's models are altered substantially when the equilibrium assumption is relaxed. Further, although Tilman places considerable emphasis on the results of plot fertilization experiments at Rothamsted, he also notes that the plant community had not yet attained a new equilibrium over a century after the initial resource perturbation! Other field observations (e.g., Pulliam and Parker, 1979; Folse, 1982; Järvinen, 1980; Rotenberry and Wiens, 1980) have also found substantial temporal variability in populations or communities that seems unrelated to variations in levels of resources (see Wiens, 1983a,b).

Thus, populations, rather than tracking variations in resource levels with the precision predicted by theory or asserted by Cody (1981), may react to these variations, but with a time lag, a "tracking inertia" (Fig. 4). Noncoincidence of many of the factors involved in resource systems may also reduce the match of populations to resources, and stochastic influences further distort the relationship. The consequence is that, although populations may at times be clearly limited by resource availability, resources may be superabundant at other times, and the patterns expected from the predictions of equilibrium-based theory may well not materialize (Wiens, 1977a). Schoener (1982) has recently argued that such "ecological crunches" occur quite often and that resource limitation and competition are thus frequent and contribute importantly to population and community patterns. Resolution of the differences between Schoener's view and my own rests on a documentation of the frequency of "ecological crunches" vs. periods of resource superabundance in nature. This will require long-term investigations of population dynamics that include careful monitoring of the abundance and availability of key resources.

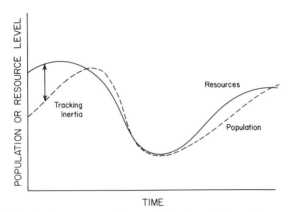

Figure 4. Variations in the resources of a population through time. The population responds to increasing resource levels with a lag, creating a "tracking inertia." When resources decline, the population may encounter resource limitation, effecting a decline in population size. Subsequent increases in resources relative to population demands prompt an increase in population size, but again with a time lag. Much of the time, then, the population size may not track resource variations closely.

4 COMMUNITY CONSEQUENCES

Aggregates of populations and underlying resource systems occurring together in some area over some period of time make up communities. One may conveniently view communities as *interactive*, structured primarily as a result of biotic interactions (primarily competition) among species, or *noninteractive*, in which the species respond to environmental features independently of one another and in which community "patterns" may largely be epiphenomena resulting from these individualistic responses (see also Lawton, Chapter 12). This dichotomy does not necessarily parallel that between structured versus "random" communities (e.g., Connor and Simberloff, 1978, 1983; Diamond and Gilpin, 1982; Gilpin and Diamond, 1982), as noninteractive communities may nonetheless exhibit clear patterns of structure (Caswell, 1976; Joern and Lawlor, 1980).

Most classical and current community theory deals with interactive rather than noninteractive communities (MacArthur, 1972; Wiens, 1983a). The constituent species of the community are anticipated to be in equilibrium with their resources and with one another, resource limitation is commonplace, and direct interactions among the species determine the resource use patterns of species and the composition of the community. Although evidence supporting this view of communities is generally obtained from comparative observations (e.g., Schoener, 1974; Diamond, 1978), experimental manipulations of resource levels or species removals have also been used to test for interactions directly (see Dhondt and Eyckerman, 1980; Minot, 1981; Alatalo, 1982). In some cases the findings have clearly demonstrated that interactions are important, but other studies have failed to find the patterns expected from theory. Schroder and Rosenzweig (1975), for example, removed individuals from populations of two coexisting *Dipodomys* species, but found that they were quickly replaced by individuals of the removed species. They concluded that the two species appeared to avoid direct competition by means of subtle differences in habitat preferences, differences that were presumably adaptive responses to competition in the past [what Connell (1980) has dubbed the "ghost of competition past"]. Noon (1981) also relied on past competition to explain the apparently inconsistent results of his studies of thrush distributions on woodland altitudinal gradients. Unfortunately, hypotheses that permit such ghosts to enter into explanations are not testable, rendering experimental results equivocal.

In noninteractive communities, the different species may respond to characteristics of the pool of available resources in an individualistic manner, independently of the presence or absence of other species that share some aspects of resource use. Some of the species may be in an equilibrium determined by resource limitation at some times, other species at other times. The species are likely to differ to varying degrees in their patterns of resource use, and it thus becomes difficult to determine whether these differences are proximate or evolved mechanisms for resource allocation or simply responses by very different species to the environment, responses that incidentally aid in coexistence (Reichman, 1975). Still, several recent studies (e.g., Rathcke, 1976; Rotenberry, 1980; Rotenberry and Wiens, 1980; Cockburn, 1981; Maurer and Whitmore, 1981; Wiens and Rotenberry, 1981; Collins et al., 1982; Strong, 1982; Wiens, 1983a) have considered the patterns of similarities and differences among species they found to be more consistent with an individualistic, noninteractive view than with the competition-based community paradigm.

The linkage between community structuring and resources has been most explicitly addressed (at least conceptually) in studies of resource partitioning between coexisting species. As Schoener (1974, p. 27) has noted, "the major purpose of resource-partitioning studies is to analyze the limits interspecific competition place on the number of species that can stably coexist." Such intents notwithstanding, a demonstration of resource partitioning alone says *nothing* about the role of competition, but is a documentation of a *pattern* in the community, not of any underlying *process* (Wiens, 1983a). A pattern of resource partitioning may be just as likely to occur in a noninteractive community as in an interactive (competitive) one. Not only does the usual interpretation of resource partitioning in communities carry an implicit premise of process that has produced the pattern, but it also assumes that resources are in fact limiting and the system equilibrial. Anderson's (1982) study of resource partitioning in carrion beetles is perhaps all too typical. Anderson used pitfall trapping to assess the patterns of between-species differences in seasonality and habitat use. Combining this information with literature statements on food use by the species and an inference that food was limiting in his system, he then concluded (1982, p.1323) that "in the silphids and other carrion-feeding insects, competition for these resources is probably the most important factor responsible for the inducement of ecological character displacement" (the habitat and seasonal differences between species). In the absence of direct measures of resource dynamics and direct documentation of competitive interactions, such conclusions represent unjustified and unsubstantiated assertions.

Cody (1979, p. 224) has cast the conventional notion of resource partitioning in a more explicit form, noting that the "unusable tautology known as the 'Competitive Exclusion Principle' translates into ecological reality via such questions as 'Which are the resource axes along which ecological segregation occurs?' 'Are the differences on a given resource axis substitutable for differences on another resource axis in simple or at least comprehensible ways?' . . . 'How are resource utilization functions modified by the presence in the same habitat or in the same territory of a competitor species with overlap in potential resource use?'" These are legitimate questions, but they must be addressed with considerable care. First, they are all basically questions about pattern, although underlying implications of the process (competition) determining the answers are strong. Second, how resources or "resource axes" are determined is critical to the answers that may be obtained. If the resources are not defined or identified correctly, any patterns that do emerge may be suspect. Finally, the interpretation of the patterns sought by these questions is to a large degree determined by how one interprets overlap in resource use ("niche overlap"). Several studies, for example, have documented that food resource overlap among coexisting species is greatest when resources are apparently abundant and least when times are lean (see Schoener, 1982), but studies of Galapagos finches that included careful documentations of resource levels (e.g., Grant and Grant, 1980; Schluter, 1982b) found greatest overlap during periods of food scarcity, as the species converged in their use of the few resources available. Rotenberry (1980) and Vitt et al. (1981) found substantial temporal variation in resource overlap among species of birds and lizards and suggested that the consumers responded independently but in parallel to changing patterns of resource availability. Unfortunately, one can tender a competitive explanation for either pattern: if overlap is great when resources are scarce, that should promote active, direct competition, whereas if overlap is low under such conditions it can be interpreted as clear evidence of differences in resource use that are the consequence of past competition. In addition to these observational and logical complications, however, there are theoretical problems in assessing resource overlap patterns (Abrams, 1980; Case, 1981, 1982; Turelli, 1981). It is simply not possible to derive meaningful interpretations of resource overlap or partitioning patterns in the absence of detailed consideration of the underlying resource systems.

One other aspect of resource-related patterns at the community level deserves mention, the notion that there can be no more species in a community than there are types of resources (see MacArthur, 1972). Quite aside from the difficulties and potential circularities in defining resources in the context of this theory, there are clear theoretical reasons to doubt its veracity. Tilman's (1982) models suggest that, for sessile organisms such as plants that feed upon essential resources, the introduction of even modest levels of spatial heterogeneity into resource systems potentially permits an almost unlimited number of species to coexist at equilibrium on a restricted set of resources. The same prediction follows when he expands his models from two to three trophic levels, even in a homogeneous environment, or when the system persists in an nonequilibrium state. There thus may be no clear relation between the number of resources and the number of coexisting species they can support, at least for sessile forms. Interestingly, Tilman's models suggest that the expectation may be totally different (and closer to that of MacArthur) for mobile animals that switch their use of resources in a frequency-dependent manner.

5   CONCLUDING COMMENTS

The foregoing discussion has largely emphasized animal consumers and their food resources. Other resources or users of resources demonstrate many general similarities and present similar problems in analysis, but there are some important differences. Tilman (1982) has discussed many of these in considerable detail. The use of space as a resource, for example, produces discrete spatial heterogeneity in its availability, altering theoretical expectations of communities at equilibrium in important ways. Animals, by consuming resources that are nutritionally substitutable for one another, can practice resource switching, and thus differ from organisms such as plants that require various essential resources that are available from different sources or require distinctly different modes of acquisition. Use of resources that involves consumption of entire organisms may have profound effects on demographic processes and patterns in the resource populations, whereas the influences of grazers or parasites that consume only parts of individuals are quite different. Despite these differences, however, the consideration of resource systems that I have presented leads to some observations that may be of general importance to deliberations on resources and their effects on individuals, populations, and communities.

There have been some good, rigorous field studies of elements of resource systems conducted [e.g., in Galapagos finches (Grant, 1981), cottonwood aphids (Whitham et al., Chapter 2)]. Despite all our theory and a great many studies, however, our knowledge of the details of the operations of resource systems in nature is really rather superficial. Some of the problem may rest with the way in which our preconceptions about ecological systems and resources have dominated our investigations. Ecologists have generally expected such systems to be at or close to equilibrium, to be more or less continuously resource limited, to be governed largely by deterministic processes, and to express the patterns expected of optimally structured systems (Wiens, 1983a). Theory, rather than posing questions to be considered, seems at times to have specified the answers that should be sought. As Birch (1979, p. 197) has observed, "We must question the questions people have asked and see if we can formulate better ones. The trouble about good answers is that people tend to look at problems in terms of answers they already know and which they expect or hope to find."

How, then, should studies of resources and ecological systems be approached? First, preconceptions about the systems must be recognized as such, and not permitted to bias the research design or interpretation of results. Coupled to this is a need to consider several alternative hypotheses of pattern or process rather than just the single hypothesis that is currently in favor. Such hypotheses should, of course, be testable; an untestable hypothesis may have heuristic value, but it cannot lead to logically satisfying explanations. Null hypotheses that incorporate stochastic processes may be especially useful, although they must be carefully and intelligently structured. Second, the methodology that is used to measure the components of resource systems, namely, resource abundance, resource availability, resource use, individual performance, and (if possible) fitness, should be unbiased and empirically rigorous. Waving a sweep net or scattering sticky boards about in an environment, for example, is not likely to produce a valid measure of resources. In addition, although inference has a place in scientific investigation, it is not as a substitute for measurement. Resource abundance or availability *cannot* simply be inferred from measures of resource use. Whenever possible, experimental manipulations of resources or of organisms that use them should be part of the research. Experiments should be conducted, however, only if they can be carefully designed, replicated, and controlled. Dumping food into a system to see what happens is an "experiment" only in the coarsest sense of the word. Attempts should also be made to develop novel approaches to measuring resource limitation (e.g., Sinclair et al., 1982), or to derive ways of calibrating the relationship between resource abundance and availability.

Obviously it will be easier to conduct careful studies of resources in some systems than in others. Such systems are likely to be characterized by resources that are easily measurable, relatively obvious, discrete, and that have a restricted set of rather specialized species using them. Resources such as nectar (Gass and Montgomerie, 1981; Brown et al., 1981), shells for hermit crabs (Bertness, 1981), or cow dung (Hughes and Walker, 1970) come to mind, but there are many others. Abiotic influences on the system should be clear. The use of resources by organisms should be amenable to observation; otherwise, many of the factors translating availability into use cannot be determined. Some means should be available for documenting the effects of differential use and allocation of resources on the performance of individuals, and ideally some measure of individual fitness under different conditions should be obtainable. If the investigations are extended to the population level, insights may be gained most readily if the population turnover rate is rapid relative to that of its resources (e.g., short-lived organisms), and if dispersal movements are minor or constrained (e.g., on islands). At a community level, systems in which population interactions with one another are potentially direct (e.g., predation, interference competition) and clearly related to an identifiable resource base will probably be more tractable for analysis than communities dominated by indirect interactions (e.g., diffuse exploitation competition). It is not likely that very many natural systems will possess very many of these characteristics, and those that do are likely to be so special or unique that it will severely limit our ability to generalize from them. Still, some consideration of these features of systems in the planning stages of investigations may lead to better studies, or at least to a clearer recognition of the constraints that a particular system will place on our ability to obtain results or draw conclusions from it.

The title of this volume implies change, challenge, and complexity: "new ecology," "novel approaches," and "interactive systems." There is little doubt that resource systems are complex and interactive; even the general form of Fig. 1 attests to that. There is really rather little new or novel in what I have discussed in this chapter, however. What *would* be new would be to conduct our investigations of populations and communities and their relations to resources in the conceptual framework of resource systems, incorporating a strong committment to conducting long-term investigations that include careful measurement of the dynamics of both populations and resources. That, indeed, is the challenge.

6 SUMMARY

Resources are of obvious importance to individuals, populations, and communities. The treatment of resources in much of the ecological literature, however, has often been rather superficial and characterized by a heavy reliance on inference. Often resources have not been defined at all, or have been considered on the basis of interspecific differences as those limiting factors that are partitioned among coexisting species. I suggest that a "resource" should be defined as an environmental factor that is directly used by an organism and that may potentially influence individual fitness.

This definition lies at the foundation of what I call "resource systems," sets of components and processes that relate to resources, their use by organisms, and the consequences of resource use to individuals, populations, and communities. The initial component of resource systems is resource *abundance* in the environment. Ideally, documentations of resource abundance should be direct or at least carefully calibrated, and they should consider the spatial and temporal dynamics (renewal rates) rather than simply record standing crops in a static fashion. Not all of a resource that in present in an environment at a given time is available to an organism, however. The translation of resource abundance into resource *availability* is influenced by a variety of features of the abiotic environment, of the resources themselves, and of the organisms that use them. Resources that are present in an area but, because of the effects of such "translators," are not available to organisms, cannot contribute to individual, population, or community patterns in a direct manner. Considerations of the ecological consequences of resource patterns must thus rest on a knowledge of resource availability, but how various factors influence the abundance–availability translation cannot be ignored.

Organisms use only a portion of the resources in an environment that are available to them. This resource *use* is constrained by features of consumer morphology, behavioral preferences, foraging or settling behavior, and so on. Once resources have been acquired by an individual, they are then internally *allocated* to various physiological functions, such as growth, maintenance, reproduction, or defense. The ways in which resources are used and allocated in relation to their availability may at times limit individual function. This is resource limitation in the individual, physiological sense, and it is through such limitation that individual performance and, eventually, fitness are affected.

Patterns of both resource use and resource allocation have generally been considered theoretically in the context of ecological optimization, with the expectation that under a given set of environmental circumstances there is a "best" way of foraging or allocating resources to different functions. If environments vary at all (and all do), however, there may be little reason to expect phenotypes to be restricted to such a single optimum in nature; rather, a wide range of alternative phenotypes may be equally fit, and optimality then will be expressed as a "plateau" rather than a single point.

These components of resource systems, and the various factors that affect them and their interrelationships, lie at the foundation of considerations of resources in relation to populations and communities. At the population level, resources have generally been presumed to be limiting, the populations in equilibrium with their resource supplies. This then sets the stage for competitive interactions among co-occurring populations using the same resources, which, in turn, are generally thought to produce the patterns of resource partitioning among community members. Most studies of resource partitioning, however, document a *pattern* of divergence between species, and say little or nothing of the underlying *processes* that might have produced the pattern. Also, there generally is not much direct evidence of resource limitation of populations available, despite the attractiveness of the proposition that populations must, ultimately, be limited. Some experimental manipulations of resources and/or their consumers have produced clearly interpretable results, but others have not. Considerable care must thus be exercised in the design and interpretation of such experiments.

Much of the evidence bearing on resource limitation, competition, and the processes underlying resource partitioning relies heavily on inference, assumption, and assertion. Little consideration is given to the direct measurement of resources, much less to gathering information on the various components of resource systems, which is really essential to an understanding of resources and their influences on populations and communities. Beginnings toward such understanding are most likely to emerge from studies that are conducted in systems of restricted resources supporting specialist consumers, and that adopt a rigorous methodology (rather than inference) to test alternative hypotheses. Above all, resource systems require the close attention of minds unfettered by preconceptions of resource limitation, equilibrium, optimization, competition, and the like.

ACKNOWLEDGMENTS

The thoughts contained in this chapter underwent gestation during several years of research with grassland and shrubsteppe bird communities, which were supported by the United States National Science Foundation. My colleagues at New Mexico provided the encouragement and the atmosphere in which to think about resources, and Rich Bradley, Jim Findley, Manuel Molles, Bob Pietruszka, Bea Van Horne, and Steve Zack offered comments on a draft manuscript. Jean Ferner prepared and improved on several drafts of the chapter.

LITERATURE CITED

Abbott, I., L. K. Abbott, and P. R. Grant. 1977. Comparative ecology of Galápagos ground finches (*Geospiza* Gould): Evaluation of the importance of floristic diversity and interspecific competition. *Ecol. Monogr.* **47**:151–184.

Abrams, P. 1980. Some comments on measuring niche overlap. *Ecology* **61**:44–49.

Alatalo, R. V. 1982. Evidence for interspecific competition among European tits *Parus* spp.: A review. *Ann. Zool. Fennici* **19**:309–317.

Anderson, R. S. 1982. Resource partitioning in the carrion beetle (Coleoptera: Silphidae) fauna of southern Ontario: Ecological and evolutionary considerations. *Can. J. Zool.* **60**:1314–1325.

Andrzejewski, R. 1975. Supplementary food and the winter dynamics of bank vole populations. *Acta Theriolog.* **20**:23–40.

Atsatt, P. R., and D. J. O'Dowd. 1976. Plant defense guilds. *Science* **193**:24–29.

Ballinger, R. E. 1977. Reproductive strategies: Food availability as a source of proximal variation in a lizard. *Ecology* **58**:528–535.

Bertness, M. D. 1981. Competitive dynamics of a tropical hermit crab assemblage. *Ecology* **62**:751–761.

Bethel, W. M., and J. C. Holmes. 1973. Altered evasive behavior and responses to light in amphipods harboring acanthocephalan cystacanths. *J. Parasitol.* **59**:945–956.

Bethel, W. M., and J. C. Holmes. 1974. Correlation of development of altered evasive behavior in *Gammarus lacustris* (Amphipoda) harboring cystacanths of *Polymorphus paradoxus* (Acanthocephala) with the infectivity to the definitive host. *J. Parasitol.* **60**:272–274.

Bethel, W. M., and J. C. Holmes. 1977. Increased vulnerability of amphipods to predation owing to altered behavior induced by larval acanthocephalans. *Can. J. Zool.* **55**:110–115.

Birch, L. C. 1979. The effect of species of animals which share common resources on one another's distribution and abundance. *Fortschr. Zool.* **25**:197–221.

Boag, P. T., and P. R. Grant. 1981. Intense natural selection in a population of Darwin's finches (Geospizinae) in the Galápagos. *Science* **214**:82–85.

Bowers, M. A. 1982. Foraging behavior of heteromyid rodents: Field evidence of resource partitioning. *J. Mammal.* **63**:361–367.

Brömssen, A. von, and C. Jansson. 1980. Effects of food addition to Willow Tit *Parus montanus* and Crested Tit *P. cristatus* at the time of breeding. *Ornis Scand.* **11**:173–178.

Brooke, M. de L. 1979. Differences in the quality of territories held by Wheatears (*Oenanthe oenanthe*). *J. Anim. Ecol.* **48**:21–32.

Brown, J. H. 1975. Geographical ecology of desert rodents. In M. L. Cody and J. M. Diamond (eds.), *Ecology and evolution of communities*, pp. 315–341. Belknap Press, Cambridge, Massachusetts.

Brown, J. H., and G. A. Lieberman. 1973. Resource utilization and coexistence of seed-eating desert rodents in sand dune habitats. *Ecology* **54**:788–797.

Brown, J. H., O. J. Reichman, and D. W. Davidson. 1979. Granivory in desert ecosystems. *Annu. Rev. Ecol. Syst.* **10**:201–227.

Brown, J. H., A. Kodric-Brown, T. G. Whitham, and H. W. Bond. 1981. Competition between hummingbirds and insects for the nectar of two species of shrubs. *Southwest. Nat.* **26**:133–145.

Bryant, D. M., and K. R. Westerterp. 1980. The energy budget of the House Martin (*Delichon urbica*). *Ardea* **68**:91–102.

Bryant, J. P., and P. J. Kuropat. 1980. Selection of winter forage by subarctic browsing vertebrates: The role of plant chemistry. *Annu. Rev. Ecol. Syst.* **11**:261–285.

Cain, A. J. 1964. The perfection of animals. In J. D. Carthy and C. L. Duddington (eds.), *Viewpoints in biology*, Vol. 3, pp. 36–63. Butterworths, London.

Calder, W. A., III. 1979. On the temperature-dependency of optimal nectar concentration for birds. *J. Theor. Biol.* **78**:185–196.

Camp, J. W., and H. W. Huizinga. 1979. Altered color, behavior, and predation susceptibility of the isopod *Asellus intermedius* infected with *Acanthocephalus dirus*. *J. Parasitol.* **65**:667–669.

Case, T. J. 1979. Character displacement and coevolution in some *Cnemidophorus* lizards. *Fortschr. Zool.* **25**:235–282.

Case, T. J. 1981. Niche separation and resource scaling. *Am. Nat.* **118**:554–560.

Case, T. J. 1982. Coevolution in resource-limited competition communities. *Theor. Popul. Biol.* **21**:69–91.

Caswell, H. 1976. Community structure: A neutral model analysis. *Ecol. Monogr.* **46**:327–354.

Charnov, E. L. 1976. Optimal foraging: The marginal value theorem. *Theor. Popul. Biol.* **9**:129–136.

Chew, F. S. 1980. Foodplant preferences of *Pieris* caterpillars (Lepidoptera). *Oecologia* **46**:347–353.

Christian, D. P. 1980. Vegetative cover, water resources, and microdistributional patterns in a desert rodent community. *J. Anim. Ecol.* **49**:807–816.

Christian, K. A. 1982. Changes in the food niche during postmetamorphic ontogeny of the frog *Pseudacris triseriata*. *Copeia* **1982**:73–80.

Cockburn, A. 1981. Population regulation and dispersion of the smoky mouse, *Pseudomys fumeus*. I. Dietary determinants of microhabitat preference. *Aust. J. Ecol.* **6**:231–254.

Cody, M. L. 1974. *Competition and the structure of bird communities*. Princeton University Press, Princeton, New Jersey.

Cody, M. L. 1979. Resource allocation patterns in Palaearctic warblers (Sylviidae). *Fortschr. Zool.* **25**:223–234.

Cody, M. L. 1981. Habitat selection in birds: The roles of vegetation structure, competitors, and productivity. *BioScience* **31**:107–113.

Cole, F. R., and G. O. Batzli. 1978. Influence of supplemental feeding on a vole population. *J. Mammal.* **59**:809–819.

Cole, F. R., and G. O. Batzli. 1979. Nutrition and population dynamics of the prairie vole, *Microtus ochrogaster*, in central Illinois. *J. Anim. Ecol.* **48**:455–470.

Coley, P. D. 1980. Effects of leaf age and plant life history patterns on herbivory. *Nature (London)* **284**:545–546.

Collier, B. D., N. C. Stenseth, S. Barkley, and R. Osborn. 1975. A simulation model of energy acquisition and utilization by the brown lemming *Lemmus trimucronatus* at Barrow, Alaska. *Oikos* **26**:276–294.

Collins, S. L., F. C. James, and P. G. Risser. 1982. Habitat relationships of wood warblers (Parulidae) in northern central Minnesota. *Oikos* **39**:50–58.

Connell, J. H. 1980. Diversity and the coevolution of competitors, or the ghost of competition past. *Oikos* **35**:131–138.

Connor, E. F., and D. Simberloff. 1978. Species number and compositional similarity of the Galápagos flora and avifauna. *Ecol. Monogr.* **48**:219–248.

Connor, E. F., and D. Simberloff. 1983. Neutral models of species' co-occurrence patterns. In D. R. Strong, Jr., D. Simberloff, and L. Abele (eds.), *Ecological communities: Conceptual issues and the evidence*. Princeton University Press, Princeton, New Jersey. (in press).

Courtney, P. A., and H. Blokpoel. 1980. Food and indicators of food availability for common terns on the lower Great Lakes. *Can. J. Zool.* **58**:1318–1323.

Crombie, A. C. 1946. Further experiments on insect competition. *Proc. R. Soc. London, Ser. B* **133**:76–109.

Dhondt, A. A., and R. Eyckerman. 1980. Competition between the Great Tit and the Blue Tit outside the breeding season in field experiments. *Ecology* **61**:1291–1296.

Diamond, J. M. 1978. Niche shifts and the rediscovery of interspecific competition. *Am. Sci.* **66**:322–331.

Diamond, J. M., and M. E. Gilpin. 1982. Examination of the "null" model of Connor and Simberloff for species co-occurrences on islands. *Oecologia* **52**:64–74.

Dunham, A. E. 1980. An experimental study of interspecific competition between the iguanid lizards *Sceloporus merriami* and *Urosaurus ornatus*. *Ecol. Monogr.* **50**:309–330.

Dunning, J. B., Jr., and J. H. Brown. 1982. Summer rainfall and winter sparrow densities: A test of the food limitation hypothesis. *Auk* **99**:123–129.

Ellis, J. E., J. A. Wiens, C. F. Rodell, and J. C. Anway. 1976. A conceptual model of diet selection as an ecosystem process. *J. Theor. Biol.* **60**:93–108.

Elton, C. 1927. *Animal ecology*. Sidgwick and Jackson, London.

Feeny, P. 1976. Plant apparency and chemical defense. *Recent Adv. Phytochem.* **10**:1–40.

Feinsinger, P. 1976. Organization of a tropical guild of nectarivorous birds. *Ecol. Monogr.* **46**:257–291.

Feinsinger, P. 1978. Ecological interactions between plants and hummingbirds in a successional tropical community. *Ecol. Monogr.* **48**:269–287.

Ferns, P. N. 1980. Energy flow through small mammal populations. *Mammal Rev.* **10**:165–188.

Fleischer, S. J., W. A. Allen, J. M. Luna, and R. L. Pienkowski. 1982. Absolute-density estimation from sweep sampling, with a comparison of absolute-density sampling techniques for adult potato leafhopper in alfalfa. *J. Econ. Entomol.* **75**:425–430.

Flowerdew, J. R. 1972. The effect of supplementary food on a population of wood mice (*Apodemus sylvaticus*). *J. Anim. Ecol.* **41**:553–566.

Folse, L. J., Jr. 1982. An analysis of avifauna-resource relationships on the Serengeti Plains. *Ecol. Monogr.* **52**:111–127.

Ford, R. G., J. A. Wiens, D. Heinemann, and G. L. Hunt. 1982. Modelling the sensitivity of colonially breeding marine birds to oil spills: Guillemot and kittiwake populations on the Pribilof Islands, Bering Sea. *J. Appl. Ecol.* **19**:1–31.

Foster, M. S. 1977. Ecological and nutritional effects of food scarcity on a tropical frugivorous bird and its fruit source. *Ecology* **58**:73–85.

Franzblau, M. A., and J. P. Collins. 1980. Test of a hypothesis of territory regulation in an insectivorous bird by experimentally increasing prey abundance. *Oecologia* **46**:164–170.

Fretwell, S. D. 1972. *Populations in a seasonal environment*. Princeton University Press, Princeton, New Jersey.

Fretwell, S. D. 1977. The regulation of plant communities by the food chains exploiting them. *Perspec. Biol. Med.* **20**:169–185.

Frye, R. J., and M. L. Rosenzweig. 1980. Clump size selection: A field test with two species of *Dipodomys*. *Oecologia* **47**:323–327.

Gass, C. L. 1978. Rufous hummingbird feeding territoriality in a suboptimal habitat. *Can. J. Zool.* **56**:1535–1539.

Gass, C. L., and R. D. Montgomerie. 1981. Hummingbird foraging behavior: decision-making and energy regulation. In A. C. Kamil and T. D. Sargent (eds.), *Foraging behavior. Ecological, ethological, and psychological approaches*, pp. 159–194. Garland STPM Press, New York.

Gass, C. L., G. Angehr, and J. Centa. 1976. Regulation of food supply by feeding territoriality in the rufous hummingbird. *Can. J. Zool.* **54**:2046–2054.

Gilbert, B. S., and C. J. Krebs. 1981. Effects of extra food on *Peromyscus* and *Clethrionomys* populations in the southern Yukon. *Oecologia* **51**:326–331.

Gill, F. B., A. L. Mack, and R. T. Ray. 1982. Competition between hermit hummingbirds Phaethorninae and insects for nectar in a Costa Rican rain forest. *Ibis* **124**:44–49.

Gilpin, M. E., and J. M. Diamond. 1982. Factors contributing to non-randomness in species co-occurrences on islands. *Oecologia* **52**:75–84.

Goss-Custard, J. D. 1977. Responses of redshank, *Tringa totanus*, to the absolute and relative densities of two prey species. *J. Anim. Ecol.* **46**:867–874.

Goss-Custard, J. D. 1980. Competition for food and interference among waders. *Ardea* **68**:31–52.

Gould, S. J., and R. C. Lewontin. 1979. The spandrels of San Marco and the Panglossian paradigm: A critique of the adaptationist programme. *Proc. R. Soc. London Ser. B* **205**:581–598.

Grant, P. R. 1981. Speciation and adaptive radiation of Darwin's finches. *Am. Sci.* **69**:653–663.

Grant, P. R., and B. R. Grant. 1980. Annual variation in finch numbers, foraging and food supply on Isla Daphne Major, Galápagos. *Oecologia* **46**:55–62.

Hairston, N. G., F. E. Smith, and L. B. Slobodkin. 1960. Community structure, population control, and competition. *Am. Nat.* **94**:421–425.

Hamilton, W. D., and M. Zuk. 1982. Heritable true fitness and bright birds: A role for parasites? *Science* **218**:384–387.

Hansen, L. P., and G. O. Batzli. 1979. Influence of supplemental food on local populations of *Peromyscus leucopus*. *J. Mammal.* **60**:335–342.

Harper, J. L. 1977. *Population biology of plants*. Academic Press, New York.

Hart, D. D. 1981. Foraging and resource patchiness: Field experiments with a grazing stream insect. *Oikos* **37**:46–52.

Hay, M. E., and P. J. Fuller. 1981. Seed escape from heteromyid rodents: The importance of microhabitat and seed preference. *Ecology* **62**:1395–1399.

Holdren, C. E., and P. R. Ehrlich. 1982. Ecological determinants of food plant choice in the checkerspot butterfly *Euphydryas editha* in Colorado. *Oecologia* **52**:417–423.

Holling, C. S. 1965. The functional response of predators to prey density and its role in mimicry and population regulation. *Mem. Entomol. Soc. Can.* **45**.

Holmes, J. C., and W. M. Bethel. 1972. Modification of intermediate host behavior by parasites. In E. U. Canning and C. A. Wright (eds.), *Behavioral aspects of parasite transmission*, pp. 123–149. Academic Press, New York.

Holmes, R. T., J. C. Schultz, and P. Nothnagle. 1979. Bird predation on forest insects: An exclosure experiment. *Science* **206**:462–463.

Huey, R. B., and E. R. Pianka. 1981. Ecological consequences of foraging mode. *Ecology* **62**:991–999.

Hughes, R. D., and J. Walker. 1970. The role of food in the population dynamics of the Australian bushfly. In A. Watson (ed.), *Animal populations in relation to their food resources*, pp. 255–269. Blackwell, Oxford.

Hutto, R. L. 1978. A mechanism for resource allocation among sympatric heteromyid rodent species. *Oecologia* **33**:115–126.

Hutto, R. L. 1981. Temporal patterns of foraging activity in some wood warblers in relation to the availability of insect prey. *Behav. Ecol. Sociobiol.* **9**:195–198.

Jaeger, R. G. 1980. Fluctuations in prey availability and food limitation for a terrestrial salamander. *Oecologia* **44**:335–341.

Jansson, C., J. Ekman, and A. von Brömssen. 1981. Winter mortality and food supply in tits *Parus* spp. *Oikos* **37**:313–322.

Järvinen, O. 1980. Dynamics of north European bird communities. *Acta 17th Int. Congr. Ornithol.*, 770–776.

Joern, A., and L. R. Lawlor. 1980. Food and microhabitat utilization by grasshoppers from arid grasslands: Comparisons with neutral models. *Ecology* **61**:591–599.

Johnson, D. H. 1980. The comparison of usage and availability measurements for evaluating resource preference. *Ecology* **61**:65–71.

Jones, P. J., and P. Ward. 1976. The level of reserve protein as the proximate factor controlling the timing of breeding and clutch-size in the red-billed quelea *Quelea quelea. Ibis* **118**:547–574.

Jones, P. J., and P. Ward. 1979. A physiological basis for colony desertion by red-billed queleas (*Quelea quelea*). *J. Zool. London* **189**:1–19.

Källander, H. 1981. The effects of provision of food in winter on a population of the Great Tit *Parus major* and the Blue Tit *P. caeruleus. Ornis Scand.* **12**:244–248.

Karr, J. R., and F. C. James. 1975. Eco-morphological configurations and convergent evolution in species and communities. In M. L. Cody and J. M. Diamond (eds.), *Ecology and evolution of communities*, pp. 258–291. Belknap Press, Cambridge, Massachusetts.

Kaufman, D. W., and G. A. Kaufman. 1982. Effect of moonlight on activity and microhabitat use by Ord's kangaroo rat (*Dipodomys ordii*). *J. Mammal.* **63**:309–312.

Krebs, J. R. 1971. Territory and breeding density in the Great Tit *Parus major* L. *Ecology* **52**:2–22.

Krebs, J. R., and N. B. Davies (eds.). 1978. *Behavioural ecology: An evolutionary approach.* Sinauer, Sunderland, Massachusetts.

Krebs, J. R., A. I. Houston, and E. L. Charnov. 1981. Some recent developments in optimal foraging. In A. C. Kamil and T. D. Sargent (eds.), *Foraging behavior. Ecological, ethological, and psychological approaches*, pp. 3–18. Garland STPM Press, New York.

Lack, D. 1954. *The natural regulation of animal numbers.* Clarendon Press, Oxford.

Lack, D. 1966. *Population studies of birds.* Clarendon Press, Oxford.

Lawhon, D. K., and M. S. Hafner. 1981. Tactile discriminatory ability and foraging strategies in kangaroo rats and pocket mice (Rodentia: Heteromyidae). *Oecologia* **50**:303–309.

Lemen, C. A. 1978. Seed size selection in heteromyids. A second look. *Oecologia* **35**:13–19.

Levinton, J. S., and G. R. Lopez. 1977. A model of renewable resources and limitation of deposit-feeding benthic populations. *Oecologia* **31**:177–190.

Lister, B. C. 1981. Seasonal niche relationships of rain forest anoles. *Ecology* **62**:1548–1560.

MacArthur, R. H. 1972. *Geographical ecology.* Harper and Row, New York.

Magnuson, J. J., L. B. Crowder, and P. A. Medvick. 1979. Temperature as an ecological resource. *Am. Zool.* **19**:331–343.

Maurer, B. A., and R. C. Whitmore. 1981. Foraging of five bird species in two forests with different vegetation structure. *Wilson Bull.* **93**:478–490.

May, R. M. (ed.). 1981. *Theoretical ecology. Principles and applications*, 2nd ed. Sinauer, Sunderland, Massachusetts.

Maynard Smith, J. 1978. Optimization theory in evolution. *Annu. Rev. Ecol. Syst.* **9**:31–56.

M'Closkey, R. T. 1981. Microhabitat use in coexisting desert rodents—the role of population density. *Oecologia* **50**:310–315.

Menge, B. A. 1972. Foraging strategy of a starfish in relation to actual prey availability and environmental predictability. *Ecol. Monogr.* **42**:25–50.

Milton, K. 1979. Factors influencing leaf choice by howler monkeys: A test of some hypotheses of food selection by generalist herbivores. *Am. Nat.* **114**:362–378.

Minot, E. O. 1981. Effects of interspecific competition for food in breeding Blue and Great tits. *J. Anim. Ecol.* **50**:375–385.

Moore, J. 1981. The ecology of the acanthocephalan (*Plagiorhynchus cylindraceus*) in the isopod (*Armadillidium vulgare*) and the starling (*Sturnus vulgaris*). Ph.D. diss., University of New Mexico, Albuquerque.

Moore, J. 1983. Responses of an avian predator and its isopod prey to an acanthocephalan parasite. *Ecology* **64:** in press.

Morse, D. H. 1980. *Behavioral mechanisms in ecology.* Harvard University Press, Cambridge, Massachusetts.

Moss, R. 1972. Food selection by red grouse (*Lagopus lagopus scoticus*) in relation to chemical composition. *J. Anim. Ecol.* **41**:411–428.

Murdoch, W. W. 1969. Switching in general predators: experiments on predator specificity and stability of prey populations. *Ecol. Monogr.* **39**:335–354.

Murdoch, W. W., and A. Oaten. 1975. Predation and population stability. *Adv. Ecol. Res.* **9**:1–131.

Myers, J. P., S. L. Williams, and F. A. Pitelka. 1980. An experimental analysis of prey availability for sanderlings (Aves: Scolopacidae) feeding on sandy beach crustaceans. *Can. J. Zool.* **58**:1564–1574.

Newton, I. 1980. The role of food in limiting bird numbers. *Ardea* **68**:11–30.

Nielsen, B. O. 1978. Above ground food resources and herbivory in a beech forest ecosystem. *Oikos* **31**:273–279.

Noon, B. R. 1981. The distribution of an avian guild along a temperate elevational gradient: The importance and expression of competition. *Ecol. Monogr.* **51**:105–124.

O'Connor, R. J., and R. A. Brown. 1977. Prey depletion and foraging strategy in the oystercatcher *Haematopus ostralegus. Oecologia* **27**:75–92.

O'Dowd, D. J., and M. E. Hay. 1980. Mutualism between harvester ants and a desert ephemeral: Seed escape from rodents. *Ecology* **61**:531–540.

Oster, G., and E. O. Wilson. 1978. *Caste and ecology in the social insects.* Princeton University Press, Princeton, New Jersey.

Owen, D. F., and R. G. Wiegert. 1976. Do consumers maximize plant fitness? *Oikos* **27**:488–492.

Pacala, S., and J. Roughgarden. 1982. Resource partitioning and interspecific competition in two two-species insular *Anolis* lizard communities. *Science* **217**:444–446.

Pearson, D. L., and S. L. Stemberger. 1980. Competition, body size and the relative energy balance of adult tiger beetles (Coleoptera: Cicindelidae). *Am. Midl. Nat.* **104**:373–377.

Pimm, S. L. 1978. An experimental approach to the effects of predictability on community structure. *Am. Zool.* **18**:797–808.

Pleasants, J. M., and M. Zimmerman. 1979. Patchiness in the dispersion of nectar resources: Evidence for hot and cold spots. *Oecologia* **41**:283–288.

Price, M. V. 1978. Seed dispersion preferences of coexisting desert rodent species. *J. Mammal.* **59**:624–626.

Pulliam, H. R. 1975. Diet optimization with nutrient constraints. *Am. Nat.* **109**:765–768.

Pulliam, H. R. 1980. Do chipping sparrows forage optimally? *Ardea* **68**:75–82.

Pulliam, H. R., and T. A. Parker, III. 1979. Population regulation of sparrows. *Fortschr. Zool.* **25**:137–147.

Puttick, G. M. 1980. Energy budgets of Curlew Sandpipers at Langebaan Lagoon, South Africa. *Estuar. Coastal Mar. Sci.* **11**:207–215.

Puttick, G. M. 1981. Sex-related differences in foraging behaviour of Curlew Sandpipers. *Ornis Scand.* **12**:13–17.

Pyke, G. H. 1980. The foraging behaviour of Australian honeyeaters: A review and some comparisons with hummingbirds. *Aust. J. Ecol.* **5**:343–369.

Pyke, G. H., H. R. Pulliam, and E. L. Charnov. 1977. Optimal foraging: a selective review of theory and tests. *Q. Rev. Biol.* **52**:137–154.

Rabinowitz, D. 1981. Buried viable seeds in a North American tall-grass prairie: The resemblance of their abundance and composition to dispersing seeds. *Oikos* **36**:191–195.

Rapport, D. J. 1980. Optimal foraging for complementary resources. *Am. Nat.* **116**:324–346.

Rapport, D. J. 1981. Foraging behavior of *Stentor coeruleus*: A microeconomic interpretation. In A. C. Kamil and T. D. Sargent (eds.), *Foraging behavior. Ecological, ethological, and psychological approaches*, pp. 77–93. Garland STPM Press, New York.

Rathcke, B. J. 1976. Competition and coexistence within a guild of herbivorous insects. *Ecology* **57**:76–87.

Real, L. A. 1975. A general analysis of resource allocation by competing individuals. *Theor. Popul. Biol.* **8**:1–11.

Reichman, O. J. 1975. Relation of desert rodent diets to available resources. *J. Mammal.* **56**:731–751.

Reichman, O. J. 1979. Desert granivore foraging and its impact on seed densities and distributions. *Ecology* **60**:1085–1092.

Reichman, O. J. 1981. Factors influencing foraging in desert rodents. In A. C. Kamil and T. D. Sargent (eds.), *Foraging behavior. Ecological, ethological, and psychological approaches*, pp.195–213. Garland STPM Press, New York.

Reichman, O. J., and D. Oberstein. 1977. Selection of seed distribution types by *Dipodomys merriami* and *Perognathus amplus*. *Ecology* **58**:636–643.

Rhoades, D. F. 1979. Evolution of plant chemical defense against herbivores. In G. A. Rosenthal and D. H. Janzen (eds.), *Herbivores. Their interaction with secondary plant metabolites*, pp. 3–54. Academic Press, New York.

Rhoades, D. F., and R. G. Cates. 1976. Toward a general theory of plant antiherbivore chemistry. *Recent Adv. Phytochem.* **10**:168–213.

Ricklefs, R. E., and G. W. Cox. 1977. Morphological similarity and ecological overlap among passerine birds on St. Kitts, British West Indies. *Oikos* **29**:60–66.

Rissing, S. W. 1981. Prey preferences in the desert horned lizard: Influence of prey foraging method and aggressive behavior. *Ecology* **62**:1031–1040.

Rosenthal, G. A., and D. H. Janzen (eds.). 1979. *Herbivores. Their interaction with secondary plant metabolites*. Academic Press, New York.

Rotenberry, J. T. 1980. Dietary relationships among shrub-steppe passerine birds: Competition or opportunism in a variable environment? *Ecol. Monogr.* **50**:93–110.

Rotenberry, J. T., and J. A. Wiens. 1980. Temporal variation in habitat structure and shrub-steppe bird dynamics. *Oecologia* **47**:1–9.

Savory, C. J. 1978. Food consumption of red grouse in relation to the age and productivity of heather. *J. Anim. Ecol.* **47**:269–282.

Schluter, D. 1982a. Seed and patch selection by Galápagos ground finches: relation to foraging efficiency and food supply. *Ecology* **63**:1106–1120.

Schluter, D. 1982b. Distributions of Galápagos ground finches along an altitudinal gradient: The importance of food supply. *Ecology* **63**:1504–1517.

Schoener, T. W. 1971. Theory of feeding strategies. *Annu. Rev. Ecol. Syst.* **2**:369–404.

Schoener, T. W. 1974. Resource partitioning in ecological communities. *Science* **185**:27–39.

Schoener, T. W. 1982. The controversy over interspecific competition. *Am. Sci.* **70**:586–595.

Schroder, G. D., and M. L. Rosenzweig. 1975. Perturbation analysis of competition and overlap in habitat utilization between *Dipodomys ordii* and *Dipodomys merriami*. *Oecologia* **19**:9–28.

Sherry, T. W., and L. A. McDade. 1982. Prey selection and handling in two neotropical hover-gleaning birds. *Ecology* **63**:1016–1028.

Sinclair, A. R. E., C. J. Krebs, and J. N. M. Smith. 1982. Diet quality and food limitation in herbivores: The case of the snowshoe hare. *Can. J. Zool.* **60**:889–897.

Singer, M. C., and P. R. Ehrlich. 1979. Population dynamics of the checkerspot butterfly *Euphydryas editha*. *Fortschr. Zool.* **25**:53–60.

Smith, C. C. 1981. The indivisible niche of *Tamiasciurus*: An example of nonpartitioning of resources. *Ecol. Monogr.* **51**:343–363.

Smith, K. G. 1982. Drought-induced changes in avian community structure along a montane sere. *Ecology* **63**:952–961.

Smith, M. H. 1971. Food as a limiting factor in the population ecology of *Peromyscus polionotus* (Wagner). *Ann. Zool. Fennici* **8**:109–112.

Sorensen, A. E. 1981. Interactions between birds and fruit in a temperate woodland. *Oecologia* **50**:242–249.

Southwood, T. R. E. 1978. *Ecological methods*, 2nd ed. Chapman and Hall, London.

Stamps, J., S. Tanaka, and V. V. Krishnan. 1981. The relationship between selectivity and food abundance in a juvenile lizard. *Ecology* **62**:1079–1092.

Strong, D. R., Jr. 1982. Potential interspecific competition and host specificity: Hispine beetles on *Heliconia*. *Ecol. Entomol.* **7**:217–220.

Taitt, M. J. 1981. The effect of extra food on small rodent populations: I. Deermice (*Peromyscus maniculatus*). *J. Anim. Ecol.* **50**:111–124.

Taitt, M. J., and C. J. Krebs. 1981. The effect of extra food on small rodent populations: II. Voles (*Microtus townsendii*). *J. Anim. Ecol.* **50**:125–137.

Taitt, M. J., J. H. W. Gipps, C. J. Krebs, and Z. Dundjerski. 1981. The effect of extra food and cover on declining populations of *Microtus townsendii*. *Can. J. Zool.* **59**:1593–1599.

Templeton, A. R., and D. A. Levin. 1979. Evolutionary consequences of seed pools. *Am. Nat.* **114**:232–249.

Tepedino, V. J., and N. L. Stanton. 1981. Diversity and competition in bee-plant communities on short-grass prairie. *Oikos* **36**:35–44.

Tepedino, V. J., and N. L. Stanton. 1982. Estimating floral resources and flower visitors in studies of pollinator-plant communities. *Oikos* **38**:384–386.

Thompson, S. D. 1982a. Microhabitat utilization and foraging behavior of bipedal and quadrupedal heteromyid rodents. *Ecology* **63**:1303–1312.

Thompson, S. D. 1982b. Structure and species composition of desert heteromyid rodent species assemblages: Effects of a simple habitat manipulation. *Ecology* **63**:1313–1321.

Thorman, S. 1982. Niche dynamics and resource partitioning in a fish guild inhabiting a shallow estuary on the Swedish west coast. *Oikos* **39**:32–39.

Tilman, D. 1982. *Resource competition and community structure*. Princeton University Press, Princeton, New Jersey.

Townsend, C. R., and P. Calow (eds.). 1981. *Physiological ecology: an evolutionary approach to resource use*. Sinauer, Sunderland, Massachusetts.

Trombulak, S. C., and G. J. Kenagy. 1980. Effects of seed distribution and competitors on seed harvesting efficiency in heteromyids. *Oecologia* **44**:342–346.

Turelli, M. 1981. Niche overlap and invasion of competitors in random environments. I. Models without demographic stochasticity. *Theor. Popul. Biol.* **20**:1–56.

Vitt, L. J., R. C. van Loben Sels, and R. D. Ohmart. 1981. Ecological relationships among arboreal desert lizards. *Ecology* **62**:398–410.

Waser, P. M., and R. H. Wiley. 1980. Mechanisms and evolution of spacing in animals. In P. Marler and J. G. Vandenberg (eds.), *Handbook of behavioral neurobiology*, Vol. 3, pp. 159–223. Plenum Press, New York.

White, T. C. R. 1978. The importance of a relative shortage of food in animal ecology. *Oecologia* **33**:71–86.

Whitham, T. G. 1978. Habitat selection by *Pemphigus* aphids in response to resource limitation and competition. *Ecology* **59**:1164–1176.

Whitham, T. G. 1980. The theory of habitat selection: Examined and extended using *Pemphigus* aphids. *Am. Nat.* **115**:449–466.

Whitham, T. G. 1981. Individual trees as heterogeneous environments: Adaptation to herbivory or epigenetic noise? In R. F. Denno and H. Dingle (eds.), *Insect life history patterns: habitat and geographic variation*, pp. 9–27. Springer-Verlag, New York.

Wiens, J. A. 1976. Population responses to patchy environments. *Annu. Rev. Ecol. Syst.* **7**:81–120.

Wiens, J. A. 1977a. On competition and variable environments. *Am. Sci.* **65**:590–597.

Wiens, J. A. 1977b. Model estimation of energy flow in North American grassland bird communities. *Oecologia* **31**:135–151.

Wiens, J. A. 1981. Scale problems in avian censusing. *Stud. Avian Biol.* **6**:513–521.

Wiens, J. A. 1982. On size ratios and sequences in ecological communities: Are there no rules? *Ann. Zool. Fennici* **19**:297–308.

Wiens, J. A. 1983a. Avian community ecology: An iconoclastic view. In A. H. Brush and G. A. Clark, Jr. (eds.), *Perspectives in ornithology*, pp. 355–403. Cambridge University Press, Cambridge.

Wiens, J. A. 1983b. On understanding a nonequilibrium world: Myth and reality in community patterns and processes. In D. R. Strong, D. Simberloff, and L. Abele (eds.), *Ecological communities: Conceptual issues and the evidence*. Princeton University Press, Princeton, New Jersey. (in press).

Wiens, J. A., and G. S. Innis. 1974. Estimation of energy flow in bird communities: a population bioenergetics model. *Ecology* **55**:730–746.

Wiens, J. A., and R. A. Nussbaum. 1975. Model estimation of energy flow in northwestern coniferous forest bird communities. *Ecology* **56**:547–561.

Wiens, J. A., and J. T. Rotenberry. 1980. Patterns of morphology and ecology in grassland and shrubsteppe bird populations. *Ecol. Monogr.* **50**:287–308.

Wiens, J. A., and J. T. Rotenberry. 1981. Habitat associations and community structure of birds in shrubsteppe environments. *Ecol. Monogr.* **51**:21–41.

Wilson, E. O. 1975. *Sociobiology*. Harvard University Press, Cambridge, Massachusetts.

Zimmerman, M. 1982. The effect of nectar production on neighborhood size. *Oecologia* **52**:104–108.

Zimmerman, M., and J. M. Pleasants. 1982. Competition among pollinators: quantification of available resources. *Oikos* **38**:381–383.

Zucker, W. V. 1982. How aphids choose leaves: The role of phenolics in host selection by a galling aphid. *Ecology* **63**:972–981.

CHAPTER 16

Mutualistic Interactions in Population and Community Processes

JOHN F. ADDICOTT
Department of Zoology
University of Alberta
Edmonton, Alberta, Canada
and
Rocky Mountain Biological Laboratory
Crested Butte, Colorado

CONTENTS

---

1   INTRODUCTION

The study of mutualism has begun to emerge as a strong and active discipline within ecology (for recent reviews see Boucher et al., 1982; Thompson, 1982; Wilson, 1980). Although ecologists are still identifying and describing mutualistic systems, particularly those in which benefits arise indirectly or in nonobvious ways, there is now a greater emphasis on analyses of the dynamics, regulation, and consequences of mutualism. This chapter is an examination of some recent trends and directions in the study of mutualism. It will emphasize what mutualism is and how it arises, the population structure of mutualistic systems, density-, time-, and location-dependent mutualism, costs vs. benefits in mutualistic systems, and the consequences of competition, mutualism, and parasitism within mutualistic systems.

1.1   Definitions

Mutualism is any interaction in which two (or more) species reciprocally benefit from the presence of the other species. There are many different factors that further define the nature of mutualistic interactions. These include the degree of association among two species (symbiotic, nonsymbiotic), the degree to which the interaction is necessary for the survival of each species (obligate, facultative), the kinds of benefits (e.g., protection, dispersal, feeding), the extent of reciprocal specialization by one species in response to the other (coevolved, noncoevolved), the temporal pattern of the association of species (permanent, intermittent), the temporal pattern of the strength of the interaction (constant, variable), the extent to which benefit arises from association in close contact (direct, indirect), the number and specificity of species involved in the interaction, and the relative size or reproductive potential of each species.

Terminology for beneficial interactions is not consistent among authors, some using the term symbiosis for reciprocally beneficial interactions among species (e.g., Hutchinson, 1978), or the terms mutualism and protocooperation for what I would call obligate and facultative mutualism, respectively (e.g., Odum, 1971). However, it is preferable to retain the term symbiosis to refer to the extent to which two species live in close association (see Hertig et al., 1937). In the long run, less confusion will arise by using the single, general term, mutualism, for reciprocally beneficial interactions, applying appropriate qualifiers as needed.

Commensalism is an interaction that benefits one species without affecting the other species. Facilitation can be a synonym for commensalism (e.g., McNaughton, 1976; Sinclair and Norton-Grifiths, 1982), but it is also a synonym for benefit, without specifying whether this occurs within mutualism or commensalism. This chapter considers both mutualism and commensalism because, in practice, it is difficult to obtain unequivocal demonstrations of reciprocal benefit.

1.2   Measurement of Benefit

The concept of benefit can apply to either an individual or population level of organization. At an individual level, benefit occurs if the fitness of those organisms capable of associating with another species is greater than those organisms not capable of associating with another species (e.g., Keeler, 1981). At the population level, species j benefits species i if the per capita growth rate of species i increases with increases in the density of species j (i.e., $\partial(dN_i/N_i\, dt)/\partial N_j > 0$).

Application of these concepts to either model or biological systems can be difficult, particularly if benefits arise indirectly (see below and Table 1). Lawlor (1979) discusses this problem for a population model of mutualism arising from competitive interactions and suggests defining the net interspecific effect, γ, as the net change in the equilibrium density of one species per change in the density of a second species. This takes into account both the direct and indirect effects of species on each other, but only at equilibrium. If the benefits are indirect, there does not appear to be an analytic method for measuring net interspecific effects at any arbitrary density, but computer simulations can approximate these effects (Addicott and Freedman, 1983). Cost–benefit models of mutualism (e.g., Roughgarden, 1975; Keeler, 1981) also encounter difficulties if indirect interactions occur.

The measurement of benefit in biological systems typically involves manipulating the density of one species and observing changes in the fitness, per capita population growth rate, or equilibrium density of a target species. Problems in choosing measures of fitness can lead to inappropriate conclusions about the existence of mutualism. For example, Hutchinson (1978, p. 148) interprets Wilbur's (1972) work on interactions among larval amphibians as an apparent kind of mutualism. Although one component of fitness (survival) does show beneficial effects of the presence of other species, there are negative effects on other components (size, development rate), making it likely that the net interaction is neutral or even detrimental. Owen and Wiegert (1976, 1981) hypothesize, and provide some evidence, that plants may have greater productivity or sexual reproductive success following grazing. However, in long-lived plants, increased aboveground production or sexual reproduction may occur at the expense of nutrient storage in the roots. This could negatively affect the ability of the plant to survive, or to reproduce asexually or sexually in future years, and could result in a net loss of fitness. Therefore, particularly for mutualism that might arise indirectly or in a "nonobvious" manner, it is important to consider trade-offs among fitness components and short- and long-term assessments of fitness.

2 PATTERNS OF MUTUALISM

2.1 Types of Benefits and Systems

Appreciating the diversity of beneficial effects in mutualistic or commensal systems will lead to a better understanding of the dynamics, regulation, development, and consequences of mutualism. Table 1 summarizes the many types of benefit, distinguishing between those that arise through direct interactions among two species and those that arise indirectly. An indirect benefit is one that requires the presence of one or more additional species besides the mutualists themselves. For example, if A benefits B by decreasing a negative effect of C on B, then the effect of A on B is indirect. Indirect benefits are common and involve modification of existing predator–prey, competitive, or mutualistic interactions. Mutualism in which benefits for each species are indirect is an indirect mutualism (Boucher et al., 1982).

Most types of benefit may occur in either mutualistic or commensal systems. However, although increasing prey availability is a common form of benefit in commensal systems (Wilson, 1980), there appear to be no verified examples of its occurrence in mutualistic systems (see Section 2.2). The variety of both direct and indirect benefits results in many different kinds of mutualism. Most are asymmetric in the sense that each species benefits in a different way, but there are exceptions (e.g., Brown and Kodric-Brown, 1979; Mares and Rosenzweig, 1978; Munn and Terborgh, 1979; Vance, 1978). Depending on the kinds of benefits, mutualism between two species may involve interactions among two (e.g., Bequaert, 1921), three (e.g., Sankurathri and Holmes, 1976), or more species (e.g., Messina, 1981). The variety of mutualisms makes it difficult to generalize among either model or natural systems (e.g., Addicott, 1981; Addicott and Freedman, 1983; Wolin and Lawlor, 1983).

2.2 On "Nonobvious" Mutualism

Mutualism is far more prevalent and important in ecological systems than would be indicated by the ecological literature prior to the mid-1970s (see Risch and Boucher, 1976). Attempts to rectify this neglect have included interpretations of ecological data and hypotheses for how mutualism might

Table 1. Types of Benefits, Hypothesized or Observed, among Species in Mutualistic and Commensal Systems

Types of Benefits	Representative References
Direct	
Nutrient transfer	Kleinfeldt (1978), Lewis (1973)
Energy transfer	Glynn (1976), Noble (1975)
Provide habitat	Bloom (1975), Vance (1978)
Disperse gametes or individuals	Bequaert (1921), Price and Waser (1979)
Lower intraspecific competition	Collins et al. (1976)
Increase survival	Valerio (1975)
Indirect	
Modified predator–prey interactions	
Protection from predation	
Feed on predator	D'Elisca (1977), Sankurathri and Holmes (1976)
Deter feeding by predator	Berger (1980), Bloom (1975), Ross (1971)
Predator satiation	Hutchinson (1978)
Common defense	Munn and Terborgh (1979), Wickler (1968)
Diverse defense (associational resistence)	Feeny (1975)
Increase prey availability	Mares and Rosenzweig (1976), Vandermeer (1980)
Modified competitive interactions	
Decrease competition on mutualist	
Niche differentiation	Friedmann and Kern (1956)
Deter competition	Osman and Haugsness (1981), Wright (1973)
Increase effects on competitor	Janzen (1969), Messina (1981), Quinlan and Cherret (1978)
Feed on competitor	Barbehenn (1969), Springett (1968)
Compete with competitor	Davidson (1980), Lawlor (1979)
Modified mutualistic interactions	
Attract mutualists	
Simultaneously	Bobisud and Neuhaus (1975), Brown and Kodric-Brown (1979)
Sequentially	Waser and Real (1979)

arise in "nonobvious" ways (e.g., Collins et al., 1976; Hutchinson, 1978; Lawlor, 1979; Mares and Rosenzweig, 1978; Owen and Wiegart, 1976, 1981; Patelle, 1981; Simberloff et al., 1978; Vandermeer, 1980; Vandermeer and Boucher, 1978).

Systems interpreted as displaying "nonobvious" mutualism (for examples, see Hutchinson, 1978; and Vandermeer and Boucher, 1978) do show "nonobvious" beneficial effects of one species on another, but in many cases, benefits are unidirectional, indicating commensalism, not mutualism. The ecological and evolutionary implications of mutualism and commensalism are substantially different; there should be a clear delimitation between them whenever possible. This should not discourage the search for novel ways in which species reciprocally benefit each other, because there are many systems once interpretated as commensal or parasitic that are now known to be mutualistic (e.g., Glynn, 1976; Smith, 1968). However, there must be a coupling of a willingness to "look for" beneficial interactions with a clear discrimination between mutualism, involving reciprocal benefits, and commensalism (= facilitation) involving a unidrectional benefit.

3 SOME PROBLEMS FOR FUTURE RESEARCH

3.1 On the Structure of Mutualistic Systems

The structure of a population affects both its dynamics and evolution. For potentially mutualistic systems, the evolution of mutualism depends upon the distribution of benefits among individuals within a population (Boucher et al., 1982; Wilson, 1980). Benefits are localized at the level of one or a few individuals in symbiotic mutualism (e.g., Taylor, 1969), and cost–benefit models usually assume this structure, with nonassociated and nonmutualistic individuals receiving no benefits (e.g., Keeler, 1981). Wilson's (1980) model of trait group structure allows benefits to be felt within a larger group of individuals, by both mutualistic and nonmutualistic types, and there are natural systems in which benefits are shared among larger groups of individuals. In pollination systems, a local population of plants may provide nectar for, and be pollinated by, a large number of individuals of one or more species of pollinators. Even nonmutualists, those producing no nectar, may benefit from the interaction. In protection mutualisms, the distribution of the benefits depend on how the density or behavior of the predator is affected. If the density of the predator is decreased, then all mutualists and nonmutualists within an ecological neighborhood would benefit. Alternatively, deterring predation from associated mutualists could cause higher levels of predation on nonassociated and nonmutualistic types.

There is a need for more analyses of the distribution of benefits, particularly in those systems with indirect benefits. This will require the simultaneous analysis of population structure of from two to four (or more) species. This presents a serious challenge for ecologists, because the population structure of each species may be different. For example, in ant–homopteran–predator systems, many local populations of aphids may be contained within the foraging range of one nest of ants, many of which are in turn contained within the movement range of predators (Addicott, 1979). Correct assessment of population structure is essential for determining whether a beneficial interaction occurs. Bierzychudek (1981) tests and rejects the hypothesis that the pollination success of two similar plant species increases when they co-occur. However, she notes that the plots used for sampling co-occurrence were much smaller than the foraging range of the pollinator, and suggests that sampling larger plots might be more appropriate and could possibly lead to a different conclusion.

The structure of mutualistic systems also affects their dynamics. Most population models of mutualism assume that populations interact as homogeneous units (see Addicott, 1981; Vandermeer and Boucher, 1978). With these models, the type of mutualistic benefit determines the dynamic properties of the system (Addicott and Freedman, 1983). However, in natural systems homogeneous interactions among mutualists are unlikely, and there is therefore a need for development and analysis of structured models. These are now common in the study of competition and predation, where structure usually increases stability.

3.2 On the Distribution and Importance of Mutualism

Ecologists are concerned with the relative frequency and importance in different ecosystems of predation, competition, mutualism, commensalism, and spatial and temporal resource division and heterogeneity. Although some kinds of obligate mutualistic systems occur more frequently in tropical and subtropical ecosystems than in temperate and arctic ecosystems, this pattern may not reflect either the overall frequency or significance of mutualism. For example, microbial ecologists have stressed the importance of mutualism in environments in which there are high intensity abiotic "problems" (e.g., Lewis, 1973). Symbiotic nitrogen fixation and mycorrhizal associations are particularly frequent in high latitude plant communities where soil nutrients are limiting (Harley, 1970), and lichens colonize the relatively "harsh" environments of bare rock surfaces.

A better understanding of this problem will involve both empirical and conceptual approaches. "What ecologists look for" (Risch and Boucher, 1976) may be very important. We need to consider not only highly coevolved, symbiotic or obligate mutualisms, but also nonsymbiotic, facultative, intermittent, or diffuse mutualisms. Recent research in temperate ecosystems has been effective in revealing mutualisms that are less obvious because of reduced specificity (e.g., Culver and Beattie, 1978; Keeler, 1981; Vance, 1978) or intermittent association (e.g., Tilman, 1978). Another problem is that systems that appear to be parasitic or commensal may on closer analysis reveal mutualistic interactions (e.g., Glynn, 1976; Smith, 1968). In many cases whole groups of organisms are classified as parasites, based on taxonomic affinities, yet we know almost nothing about their effects, positive or negative, on their hosts (J. C. Holmes, personal communication). We also need to know not just the absolute but also the relative frequency of mutualistic interactions in different ecosystems.

A conceptual approach involves asking under what circumstances an organism would be more successful as a mutualist than as a nonmutualist. This requires focusing on how mutualism arises, and on the intensity and variability of "problems" associated with obtaining energy, nutrients, or mutualists, or avoiding predators or competitors. Where an organism encounters a constant, high intensity biotic "problem," mutualists can provide a "solution" that is consistent and "immune to evolutionary changes" (Schemske, 1980). Alternatively, where the intensity of the biotic "problem" is variable in time or space, mutualism can provide a flexible, facultative "solution" (Culver and Beattie, 1978; Stephenson, 1982). The questions that need to be answered are, ecologically, when will a mutualist be able to "solve" a "problem" more effectively, and, evolutionarily, when will a second genome make the "solution" more stable to evolutionary change. This will probably require the use of evolutionary stable strategies (ESS) and related modeling techniques.

3.3 Probability of Extinction of Mutualists

Another factor that could influence the distribution of mutualisms is their persistence in the face of fluctuating environments. Using a two-species graphic model of obligate mutualism, May (1976) argues that there will be a region of densities within which one or both mutualists would necessarily go extinct, and that given a fluctuating environment, obligate mutualists would eventually enter this region. This hypothesis has been invoked as at least a partial explanation for geographic (May, 1976) and taxonomic (Briand and Yodzis, 1982) distributions of obligate mutualism. However, the hypothesis suffers from the absence of an appropriate null model against which to compare the dynamics of obligate mutualists (Addicott, 1981). An alternative hypothesis is that mutualism leads to lower extinction rates. Mutualism could increase the resiliency of species to ecological perturbations by isolating them from the effects of factors that might fluctuate or become limiting, or it could lead to a faster rate of return to equilibrium conditions following a perturbation. Providing appropriate null comparisons for any of these hypotheses is difficult, because we do not know enough about what biological changes occur in going from a nonmutualist to a facultative mutualist to an obligate mutualist. One approach to getting around this problem is to examine facultative mutualism instead of obligate mutualism, assuming that differences between mutualists and nonmutualists are small (Addicott, 1981).

However, differentiating between these views will ultimately require more careful observation and analysis of mutualisms in nature. There are relatively few studies of the dynamics of mutualism (see Gilbert, 1977), and more research effort should be directed toward comparisons among mutualistic and nonmutualistic populations in similar environments. Another approach is to analyze how a species changes as it becomes more and more dependent on another species (Roughgarden, 1975), and then analyze how these changes influence the resiliency and persistence of model systems. Close examination of the structure of mutualistic systems may reveal how the mechanism of an interaction may lead to variable mutualism "coefficients" that increase as density decreases, providing a strong stabilizing influence (see Section 3.5).

3.4 The Costs of Being a Mutualist

Mutualism only occurs if there is a net benefit of an association among species, with costs being less than (gross) benefits. There is relatively little attention paid to the costs of being a mutualist (Janzen, 1979), despite the importance in model systems of cost terms on the evolution (Keeler, 1981; Roughgarden, 1975) and dynamics of mutualism (Addicott and Freedman, 1983). Ecologists generally measure net benefits, ignoring the component cost and benefits, because costs may be difficult to measure or directly compare with gross benefits. It is relatively easy to measure some costs, such as the production of floral or extrafloral nectar in pollination or ant–plant mutualisms. However, it is a much more demanding task to measure the costs of a floral display used to attract pollinators, or the pits on a beetle used to carry phoretic mites. Even when it is possible to measure costs, it is necessary to be able to compare costs and (gross) benefits in similar terms. This is difficult to accomplish because of another asymmetry of mutualistic systems: for any one species the "currency" in which it receives benefits is usually different from the "currency" in which costs arise. For example, in protection mutualisms a species may benefit by decreased losses to predators at the cost of some energy transferred to its mutualist. One solution is to attempt to calculate the cost of being a mutualist as a proportion of the total energy budget of the organism (see O'Dowd, 1979; O'Dowd and Hay, 1980), and to compare this with the benefits of mutualism measured in a similar fashion (Janzen, 1979). A more direct solution is found in those systems in which *most* of the costs and benefits can be compared directly, preferably in terms of a component of fitness, such as survivorship. A predator may feed upon members of a population but survivorship may be higher with predation than without (e.g., Collins et al., 1976; Valerio, 1975). In a few pollination systems the pollinators are also seed predators (e.g., Janzen, 1969). A predator associated with a plant may decrease seed production by feeding on pollinators or increase seed survival by feeding on seed predators (Louda, 1982). Mutualisms based on "sharing" predators or mutualists (e.g., Brown and Kodric-Brown, 1979; Munn and Terborgh, 1979) may also be appropriate. These kinds of systems deserve further exploration.

Another problem associated with the cost of being a mutualist is the extent to which costs occur, or are constant or facultative, depending on the presence or absence of an associated mutualist. If there were reduced costs or no costs to a nonassociated but potentially mutualistic individual, this would increase the probability of mutualists being favored over nonmutualists (Keeler, 1981). There may be no cost to a mutualist involved in an indirect mutualism (Lawlor, 1979; Vandermeer, 1980), where each species appears to act as it would in the absence of the other, yet each benefits from the presence of the other species. However, these systems may not necessarily be stable to evolutionary change, and costs might be incurred to obtain additional benefits (see Wilson, 1980). There are examples of both fixed and variable costs in natural systems. The production of nectar by most plants appears to be fixed, regardless of whether it is utilized by a mutualist or not (Cruden et al., 1983), but there are some plants in which the production of nectar increases or decreases with utilization (Cruden et al., 1983; McDade and Kinsman, 1980). The food bodies present on most ant plants are produced whether or not ants are present, but in one case the production is facultative (Risch and Rickson, 1981). In some ant–plant systems the costs to a plant of producing extrafloral nectar vary in response to the potential benefits achieved by associating with ants, increasing in response to herbivore damage (Stephenson, 1982) or decreasing when ants are no longer effective in removing herbivores (Tilman, 1978).

3.5   Nonconstant Interactions: Space, Time, and Density

A mechanistic view of mutualism (Section 2) makes it easier to understand and model how the strength of a mutualistic interaction may vary. Both the "problem" (e.g., predation, competition, nutrient availability) and the ability of a mutualist to "solve" that "problem" may vary either in time (e.g., Culver and Beattie, 1978; Laine and Niemela, 1980; Messina, 1981; Sinclair and Norton-Griffiths, 1982; Tilman, 1978) or space (e.g., Bach and Herrnkind, 1980; Cruden et al., 1976; Losey, 1978; Smith, 1968). In protection mutualisms, the density of predators or competitors of one of the mutualists may vary independently of the densities of the mutualists. Thus, the maximum benefit that a mutualist provides could also vary. Some studies of cleaning symbioses show beneficial effects of the cleaners, but others show no effects or detrimental effects (Losey, 1978). This is probably the result of variation in the densities of parasites of fishes in space and time and the difficulty of observing these systems for long periods of time. Herbivory on a plant may become so intense that a mutualist can no longer provide effective protection (Tilman, 1978), leading to a temporary breakdown of an association. With mycorrhizal associations between vascular plants and fungi, the association is beneficial for the vascular plant when soil nutrients are low. But, when soil nutrients are high, plants can obtain all the necessary nutrients in the absence of fungi, and fungi are essentially parasitic on the plants (Crush, 1975). The abiotic environment may limit the distribution and activity of a potential mutualist, thereby limiting advantages of associating with that mutualist (Cruden et al., 1976). Louda's (1982) study of the effects of spiders on the pollinators and seed predators of a plant illustrates that variation in the temporal pattern of association can affect whether or not the association is beneficial.

The degree of mutualism will vary not only as the "problem" and potential "solutions" vary, but also as the density of the mutualists themselves vary. Mutualism coefficients may be density- or frequency-dependent functions (e.g., Gause and Witt, 1935), leading to curvilinear isoclines and increased stability of mutualistic interactions (e.g., Vandermeer and Boucher, 1978). Mechanistic modeling of mutualism shows how density-dependent variation in mutualism may occur. Bobisud and Neuhaus (1975) model three plant species and a pollinator, showing that two plant species, each at low frequency, may benefit each other if the pollinator does not distinguish between them. However, as the frequency of one of these species increases, the advantage it gains from the other species decreases. Predator–prey and competition models, modified by the presence of a mutualist, also show declines in the benefits of association as density increases (Addicott and Freedman, 1983).

Studies of natural systems show both positive and negative density dependent variation in the effects of mutualism. The interaction among aphids and ants on fireweed (Addicott, 1979) shows that at high aphid densities, ants no longer have a positive effect on aphids. In a system that may be mutualistic, isopod and insect root borers increase root production of mangroves, but this may only occur at low densities of borers (Simberloff et al., 1978). In pollination systems, seed production may increase with pollinator densities. However, at high pollinator densities, seed production may be independent of pollinator density, being limited instead by energy available for seed maturation (Udovic, 1981). There are many other mutualisms in which positive density effects occur (e.g., Platt et al., 1974; Sankurathri and Holmes, 1976; Schemske, 1981; Silander, 1981).

More research should be directed to the study of temporal, spatial, and density-dependent variation in mutualism in both model and natural systems. Model systems can be used to examine the development and stability of mutualism under variable costs and benefits, or to explore under what conditions there will be an optimal density for one or both mutualists. We also need to incorporate variation in the strength of mutualistic interactions into our concept of mutualism. An interaction may be mutualistic and yet not show beneficial effects at all times, places or densities.

3.6 Nonequilibrium Interactions in Mutualism

In thinking of mutualism as a population phenomenon, there is a heavy emphasis placed on the effects of mutualism on equilibrium density. However, a consequence of negative density-dependent interactions in mutualism is that mutualism may have no effect or even a negative effect on equilibrium densities (Addicott, 1981; Addicott and Freedman, 1983; Wolin and Lawlor, 1983). Similarly, mutualism may affect parameters that have no effect upon the equilibrium density of a species (Gilbert, 1977). Nonetheless, the interaction would be beneficial if the system rarely reaches equilibrium, or the fitness of individuals is determined primarily during initiation of an association, or mutualists make a greater contribution to the equilibrium population than nonmutualists. In these situations conceptual or experimental emphasis on the effects of mutualism on equilibrium density will lead to inappropriate conclusions.

With facultative mutualisms in varying environments, the densities of mutualists may fluctuate independently, with equilibrium conditions being rare. With ant–homopteran interactions, the densities of homopterans available for tending in any given year are strongly influenced by weather conditions during the previous winter when there is usually no association between ants and homopterans. Successional plant species involved in mutualistic interactions (e.g., Harley, 1970) may never achieve an equilibrium density.

Under these conditions what is important is whether and how soon an association is established and how strong that association is at low densities. For example, the interaction between aphids and ants on fireweed (Addicott, 1978b, 1979) shows that once an aphid population reaches a certain size, it is very unlikely to go extinct even in the absence of tending ants. However, ants decrease the probability of extinction of newly established populations by increasing fecundity and maturation rate of aphids, and by decreasing the impacts of predators or parasitoids. The effects of ants on aphids decrease with higher densities of aphids. Viewing mutualism as an individual phenomenon, the time required for an association to occur may determine whether or not an individual is successful (e.g., Collins et al., 1977). This is often a part of the natural history of a system about which very little is known.

Mutualism may have little effect on equilibrium density, as suggested by Gilbert (1977) for the *Heliconius–Passiflora* system. Rather, mutualism may be a governing factor, not setting the equilibrium population density, but maintaining it at that level with very little fluctuation. In pollination mutualisms it is very unlikely that seed production will determine the density of mature plants, other compensatory factors playing more important roles. Knowledge of the dynamics of mutualism in nature is relatively poor, as most research has been directed at identifying the existence of mutualism. There is some information on the localization of density-dependent population regulation (e.g., Addicott, 1979; Dimock, 1974; Glynn, 1976; Taylor, 1969; Tilman, 1978). Considerably more research needs to be directed towards studying the dynamics of mutualism, particularly during the low density or establishment phases of an interaction.

3.7   Intermittent, Sequential, and Diffuse Mutualism

Some mutualisms involve tight symbiosis and coevolution leading to persistent, species-specific interactions. However, there are mutualisms in which interactions are intermittent, occurring only for limited periods of time but reoccurring at predictable intervals, or are diffuse, occurring between large assemblages of species. These kinds of mutualism deserve more study, to increase both our understanding of the distribution of mutualism (see Section 3.2) and stability of mutualism that may arise from lack of specificity.

The extent to which mutualism is intermittent reflects the extent to which ecological "problems" or their potential "solutions" are temporary. An example of this phenomenon is the interaction between ants, wild cherry and tent caterpillars (Tilmann, 1978). Ants are attracted to the wild cherry by the production of extrafloral nectar, and as long as the caterpillar larvae are small enough, ants remove them from the wild cherry. The interaction is intermittent, because in any one year ants associate with wild cherry for only a few weeks. This is similar to those pollination systems in which flowering seasons are relatively short.

Where either a "problem" or its potential "solution" leads to an intermittent association, one or both mutualists must be able to establish a new association with another mutualist or revert to a nonmutualistic pattern. Establishing new associations can lead to sequential mutualism (or commensalism) among species that utilize the same mutualist but at different times (Waser and Real, 1979). This pattern may be common not only in pollination systems (Waser and Real, 1979), but also in ant–plant and ant–homopteran interactions where ants associate with plants with extrafloral nectaries early in the season and then with homopterans (personal observations). One consequence of the pattern of intermittent and sequential association is the difficulty of recognizing these systems as mutualistic, which can lead to an underestimation of the importance of mutualism (Tilman, 1978).

Diffuse mutualism, in analogy to diffuse competition, arises when any of a number of species singly or simultaneously interact with a mutualist. Vance (1978) describes an interaction among sessile bivalve molluscs and their associated epifauna. The encrusting forms apparently benefit from the hard substrate formed by the bivalves, which is otherwise in short supply, while the bivalves are protected from their starfish predators by the epifauna. The association is nonspecific, occurring with a large number of different species. Similar interactions can be found in pollination, ant–plant, seed dispersal, and ant–homopteran mutualisms (e.g., Culver and Beattie, 1978; Schemske, 1982). With such loose associations it is unlikely that there will be strong coevolution (Janzen, 1980) among any of the species. Instead, there are likely to be generalized responses (Keeler, 1981) that can allow successful interactions even among species newly introduced to an area (Koptur, 1979).

Because of the lack of specificity in the association of species, diffuse mutualism may also be difficult to detect (Vance, 1978). The lack of specificity of the interaction and the availability of a variety of species that can effectively act as mutualists may be an important component leading to the long-term stability of mutualistic interactions (e.g., Culver and Beattie, 1978; Schemske, 1982). This assertion needs verification in both model and natural systems.

3.8   Regulation of Diversity in Mutualistic Systems

Diversity remains a central problem in the study of ecological communities, but there has been relatively little work on the regulation of diversity in mutualistic systems and the effect of mutualism on diversity in more broadly defined communities. Three factors will influence the potential of mutualism to add species to communities: parasitism of mutualism, mutualism for mutualists, and competition for mutualists. A parasite or commensal of mutualism utilizes the same benefits normally provided to a true mutualist but without providing any benefits in return. The best known examples are from pollination systems where nectar robbers and nector thieves obtain nectar without providing any pollination service (Inouye, 1983), or where plants obtain pollination service without providing any reward (Brown and Kodric-Brown, 1979). Further research on the occurrence of parasitism of mutualism would be appropriate. Species that exist exclusively or primarily as parasites of mutualism (e.g., Colwell et al., 1974; Schemske, 1981) increase diversity in a system, but they are a strong selective pressure on the parasitized mutualism. This could lead to strong mechanisms for limiting losses to the parasites, such as long, thick or noxious corolla tubes in pollination systems. Or, it could lead to a dissolution of mutualism. This raises interesting modeling problems common to the study of exploiter–victim systems, with the added complication of another species. Depending on whether a coevolutionary equilibrium were established or not, parasitism could either increase or ultimately decrease diversity.

Mutualism can also increase diversity, if mutualism becomes more effective with more species. There are two major patterns of this type of interaction. In Mullerian mimicry or interspecific flocking, two or more species may benefit by decreased predation. In another type of system two or more species may be more effective in maintaining a common mutualist (Brown and Kodric-Brown, 1979; Waser and Real, 1979). Since the equilibrium density of a species is not necessarily determined by the magnitude of a mutualistic interaction, the density of one mutualist may become limiting for its partner(s), resulting in competition for mutualists. This is known to occur in pollination systems (Waser, 1978), but is a likely occurrence in other facultative mutualistic systems (Addicott, 1978a).

Parasitism of mutualism and mutualism for mutualists should increase diversity, whereas competition for mutualists should limit diversity. A major challenge for ecologists studying both model and natural systems is to understand how these three factors interact with each other to control diversity in both ecological and evolutionary time. There have been some attempts at this (e.g., King et al., 1975), but there should be more.

The role of a species in the structure and function of a community may not be adequately reflected by its density. Gilbert (1980) uses the term keystone mutualists for those species that maintain much of the diversity of a community by maintaining a pool of mutualists that are in turn responsible for the seed dispersal, pollination, or protection of many other species. For example, a tree that provides nectar not just to its own obligate and effective pollinator, but also to a series of "minor parasites" (Baker et al., 1971), may enable those pollinators to persist in an area without having to emigrate when their primary nectar sources, of which they are obligate pollinators, are not available. Atsatt and O'Dowd (1976) have made similar suggestions for the role of nectar plants in the maintenance of parasitoids. These kinds of indirect but far-reaching effects deserve more attention.

4   SUMMARY

In this chapter I have identified a series of problems which should provide the focus for the study of mutualism. There is still the need to identify mutualistic systems, particularly those in which interactions are indirect, diffuse, or intermittent. Analyses of the evolution of mutualism have emphasized particular systems, usually symbiotic, but a more general analysis would be appropriate, with particular emphasis upon when mutualism provides an evolutionarily stable "solution" to an ecological "problem." The dynamics of mutualism needs more attention, with the development of population models incorporating more realistic assumptions about the structure and mechanisms of mutualistic interactions, and with closer analysis of the dynamics of natural systems. The regulation and variation of costs and benefits should be addressed in both model and natural systems. Finally, there needs to be further consideration of the effects of mutualism on diversity.

ACKNOWLEDGMENTS

I wish to thank Carol Addicott, John Aho, Mike Antolin, John Holmes, and Paul Marino for their comments on the manuscript.

LITERATURE CITED

Addicott, J. F. 1978a. Competition for mutualists: Aphids and ants. *Can. J. Zool.* **56**:2093–2096.

Addicott, J. F. 1978b. The population dynamics of aphids on fireweed: A comparison of local and metapopulations. *Can. J. Zool.* **56**:2554–2564.

Addicott, J. F. 1979. A multispecies aphid-ant association: Density dependence and species-specific effects. *Can. J. Zool.* **57**:558–569.

Addicott, J. F. 1981. Stability properties of 2-species models of mutualism: Simulation studies. *Oecologia (Berlin)* **49**:42–49.

Addicott, J. F., and H. I. Freedman. 1983. On the structure and stability of mutualistic systems: Analysis of predator-prey and competition models as modified by the action of a slow-growing mutualist. *Theor. Popul. Biol.* (in review).

Atsatt, P. R., and D. J. O'Dowd. 1976. Plant defense guilds. *Science (Wash. D.C.)* **193**:24–29.

Bach, C. E., and W. F. Herrnkind. 1980. Effects of predation pressure on the mutualistic interaction between the hermit crab, *Pagurus pollicaris* Say, 1817, and the sea anemone, *Calliactis tricolor* (Lesueur, 1817). *Crustaceana* **38**:104–108.

Baker, H. G., R. Cruden, and I. Baker. 1971. Minor parasitism in pollination biology and its community function: The case of *Ceiba acuminata*. *Bioscience* **21**:1127–1129.

Barbehenn, K. R. 1969. Host-parasite relationships and species diversity in mammals: An hypothesis. *Biotropica* **1**:29–35.

Bequaert, J. 1921. On the dispersal by flies of the spores of certain mosses of the family Splachnaceae. *Bryologist* **24**:1–4.

Berger, J. 1980. Feeding behaviour of *Didinium nasutum* on *Paramecium bursaria* with normal or apochlorotic zoochlorellae. *J. Gen. Microbiol.* **118**:397–404.

Bierzychudek, P. 1981. *Asclepias, Lantana* and *Epidendrum*: a floral mimicry complex? *Biotropica* **13**(Suppl.):54–58.

Bloom, S. A. 1975. The motile escape response of a sessile prey: A sponge-scallop mutualism. *J. Exp. Mar. Biol. Ecol.* **17**:311–321.

Bobisud, L. E., and R. J. Neuhaus. 1975. Pollinator constancy and survival of rare species. *Oecologia (Berlin)* **21**:263–272.

Boucher, D. H., S. James, and K. H. Keeler. 1982. The ecology of mutualism. *Annu. Rev. Ecol. Syst.* **13**:315–347.

Briand, F., and P. Yodzis. 1982. The phylogenetic distribution of obligate mutualism: Evidence of limiting similarity and global instability. *Oikos* **39**:273–275.

Brown, J. H., and A. Kodric-Brown. 1979. Convergence, competition, and mimicry in a temperate community of hummingbird pollinated flowers. *Ecology* **60**:1022–1035.

Collins, J. P., R. C. Berkelhamer, and M. Mesler. 1977. Notes on the natural history of the mangrove *Pelliciera rhizophorae* Tr. & Pl. (Theaceae). *Brenesia* **10**:17–29.

Collins, N. C., R. Mitchell, and R. G. Wiegert. 1976. Functional analysis of a thermal spring ecosystem, with an evaluation of the role of consumers. *Ecology* **57**:1221–1232.

Colwell, R. K., B. J. Betts, P. Bunnell, F. L. Carpenter, and P. Feinsinger. 1974. Competition for the nectar of *Centropogon valerii* by the hummingbird *Colibri thalassinus* and the flower-piercer *Diglossa plumbea*, and its evolutionary implications. *Condor* **76**:447–452.

Cruden, R. W., S. Kinsman, R. E. Stockhouse, II, and Y. B. Linhart. 1976. Pollination, fecundity, and the distribution of moth-flowered plants. *Biotropica* **8**:204–210.

Cruden, R. W., S. M. Hermann, and S. Peterson. 1983. Patterns of nectar production and plant-pollinator coevolution. In B. Bentley and T. Elias (eds.), *The biology of nectaries*, pp. 80–125. Columbia University Press, New York.

Crush, J. R. 1975. Occurrence of endomycorrhizas in soils of the Mackenzie Basin, Cantebury, New Zealand. *N.Z. J. Agric. Res.* **18**:361–364.

Culver, D. C., and A. J. Beattie. 1978. Myrmecochory in *Viola*: Dynamics of seed-ant interactions in some West Virginia species. *J. Ecol.* **66**:53–72.

D'Elisca, P. N. 1977. *Endosphaerium funiculatum*, gen. nov., sp. nov., a new predaceous fungus mutualistic in gills of freshwater pelecypods. *J. Invert. Pathol.* **30**:418–421.

Davidson, D. W. 1980. Some consequences of diffuse competition in a desert ant community. *Am. Nat.* **116**:92–105.

Dimock, R. V., Jr. 1974. Intraspecific aggression and the distribution of a symbiotic polychaete on its host. In W. B. Vernberg (ed.), *Symbiosis in the sea*, pp. 29–44. University of South Carolina Press, Columbia.

Feeny, P. 1975. Biochemical coevolution between plants and their insect herbivores. In L. E. Gilbert and P. H. Raven (eds.), *Coevolution of animals and plants*, pp. 3–19. University of Texas Press, Austin.

Friedmann, H., and J. Kern. 1956. *Micrococcus cerolyticus*, nov. sp., an aerobic lypolytic organism isolated from the African honey-guide. *Can. J. Microbiol.* **2**:515–517.

Gause, G. F., and A. A. Witt. 1935. Behaviour of mixed populations and the problem of natural selection. *Am. Nat.* **69**:596–609.

Gilbert, L. E. 1977. The role of insect-plant coevolution in the organization of ecosystems. In V. Labyrie (ed.), *Comportement des insectes et milieu trophique*, pp. 399–413. C.N.R.S., Paris.

Gilbert, L. E. 1980. Food web organization and the conservation of neotropical diversity. In M. E. Soule and B. A. Wilcox (eds.), *Conservation biology: An evolutionary-ecological perspective*, pp. 11–33. Sinauer, Sunderland, Massachusetts.

Glynn, P. W. 1976. Some physical and biological determinants of coral community structure in the eastern Pacific. *Ecol. Monogr.* **46**:431–456.

Harley, J. L. 1970. The importance of micro-organisms to colonizing plants. *Trans. Bot. Soc. Edinb.* **41**:65–70.

Hertig, M., W. H. Taliaferro, and B. Schwartz. 1937. Report of the Committee on Terminology. *J. Parasitol.* **23**:325–329.

Hutchinson, G. E. 1978. *An introduction to population ecology*. Yale University Press, New Haven, Connecticut. 260 pp.

Inouye, D. W. 1983. The ecology of nectar robbing. In B. Bentley and T. Elias (eds.), *The biology of nectaries*, pp. 153–173. Columbia University Press, New York.

Janzen, D. H. 1969. Allelopathy by myrmecophytes: The ant *Azteca* as an allelopathic agent of *Cecropia*. *Ecology* **50**:147–153.

Janzen, D. H. 1979. How many babies do figs pay for babies? *Biotropica* **11**:48–50.

Janzen, D. H. 1980. When is it coevolution? *Evolution* **34**:611–612.

Keeler, K. H. 1981. A model of selection for facultative nonsymbiotic mutualism. *Am. Nat.* **118**:488–498.

King, C. E., E. E. Gallaher, and D. A. Levin. 1975. Equilibrium diversity in plant-pollinator systems. *J. Theor. Biol.* **53**:263–275.

Kleinfeldt, S. E. 1978. Ant-gardens: The interaction of *Codonathe crassifolia* (Gesneriaceae) and *Crematogaster longispina* (Formicidae). *Ecology* **59**:449–456.

Koptur, S. 1979. Facultative mutualism between weedy vetches bearing extrafloral nectaries and weedy ants in California. *Am. J. Bot.* **66**:1016–1020.

Laine, K. J., and P. Niemela. 1980. The influence of ants on the survival of mountain birches during an *Oporinia autumnata* (Lep., Geometridae) outbreak. *Oecologia (Berlin)* **47**:39–42.

Lawlor, L. R. 1979. Direct and indirect effects on *n*-species competition. *Oecologia (Berlin)* **43**:355–364.

Lewis, D. H. 1973. The relevance of symbiosis to taxonomy and ecology, with particular reference to mutualistic symbioses and the exploitation of marginal habitats. In V. H. Heywood (ed.), *Taxonomy and ecology*, pp. 151–172. Academic Press, London.

Losey, G. S., Jr. 1978. The symbiotic behavior of fishes. In D. I. Mostofsky (ed.), *The behavior of fish and other aquatic animals*, pp. 1–31. Academic Press, New York.

Louda, S. M. 1982. Inflorescence spiders: A cost/benefit analysis for the host plant, *Haplopapus venetus* Blake (Asteraceae). *Oecologia (Berlin)* **55**:185–191.

McDade, L., and S. Kinsman. 1980. The impact of floral parasitism in two neotropical hummingbird-pollinated plant species. *Evolution* **34**:944–958.

McNaughton, S. J. 1976. Serengeti migratory wildebeest: Facilitation of energy flow by grazing. *Science (Wash. D.C.)* **191**:92–94.

Mares, M. A., and M. L. Rosenzweig. 1978. Granivory in North and South American deserts: Rodents, birds, and ants. *Ecology* **59**:235–241.

May, R. M. 1976. Models for two interacting populations. In R. M. May (ed.), *Theoretical ecology, principles and applications*, pp. 49–70. Saunders, Philadelphia.

Messina, F. J. 1981. Plant protection as a consequence of an ant-membracid mutualism: Interactions on goldenrod (*Solidago* sp.). *Ecology* **62**:1433–1440.

Munn, C. A., and J. W. Terborgh. 1979. Multi-species territoriality in neotropical foraging flocks. *Condor* **81**:338–347.

Noble, J. C. 1975. The effects of emus (*Dromaius novae-hollandiae* Latham) on the distribution of the nitre bush (*Nitraria billardieri* D.C.). *J. Ecol.* **63**:979–984.

O'Dowd, D. J. 1979. Foliar nectar production and ant activity on a neotropical tree, *Ochroma pyramidale*. *Oecologia (Berlin)* **43**:233–248.

O'Dowd, D. J., and M. E. Hay. 1980. Mutualism between harvester ants and a desert ephemeral: Seed escape from rodents. *Ecology* **61**:531–540.

Odum, E. P. 1971. *Fundamentals of Ecology*, 3rd ed. Saunders, Philadelphia. 574 pp.

Osman, R. W., and J. A. Haugsness. 1981. Mutualism among sessile invertebrates: A mediator of competition and predation. *Science (Wash. D.C.)* **211**:846–848.

Owen, D. F., and R. G. Wiegert. 1976. Do consumers maximize plant fitness? *Oikos* **27**:488–492.

Owen, D. F., and R. G. Wiegert. 1981. Mutualism between grasses and grazers: An evolutionary hypothesis. *Oikos* **36**:376–378.

Patelle, M. 1981. More mutualisms between consumers and plants. *Oikos* **38**:125–127.

Platt, W. J., G. R. Hill, and S. Clark. 1974. Seed production in a prairie legume (*Astragalus canadensis* L.). Interactions between pollination, predispersal seed predation, and plant density. *Oecologia (Berlin)* **17**:55–63.

Price, M. V., and N. M. Waser. 1979. Pollen dispersal and optimal outcrossing in *Delphinium nelsoni*. *Nature (London)* **277**:294–297.

Quinlan, R. J., and J. M. Cherret. 1978. Aspects of the symbiosis of the leaf-cutting ant *Acromyrmex octospinosus* (Reich) and its fungus food. *Ecol. Entomol.* **3**:221–230.

Risch, S., and D. H. Boucher. 1976. What ecologists look for. *Bull. Ecol. Soc. Am.* **57**(3):8–9.

Risch, S. J., and F. R. Rickson. 1981. Mutualism in which ants must be present before plants produce food bodies. *Nature (London)* **291**:149–150.

Ross, D. M. 1971. Protection of hermit crabs (*Dardanus* spp.) from octopus by commensal sea anemones (*Calliactis* spp.), *Nature (London)* **230**:401–402.

Roughgarden, J. 1975. Evolution of marine symbiosis—A simple cost-benefit model. *Ecology* **56**:1201–1208.

Sankurathri, C. S., and J. C. Holmes. 1976. Effects of thermal effluents on parasites and commensals of *Physa gyrina* Say (Mollusca: Gastropoda) and their interactions at Lake Wabamun, Alberta. *Can. J. Zool.* **54**:1742–1753.

Schemske, D. W. 1980. Evolutionary significance of extrafloral nectar production by *Costus woodsonii* (Zingiberaceae): An experimental analysis of ant protection. *J. Ecol.* **68:**959–967.

Schemske, D. W. 1981. Floral convergence and pollinator sharing in two bee-pollinated tropical herbs. *Ecology* **62:**946–964.

Schemske, D. W. 1982. Ecological correlates of a neotropical mutualism: Ant assemblages at *Costus* extrafloral nectaries. *Ecology* **63:**932–941.

Silander, J. A., Jr. 1978. Density-dependent control of reproductive success in *Cassia biflora*. *Biotropica* **10:**292–296.

Simberloff, D., B. J. Brown, and S. Lowrie. 1978. Isopod and insect root borers may benefit Florida mangroves. *Science (Wash. D.C.)* **201:**630–632.

Sinclair, A. R. E., and M. Norton-Griffiths. 1982. Does competition or facilitation regulate migrant ungulate populations in the Serengeti? A test of hypotheses. *Oecologia (Berlin)* **53:**364–369.

Smith, N. G. 1968. The advantage of being parasitized. *Nature (London)* **219:**690–694.

Springett, B. P. 1968. Aspects of the relationship between burying beetles, *Necrophorus* spp. and the mite, *Poecilochirus necrophori* Vitz. *J. Anim. Ecol.* **37:**417–424.

Stephenson, A. G. 1982. The role of extrafloral nectaries of *Catalpa speciosa* in limiting herbivory and increasing fruit production. *Ecology* **63:**663–669.

Taylor, D. L. 1969. On the regulation and maintenance of algal numbers in zooxanthellae-coelenterate symbiosis, with a note on the relationship in *Anemonia sulcata*. *J. Mar. Biol. Assoc. U.K.* **49:**1057–1065.

Thompson, J. N. 1982. *Interaction and coevolution.* Wiley, New York. 179 pp.

Tilman, D. 1978. Cherries, ants and tent caterpillars: Timing of nectar production in relation to susceptibility of caterpillars to ant predation. *Ecology* **59:**686–692.

Udovic, D. 1981. Determinants of fruit set in *Yucca whipplei:* Reproductive expenditure vs. pollinator availability. *Oecologia (Berlin)* **48:**389–399.

Valerio, C. E. 1975. A unique case of mutualism. *Am. Nat.* **109:**235–238.

Vance, R. R. 1978. A mutualistic interaction between a sessile marine clam and its epibionts. *Ecology* **59:**679–685.

Vandermeer, J. 1980. Indirect mutualism: variations on a theme by Stephen Levine. *Am. Nat.* **116:**441–448.

Vandermeer, J. H., and D. H. Boucher. 1978. Varieties of mutualistic interaction in population models. *J. Theor. Biol.* **74:**549–558.

Waser, N. M. 1978. Competition for hummingbird pollination and sequential flowering in two Colorado wildflowers. *Ecology* **59:**934–944.

Waser, N. M., and L. A. Real. 1979. Effective mutualism between sequentially flowering plant species. *Nature (London)* **281:**670–672.

Wickler, W. 1968. *Mimicry in plants and animals.* McGraw-Hill, New York. 255 pp.

Wilbur, H. M. 1972. Competition, predation and the structure of the *Ambystoma-Rana sylvatica community. Ecology* **53:**3–21.

Wilson, D. S. 1980. *The natural selection of populations and communities.* Benjamin/Cummings, Menlo Park, California. 186 pp.

Wolin, C. L., and L. R. Lawlor. 1983. Models of facultative mutualism: Density Effects. *Am. Nat.* (in press).

Wright, H. O. 1973. Effect of commensal hydroids on hermit crab competition in the littoral zone of Texas. *Nature (London)* **241:**139–140.

CHAPTER 17

The Importance of Scale
in Community Ecology:
A Kelp Forest Example
with Terrestrial Analogs

PAUL K. DAYTON

MIA J. TEGNER

Scripps Institution of Oceanography
La Jolla, California

CONTENTS

1 INTRODUCTION

The scale of resolution chosen by ecologists is perhaps the most important decision in their research program, because it largely predetermines the questions, the procedures, the observations, and the results. This chapter summarizes aspects of our long-term studies in southern California kelp forests which have focused on (1) community structure as modulated by patch dynamics and (2) the population dynamics of a few important species. Here we discuss our results and argue that larger scales would improve our understanding and that this is generally true in ecological research, a position we document with literature from the American Southwest.

There seems to be a tendency for ecologists to reduce the scale of their research both to make it more tractable and to ensure prompt and clean results. Marine examples and good summary papers include boundary layer hydrodynamics of algae and their role in nutrient transfer and irradiance (Wheeler and Neushul, 1981) and soft substrata (Eckman, 1979), larval behavior (Thorson, 1946; Wilson, 1952), adult polychaete behavior (L. A. Levin, 1981, 1982), adult–larval interactions (Woodin, 1976), resistance of fouling communities to invasion (Sutherland, 1981), and so on. Research continues on such popular questions as competition (Connell, 1983) and predation/disturbance (Paine and Levin, 1981). Here we argue that populations of coastal marine species, especially those contributing the "strong interactions" (*sensu* Paine, 1980), are characterized by episodic recruitment patterns unpredictable in time and/or space (see summary in Dayton, 1983). Because most such species have long-lived individuals and slowly declining populations, recruitment pulses result in long-lasting spatial and temporal patterns. The importance of year-class phenomena is well known in pelagic fishery literature (Cushing, 1975), where much research focuses on large (1000s of km) and mesoscale (100s of km) oceanic patterns (Cushing and Dickson, 1976; Parrish et al., 1981, for summaries) and climatology (Lasker, 1978). Many ecologists, on the other hand, focus on their small scale questions amenable to experimental tests and remain oblivious to the larger scale processes which may largely account for the patterns they study.

2 PATCH DYNAMICS OF KELP FORESTS

Like most communities kelp forests are distinctly patchy. The patch types are characterized by vegetation layers which are defined by distinct morphological adaptations including (1) a floating morphological structure in which the canopy is at or near the surface, (2) a stipitate morphology that supports the canopy above the substratum, (3) a prostrate structure in which the fronds lie on the substratum attached by a short stipe, (4) a turf composed of many species of coralline, foliose, and siphonious red algae, and (5) a pavement of encrusting coralline algae. There is a convergence into these morphological types as most kelp communities have similar canopy guilds, but the actual species differ. In California kelp forests the above categories often include more than one species within a given patch.

The persistence, resistance, and resilience stabilities of these patches have been evaluated in three California habitats: Pt. Loma, San Diego County; Bird Rock, Santa Catalina Island in Los Angeles County; and Pt. Piedras Blancas, San Luis Obispo County (Dayton et al., 1984, in press). The three habitats represent different types of physical stress. The persistence of individual patches was evaluated when possible by observing their edges over time and by evaluating survivorship curves which document that the patches have persisted more than one generation. Indeed, some patches in the Pt. Loma forest are thought to have persisted since 1946 (Ron McPeak, personal communication). Resistance to invasion can be evaluated by canopy removal and transplantation experiments, whereas resistance to disturbance must be evaluated by observation and, when appropriate, by manipulation of disturbance agents. Resilience refers to the recoverability rate following a perturbation sufficient to allow colonization by different species.

The taller perennial canopy guilds dominate the competition for light but are more susceptible to wave stress, and the dominance hierarchies appear reversed in areas exposed to increasing wave stress from storms. There is a threshold (at which the plant is ripped free) beyond which the intensity of the storm is irrelevant to the plant. More important, however, is the frequency of storms sufficient to rip plants free. The main source of mortality at Pt. Loma is from entanglement with storm-dislodged *Macrocystis pyrifera* plants. But normally the disturbed area is sufficiently small that the previous patch type recovers. At Pt. Piedras Blancas, where storms are much more frequent and intense, there are different patterns: in some areas *Macrocystis* fronds are ripped free but the plants (and the patch) persist, but in other areas (successful transplantations notwithstanding) this species simply does not persist. In more wave-stressed situations, the long-lived stipitate species (*Pterygophora californica, Eisenia arborea*, and *Laminaria digitata*) seem more tolerant. Finally, the prostrate *Dictyoneurum californica* and turf-forming articulated coralline algae seem extremely tolerant to wave stress. In these more exposed situations the lower canopy guilds become the respective dominants. In other areas, often with unstable substrata or patchy but intense sea urchin grazing, annual species such as *Nereocystis luetkeana* or *Desmarestia* spp. are most abundant.

Sea urchin grazing is another well-known form of disturbance in kelp forests (Lawrence, 1975); and Vadas (1968), working in Puget Sound, Washington, has documented the tradeoffs between unpalatability (*Agarum* spp.), ephemeral tactics with high palatability and low competitive ability (*Nereocystis*), and high palatability and competitive ability (*Laminaria* spp.). In southern California sea urchins are occasionally responsible for denuding large areas of algae, but predators such as spiny lobsters and fishes (Tegner, 1980; Tegner and Dayton, 1981; Cowen, 1983; Tegner and Levin, 1983) or, north of Pt. Conception, sea otters (Lowry and Pearse, 1973; Woodhouse *et al.*, 1977) usually prevent sea urchin populations from devastating entire kelp forests as they do in Nova Scotia (Mann, 1977) or sea otter-free areas of the northeast Pacific (Estes and Palmisano, 1974; Simenstad et al., 1978). The scale of urchin disturbance varies from the small <1 m ambit from a refuge to a few square meter area in a boulder area patch to a few hectares to an entire coastline. Obviously it is extremely important to understand the processes determining the actual scale; and probably in most cases the processes involve dispersal, which immediately implies a large scale oceanographic study.

Resistance to invasion is another important asset of each patch type in southern California (Dayton et al., 1984, in press). In most cases distinct patches resisted invasion for over 10 years. The mechanisms of resistance involve competition for light and limits to spore dispersal as demonstrated by clearance experiments and seeding with fertile sporophylls. The area of the disturbance is an integral component of patch stability: small disturbances killing only a few plants usually do not alter the patch composition because the disturbed area is almost certain to be recolonized by members of the existing species, both from preceding sporulations and from swamping by spores from nearby plants. Larger disturbances involve the relative dispersal abilities of the different species. The mechanisms of kelp dispersal involve very limited spore dispersal or most commonly, drifting plants leaving a trail of recruits or fragments of fertile material which need to be snagged and held against the substratum long enough to inoculate an area with a high density of spores. Furthermore, the different species (especially their gametophytes) have very different physiological thresholds (Lüning and Neushul, 1978). Finally, the reproduction of most kelps is distinctly seasonal, with the spores being released in the winter. For these reasons there are different thresholds affecting the probabilities of invasion into different-sized disturbances. The actual observed spatial scales range from that of the attachment site of single plants to very large areas resulting from catastrophes such as overgrazing, unusual storms, or severe climatic fluxes such as the northern "El Niño" of the late 1950s, which devastated the kelp forests (IMR, 1963; Chelton et al., 1982).

3 ROLE OF SEA URCHINS

As mentioned above, sea urchins are particularly important grazers; in southern California *Strongylocentrotus purpuratus* and especially the large, motile *S. franciscanus* are the most important. For reasons not completely understood but probably relating to hunger, they occasionally move out of their crevices and forage in large aggregations; when this happens large areas of a kelp forest can be denuded of macroalgae, leaving only a pavement of encrusting coralline algae (the "barren areas" of Lawrence, 1975). Enough urchins remain foraging in these "barren areas" to preclude recruitment of macroalgae so that these patches, too, are stable for months and sometimes years.

Even when restricted to its crevice habitat, *S. franciscanus* has important community roles. It has a nocturnal foraging pattern in which it grazes out about a meter from its crevice or boulder; this results in a "barren" zone free of macroalgae, and in some areas this barren zone offers a superior habitat to a cup coral, *Balanophylia elegans* (Gerrodette, 1981; Dayton et al., 1984, in press). Perhaps more important, the encrusting coralline algae offer important inducements for abalone larvae to settle (Morse et al., 1980); thus urchins maintain recruitment areas for abalone. In addition, the spine canopy of *S. franciscanus* is a refuge from predators to recently metamorphosed individuals of several species, including abalones and the *S. franciscanus* themselves (Tegner and Dayton, 1977). Another important role of sea urchins is that of snagging and holding drift kelp, much of which is fertile, long enough to allow a local inoculation of spores, especially along the ridges and ledges above the urchins. This is a critical factor in the dispersal of kelp.

4 KELP FOREST EFFECTS ON CURRENTS: NUTRIENTS AND LARVAL TRANSPORT

Dominant current flows in southern California coastal areas are in the longshore directions (Winant and Bratkovich, 1981). The terrestrial boundary constrains cross-shore currents to be weaker and generally to have long-period motions. The main sources of cross-shore transport are internal tides, the strength of which vary with the seasonal stratification of the water column. These currents supply nutrients and planktonic larvae. Just as terrestrial forests affect the wind (Raupach and Thom, 1981), the drag of the kelp forest strongly alters the water flow (Jackson and Winant, 1983). Jackson and Winant's preliminary data show that the mean current inside the kelp forest at Pt. Loma is 0.1 cm/sec compared to 0.3 cm/sec outside the forest. The magnitude of the calculated drag implies that the most important transition from the outside to the inside of the forest in the longshore direction should occur within approximately 100 m of the upstream edge. While organisms on the outer edge get nutrient- and larval-rich water, those on the inside are exposed to water from which nutrients and plankton are filtered. The longshore within-kelp-forest water movement along the length of the 8 km kelp forest at Pt. Loma is sufficiently slow that the transit time is about 220 hr. The biological uptake of nitrate can occur at 23%/hr (Jackson, 1980), so nutrient stress may be rather common. Similarly, once presented with appropriate clues, at least some larvae begin settlement behavior very quickly (Hinegardner, 1975). It is clear that the kelp-induced drag is sufficient to much reduce or eliminate the transport of nutrients and many larvae from the interior of the kelp forest. Cross-shore currents provide an alternate mechanism to reach interior parts of the kelp bed. Such currents are weaker and more sporadic than the longshore, but the shorter distance may compensate. Because an important source of cross-shore transport is internal tide motion, this depends on the thermal stratification of the water column, which is much stronger in the summer.

The time that larvae spend in the water column is variable but obviously important to the scale of study. Larvae known to have long (weeks to months) planktonic lives (the cyphonautes larvae of the bryozoan *Membranipora membranacea* and the plutei larvae of the sea urchins) tend to have large (often order of magnitude) decreases in density from upstream to the middle to the downstream edge of the kelp (Bernstein and Jung, 1979; personal observation). Recruitment of *S. franciscanus* over a 3-year period was consistently about two times higher on the outside than the inside edge of the Point Loma kelp forest (Tegner and Dayton, 1981). Animals with somewhat shorter larval life-spans such as green abalone, *Haliotis fulgens*, which have larval spans of 4–9 days (Leighton et al., 1981), are strongly influenced by current flow rates and distances between appropriate habitats. The depression of the current by the kelp forest can act to "trap" the larvae within a large kelp forest such as Pt. Loma. Finally, many of the invertebrate species within the kelp forests—such as sponges, bryozoans, and tunicates—have larvae with very short larval life spans (minutes to a few hours). These larvae are likely to be retained within the kelp forest. Thus, physical events on the scale of a kelp forest have important consequences regarding gene flow and population dynamics.

5   EDGE EFFECTS AND THE CRITICAL SIZE OF A KELP FOREST

Edge effects, such as the apparent aggregation of organisms at the periphery of the patches, are another general ecological feature; all naturalists are aware of the importance of this environmental heterogeneity (Elton, 1966). The mechanisms contributing to edge effects are varied. They include the simple physical effect described above and the aggregation of animals that use the shelter as a refuge from predation such as seen for sparrows (Pulliam, personal communication), rodents (Bartholomew, 1970), and herbivorous fish (Randall, 1965). The edge facilitates the filtration of food (perhaps by species including web spiders or insectivorous birds). These mechanisms contribute to a pronounced edge effect in larger kelp forests. This has been described by Hobson and Chess (1976), and Bray (1981) experimentally documented the filtering of plankton by planktivorous fishes. Hobson (1976) and Bernstein and Jung (1979) found this edge effect for a picker fish, which cleans other fish and forages on the bottom and on kelp fronds. The kelp forest may offer shelter from predation by sharks (Tricas, 1979) or seals or simply provide structure for fish to orient to; we suggest also that the plankton accumulates at the upper edge of the kelp forest as the kelps slow and divert the currents and that those larvae competent to settle do so, either on the bottom or on the fronds directly.

Another scale-related complication of an edge effect is that it can determine size thresholds of kelp forests. Kelp fronds are grazed by fishes such as the half moon, *Medialuna californiensis*; the opaleye, *Girella nigricans*; and the señorita, *Oxyjulis californica*, which may consume the frond directly or, more commonly, incidentally as they pick encrusting organisms (North, 1971; Bernstein and Jung, 1979). Isolated or small patches of *Macrocystis* plants too small to have an edge effect often attract unusually high densities of encrusting organisms throughout the patch and are thus especially attractive to the fishes that graze the fronds at a rate much higher than the frond growth rate. This sets a size threshold for the establishment of a stable kelp forest (North, 1971). At the other extreme, it is conceivable that the absorption of nutrients by the outer plants can decrease nutrients sufficiently that plants in the inner areas cannot survive. Although this has not been observed, inner kelp plants often are much less healthy (Bernstein and Jung, 1979; Dayton et al., 1984, in press), and this could, at least in theory, establish the maximum dimensions of a kelp forest.

6   THE IMPORTANCE OF LARGER SCALES IN SPACE

Kelp forests are not isolated systems driven entirely by local processes; they are also driven by much larger mesoscale physical forces which may involve all of the Southern California Bight and the very much larger California Current System (Fig. 1). Large but poorly understood oceanic processes in the Pacific cause large scale, low frequency changes in the California Current leading to highly significant interannual variability in physical and biological parameters throughout the system (Bernal, 1981; Chelton et al., 1982; Fig. 2). These and other more local (mesoscale of 100s of km) oceanographic anomalies are probably responsible for the episodic recruitment events which seem to characterize most of the world's coastal fisheries. These patterns are also true for many coastal benthic populations (summarized in Dayton and Oliver, 1980).

An example of a large scale event was the "El Niño" of 1957–1959 when the California Current system was marked by abnormally high water temperatures (Fig. 2), an increase in salinity, decreased southward flow, raised sea level, decreased abundance of zooplankton, and a virtual collapse of all the kelp forests, even those of the offshore islands (IMR, 1963; Chelton et al., 1982). In addition, many southern species were observed far north of their usual range and some successfully spawned off southern California (Radovitch, 1961). It is clear that a large amount of tropical warm water intruded into the California system in the same manner as the more famous southern "El Niño" off Peru. There was little documentation of effects on kelp communities; however, it is clear that most of the kelp forests virtually

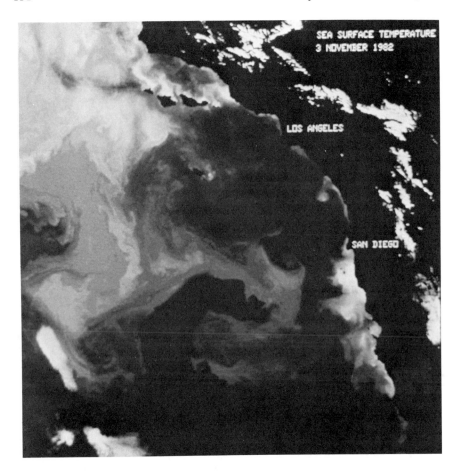

Figure 1. Thermal image of the Southern California Bight on 3 November 1982. The warmest surface temperatures, black, grade to the coldest, white; however, the white masses over the mainland and in the lower left are clouds. Conspicuous processes include the cold California Current moving south from Point Conception. There is strong upwelling along the coast from San Diego south. Tongues of warm water are visible in black. The one closest to the coast, called the southern California eddy, brings warmer water from the south and mixes with the California Current south of the northern Channel Islands. Courtesy of Mark Abbott, Scripps Institution of Oceanography.

disappeared and sea urchin densities vastly increased (IMR, 1963). In some areas such as Palos Verdes, Los Angeles County, the kelp only started returning in the late 1970s; on the other hand, kelp forests returned within a few years in other areas such as the southern Channel Islands. Unfortunately, the patterns are confused by improvements in coastal sewage disposal systems and heavy increases in fisheries, especially of lobsters and abalones, both of which affect sea urchins (Tegner, 1980) and active kelp restoration.

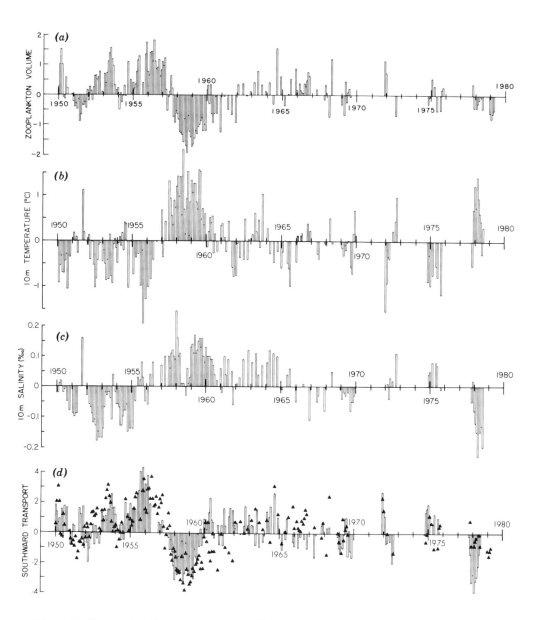

Figure 2. Time series of nonseasonal values of four parameters in the California Current: (*a*) the average zooplankton volume (10^{-3} ml m^{-3}) time series for CalCOFI lines 60-136; (*b*) the average 10 m temperature over 150 hydrographic stations in the same area; (*c*) the average 10 m salinity over 150 hydrographic stations in the same area; (*d*) southward transport as determined by principal component analyses of 0/500 steric height. The triangles in (*d*) represent the zooplankton time series shown in (*a*). From Chelton et al. (1982).

Populations with long-term planktonic larvae have been troublesome to naturalists and theoreticians alike (Thorson, 1950; Scheltema, 1974; Strathmann, 1974; S. Levin, 1976; Jackson and Strathmann, 1981). We argue that larval dispersal is a key to understanding the response of communities to such disturbances, but despite the obvious importance to most nearshore communities, larval ecology and dispersal remain mysteries that are rarely acknowledged. This is surprising because important species in even such well-known systems as the rocky intertidal commonly exhibit rare or episodic recruitment events that probably relate to larval parameters but have important community consequences. For example, *Mytilus californianus* is the dominant species in the Pacific Northwest, yet its recruitment is highly episodic (Dayton, 1971; Paine, 1974, 1980). Similarly, the lower intertidal is strongly influenced by *Strongylocentrotus purpuratus* (Paine and Vadas, 1969; Dayton, 1975), the recruitment of which is also episodic and unpredictable (Ebert, 1968; Dayton, 1975a; Paine, 1980). Clearly larval dispersal is a major element of episodic recruitment and is critical to the structure of most benthic communities.

We have observed large scale spatial variation in recruitment of the giant red sea urchin, *Strongylocentrotus franciscanus*. *S. franciscanus* has a relatively long-lived pelagic larval stage (from 50 to 130 days; Strathmann, 1978; Cameron and Schroeter, 1980), and these patterns are relatively consistent over time, suggesting that large scale oceanic processes may be mediating recruitment success. Recruitment of sea urchins has been evaluated in many locations, including the Channel Islands and other parts of the Southern California Bight (Tegner, personal observation). The northwestern San Miguel and Santa Rosa Islands had recruitment rates averaging 16% and 9%, respectively, of the total populations per year from 1976 to 1980. In contrast, Santa Barbara and San Clemente Island which are further south and east in the bight, had average recruitment rates of 55% and 33% per year. The pattern observed on the northwestern islands appears to be similar to that observed along the coast of central California north of Point Conception, where recruitment rates are very low; and the southeastern islands compare with the coastline from San Diego south (Tegner, personal observation).

A number of parameters such as temperature, larval food and current patterns might be expected to affect recruitment on large spatial scales. Temperature probably has little effect as *S. franciscanus* is a northern species and southern California is near the southern limit of its range, yet the colder areas of our study (the Pt. Conception area and the northern Channel Islands) have lower recruitment. Larval food availability is undoubtedly higher in the upwelling, plankton-rich region around the northwestern Channel Islands and along the coast of central California (e.g., Smith, 1978; Parrish et al., 1981), where recruitment is lower, suggesting that larval food is not limiting recruitment success. Parrish et al. (1981) have recently proposed a relationship between large scale seasonal patterns of ocean surface drift along the California coast and reproductive strategies of successful coastal fishery species. They describe a region of maximum upwelling from Cape Blanco to Point Conception along the coast, which includes San Miguel and Santa Rosa Islands. *S. franciscanus* spawns in April–May in central California (Bennett and Giese, 1955), during the height of the offshore Ekman transport when prevailing winds cause seaward drift of surface waters (Parrish et al., 1981); thus, the larvae are spawned into a nutrient-rich environment but dispersed offshore into the south-flowing California Current. In contrast, the Southern California Bight is characterized by minimal offshore Ekman transport and a geostrophic flow pattern that features a closed gyral circulation near the coast (the Southern California Eddy; Owen, 1980) and an onshore component of flow further offshore (Parrish et al., 1981). This circulation pattern would tend to maintain the larvae in the areas of high recruitment rates.

The length of the pelagic period of larval development (a temporal scale) is a factor in limiting distributions. Over long time periods, these effects may not be evident; however, recolonization of an isolated area that has undergone a severe disturbance suggests that the interaction between time and space scales is important. Green abalones, *Haliotis fulgens*, which inhabit the shallow inner margins of kelp forests, are an example of a species with a relatively short larval life [ranging from 4 to 9 days as a function of temperature (Leighton et al., 1981)]. Our study of the recovery of green abalone populations on the once badly disturbed Palos Verdes Peninsula and a drift bottle experiment (Tegner and Butler, personal observation) suggest that the combination of the interaction of this species' location in a low current region of the forest with the currents during its spawning season and its short larval life make *H. fulgens* poor long-distance dispersers. The results from drift bottles released into green abalone habitat indicate that transport between isolated kelp forests within appropriate time periods is probably rare. They underscore the importance of endemic brood stock for the maintenance of local populations of animals with limited dispersal capabilities (Tegner and Butler, personal observation).

7 SCALES IN TIME

Developing a proper temporal perspective may be one of the most important unresolved problems in ecology, because none of us has more than one life-span to study patterns that often have longer duration. More realistically, funding, constraints on duration of study, and academic promotional procedures force ecologists into 2- to 3-year projects; few natural patterns are of such short duration. Eventually it is only the old, often tired voices that call for better perspective; these pleas seem easily ignored. Nevertheless, the many examples of episodic natural disasters and/or recruitment and usually long-term population decline demonstrate that the natural patterns we see are "footprints" of previous events (see also Colwell, Chapter 14). This emphasizes the importance of perceiving and appreciating temporal patterns even if only by inference.

The San Diego kelp forests have been reasonably stable since 1911 when the first records were made (North, 1971). They were, however, declining in the early 1950s for various reasons including pollution, overfishing, and so on (Tegner, 1980), and they were reduced to 1% of their former cover (IMR, 1963; North, 1974; Dayton et al., 1984, in press) by the El Niño. Since their recovery from the El Niño, the percent cover has varied considerably with major storms, but the canopy has usually recovered rapidly. The data and observations in Dayton et al. suggest that 10 years is a minimal time to study patch dynamics at Pt. Loma, but cursory study of mosaics of air photos of the Pt. Loma forest taken by Wheeler North and Kelco Company suggest that 20 years is needed to see major changes in *Macrocystis* cover. The mean generation time of *Macrocystis* which survive to adulthood is less than 6 years, but stipitate kelps have much longer life-spans. Other long-term patterns observed or inferred for shellfish (Coe, 1956) and echinoderms (Merrill and Hobson, 1970; Warner, 1971) suggest that several decades are necessary for an appreciation of the temporal scale important to large invertebrates, and these scales are likely to be representative of many such populations. Some pelagic fish species have decade- to century-long fluctuations (Soutar and Isaacs, 1974; Cushing and Dickson, 1976).

Historical records in southern California show that within the last 200 years there have been several periods of very important climatological changes (Kuhn and Shepard, 1981). For example, extremely violent hurricanes from the southeast (described by mariners as being worse than the storms of Cape Horn) were relatively common in the early 1800s. The Pacific Railway Survey of 1853–1857 collected a suite of fish species off San Diego that was more tropical than any seen in subsequent decades (Hubbs, 1948). At the same time, much of the fish fauna of Monterey was made up of species now characteristic of the warmer waters south of Point Conception. Hubbs (1948), demonstrating a close correlation between average air and sea temperatures along the Pacific coast, used records of coastal air temperatures to argue that 1850 and 1860 decades were a prolonged warm period. In the early 1860s the state of California went bankrupt when the San Joaquin Valley became a lake of 30 × 300 miles while coastal areas were ravaged by floods. Another warm, wet period in 1884–1891 (including a measured ppt. of 11.5 inches in 80 min) resulted in devastating floods in southern California. These rains certainly included the Colorado Plateau, as the Colorado River at Yuma was reported to be 20 miles wide! (See Kuhn and Shepard, 1981, for more details and references.) While the El Niño of 1957–1959 was not associated with heavy rainfall, there is abundant evidence that these wet periods in the 1800s were associated with El Niños (Hubbs, 1948; Kuhn and Shepard, 1981). Thus, the massive marine shifts documented by the 1963 IMR report, Radovitch (1961), and Chelton et al. (1982) may be relatively common. Douglas (1976) correlated tree ring patterns and seawater

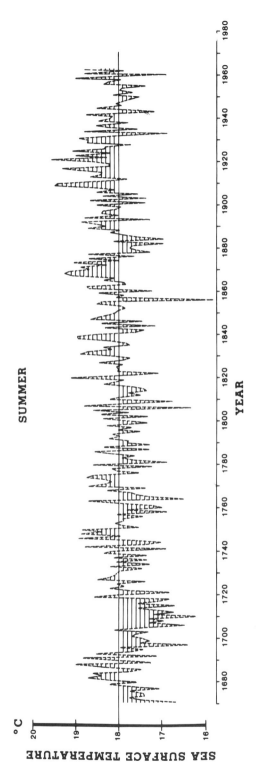

Figure 3. Summer sea temperatures 5 m below sea level at La Jolla, California, reconstructed from standardized tree ring indices for *Pseudotsuga macrocarpa* from 1214 m on the Santa Ana Mountains, California. Data from Douglas (1976, p. 188) were put in graphic form by Gerry Kuhn.

temperature. Then, using these correlations, he recalculated the sea temperature back to 1680 (Fig. 3). Although these calculations lack the precision of the anomalies evaluated by Chelton et al. (1982), they do show that major oceanic changes occur regularly in southern California.

Vermeij (1982) and Wethey (1983, in press) have compiled very convincing historical, archaeological, and paleontological data buttressing their argument that climatic shifts (over centuries) have resulted in major biogeographic shifts and even local extinctions. Several types of biogeographical shifts, each sensitive to different types of climatological change such as rainfall, temperature highs, duration of freezes, and so on, are well known in terrestrial systems; in addition, these climate changes often had catastrophic effects on human populations (see summary in Ford, 1982). Finally, at a much larger scale, there were major sea level lowerings during the Pleistocene, including a drop of several hundred meters below the present level. This would reduce the rocky habitat area of kelps because most of the shelf would be above sea level; the remaining habitat would be mostly sand and mud. Furthermore, the northern Channel Islands merged into a very large island (Johnson, 1978). What was left of the Southern California Bight would have been a very different habitat, as it would have lost its shelf and would be in the lee of the island and thus protected from most sea conditions (Johnson, 1978). The narrow channel may, however, have had large tidal bores such as is today seen between Tiburon Island and mainland Sonora, Mexico.

Which time scale is appropriate to community ecologists? Many regularly occurring processes (competition, predation, etc.) can be understood in short term (<10 years) studies. The mean generation times of kelps range from 1 to 15 years; hence, these would be minimal time spans to observe their population dynamics. But because the patches of particular kelps can persist for several generations, evaluation of the distribution and abundance of kelps necessitates decades, at a minimum. The same seems to be true of most large invertebrates, which have episodic recruitment and potentially are very long-lived. The historical records of southern California (Hubbs, 1948, 1960; Kuhn and Shepard, 1981) show that climatological events which happen at even 100-year intervals may involve habitat alterations that could cloud the understanding of present-day patterns. It is impossible to know the climatological and sea level effects of the Pleistocene on present day kelp and other coastal communities, but they may have had important evolutionary consequences, and it is important at least to be cognizant of these dramatic changes of thousand-year time scales.

8   THE ROLE OF MAN IN COMMUNITY PATTERNS

We all agree that it is important to preserve as much habitat as possible in as "natural" a state as is practical. However, it is difficult to objectively define "natural." There is a tendency to describe it as pre-white man, but this ignores what may have been substantial effects by native man. It is important to realize that we live in a considerable ecological shadow of primitive man. We make this case for the California kelp system as dominated by the sea otter, but the role of early man as a major influence in natural systems is probably general.

By eating sea urchins, the herbivore from which most kelps have no defense, the sea otter is known to have a keystone role in all the kelp communities where it occurs. This is especially true in Alaska, where sea urchins have no other important predators (for the Aleutian Islands: Estes and Palmisano, 1974; Dayton, 1975b; for southeastern Alaska: Duggins, 1980); it is probably also true in central California (McLean, 1962) but much less clear-cut as physical disturbances, substrate instability, and other environmental factors cloud the sea otter–sea urchin–kelp relationships (Miller and Giebel, 1973; Cowen et al., 1982). In southern California the sea otter is not the key predator that it is north of Pt. Conception because there are effective alternate predators. For example, sea otters have been functionally absent for at least 150 years, during which time there have remained luxuriant kelp forests while urchins have largely been controlled by the predation from spiny lobsters and sheephead (Tegner, 1980; Tegner and Dayton, 1981; Cowen, 1983; Tegner and Levin, 1983).

But have sea otters always had such a prominent role? We know that the presence of sea otters at their carrying capacity excludes commercial or sport shell fisheries (Miller and Giebel, 1973) and that remaining sea urchins and abalones are forced to live in deep crevices and cracks (Lowry and Pearse, 1973). Yet Indian middens in central California and the Channel Islands are ubiquitous, and dominated by the shells of large abalone, and large Aristotle's lanterns of sea urchins are common (Woodhouse et al., 1977; personal observations); this runs counter to the recent evidence that sea otter predation is important. There are also large sheephead in the middens, again indicating that otters were not functionally dominating the system. Because we have no estimates of the rate of deposition of the shell material in the middens, we cannot exclude the hypothesis that the large shellfish were the very rare collection from deep crevices. However most of the big shells we have seen do not have the degree of shell erosion usually seen on abalones from deep crevices. We know that the Indians harvested sea otters, and we feel that the situation in California paralleled that of Amchitka Island, Alaska, where the otters were apparently eliminated for a period by early Aleuts, thereby releasing shellfish (Simenstad et al., 1978). Unanswered is the question of what is a "natural" sea otter–sea urchin–kelp relationship in California. If early man has been in California >50,000 years (Bada et al., 1974; but c.f. Taylor, et al., 1983), it is possible that he has strongly impacted otter populations for much of that time and this impact could have been especially important during the lowered sea level periods at which time the kelp–otter habitat may have been much diminished. Furthermore, we can only guess what relationship existed between the huge, herbivorous, Steller's sea cow and the kelp community (Dayton, 1975b; Domning, 1978), but Steller found it in what must have been an island refuge from Aleuts. It seems clear that primitive man could have rendered such a large, defenseless animal locally extinct as readily as the Russian fur hunters, who rapidly hunted it to total extinction.

9 DISCUSSION

The proper integration of spatial and temporal scales is perhaps the most important challenge in ecology today. Certainly it is evident that organisms have patchy and aggregated distributions on a wide variety of both scales, and their integration must involve the more complex larger scales as well as the relatively well-studied smaller scales. Still lacking but perhaps obtainable in the kelp system is an understanding of the relationship between the small scale spatial patterns (centimeters to 10s of meters and <10 years) and the larger, coarse scales of 100 m to 100 km and 10–100 years. On the larger mesoscale level (100s of km) there has been recent progress from preliminary attempts to wed oceanography and climatology, especially with regard to plankton and fishery questions (Cushing and Dickson, 1976; Lasker, 1978; Parrish et al., 1981; Colebrook, 1982). Some of the approaches can be utilized in nearshore ecosystems such as kelp communities, but this has yet to be done. Time series analyses are available for canopy area and relative harvest of *Macrocystis* from 1911. Again, the fishery literature has time series data of catch records. But for marine species of no commercial interest, time series data of >10 years are extremely rare.

Although details differ, the same problems of scale exist for all systems. Marine systems are especially difficult because of their serious sampling problems, but terrestrial systems also integrate all levels of scales, and terrestrial parallels abound. Spatial scales integrate dispersal patterns which range from slow lateral growth to low dispersing seeds to widely dispersing wind and animal-borne seeds to animals with long migrations. Temporal scales involve everything from insects with very short (weeks) turnover times to annual grasses to the rare mast years of many trees to the same long-period climatological shifts discussed above. Terrestrial examples of long-term patch persistence are common in the American southwest. For example, Vasek (1980) speculates that clones of the creosote bush (*Larrea tridentata*) are over 11,000 years old. The photographs in Hastings and Turner (1965) and Gehlbach (1981) document numerous patches of plants persisting many decades (in most cases simply reflecting plant longevity), and Gehlbach also mentions examples of plant and animal changes and persistence over one to two decades. In the absence of fire, sage and chaparral patches remain distinct for many decades; a 1931 photograph in Mooney (1977) of Banner Grade, San Diego Co, shows patches which are still unchanged in 1983 (personal observation). Fire disturbance is an integral component of most plant associations in the Southwest (for reviews see Humphrey, 1958; Wright and Heinselman, 1973; Keeley et al., 1981; Minnich, 1983). This disturbance parallels that of major storms in kelp systems, as the affected area is relatively large and the impact varies for particular species. More local but still selective disturbances in terrestrial systems paralleling sea urchin grazing and smaller storms in kelp systems include the foraging of gophers, badgers, peccaries, rodents, rabbits, cervids, and so on.

It is difficult to think of many terrestrial parallels for the long-distance (100s of km) dispersal characteristic of some marine invertebrates. However, smaller scale dispersal mechanisms such as wind and sheet flooding (Reichman, 1979) are common, as is bird, rodent, cervid, ant, and even coyote seed dispersal in feces. Parallels to kelp edge effects are also probably common, as rodents and birds forage from shelter (Bartholemew, 1970; Halligan, 1974; Pulliam, personal communication). Other interactions roughly paralleling those of sea urchin–coralline algae–abalone interactions in kelp habitats can be found in such habitat modifications as nurse trees (Niering et al., 1963) and bacterial–fungal relationships (Kaminsky, 1981).

There are numerous examples of short-term temporal fluxes in the Southwest including many studies of the variation in seed predation (summarized in Brown et al., 1979) and of the duration of freezes (Niering et al., 1963). Longer-term and relatively important ecological changes over the last 100 years have been attributed to climate changes (Hastings and Turner, 1965), but for some reason southwestern ecologists have not considered the importance of relatively rare but catastrophic climatological events which are becoming appreciated in marine systems. For example the focus on overgrazing, predator control, and fire to explain the heavy erosion and gully formation of the 1880s (Hastings and Turner, 1965; Humphrey, 1958), may not sufficiently appreciate the extremely heavy rains of the 1860s, late 1880s and early 1890s (Brewer, 1974; Kuhn and Shepard, 1981). Finally, on a still longer time scale, the remarkable records of tree rings and especially packrat middens offer surprisingly precise records of climate and vegetation shifts through the Holocene (Phillips, 1977; Betancourt and Van Devender, 1981; Cole, 1982). These records are sufficiently precise to identify periods of early and late summer rains and slight species elevational shifts in some components of the life zones (A. Phillips, personal communication). Certainly the biological consequences of changing climates are much appreciated (Ford, 1982), but the issue of the integration of space and time scales is rarely explicitly addressed.

Another very difficult problem is that of integrating the ecological consequences of early man's activities with our concept of ''natural.'' We try to minimize the effects of modern man in our studies by identifying old fields or logged areas, land subject to livestock grazing, introduced species, and so on. Yet it is rare to find ecologists cognizant of the effects of early man, the hunter and the farmer. There are probably many southwestern parallels of our scenario concerning the interwoven effects of early man on sea cows and sea otters and the consequences for kelps, abalones, sea urchins, and probably lobsters. One gazes with wonder at the list of some 200 genera of animals that became extinct during the Pleistocene (Martin and Guilday, 1967), many of which almost certainly went extinct directly or indirectly at the hands of early man (Deevey, 1967; Martin, 1967). In most cases the animals apparently lost to primitive hunters were "big game" mammalian and avian herbivores; collapses of dependent carnivores, scavengers, and commensals apparently followed. The ecological consequences of foraging by several species of mastodons and mammoths will never be known (but have been inferred; Johnson, 1980), but judging from the role of African elephants they must have had important effects on plant communities; mesquite thickets and other riparian habitats in particular must have been very different than they are now. Similarly, foraging by giant sloths, peccaries, bears, tapirs, and armadillos must have created patchy disturbances important to many species of weedy plants and birds. Horses, camels, pronghorns, cervids, bovids, and beavers also must have had important community roles in addition to seed dispersal (see Janzen and Martin, 1982). These roles have not been replaced; the "niches" are empty and in a real sense our perception of "natural" is illusory.

The activities of early man, the farmer, are perhaps less dramatic but still important. The importance of "wildfire" is becoming known, but early man's agricultural use of fire was extensive (Pyne, 1982) and its importance must also be recognized. In a similar vein, it seems likely that by stripping their environment of firewood, the Anasazi may have contributed importantly to erosion (see Betancourt and Van Deventer, 1981). Certainly, early farmers had intimate relationships with the earth, the effects of which persist today (see references in Nabhan, 1982, for desert people). Nabhan describes a large variety of desert native plants which were at least partially domesticated and can now be found growing with their wild forms. The desert people had many ingenious means of domesticating the land as well; Nabhan describes subtle but effective soil and drainage improvements. In some cases irrigation was practiced on a much grander scale; for example, the Hohokam constructed over 2000 km of canals in the Salt River Valley of Arizona; some of the canals were 11 m wide (26 m between crests) and 3 m deep (Masse, 1981). Many of these canals can still be seen in aerial photographs of modern farms. Undoubtedly, "primitive" irrigation systems significantly altered many additional terrestrial habitats we consider "natural."

Attempting to summarize the message of this essay brings to mind the parable of the blind men and the elephant. In their initial encounter with the proboscidian, each "sampled" a different location and came to vastly different conclusions about the nature of the beast. Our experience with the kelp forest has illustrated how easy it is to study an isolated component of the system without appreciation of larger temporal and spatial scales that are critical to an integrated understanding of the system. Events that seem random on one scale compose patterns on another. Certainly these conclusions are generally true in terrestrial ecology, where large spatial scales and especially long-term temporal scales are particularly important to a general evolutionary understanding of nature. It is especially important for evolutionary ecologists studying terrestrial systems to explicitly consider the important roles of early hunters and farmers, because even our most "pristine" sites are haunted by the empty niches of recently extinct species, by the genes of domesticated and partially domesticated crops, by the vestiges of fire, and centuries of agriculture. Clearly, it is not possible for each ecologist to work among all these scales; indeed, proper scientific methods demand more restricted scales amenable to challenging specific hypotheses. Nevertheless, it is very realistic and important to plead for an enlarged perspective in interpreting and generalizing the results of more specific research.

10 SUMMARY

Our research in California kelp communities has utilized classical approaches to the study of community structure and the demography of important kelp and invertebrate species. We have discussed the stability of small patches (square meters to hectares) in the kelp forest with regard to invasion, competition, grazing, and different types of disturbance. Sea urchins are important grazers, and we have studied their community roles and patterns of recruitment and mortality. These approaches are, however, inadequate to obtain a complete understanding of the system because many important patterns depend on processes that work on much larger scales in space and time. For example, the drag of the kelp forest itself slows the currents and substantially reduces the transport of nutrients and long-lived larvae into the forest. In addition, the reduction in current velocity can act to "trap" short-lived larvae that originate within the forest. Certainly the kelp forest has many important, usually nonlinear, intermediate scale effects on its environment.

Much larger scale oceanographic processes such as eddies spinning off the California current bring nutrients and larvae to kelp forests. Alternatively, warm water intrusions from the south can devastate kelp forests. Finally, man too has impacted kelp habitats in a myriad of ways. In recent years fishing and waste disposal have had many effects on kelp forests. But the effects of humans are not necessarily only recent. For example, sea otters are known to be important predators in some kelp communities and there is indirect evidence that their populations may have been limited by effective hunting by native human populations. Early man may also have much reduced the range of Steller's sea cow, a large herbivore that probably had a substantial impact on the nearshore environment.

Focusing only on the southwestern United States, we find that there are numerous parallels to our kelp story. There are many examples of patch stability and various dispersal patterns dependent upon physical and biological processes. There is abundant evidence that shifts in climate have had significant and long-lasting consequences. Finally, the role of early man structuring our "natural" ecosystems is even more dramatic in the terrestrial system; examples are the efficacy of his hunting with regard to the direct and indirect extinction of many important species, the far-ranging ecological effects of man-caused fire used for agricultural and hunting purposes, and long-lasting genetic and habitat effects of many agricultural procedures.

ACKNOWLEDGMENTS

We thank J. Barry, R. Cowen, L. Dayton, J. Estes, A. Genin, L. Haury, and P. Price for reading and commenting on the manuscript. We appreciate the support of G. Kuhn, W. J. North, R. McPeak, and Kelco Co. This work is a result of research sponsored by the National Science Foundation, Scripps Industrial Associates, the Marine Life Research Group, and NOAA, National Sea Grant College Program, Department of Commerce, under grant number NA80AA-D-00120, and by the California State Resources Agency, project numbers R/F-36 and R/F-73. The U.S. Government is authorized to produce and distribute reprints for governmental purposes.

LITERATURE CITED

Bada, J. L., R. A. Schroeder, and G. F. Carter. 1974. New evidence for the antiquity of man in North America deduced from aspartic acid racemization. *Science* **184**:791–793.

Bartholomew, B. 1970. Bare zone between California shrub and grassland communities: The role of animals. *Science* **170**:1210–1212.

Bennett, J., and A. C. Giese. 1955. The annual reproductive and nutritional cycles in two western sea urchins. *Biol. Bull.* **109**:226–237.

Bernal, P. A. 1981. A review of the low-frequency response of a pelagic ecosystem in the California current. *Calif. Coop. Oceanic Fish. Invest. Rep.* **22**:49–64.

Bernstein, B. B., and N. Jung. 1979. Selective processes and coevolution in a kelp canopy community in southern California. *Ecol. Monogr.* **49**:335–355.

Betancourt, J. L., and T. R. Van Deventer. 1981. Holocene vegetation in Chaco Canyon, New Mexico. *Science* **214**:656–658.

Bray, R. N. 1981. Influence of water currents and zooplankton densities on daily foraging movements of blacksmith, *Chromis punctipinnis*, a planktivorous reef fish. *Fish. Bull.* **78**:829–841.

Brewer, W. H. 1974. *Up and down California in 1860–1864.* University of California Press, Berkeley. 583 pp.

Brown, J. H., O. J. Reichman, and D. W. Davidson, 1979. Granivory in desert ecosystems. *Annu. Rev. Ecol. Syst.* **10**:201–227.

Cameron, R. A., and S. C. Schroeter. 1980. Sea urchin recruitment: Effect of substrate selection on juvenile distribution. *Mar. Ecol. Prog. Ser.* **2**:243–247.

Chelton, D. B., P. A. Bernal, and J. A. McGowan. 1982. Large-scale interannual physical and biological interactions in the California Current. *J. Mar. Res.* **40**:1095–1125.

Coe, W. R. 1956. Fluctuations in populations of littoral marine invertebrates. *J. Mar. Res.* **15**:212–232.

Cole, K. 1982. Late quaternary zonation of vegetation in the eastern Grand Canyon. *Science* **217**:1142–1145.

Colebrook, J. M. 1982. Continuous plankton records: phytoplankton, zooplankton and environment, north-east Atlantic and North Sea, 1958–1980. *Oceanologia Acta* **5**:473–480.

Connell, J. H. 1983. On the prevalence and relative importance of interspecific competition: Evidence from field experiments. *Am. Nat.* (in press).

Cowen, R. K. 1983. The effect of sheephead (*Semicossyphus pulcher*) predation on red sea urchin (*Strongylocentrotus franciscanus*) populations: An experimental analysis. *Oecologia* (in press).

Cowen, R. K., C. R. Agegian, and M. F. Foster. 1982. The maintenance of community structure in a central California giant kelp forest. *J. Exp. Mar. Biol. Ecol.* **64**:189–201.

Cushing, D. H. 1975. *Marine ecology and fisheries.* Cambridge University Press, Cambridge. 278 pp.

Cushing, D. H., and R. R. Dickson. 1976. The biological response in the sea to climatic changes. *Adv. Mar. Biol.* **14**:1–122.

Dayton, P. K. 1971. Competition, disturbance and community organization: The provision and subsequent utilization of space in a rocky intertidal community. *Ecol. Monogr.* **41**:351–389.

Dayton, P. K. 1975a. Experimental evaluation of ecological dominance in a rocky intertidal algal community. *Ecol. Monogr.* **45**:137–159.

Dayton, P. K. 1975b. Algal canopy interactions in a sea otter-dominated kelp community at Amchitka Island, Alaska. *Fish. Bull.* **73**:230–237.

Dayton, P. K. 1983. Processes structuring some marine communities: Are they general? In L. G. Abele, D. S. Simberloff, and D. R. Strong, (eds.), *Ecological communities: Conceptual issues and the evidence.* Princeton University Press, Princeton, New Jersey (in press).

Dayton, P. K., and J. S. Oliver. 1980. An evaluation of experimental analyses of population and community patterns in benthic marine environments. In K. R. Tenore and B. C. Coull (eds.), *Marine benthic dynamics*, pp. 93–120. University of South Carolina Press, Columbia.

Dayton, P. K., et al. 1984. Patch dynamics and stability of southern California kelp communities. *Ecol. Monogr.* (in press).

Deevey, E. S. 1967. Introduction. In P. S. Martin and H. E. Wright (eds.), *Pleistocene extinctions, the search for a cause*, pp. 61–72. Yale University Press, New Haven, Connecticut.

Domning, D. P. 1978. Sirenian evolution in the North Pacific Ocean. *University of California Publications in Geological Sciences* **118**:1–176.

Douglas, A. V. 1976. Past air-sea interactions over the eastern north Pacific Ocean as revealed by tree ring data. Ph.D. thesis, University of Arizona, Tucson, 196 pp.

Duggins, D. O. 1980. Kelp beds and sea otters: An experimental approach. Ecology **61**:447–453.

Ebert, T. A. 1968. Growth rates of the sea urchin *Strongylocentrotus purpuratus* related to food availability and spine abrasion. *Ecology* **49**:1075–1091.

Eckman, J. E. 1979. Small scale patterns in a soft-substratum, intertidal community. *J. Mar. Res.* **37**:437–457.

Elton, C. S. 1966. *The pattern of animal communities.* Methuen, London. 432 pp.

Estes, J. A., and J. F. Palmisano. 1974. Sea otters: Their role in structuring nearshore communities. *Science* **185**:1058–1060.

Ford, M. J. 1982. *The changing climate: Respones of the natural fauna and flora.* George Allen and Unwin Press, London. 190 pp.

Gehlbach, F. R. 1981. Mountain islands and desert seas. Texas A&M University, College Station. 298 pp.

Gerrodette, T. 1981. Dispersal of the solitary coral *Balanophyllia elegans* by demersal planular larvae. *Ecology* **62**:611–619.

Halligan, J. 1974. Relationship between animal activity and bare areas associated with California sagebrush in annual grassland. *J. Range Manage.* **27**:358–362.

Hastings, J. R., and R. M. Turner. 1965. *The changing mile.* University of Arizona Press, Tucson, Arizona. 317 pp.

Hinegardner, R. T. 1975. Morphology and genetics of sea urchin development. *Am. Zool.* **15**:679–689.

Hobson, E. S. 1976. The rock wrasse, *Halichoeres semicinctus*, as a cleaner fish. *Calif. Fish Game* **62**:73–78.

Hobson, E. S., and J. R. Chess. 1976. Trophic interactions among fishes and zooplankters near shore at Santa Catalina Island, California. *Fish. Bull. U.S.* **74**:567–598.

Hubbs, C. L. 1948. Changes in the fish fauna of western North America correlated with changes in ocean temperature. *J. Mar. Res.* **7**:459–482.

Hubbs, C. L. 1960. Quaternary paleoclimatology of the Pacific Coast of North America. *CalCOFI Rep.* **7**:105–112.

Humphrey, R. R. 1958. The desert grassland. *Bot. Rev.* **24**:1–74. (Reprinted by University of Ariz. Press, Tucson.)

IMR, 1963. An investigation of the effects of discharged wastes on kelp. Final Rept. Univ. Calif. Inst. Mar. Res. IMR Ref. 63–6. 305 pp.

Jackson, G. A. 1980. Marine biomass production through seaweed aquaculture. In A. San Pietro (ed.), *Biochemical and photosynthetic aspects of energy production*, pp. 31–80. Academic Press, New York.

Jackson, G. A., and R. R. Strathmann. 1981. Larval mortality from offshore mixing as a link between precompetent and competent periods of development. *Am. Nat.* **118**:16–26.

Jackson, G. A., and C. W. Winant. 1983. Effects of a kelp forest on coastal currents. *Continental Shelf Res.* (in press).

Janzen, D. H., and P. S. Martin. 1982. Neotropical anachronisms: The fruits the Gomphotheres ate. *Science* **215**:19–27.

Johnson, D. L. 1978. The origin of island mammoths and the quaternary land bridge history of the northern Channel Islands, California. *Q. Res.* **10**:204–225.

Johnson, D. L. 1980. Episodic vegetational stripping, soil erosion, and landscape modification in prehistoric and recent historic time, San Miguel Island, California. In D. M. Power (ed.), The california islands, pp. 103–121. Santa Barbara Museum of Natural History, Santa Barbara, California.

Kaminsky, R. 1981. The microbrial origin of the allelopathic potential of *Adenostoma fasciulatum. Ecol. Monogr.* **51**:365–382.

Keeley, S. C., J. E. Keeley, S. M. Hutchinson, and A. W. Johnson. 1981. Postfire succession of the herbaceous flora in southern California chaparral. *Ecol. Monogr.* **62**:1608–1621.

Kuhn, G. G., and F. P. Shepard. 1981. Should southern California build defenses against violent storms resulting in lowland flooding as discovered in records of past century. *Shore Beach* **49**:2–10.

Lasker, R. 1978. The relationship between oceanographic conditions and larval anchovy food in the California Current: identification of factors contributing to recruitment failure. *Rapp. P.-V. Réun. Cons. Int. Explor. Mer.* **173**:212–230.

Lawrence, J. M. 1975. On the relationship between marine plants and sea urchins. *Oceanogr. Mar. Biol. Annu. Rev.* **13**:213–286.

Leighton, D. L., M. J. Byhower, J. C. Kelly, G. N. Hooker, and D. E. Morse. 1981. Acceleration of development and growth in young green abalone (*Haliotis fulgens*) using warm effluent seawater. *J. World Maricul. Soc.* **12**:170–180.

Levin, L. A. 1981. Dispersion, feeding behavior and competition in two spionid polychaetes. *J. Mar. Res.* **39**:99–117.

Levin, L. A. 1982. Interference interactions among tube-building polychaetes in a dense infaunal assemblage. *J. Exp. Mar. Biol. Ecol.* **65**:107–119.

Levin, S. A. 1976. Population dynamics in heterogeneous environments. *Annu. Rev. Ecol. Syst.* **7**:287–310.

Lowry, L. F., and J. S. Pearse. 1973. Abalones and sea urchins in an area inhabited by sea otters. *Mar. Biol.* **23**:213–219.

Lüning, K., and M. Neushul. 1978. Light and temperature demands for growth and reproduction of Laminarian gametophytes in southern and central California. *Mar. Biol.* **45**:297–309.

Mann, K. H. 1977. Destruction of kelp beds by sea urchins: A cyclical phenomena or irreversible degradation? *Helgolander Wiss. Meeresunters* **30**:455–467.

Martin, P. S. 1967. Prehistoric overkill. In P. S. Martin and H. E. Wright (eds.), *Pleistocene extinctions: The search for a cause*, pp. 75–120. Yale University Press, New Haven, Connecticut. 453 pp.

Martin, P. S., and J. E. Guilday. 1967. A bestiary for Pleistocene biologists. In P. S. Martin and H. E. Wright (eds.), *Pleistocene extinctions: The search for a cause*, pp. 1–62. Yale University Press, New Haven, Connecticut.

Masse, W. B. 1981. Prehistoric irrigation systems in the Salt River Valley, Arizona. *Science* **214**:408–415.

McLean, J. H. 1962. Sublittoral ecology of kelp beds of the open coast near Carmel, California. *Biol. Bull.* **122**:95–114.

Merrill, R. J., and E. S. Hobson. 1970. Field observations of *Dendraster excentricus*, a sand dollar of western north America. *Am. Midl. Nat.* **83**:595–624.

Minnich, R. A. 1983. Five mosaics in southern California and northern Baja California. *Science* **219**:1287–1294.

Miller, D. J., and J. J. Giebel. 1973. Summary of blue rockfish and lingcod life histories; a reef ecology study; and giant kelp, *Macrocystis pyrifera*, experiments in Monterey Bay, California. *Dept. Fish Game Fish Bull.* **158**:1–137.

Mooney, H. A. 1977. Southern coastal scrub. In M. G. Barbour and J. Major (eds.), *Terrestrial vegetation of California*, pp. 471–489. Wiley, New York.

Morse, D. E., M. Tegner, H. Duncan, N. Hooker, G. Trevelyan, and A. Cameron. 1980. Induction of settling and metamorphosis of planktonic molluscan (*Haliotis*) larvae. III. Signaling of metabolites of intact algae is dependent on contact. In D. Muller-Schwarze and R. M. Silverstein (eds.), *Chemical signals*. Plenum, N.Y.

Nabham, G. P. 1982. *The desert smells like rain*. North Point Press, San Francisco. 148 pp.

Niering, W., R. Whittaker, and C. Lowe. 1963. The Saguaro: A population in relation to environment. *Science* **142**:15–23.

North, W. J. 1971. The biology of giant kelp beds (*Macrocystis*) in California. Beihefte Zur Nova Hedwigia Heft 32, 600 pp.

North, W. J. 1974. Kelp habitat improvement project. Annual Report, 1 July 1973–30 June 1974. W. M. Keck Laboratory of Environmental Health Engineering, California Institute of Technology, Pasadena, 137 pp.

Owen, R. W. 1980. Eddies of the California Current system: Physical and ecological characteristics. In D. M. Power (ed.), *The California islands* pp. 237–263. Santa Barbara Museum of Natural History, Santa Barbara, California.

Paine, R. T. 1974. Intertidal community structure, experimental studies on the relationship between a dominant competitor and its principal predator. *Oecologia* **15**:93–120.

Paine, R. T. 1980. Food webs: Linkage, interaction, strength and community infrastructure. *J. Anim. Ecol.* **49**:667–685.

Paine, R. T., and S. A. Levin. 1981. Intertidal landscapes: disturbance and the dynamics of pattern. *Ecol. Monogr.* **51**(2):145–198.

Paine, R. T., and R. Vadas. 1969. The effects of grazing by sea urchins, *Strongylocentrotus* spp. on benthic algal populations. *Limnol. Oceanogr.* **14**:710–719.

Parrish, R. H., C. S. Nelson, and A. Bakun. 1981. Transport mechanisms and reproductive success of fishes in the California Current. *Biol. Oceanogr.* **1**:175–203.

Phillips, A. M. 1977. Packrats, plants, and the Pleistocene in the lower Ground Canyon. Ph.D. thesis, University of Arizona, Tucson.

Pyne, S. J. 1982. *A cultural history of wildland and rural fire.* Princeton University Press, Princeton, New Jersey. 656 pp.

Radovitch, J. 1961. Relationships of some marine organisms of the northeast Pacific to water temperatures particularly during 1957 through 1959. *Calif. Dept. Fish Game Fish Bull.* **112**:1–62.

Randall, J. E. 1965. Grazing effects on sea grasses by herbivorous reef fishes in the West Indies. *Ecology* **46**:255–260.

Raupach, M. R., and A. S. Thom. 1981. Turbulence in and above plant canopies. *Annu. Rev. Fluid Mech.* **13**:97–129.

Reichman, O. J. 1979. Desert granivore foraging and its impact on seed densities and distributions. *Ecology* **60**:1085–1092.

Scheltema, R. S. 1974. Biological interactions determining larval settlement of marine invertebrates. *Thalassia Jugoslavica* **10**:263–396.

Simenstad, C. A., J. A. Estes, and K. W. Kenyon. 1978. Aleuts, sea otters, and alternate stable-state communities. *Science* **200**:403–411.

Smith, P. E. 1978. Biological effects of ocean variability: Time and space scales of biological response. *Rapp. P.-V. Réun Cons. Int. Explor. Mer.* **173**:117–127.

Soutar, A., and J. D. Isaacs. 1974. Abundance of pelagic fish during the 19th and 20th centuries as recorded in anaerobic sediments off the Californias. *Fish. Bull.* **72**:257–273.

Strathmann, R. 1974. The spread of sibling larvae of sedentary marine invertebrates. *Am. Nat.* **108**:28–44.

Strathmann, R. 1978. Length of pelagic period in echinoderms with feeding larvae from the northeast Pacific. *J. Exp. Mar. Biol. Ecol.* **34**:23–27.

Sutherland, J. P. 1981. The fouling community at Beaufort, North Carolina: a study in stability. *Am. Nat.* **118**:449–519.

Taylor, R. E., L. A. Payen, B. Gerow, D. J. Donahue, T. H. Zabel, A. J. T. Jull, P. E. Damon. 1983. Middle holocene age of the Sunnyvale human skeleton. *Science* **220**:1271–1273.

Tegner, M. J. 1980. Multispecies considerations of resource management in southern California kelp beds. *Can. Tech. Rep. Fish. Aquat. Sci.* **945**:125–143.

Tegner, M. J., and P. K. Dayton. 1977. Sea urchin recruitment patterns and implications of commercial fishing. *Science* **196**:324–326.

Tegner, M. J., and P. K. Dayton. 1981. Population structure, recruitment and mortality of two sea urchins (*Strongylocentrotus franciscanus* and *S. purpuratus*) in a kelp forest near San Diego, California. *Mar. Ecol. Prog. Ser.* **5**:255–268.

Tegner, M. J., and L. A. Levin. 1983. Spiny lobsters and sea urchins: Analysis of a predator-prey interaction. *J. Exp. Mar. Biol. Ecol.* (in press).

Thorson, G. 1946. Reproduction and larval development of Danish marine bottom invertebrates with special reference to the planktonic larvae in the Sound (Oresund). Meddeelelser fra

Kommissionen for Danmarks Fisherrog Havundersolgelser Serie: Plankton, Bind IV, Nrl. 523 pp.

Thorson, G. 1950. Reproductive and larval ecology of marine bottom invertebrates. *Biol. Rev.* **25**:1–45.

Tricas, T. C., 1979. Relationships of blue shark, *Prionace glauca*, and its prey species near Santa Catallina Island, California. *Fish. Bull.* **77**:175–182.

Vadas, R. L. 1968. The ecology of *Agarum* and the kelp bed community. Ph.D. thesis, University of Washington, Seattle. 282 pp.

Vasek, F. C. 1980. Creosote bush: Long lived clones in the Mojave Desert. *Am. J. Bot.* **67**:246–255.

Vermeij, G. J. 1982. Environmental change and the evolutionary history of the periwinkle (*Littorina littorea*) in North America. *Evolution* **36**:561–580.

Warner, G. 1971. On the ecology of a dense bed of the brittle star *Ophiothrix fragilis*. *J. Mar. Biol. Assoc.* **51**:267–282.

Wethey, D. S. 1983. Catastrophe, extinction and species diversity: A rocky intertidal example. *Ecology* (in press).

Wheeler, W. W., and M. Neushul. 1981. The aquatic environment. In O. L. Lange, P. S. Nobel, C. B. Osmond, and H. Ziegler (eds.), *Encyclopedia of plant physiology*, New Series, Vol. 12A, *Physiological Plant Ecology I*, pp. 229–247, Springer-Verlag, Berlin.

Wilson, D. P. 1952. The influence of the nature of the substratum on the metamorphosis of the larvae of marine animals, especially the larvae of *Ophelia bicornis* Savigny. *Ann. Inst. Oceanogr., Monaco,* **27**:49–156.

Winant, C. D., and A. W. Bratkovich. 1981. Temperature and currents on the southern California shelf: A description of the variability. *J. Phys. Oceanogr.* **11**:71–86.

Woodhouse, C. D., R. K. Cowen, and L. R. Wilcoxon. 1977. A summary of knowledge of the sea otter *Enhydra lutris*, L., in California and an appraisal of the completeness of biological understanding of the species. Report No. MMC-76/02, U.S. Technical Information Service, PB 270–374.

Woodin, S. A. 1976. Adult-larval interactions in dense infaunal assemblages: patterns of abundance. *J. Mar. Res.* **34**:25–41.

Wright, H. E., and M. L. Heinselman. 1973. The ecological role of fire in natural conifer forests of western and northern America. *Q. Res.* **3**:317–513.

Taxonomic Index

Author Index

493

Subject Index

Abscission:
  premature, 27–28
Adaptation, 58, 76, 156, 276
Adaptedness, 254–263
  exploiter, 54
  masking, 257, 259, 263
  morphological, 78
  organismal, 392
Adaptive Radiation, 73
Adult phase, 127
Age:
  at first reproduction, 180–181, 184–185
  regulative value, 154
  structure, 148, 153, 390
Aggression, 237
  costs and benefits, 241
Alarm signal, 284
Alkaloid, 31, 75, 77
Allee effect, 205, 316, 323
Allelochemical, 32, 54, 70, 344
  inhibition, 75
  resistance, 77–78
Altitudinal gradient, 337, 341
Amensalism, 6, 376
Amino acid, 34, 79
  sulfur, 60, 62
  synthesis, 60
Amylase, alpha-, 65
Anemia, sickle-cell, 274
Angiosperm, 332
Area effect, 369
Association:
  extracellular, 67
  insect-microbial, 54, 75–76
  intracellular, 67
  loss, 76
  pattern, 67, 69
Attack, 26
  rate, 283, 346

Barren zone, 461
Behavior, 393

  social, 3, 227, 253, 265, 279
  territorial, 37
Biochemistry, 393
Biogeography, 369, 392
  shift, 470
Biological warfare, 372
Biometrician, 146
Bird:
  census, 394
  community, 395
  wintering, 392
Body:
  size distribution, 203, 209, 213, 215–216
Breakdown, 60
  physical, 60, 68
Broad spectrum activity, 54, 74, 76
Brood, reduction, 208
Bud population, 19

Cannibalism, 37, 207
Cardenolide, 32
Carrying capacity, 342
Caste:
  evolution, 243–245
  formation, 244
Casualty, exotic, 392
Catastrophe, 392, 467, 473
Cellulase, 60, 65, 68, 73
Cell wall, bacterial, 27
Chance, 392
Change, 128, 473
  environmental, 153
  herbivore-induced, 129–130
  short-term temporal, 473
Character:
  displacement, 8, 366
  polygenic, 171
Chimera, 18, 23–24
  genetic, 36
Chitinase, 65
Choline, 63
Chromosome, 150
  number, 20–21